Plants for Human Surie

About the Editor

Bikarma Singh born as Indian, and currently working as Scientist in CSIR-Indian Institute of Integrative Medicine Jammu, one of the pioneer institute of Council of Scientific and Industrial Research (CSIR) known for the largest research and development activities coming under Ministry of Science and Technology, Govt. of India, as well as engaged and recognizeed as Assistant Professor of AcSIR, New Delhi, India. Dr. Singh Graduated up as an Honours and Gold Medalist in Botany from North-Eastern Hill University Shillong, and pursued research for his Doctorate Degree in Botany from Gauhati University Assam and Botanical Survey of India (Shillong) Meghalaya. Prior to joining CSIR-IIIM Jammu, he was working as Scientist-Ecologist in WAPCOS Limited, Gurgaon and was involved in many consultancy projects related to studies on ecology and biodiversity of areas affected due to the developmental projects such as hydroelectric dam construction, irrigation canal and coal-mining activities. Currently, Dr. Singh pursued research with passion and has more than eleven year's research experience in the area of plant sciences, having expertise in systematic-taxonomy and biotechnology, ethnobotany, value addition from essential oils, ecology, biodiversity germplasm conservation and environmental impact assessment aspects. In addition to this book, he has published six books and recent ones are 'Himalayan Orchids-distribution and taxonomy', 'Plants for Wellness and Vigour', Plants of Commercial Values and few more. Dr. Singh has more than seventy peer-reviewed scientific research articles and experimental findings published in various reputed national and international SCI, and UGC-recognized peer-reviewed journals. Besides these, he has two patents filed credited to his account for research work on medicinal aspects, and worked as project investigator or team member for more than forty research projects funded by Govt. and Private organization on R&D work and rural socio-economic upliftment activities. At present, he is also serving as a reviewing member of several National and International Scientific SCI journals, undertaken as lead leader for organizing many seminars and workshops, and also invited by several other institute as key-note speaker/ lecturer under various themes related to plant sciences and societal activities. His current research interests focused on systematics, forest dynamics, applied biotechnology, molecular tools, sustainable use of medicinal and aromatic plants, plants inventory from Himalayas, value addition of essential oils, product development from medicinal plants and IUCN categorization of Indian Flora at regional level. He is also supervising and igniting young talents pursuing research as career enrolled in various research organizations and helping Ph.D. research scholars for defending their doctorate, as well as linking research, scientist and society for economy growth of the country.

Plants for Human Survival and Medicine

Editor

Bikarma Singh

(B.Sc. Honours, M.Sc. Gold Medalist, Ph.D., NABET)
CSIR-Indian Institute of Integrative Medicine
(Council of Scientific & Industrial Research)
Ministry of Science and Technology, Govt. of India
Canal Road, Jammu-180001, Jammu & Kashmir State

CRC Press
Taylor & Francis Group
Boca Raton London New York

CRC Press is an imprint of the
Taylor & Francis Group, an **informa** business

NEW INDIA PUBLISHING AGENCY
New Delhi – 110 034

CRC Press
Taylor & Francis Group
6000 Broken Sound Parkway NW, Suite 300
Boca Raton, FL 33487-2742

First issued in paperback 2023

International Standard Book Number-13: 978-1-03-265445-4 (pbk)
International Standard Book Number-13: 978-0-3678-1894-4 (hbk)
International Standard Book Number-13: 978-1-00-301068-5 (ebk)

DOI: 10.1201/9781003010685

Publisher's Note
The publisher has gone to great lengths to ensure the quality of this reprint but points out that some imperfections in the original copies may be apparent.

Print edition not for sale in South Asia (India, Sri Lanka, Nepal, Bangladesh, Pakistan or Bhutan)

Library of Congress Cataloging-in-Publication Data
A catalog record has been requested

Visit the Taylor & Francis Web site at
http://www.taylorandfrancis.com

and the CRC Press Web site at
http://www.crcpress.com

This book entitled '**Plants for Human Survival and Medicine**' is dedicated to CSIR-IIIM Jammu, Authors of Fifteen different renowned research organizations, and family members of Editor (father: Jammu Singh, mother: Sam Devi, brother: Bishander Singh, sister: Simla, wife: Manisha Singh, daughter: Aditi Singh, son: Aryan Singh) who allowed untimely to carry out research work late night in Office and at Home to complete the assigned task and research activities

Preface

Human, animal and plant living are inter-twined since the beginning of human civilization. The usages of plants for food, shelter and medicine is an age-old practice of human. In ancient time, humans learned to cultivate plants for food as well as to live alongside with them as natural gifted companions. Traditional herbal formulations and local home-made medicine have been used throughout history and within all welfare to prevent and cure diseases. People coming using plants that were available within their geographical boundary to cure diseases addressing their local health related issues. Actually cultural tradition were exposed due to people immigration for trade and business, and gets often overwhelmed by modern scientific concepts and medications relates to various culture of different country. Since time immemorial, people believe that a plant keep the mind in tune with nature and maintains proper balance of thinking and health. As we enter the new decades of twenty-first century, drugs and medicines continue to constitute one of the essential components of health care system in promoting health and preventing illness. Usually a severe disease causes staggering amount of suffering and death in humans and animals, and with time proceeds humans committed itself to alleviate the suffering caused by pathogens and microbes.

Scientific research on plants and their application for human health care is once again assuming a prominent position. This book deals with '**Plants for Human Survival and Medicine**' is the outcome of ongoing R&D for discovering new molecules, new drugs, new leads, ethnobotany and nutraceuticals in relation to tribe, nature and climate change directly or indirectly leads to affect or effect human population. The main objectives of the present book is to provide a baseline data and information on plants and their hidden secret relates to human health and survival. Scientific data obtained from the plants always remain the basis for commercial medication and product formulation for the treatment of chronic diseases such as heart problem, blood pressure, inflammation, arthritis, rheumatism and other associated human related issues.

Plants have been recognized as a rich source of novel drugs that form the ingredients in traditional system of medicine, and approximately 90% raw botanicals of ayurveda, amchi, siddha and unani formulation were manufactured and prepared by using plants as a main source of ingredient. It is proven that the use and the search for medicine, and nutrients supplements derived and prepared from plant have accelerated the discovery. In recent decades, the discovery of several molecules such as artimisinin from *Artemisia annua*, tetrahydrocannabine (THC) and cannabidiol (CBD) from *Cannabis sativa*, diosgenin (phytosterol sapogenin) from *Costus speciosus*, trihydroxy benzoic acid glycoside (bergenin) from *Bergenia ciliata*, foskolin from *Coleus forskohlii*, withanolides and withaferin from *Withania somnifera*, santonin from *Artemisia maritima*, morphine from *Papaver somnifera*, n-triacontane from *Colebrookea oppositifolia*, taxol from *Taxus baccata*, campothecin from *Campotheca acuminata*, *Nothapodytes foetida* and *Nothapodytes nimmoniana*, curcuminin from *Curcuma longa*, and glycyrrhizin and licorice from *Glycyrrhiza glabra*, galegine from *Galega officinalis*, spilanthol from *Heliopsis longipes* and *Spilanthes acmella*, indole alkaloid (yuehchukene) from *Murraya paniculata*, and several others bioactive molecules from ethnobotanical plants which has long history of use for human needs and wants are the best examples of drug discovered for human health and survival.

Illustrating further, the most widely sold and consumed dietary supplement chyawanprash is the best example of plants serving as source for human survival and energy boost ingredient. The product so called chyawanprash is prepared as per the instruction suggested in ayurvedic text. Historically, various Indian holy books such as Mahabharat and Puranas mentioned that Ashwini Kumar brothers who were Raj Vaidya to devas during the Vedic age first prepared chyawanprash formulation for Chyawan rishi in his ashram at Dhosi hill situated near Narnaul (present Haryana), and hence the name of chyawanprash comes in focus. Amla, Astavarga plants, Ashwagandha, Satawari, Bhumi-amalaki, Long Pepper, Malabar nuts, Punarava, sugar, honey and other total 30-100 ingredients are usually used in this formulation and today serving as the most life saving product for humans.

Scrutiny of literature reveals that in traditional Ayurvedic system of medicine, herbal botanicals were usually prepared by mixing different plants, which used to be flowering twigs, leaves, leaf exudates, gums, fruits, barks, roots, rhizomes or tubers. The first chapter of this book starts with Himalayan herbs used by human as food and medicine followed by chapter that focused on Ashtavarga plants, heptatoprotective properties of plants in treatment to liver toxicity. It is then followed by chapters shift on focus to the clinical arena, and the usages of plants in relation to cancer, diabetes and skin disorders. There is published report that 122 compounds of defined structure, obtained from only 94 species of plant, that are used globally as drugs and demonstrate that 80% of these have ethnomedical applications identical or related to the current use of the active elements of the plant (Daniel S. Fabricant and Norman R. Farnsworth. 2001. *The Value of Plants Used in Traditional Medicine for Drug Discovery*. Environmental Health Perspective 109: 69-75). Therefore, this book also contains few chapters on plants that have tremendous potentials as ethnomedicine used by different tribes spreading in different geographic locations from Kashmir to Arunachal Pradesh and *vice-versa*.

Today traditional herbal formulations, drugs, food supplements, nutraceuticals, pharmaceutical intermediates, bioactive natural products and lead compounds derived from synthetic drugs are of high demand. Herbs contain many compounds with powerful antioxidant properties as evidence from the scientific data, and herb-induced change in biomarkers that assess antioxidant status and oxidative stress are of interest in relation to the mechanism of herbal protection. In cell culture studies, direct cytotoxicity, gene expression, protein synthesis and transport mechanism can be measured, and the morphology and the growth of cells could be assessed, which is a great achievement of science today. In animal studies, tumor occurrence and size can be examined which can prevents deadly diseases such as cancer and AIDS.

This book also focuses on presenting the current scientific evidence of bio-molecular effects, agrotechnology and other parameters of medicinal and nutraceutical aspects of selected plants such as *Boswellia serrata, Butea monosperma, Colebrookea oppositifolia, Cymbopogon khasianus, Dendrophthe falcata, Dysoxylum binectiferum, Echinacea purpurea, Grewia asiatica*, Northeast *Panax species, Picrorrhiza kurroa, Saussurea costus, Withania somnifera, Zanthoxylum armatum*, and most important *Ashtarvarga* group of plants such as *Habenaria intermedia, Plantathera edgeworthii, Malaxis acuminata, Crepidium muscifera, Lilium polyphyllum, Polygonatum verticillatum, Polygonatum cirrhifolium* and *Roscoea purpurea* were discussed and presented in this book for human needs and wants.

The present book is based on twenty five excellent research articles provided by fifteen topmost research organizations of India. I am sure and confident that this book will serve as the baseline information server for future research in the field of drug discovery and new nutraceuticals for human health. The biological, chemical pharmacological and clinical studies mentioned in this book will add and contribute in discovering quick leads for medicine formulations and products development relates to pharmaceutical industries. In addition to research suggestions and value of plants for human survival contained in each of chapter, an introduction section emphasizes particular research avenues for attention in the drug development programmes, and role of plants in human survival.

The Editor tried to convey maximum knowledge through this book **'Plants for Human Survival and Medicine'** regarding potential plants for human needs and medicine discovery. Readers are considered as the best panel of judges to evaluate the content of any writing and also applies to this particular book. I am sure and full confident that the readers have a moral obligation to convey their suggestions on this book in near future for its better improvement. It would be a great pleasure for me if this book could attract civilians, scientists, research scholars, strategies planners like forest department, tourists and industries who have ideas in their mind and take forward the economic plants for human survival and looks for local traditional medicines that can prepared from nearby their surroundings and forests to heal diseases. It would be worthless to mention that this book will have a way of providing a new level for future perspectives in understanding different areas of sciences and humans.

Bikarma Singh
Scientist & Editor
CSIR-Indian Institute of Integrative Medicine Jammu
(Council of Scientific & Industrial Research)
Ministry of Science and Technology
Govt. of India

Acknowledgements

Book '**Plants for Human Survival and Medicine**' must inevitably be deeply indebted to scientific contributions presented in the form of high quality research. My heartiest thanks and gratitude goes to my worthy Director IIIM Jammu, Dr. Ram A. Vishwakarma, and Director NBRI Lucknow, Professor Saroj Kanta Barik, as they continuously encouraged me for my good works and provide guidance as Guru. I owe especial debt of gratitude to sixty seven authors for giving their best research findings for this publication. I thank to Er. Rajneesh Anand, Head Chatha, and Deputy Director, IIIM Jammu; Dr. Suresh Chandra, Ex-Chief Scientist and my Head Department of Plant Sciences, and my other Departmental Collegues who continuously encouraged me for my hardwork and supported me whenever required.

I give credit of this book to all authors, scientists, professors and research scholars of India's topmost scientific organizations and institutes such as CSIR-Indian Institute of Integrative Medicine (IIIM Jammu, J&K), CSIR-Institute of Himalayan Bioresource Technology (IHBT Palampur, Himachal Pradesh), CSIR-Central Drug Research Institute (CDRI Lucknow, Uttar Pradesh), CSIR-National Botanical Research Institute (NBRI Lucknow, Uttar Pradesh), CSIR-Central Institute of Medicinal and Aromatic Plant (CIMAP, Lucknow), Assam University (Silchar, Assam), Govt. Degree College (Kathua, J&K), GB Pant National Institute of Himalayan Environment and Sustainable Development (Itanagar, Arunachal Pradesh and Almora, Uttarakhand), Indian Institute of Technology (Varanasi, Uttar Pradesh), North-Eastern Hill University (Shillong, Meghalaya), University of Jammu (Jammu, J&K), Vidyasagar University (Midnapore, West Bengal), Mahant Bachittar Singh College of Engineering and Technology (Babliana, J&K), Sikkim University (Gangtok, Sikkim) and University of Delhi (New Delhi). All these institutes/organizations are internationally known for their quality research in the field of basic sciences, drug discovery, value addition and product development. These organizations are also responsible for grooming young minds and encouraging new generation people for pursuing career in science and allied areas, and helping scholars to obtain their highest degree of study in the form of doctorate degree.

Presented invaluable contribution has come from various themes on which research articles were invited from various authors such as wild food plants, scope of medicinal and aromatic plants, product development and value addition from plants, importance of plants in relation to climate change and allied area focusing plants suitable for human survivals and wants of medicine. I am sure this publication will greatly help different workers in the task of new research activities, and will come beyond expectation of different thematic areas in which various research organizations are currently working in new era of society and science of technology transfer.

Heartliest thanks to Dr. Sougat Sarkar, Research Associate, CSIR-IIIM Jammu who encouraged me to write this book. I also thank my staff members, Rajesh and Gara associated with me during my hard time.

I would like to extend my gratitude to New India Publishing Agency, New Delhi (India) who agreed to publish these research outcomes. I appreciate all the excellent help that I received from all team members especially, NIPA during the production of this book. Thanks for their valuable time in correction of grammatical mistakes and their advised on format design of this book.

I am sure this book will attract civilians, academicians, students, scholars, industrialist and pharmaceutical companies those who are willing to understand the importance of plants for their daily needs, and looking for product development in the form of medicine and nutraceuticals. I would like to express my gratitude to many people who saw me through this book, to all those who provided support, talked things over and behind, offered comments, criticize and side-wise assisted in editing, proof-reading and designing of this book.

I would like to thank my wife, Manisha Singh, for standing beside me throughout my career and helped me indirectly in compiling this book **'Plants for Human Survival and Medicine'**. I also dedicate this book to my wonderful children, Aryan Singh and Aditi Singh for always making me smile and for understanding on those weekend mornings when I was busy in compiling and editing this book instead of playing with them, and not enjoying Park Avenue. I hope that one day they can read this book and understand why I spent so much time in front of my laptop and come home late from office and not giving much time.

Least, but not the end, I beg forgiveness from all those people who have been with me over the course of the years and whose names I have failed to mention in this book.

Bikarma Singh
Scientist & Editor
CSIR-Indian Institute of Integrative Medicine Jammu
(Council of Scientific & Industrial Research)
Ministry of Science and Technology
Govt. of India

Contents

Editor, Authors and Addresses
Number in parentheses indicate chapter number

Plants for Human Survival and Medicine

Editor and Address
Dr. Bikarma Singh (B.Sc. Honours, M.Sc. Gold Medalist, Ph.D., NABET)
Scientist, Plant Sciences (Biodiversity and Applied Botany Division), Section Incharge-Value Addition Centre, CSIR-Indian Institute of Integrative Medicine Jammu 180001, Jammu & Kashmir, INDIA

Authors and Addresses
Bikarma Singh *(1,3,5,6,8,13,17,19,21,23,24)*
Plant Sciences (Biodiversity and Applied Botany Division & Value Addition Centre) CSIR-Indian Institute of Integrative Medicine, Jammu-180001, Jammu & Kashmir INDIA; drbikarma@iiim.ac.in, drbikarma@iiim.res.in

Saroj Kanta Barik *(7)*
CSIR-National Botanical Research Institute, Lucknow-226001, Uttar Pradesh, INDIA sarojkbarik@gmail.com

Rajneesh Anand *(19)*
Instrumentation Division, CSIR-Indian Institute of Integrative Medicine Jammu-180001, Jammu-180001, Jammu & Kashmir, INDIA; ranand@gmail.com

Prasoon Gupta *(2)*
Natural Product Chemistry Division, CSIR-Indian Institute of Integrative Medicine Jammu-180001, Jammu & Kashmir, INDIA; guptap@iiim.ac.in

Inshad Ali Khan *(3)*
Clinical Microbiology Division, CSIR-Indian Institute of Integrative Medicine Jammu-180001, Jammu & Kashmir, INDIA; iakhan@iiim.ac.in

Anil Kumar Katare *(3,6)*
cGMP Division, CSIR-Indian Institute of Integrative Medicine, Jammu-180001 Jammu & Kashmir, INDIA; akkatare@iiim.ac.in

Gurdarshan Singh *(3,20)*
PK-PD Toxicology and Formulation Division, CSIR-Indian Institute of Integrative Medicine, Jammu-180001, Jammu & Kashmir, INDIA

Brijesh Kumar *(8)*
Sophisticated Analytical Instrument Facility Division, CSIR-Central Drug Research Institute, Lucknow-226001, INDIA; brijesh_kumar@cdri.res.in.

Yash Pal Sharma *(10)*
Department of Botany, University of Jammu, Jammu-180006, Jammu & Kashmir INDIA; yashdbm3@yahoo.co.in.

Govind Yadav *(13,19)*
Mutagenesis laboratory (Animal House Division), CSIR-Indian Institute of Integrative Medicine, Jammu-180001, Jammu & Kashmir, INDIA; gyadav@iiim.ac.in

Rajendra Bhanwaria *(6,17)*
Genetic Resources and Agrotechnology Division, CSIR-Indian Institute of Integrative Medicine, Jammu-180001, Jammu & Kashmir, INDIA rbhanwaria@iiim.ac.in.

Prem N. Gupta *(16)*
PK-PD-Toxicology and Formulation Division, CSIR-Indian Institute of Integrative Medicine, Jammu-180001, Jammu & Kashmir, INDIA; pngupta@iiim.ac.in.

Shreyans Kumar Jain *(4)*
Department of Pharmaceutical Engineering and Technology, Indian Institute of Technology (BHU), Varanasi -221005, Uttar Pradesh, India; sjain.phe@iitbhu.ac.in.

Krishna Upadhaya *(9)*
Department of Basic Sciences and Social Sciences, North-Eastern Hill University Shillong-793001, Meghalaya, INDIA; upkri@yahoo.com

Ratul Baishya *(14)*
Department of Botany, University of Delhi, Delhi-110007, INDIA rbaishyadu @gmail.com.

Wishfully Mylliemngap *(15)*
GB Pant National Institute of Himalayan Environment and Sustainable Development, North-East Regional Centre, Vivek Vihar, Itanagar, Arunachal Pradesh-791113, INDIA; wishm2015@gmail.com.

Amal Kumar Mondal *(18)*
Plant Taxonomy, Biosystematics and Molecular Taxonomy Laboratory, Department of Botany and Forestry, Vidyasagar University, Midnapore-721102, West Bengal INDIA; amalcaebotvu@gmail.com.

Mahendra Kumar Verma *(19)*
Medicinal Chemistry Division, CSIR-Indian Institute of Integrative Medicine Jammu 180001, Jammu-180001, Jammu & Kashmir, INDIA.

Arun Chettri *(22)*
Department of Botany, Sikkim University, 6th mile, Samdur, Tadong-737102, Gangtok Sikkim, INDIA; sentichettri@gmail.com.

Sumit G. Gandhi *(24, 25)*
Plant Biotechnology Division, CSIR-Indian Institute of Integrative Medicine, Jammu Tawi-180001, Jammu & Kashmir, INDIA; sumit@iiim.ac.in

Harish Chander Dutt *(24)*
Department of Botany, University of Jammu, Jammu-180006, Jammu & Kashmir INDIA.

Abdul Rahim *(24)*
PME Division, CSIR-Indian Institute of Integrative Medicine, Jammu Tawi-180001 Jammu & Kashmir, INDIA.

Yashbir Singh Bedi *(24, 25)*
Plant Biotechnology Division, CSIR-Indian Institute of Integrative Medicine, Jammu Tawi-180001, Jammu & Kashmir, INDIA.

Tapati Das *(12)*
Department of Ecology and Environmental Science, Assam University Silchar-788011, INDIA; das.tapati@gmail.com.

Venugopal Singamanenia *(2)*
Natural Product Chemistry Division, CSIR-Indian Institute of Integrative Medicine Jammu-180001, Jammu & Kashmir, INDIA.

Upasana Sharma *(2)*
Department of Applied Chemistry, Mahant Bachittar Singh College of Engineering and Technology, Babliana, Jammu-181101, Jammu and Kashmir, INDIA.

Durga Prasad Mindala *(3)*
cGMP Division, CSIR-Indian Institute of Integrative Medicine Jammu-180001 Jammu & Kashmir, INDIA.

Naresh Kumar Satti *(3)*
Natural Product Chemistry Division, CSIR-Indian Institute of Integrative Medicine Jammu-180001, Jammu & Kashmir, INDIA.

Bala Krishan Chandan *(3)*
Inflammation Pharmacology Division, CSIR-Indian Institute of Integrative Medicine, Jammu-180001, Jammu & Kashmir, INDIA.

Mowkashi Khullar *(3)*
Inflammation Pharmacology Division, CSIR-Indian Institute of Integrative Medicine, Jammu-180001, Jammu & Kashmir, INDIA.

Neelam Sharma *(3)*
Inflammation Pharmacology Division, CSIR-Indian Institute of Integrative Medicine, Jammu-180001, Jammu & Kashmir, INDIA.

Sushil Kumar *(5)*
Department of Botany, Government Degree College, Ramnagar-182122, Jammu and Kashmir, INDIA.

Rajendra Gochar *(6,17)*
Chatha Experimental Centre, Genetic Resources and Agrotechnology Division
CSIR-Indian Institute of Integrative Medicine, Jammu 180001, Jammu & Kashmir
INDIA.

Mamta Gochar *(6)*
Chatha Experimental Centre, Genetic Resources and Agrotechnology Division
CSIR-Indian Institute of Integrative Medicine, Jammu-180001, Jammu & Kashmir
INDIA.

Lucy B. Nongbri *(7)*
Department of Botany, North-Eastern Hill University, Shillong-793022, East Khasi
Hills, Meghalaya, INDIA.

Ladaplin Kharwanlang *(7)*
Department of Botany, North-Eastern Hill University, Shillong-793022, East Khasi
Hills, Meghalaya, INDIA.

Vikas Bajpai *(8)*
Sophisticated Analytical Instrument Facility Division, CSIR-Central Drug Research
Institute, Lucknow 226001, INDIA.

Aabid Hussain Mir *(9)*
Department of Environmental Studies, North-Eastern Hill University, Shillong-793022
Meghalaya, INDIA.

Licha Jeri *(9)*
Centre for Advanced Studies in Botany, North-Eastern Hill University
Shillong-793022, Meghalaya, INDIA.

Nazir Ahmad Bhat *(9)*
Centre for Advanced Studies in Botany, North-Eastern Hill University
Shillong-793022, Meghalaya, INDIA.

Rajib Borah *(9)*
Department of Basic Sciences and Social Sciences, North-Eastern Hill University
Shillong-793022, Meghalaya, INDIA.

Hiranjit Choudhury *(9)*
Department of Basic Sciences and Social Sciences, North-Eastern Hill University
Shillong-793022, Meghalaya, INDIA.

Yogendra Kumar *(9)*
Centre for Advanced Studies in Botany, North-Eastern Hill University
Shillong-793022, Meghalaya, INDIA.

Harpreet Bhatia *(10)*
Department of Botany, University of Jammu, Jammu-180006, Jammu & Kashmir
INDIA, harpreetbhatia2@gmail.com

Rajesh Kumar Manhas *(10)*
Department of Botany, Government Degree College, Kathua-184104, Jammu & Kashmir, INDIA.

Simmi Sharma *(11)*
CSIR-Institute of Himalayan Bioresource Technology, Palampur-176061, Himachal Pradesh, INDIA; Simmi17jan@gmail.com

Miniswrang Basumatary *(12)*
Department of Ecology and Environmental Science, Assam University Silchar –788011, INDIA.

Rakesh Kumar Nagar *(13)*
Mutagenesis laboratory (Animal House Division), CSIR-Indian Institute of Integrative Medicine, Jammu-180001, Jammu & Kashmir, INDIA.

Rhituporna Saikia *(14)*
Department of Botany, University of Delhi-110007, Delhi, INDIA.

Anup Kumar Das *(15)*
GB Pant National Institute of Himalayan Environment and Sustainable Development, North-East Regional Center, Vivek Vihar, Itanagar-791113, Arunachal Pradesh, INDIA.

Nako Laling *(15)*
GB Pant National Institute of Himalayan Environment and Sustainable Development, North East Regional Center, Vivek Vihar, Itanagar-791113, Arunachal Pradesh-791113, INDIA.

Om Prakash Arya *(15)*
GB Pant National Institute of Himalayan Environment and Sustainable Development, North East Regional Center, Vivek Vihar, Itanagar, Arunachal Pradesh-791113, INDIA.

RC Sundriya *(15)*
GB Pant National Institute of Himalayan Environment and Sustainable Development, Kosi-Katarmal, Almora-263601, Uttarakhand, INDIA.

Gourav Paudwal *(16)*
PK-PD-Toxicology and Formulation Division, CSIR-Indian Institute of Integrative Medicine, Jammu-180001, Jammu & Kashmir, INDIA; gouravpaudwal@gmail.com

Ayan Kumar Naskar *(18)*
Plant Taxonomy, Biosystematics and Molecular Taxonomy Laboratory, Department of Botany and Forestry, Vidyasagar University, Midnapore-721102, West Bengal INDIA.

Souradut Ray *(18)*
Plant Taxonomy, Biosystematics and Molecular Taxonomy Laboratory, Department of Botany and Forestry, Vidyasagar University, Midnapore-721102, West Bengal INDIA.

Sumit Singh *(19)*
Plant Sciences (Biodiversity and Applied Botany Division & Value Addition Centre)
CSIR-Indian Institute of Integrative Medicine, Jammu-180001, Jammu & Kashmir
INDIA.

Kiran Koul *(19)*
Plant Sciences (Biodiversity and Applied Botany Division & Value Addition Centre)
CSIR-Indian Institute of Integrative Medicine, Jammu-180001, Jammu & Kashmir
INDIA.

Surinder Kitchlu *(19)*
Plant Sciences (Biodiversity and Applied Botany Division & Value Addition Centre)
CSIR-Indian Institute of Integrative Medicine, Jammu-180001, Jammu & Kashmir
INDIA.

Bishander Singh *(19)*
Department of Botany, Veer Kunwar Singh University, Ara, INDIA.

Pooja Goyal *(19)*
Plant Biotechnology, CSIR-Indian Institute of Integrative Medicine, Jammu-180001
Jammu & Kashmir, INDIA.

Anjna Sharma *(20)*
PK-PD Toxicology and Formulation Division, CSIR-Indian Institute of Integrative
Medicine, Canal Road, Jammu-180001, Jammu & Kashmir, INDIA.
sharmaanjna 87@gmail.com

Javaid Fayaz Lone *(21)*
Plant Sciences (Biodiversity and Applied Botany Division & Value Addition Centre)
CSIR-Indian Institute of Integrative Medicine, Jammu-180001, Jammu & Kashmir
INDIA.

Sneha *(23)*
CSIR-Central Institute of Medicinal and Aromatic Plant, P.O. CIMAP
Lucknow-226015, INDIA.

Rajan Sachdev *(24)*
Plus Automation, 328-Nw Plot, Jammu-180005, Jammu & Kashmir, INDIA.

Sabeena Ali *(25)*
Microbial Biotechnology Division, CSIR-Indian Institute of Integrative Medicine
Srinagar – 190005, Jammu & Kashmir, INDIA.

Qazi Parvaiz Hassan *(25)*
Microbial Biotechnology Division, CSIR-Indian Institute of Integrative Medicine
Srinagar – 190005, Jammu & Kashmir, INDIA.

Introduction

Bikarma Singh

CSIR-Indian Institute of Integrative Medicine, Jammu
Jammu and Kashmir, INDIA
*Email: drbikarma@iiim.res.in

Humans are utilizing gifted natural resources such as plants, animals, sunlight, air, minerals, fossil fuels, water and microbes as the basic raw material for growth as well as for living on the earth. Human cannot create these natural resources, but, they modify little bit and use them accordingly as per their need. These natural resources are being exploited for food, medicine, urbanization, technological advancement and for industrialization since the start of human life. As plants are the basic requirement for humans, but people is involved in cutting plants, and destroying forests to fulfill their need of food, medicine, fuel, building materials for shelter and for developmental activities. Since time immemorial, a product that satisfies the human wants is intrinsic to flavour, fragrance, pharmaceutical, nutraceuticals and confectionery food industries, and these gets triggered due to industrialization coupled with globalization in recent few decades. Now, the people wants and needs have increased global demand for fine chemicals and various herbal products derived from medicinal and aromatic plants.

Whole plants or parts of plant such as roots, barks, stems, leaves, flowers, fruits and seeds are reported to be used as food and medicine. Since the beginning of human civilization, different human societies are harnessing these natural resources for their curative properties and for various pharmacological importance. Medicinal and aromatic plants have high demand in the global market. The Ministry of Environment, Forest and Climate Change of India and elsewhere across the world is continuously soughting to understand the nature and the diversity of genetically engineered crops which can be utilized to make various value added products that can be useful for commercialization (Singh 2019). This is the only reason which leads to development of new value added products and now a day herbal botanicals are replacing the existing market and the consumers select these products based on their quality, price and fewer side effect to humans. The traditional knowledge mentioned in Ayurveda, Unani and Chinese system of medicines is playing a significant role for development of new drugs and medicines (Chopra and Vishwakarma 2018). According to Fabricant and Farnsworth (2001), a total 122 compounds of defined structure have been isolated and characterized from 94 species of plants, and currently these molecules are globally used as drugs. Table 1 is provided with different drugs discovered from plants of ethnobotanical importance, plant's family and clinical usages.

History and Current Prespectives of Indian Medicinal Plants Research

Published information indicates that the usages of plants for treating human and animal diseases could be traced from 4500-1600 BC as mentioned in Hindus and Rigveda Scriptures (Meulenbeld 1999-2002). The juice of legendary mushroom, *Amanita muscaria*, often called as 'soma', was described as 'oshadhi' in literature and this plant is used as a heat-producing agent for human and animal health. While exploring the therapeutic potential of *A. muscaria*, the Indo-Aryans people discovered more plants having medicinal properties. During Vedic Period, Indo-Aryans Communities were familiar with several medicinal plants which gets described in Atharvaveda. Following this root of medicine discovery, Charak Samhita (1000-800BC), Sushrut-samhita (800-700 BC), Ashtangasangrah (heart of medicine) and Ashtangahridayasamhita (compedium of medicine) were written and described (Wujastyk 2003). According to Charaka (6th-2nd Century BC), 85% of human disease can be cured without a doctor, and only 15% of deadly disease requires a doctor (Meulenbeld 1999-2002). Unani system of medicine, originated in Greek (400 BC), came to India through Arab Physicians, accompanied by Mongols invaders and later on declined with fall of the Mongols in ancient Tamilakam in South India and Sri Lanka. The traditional Siddha system believed to have originated from Lord Shiva and passed through Parvati (wife of Lord Shiva) to a number of disciples through the use of medicinal plants popularized in Dravidian age.

As we proceeds and go deep into the history documentation, the scope of Hindu traditional system of medicine declined due to influence of

Table 1: Ethnobotanical plants, their active drug molecules and mode of clinical usages (modified after Fabricant and Farnsworth 2001)

Botanical name/Family	Drug	Action/Clinical Usages
Adonis vernalis L./ Ranunculaceae	adoniside	cardiotonic
Aesculus hippocastanum L. / Sapinadaceae	aescin	anti-inflammatory
Agrimonia eupatoria L./ Rosaceae	agrimophol	anti-helmintic
Ammi majus L./ Apiaceae	xanthotoxin	for vitilago and leukoderma
Ammi visnaga (L.) Lam./ Apiaceae	khellin	bronchodilator
Anamirta cocculus (L.) Wight & Arn./ Menispermaceae	picrotoxin	barbiturate antidote and analeptic
Ananas comosus (L.) Merr./Bromeliaceae	bromelain	anti-inflammatory, proteolytic agent
Andrographis paniculata (Burm.f.) Nees/Acanthaceae	andrographolide	bacillary dysentery
Anisodus tanguticus (Maxim.) Pascher/ Solanaceae	anisodamine	anti-cholinergic
Ardisia japonica (Thunb.) Blume / Primulaceae	bergenin	antitussive
Areca catechu L./Areaceae	arecoline	anthelmintic
Artemisia annua L./ Asteraceae	artimisinin	anti-malarial
Artemisia maritima L./ Asteraceae	santonin	ascaricide
Atropa belladonna L./ Solanaceae	atropine	pupil dilator and anticholinergic
Berberis vulgaris L./Berberidaceae	berberine	bacillary dysentery
Bergenia ciliata (Haw.) Sternb./ Saxifragaceae	bergenin	anti-inflammatory
Brassica nigra (L.) K.Koch/ Brassicaceae	allyl isothiocyanate	bubefacient
Camellia sinensis (L.) Kuntze/Theaceae	caffeine, theophy	stimulant, diuretic, anti-asthmatic and
Camptotheca acuminata Decne/ Cornaceae	llinecamptothecin	bronchodilator anti-cancerous
Cannabis sativa L./Cannabaceae	tetrahydrocannabinol (THC)	anti-emetic
Carapichea ipecacuanha (Brot.) A.Rich./ Rubiaceae	emetine	amoebicide, emetic
Carica papaya L./Caricaeae	papain	attenuator of mucus, proteolytic and mucolytic
Catharanthus roseus (L.) G.Don / Apocynaceae	vinblastine, vincristine	Hodgkin disease paediatric leukaemia
Centella asiatica (L.) Urb./Apiaceae	asiaticoside	vulnerary

Contd.

Plant / Family	Compound	Use
Chondrodendron tomentosum Ruiz & Pav./Menispermaceae	tubocurarine	skeletal muscle relaxant
Cinchona calisaya Wedd./ Rubiaceae	quinine	anti-malarial
Cinchona pubescens Vahl/Rubiaceae	quinine, quinidine	for malaria prophylaxis, cardiac arrhythmia
Cinnamomum camphora (L.) J.Presl. / Lauraceae	camphor	for rheumatic pain
Cissampelos pareira L./ Menispermaceae	cissampeline	skeletal muscle relaxant.
Colchicum autumnale L. / Colchicaceae	Colchicines	anti-gout, anti-tumor agent
Combretum indicum (L.) De Filipps/combretaceae	Trimethylindium	anthelmintic
Combretum indicum (L.) DeFilipps	quisqualic acid	anthelmintic
Convallaria majalis L./ Asparagaceae	convallotoxin	cardiotonic
Coptis japonica (Thunb.) Makino/ Ranunculaceae	palmatine	anti-pyretic, detoxicant
Corydalis ambigua Cham. & Schltdl./ Ranunculaceae	tetrahydropalmatine	analgesic, sedative
Crotalaria sessiliflora L./Fabaceae	monocrotaline	anti-tumor agent
Cullen corylifollium (L.) Medik. /Caesalpinaceae	psoralen	for vitilago
Curcuma longa L./Zingiberaceae	curcumin	choleretic
Cynara scolymus L./ Asteraceae	cynarin	choleretic
Cytisus scoparius (L.) Link/ Fabaceae	sparteine	oxytocic
Daphne genkwa Seibold & Zucc./ Thymelaeaceae	yunhuacine	abortifacient
Datura metal L./ Solanaceae	scopolamine	sedative
Datura stramonium L./ Solanaceae	scopol	for motion sickness
Digenea simplex (Wulf.) C.Agardh/ Rhodomelaceae	aminekainic acid	ascaricide
Digitalis lanata Ehrh./ Plantaginaceae	acetyldigoxin	cardiotonic
Digitalis purpurea L./ Plantaginaceae	digoxin, emetine	for atrial fibrillation and atrial fibrillation
Dysoxylum gotadhora (Buch.–Ham) Mabb./ Meliaceae	flavopiridol, rohitukine	anti-cancer, modulating apoptosis pathways
Ephedra sinica Stapf/Ephedraceae	ephedrine, pseudoephedrine	bronchodilator and rhinitis
Erythroxylum coca Lam./ Erythroxylaceae	cocaine	local anaesthetic

Contd.

Species/Family	Compound	Use
Filipendula ulmaria (L.) Maxim./Rosaceae	aspirin	analgesic and anti-inflammatory
Frangula purshiana Cooper/ Rhamnaceae	cascara	purgative
Fraxinus chinensis subsp. *rhynchophylla* (Hance) A.E.Murray/ Oleaceae	aesculetin	anti-dysenteric
Gaultheria procumbens L./ Ericaceae	methyl salicylate	rubefacient or skin allergic cause skin redness
Glaucium flavum Crantz/ Papaveraceae	glaucine	
Glycyrrhiza glabra L./Fabaceae	glycyrrhizin	sweetener, Addison's syndrome
Hamamelis virginia L./ Hamamelidaceae	gallotanins	haemorrhoid suppository
Hemsleya amabilis Diels/ Cucurbitaceae	hemsleyadin	bacillary dysentry
Huperzia serrata (Thunb.) Trevis/ Lycopodiaceae	huperzine A	enhancer, neuroprotective
Hydrangea macrophylla (Thunb.) Ser./Hydrangeaceae	phyllodulcin	sweetener
Hydrastis canadensis L./ Ranunculaceae	hydrastine	hemostatic, astringent
Hyoscyamus niger L./ Solanaceae	hyoscyamin	anti-cholinergic and bronchodilator
Justicia Adhatoda L. / Acanthaceae	vasicine	cerebral stimulant
Larrea divarticata Cav./ Zygophyllaceae	norihydroguaiartic acid	anti-oxidant
Lobelia inflata L./ Campanulaceae	lobeline	smoking deterrent, respiratory stimulant
Lonchocarpus nicou (Aubl.) DC./ Fabaceae	rotenone	piscicide
Lycoris squamigera Maxim./ Amaryllidaceae	galanthamine	cholineserase inhibitor
Melilotus officinalis (L.) Pall. /Fabaceae	dicoumarol	anti-thrombotic
Mentha arvensis L./ Lamiaceae	menthol	rubefacient
Mucuna pruriens (L.) DC. / Fabaceae	demecolcine	anti-tumor agent
Nicotiana tabacum L./ Solanaceae	nicotine	insecticides
Ocotea glazovii Mez/ Lauraceae	glaziovine	anti-depressant
Papaver somniferum L./ Papaveraceae	codeine, morphine, noscapine, papaverine	analgesic, anti-tussive, anti-tussive, and anti-spasmodic
Pausinystalia yohimbe Pierre ex Beille	yohimbine	aphrodisiac
Physostigma venenosum Balf. / Fabaceae	physostigmine, sigmasterol	for glaucoma, steroidal precursor
Pilocarpus jaborandi Holmes/ Rutaceae	pilocarpine	parasympathomimetic
Piper methysicum G.Forst. / Piperaceae	kawain	tranquilizer

Contd.

Species / Family	Compound	Use
Podophyllum peltatum L./Berberidaceae	etoposide	anti-tumor agent
Potentilla fragaroides L./ Rosaceae	(+)-catechin	haemostatic
Rauvolfia serpentina (L.) Benth. ex Kurz /Apocynaceae	rescinnamine, reserpine	anti-hypertensive
Rauvolfia tetraphylla L. /Apocynaceae	deserpidine	anti-hypertensive
Rhododendron molle G.Don/ Ericaceae	rhomitoxin	anti-hypertensive
Rorippa indica (L.) Hiern/Brassicaceae	rorifone	anti-tusssive
Salix alba L./ Salicaceae	salicin	analgesic
Sanguinaria candensis L./ Papaveraceae	sanguinarine	dental plague inhibitor
Senna alexandrina Mill./ Fabaceae	sennoside A & B	laxative
Silybum marianum (L.) Gaertn./Asteraceae	silymarin	anti-hepatotoxic
Simarouba amara DC./ Simarobaceae	glaucarubin	amoebicide
Sinopodophyllum hexandrum (Royle) T.S.Ying /Berberidaceae	podophyllotoxinrotundine	for condyloma acuminatum, analgesic, sedative
Sophora pachycarpa C.A.Mey./Fabaceae	pachycarpine	oxytocic
Stephania sinica Diels/ Menispermaceae	rotundine	analgesic, sedative, traquillizar
Stevia rebaudiana (Bertoni) Bertoni/ Asteraceae	stevioside	sweetener
Strophanthus gratus (Wall. & Hook.) Baill./ Apocynaceae	ouabain	cardiotonic
Strychnos guianensis (Aubl.) Mart./ Loganiaceae	toxiferine	relaxant in surgery
Strychnos nux-vomica L./ Loganiaceae	strychnine	CNS stimulent
Syzygium aromaticum (L.) Merr. & L.M.Perry /Myrtaceae	eugenol	for toothache
Theobroma cacao L./ Malvaceae	theobromine	diuretic, bronchodilator
Thymus vulgaris L./Lamiaceae	trichosanthin	abortifacient
Valeriana officinalis L./ Caprifoliaceae	valepotriates	sedative
Vinca minor L./Mabb./ Meliaceae	vincamine	cerebral stimulart
Veratrum album L./ Melanthiaceae	protoveratrines	anti-hypertensive
Withania somnifera (L.)Mabb./ Meliaceae	withanolides, sitaindoside VII withaferins, anaferine	anti-oxidant, anti-inflammator, immunomodulatory, anti-tumor and anti-arthritic

Greeks, Scythians, Huns, Mongols and Europeans. During British rule in India, there were many intermingling and several new medicinal plants come into the lime-light. Several english literatures published in the form of catalogues, dispensatories and pharmacopoeias at the end of 19th century. In addition Hooker's Flora (1872-1896) helped a lot in understanding the identity and the occurrence of more than 15,000 species of plants distributed in different agro-climatic zones of earnest India.

The end of 19th century and the beginning of 20th century witnessed research in various aspects of medicinal plants required for curing diseases and this emergence were due to the use of plants by different community people for treatment of different diseases based on their believes and application of it in their culture and tradition. The best examples are the establishment of School of Tropical Medicine at Calcutta in 1921, and the study of poisonous plants of India by Indian Council of Agricultural Research (ICAR) leading to the establishment of a Drug Research Laboratory at Jammu (presently CSIR-IIIM Jammu) in 1941. These organizations were involved in several All India Coordinated Research Project on medicinal and aromatic plants with 9 centre team from different parts of India, and the outcome of these research projects leads to various contribution in the field of discovery of many leads for medicines and drugs.

After India Independence in 1947, herbal botanicals revived and several new agencies were established by Government of India for research works to be carried out. The Botanical Survey of India with headquarters at Calcutta, the Forest Research Institute (FRI) at Dehradun, and the National Botanical Research Institute (NBRI) at Lucknow were established to improve the understanding of taxonomy and distribution of the indigenous flora of India and adjoining countries. The Ministry of Ayurveda, Yoga and Naturopathy, Unani, Siddha and Homeopathy of the Government of India coordinates and promotes research in the field of Ayurveda and Siddha system of medicine.

The discovery of drug reserpine (trade names: raudixin, serpalan, serpasil) from Himalayan herb, *Rauvolfia serpentina* (L.) Benth. ex Kurz (Indian snake root or Sarpagandha) in 1952 (total synthesis of *R. serpetina* were accomplished by R.B. Woodward in 1958) has brought revolution in drug discovery area from Indian plants. The reserpine is an indole alkaloid (Figure 1), used for controlling the high blood pressure, and provide relief from psychotic symptoms. If we look into the early history, the plant species were used for several decades for the treatment of fever and snake bites.

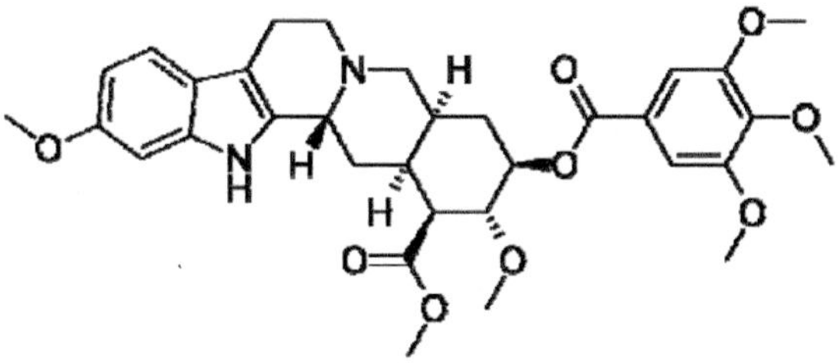

Fig. 1: Reserpine drug

It was then followed by discovery of vinblastine and vinablastin (Figure 2) from *Catharanthus roseus* (L.) G.Don. which was used to cure several types of cancer such as Hodgkin's lymphoma, lung cancer, brain cancer, bladder cancer, melanoma and testicular cancer, and the credit of its first isolation and pharmacological presentation goes to Robert Noble and Charles Thomas Beer at the University of Western Ontario of Madagascar.

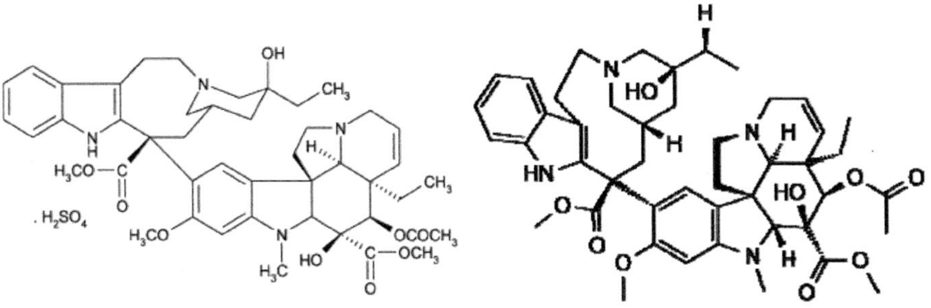

Fig. 2: Vinblastine (left) and Vinablastin (right) drug

The isolation and characterization of diosgenin from the rhizome of *Cheilocostus speciosus* (J.Koenig) C.D.Specht (synonym: *Costus speciosus* J.Koenig) carried forward the history of drug discovery for search of new compounds for human health care. Diosgenin is a polysteriod sapogenin (Figure 3) which has tremendous potential in pharmaceutical industries, and now-a-day several medicinal formulations available using this compound or plant parts extract as one of the main ingradient. This acts as a precursor for many hormones starting with marker degradation and synthesis of progesterone. As decades passes, the hunt for discovery of new molecules from plants as natural product speeds up, and the best examples are isolation and characterization of same diosgenin compounds which now a day gets isolated from several other species such as *Dioscorea composita* Hemsl.,

D. floribunda M.Martens & Galeotti, *D. alata* L., *D. hispida* Dennst, *Smilax menispermoides* Hayata, *Trillium govanianum* Wall. ex D.Don and *Trigonella foenum-graecum* L. Besides, several new compunds gets characterized and now this had wide application in product development. Currently diosgenin is used as an important bioactive phytochemical for preparation of several sterioidal drugs in pharmaceutical industry, and also used in treatment of cancer, hypercholesterolemia, inflammation and several other types of infections. After these decades, lots of new publications appeared as drugs such as artemisinin, bergenin, forskolin, CBD, THC, etc. which is serving as a direct medicine for human. Glossary of Indian Medicinal Plants and Indigenous Drugs of India by Chopra et al. (1956-1992) and Chopra (1933), respectively were some important contributions gifted to the world by Indian Scientists. After that a lot of drugs discovered and important leads were developed.

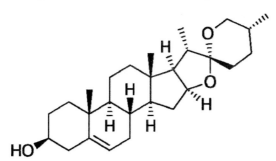

Fig. 3: Diosgenin drug

Traditional Knowledge, Plant and Drug Discovery

Chemo-systematics and morphological taxonomy applies to classify and identify the plants by studying and investigating the similarities and the differences of their physical appearance and biochemical compositions, respectively. The chemical science and the pharmacological investigation has contributed greatly to the pharmaceutical companies for discovering new drugs for curing deadly diseases. Ethnobotany which deals with the scientific study of indiginous knowledge and customs of people concerning plants and their medical, religious and other usages are important component of drug discovery programme. In the recent years, documentation of traditional knowledge of tribals and rural people has contributed lots in discovering new molecules from plants. Table 2 provide plant genera which has high value for medicinal use, their phytochemical constituents and associated biological activities.

Table 2: Genera represented by fifteen or more medicinal species having ethnobotanical usages and scientifically validated therapeutic properties

Genus / total species Worldwide	Family	Habit	Major phytochemical constituents	Pharmacological aspects
Aconitum L. / 337 spp. (Type specimen-CT: *Aconitum napellus* L.)	Ranunculaceae	herb	diterpene alkaloids (e.g. aconitine, lycoctonine, atisine, hypaconitine) and flavonoids (e.g.quercetin, kaempferol)	anodyne, anti- rheumatic, dispels internal cold, an arrow poison, anti- proliferative activity
Aralia L. / 74 spp. (Type specimen-LT: *Aralia racemosa* L.)	Araliaceae	shrub or trees	triterpenoid saponins (e.g. oleanolic acid), acetylenic lipids, triterpene carboxylic acid, flavonoids, tannins, phenols, glycosides, volatile oil.	tonic, anti-rheumatic; anti-inflammatory, anti-diabetic, anti-hyperlipidemic, hepatoprotective and anti-proliferative activities
Aristolochia L. / 485 spp. (Type specimen-LT: *Aristolochia rotunda* L.)	Aristolochiaceae	climber or liana	aristolochic acid, ceryl alcohol, â- sitosterol, Ishwarol, phenanthrene, stigmast-4-en-3-one, friedelin, cycloeucalenol and rutin	ant-ipyretic, anti-dotary for various infections, treatment of abdominal distension with pain and anti-fertility effect
Artemisia L. / 481 spp. (Type specimen-LT: *Artemisia vulgaris* L.)	Asteraceae	herb or shrub	hydrocarbons, oxygenated terpenes, acyclic monoterpenes (citronellol, myrcenol, linalool, artemisia ketone, artemisia alcohol), bicyclic monoterpenes (e.g. borneol, camphor), flavonoids (e.g. apigenin, artemisinin, luteolin, cirsiliol, kaempferol, rhamnocitrin,	anti-inflammatory, anti-microbial, hepato-protective, analgesic, anti-parasitic, anti-oxidant, hypolipidemic, anti-cancer, anti-ulcerogenic, anti-leishmanial, anti-tumor, anti-malarial, anti-

Contd.

Botanical name	Family	Habit	Chemical constituents	Pharmacological activities
			quercetin, mikanin, casticin, cirsilineol, eupatin, mearnsetin) coumarins, caffeoylquinic acids, sterols and acetylenes, volatile oils such as â-thujone, herniarin, 1,8-cineol, estragole, sabinyl acetate, cis chrysanthenyl acetate, davanone oil and terpineol	diabetic, anxiolytic, anti-convulsant, anti-promastigote, anti-depressant and anti-convulsant
Berberis L. / 580 spp. (Type specimen-LT: *Berberis vulgaris* L.)	Berberidaceae	shrub or tree	alkaloids (e.g. berberamine, palmatine, berberine, benzyl isoquinoline, berbamine, aromoline, quinoline), tannins, phenolic compounds, sterols and triterpenes	anti-microbial, anti-pyretic, anti-emetic, anti-oxidant, anti-inflammatory, anti-arrhythmic, sedative, anti-cholinergic, cholagogic, anti-leishmaniasis and anti-malarial
Clematis L. / 373 spp. (Type specimen -LT: *Clematis vitalba* L.)	Ranunculaceae	climber or liana	triterpenoid saponins (e.g. oleanane type, including oleanolic type, olean3â, 28-diol type, hederagenin type, hederagenin11,13-dien type, clematiunicinoside), flavonoids (e.g. apigeninÿ kaempfemlÿ luteolin, quercetin), steroids, tannins, proteins, fixed oils, carbohydrates.	anti-rheumatic, promotes blood circulation, anodyne; anti-inflammatory, anti-tumor activities, nervous disorders, syphilis, gout, malaria, dysentry, rheumatism, asthma, as analgesic, anti-bacterial and anti-cancer.
Codonopsis Wall. / 53 spp. (Type specimen-LT: *Codonopsis viridis* Wall.)	Campanulaceae	herb	polyacetylenes (e.g. lobetyolin, lobetyolinin, cordifolioidyne lobetyol), phenylpropanoids	revitalize the function of digestive system, relieves weakness; strengthen

Contd.

Plant	Family	Habit	Chemical constituents	Uses
			(e.g. tangshenosides, cordifoliketones), alkaloids (e.g. codonopsine, codonopsinine, codonopsinol, codonopsinol, syringin, codonopyrrolidium, tryptophan, tangshenoside, codonoside, nervolan), triterpenes (e.g. friedelin, codonopilate, taraxerol, á-spinasterol, taraxeryl acetate, codonolaside, lancemaside), flavones (e.g. luteolin, tricin, kaempferol, hesperidin), organic acids (e.g. succinic acid, caffeoylquinic acid, linoleic acid) and polysaccharides; volatile oils (e.g. 1,2-benzonedicarboxylic acid dibatyl-ester, heptedecanoic acid, 2,4,5-triisopropyl styrene, palmitic acid, linolic acid).	immune system, improve poor gastrointestinal function, gastric ulcer and appetite, decrease blood pressure, anti-tumor activity, anti-diabetic activity, hepatoprotective activity, anti-inflammatory activity.
Corydalis DC. / 586 spp. (Type specimen-LT: *Corydalis bulbosa* DC.)	Papaveraceae	herb	isoquinoline alkaloids (e.g. protopine, 13-oxoprotopine, 13-oxocryptopine, stylopine, coreximine, rheagenine, ochrobirine, sibiricine, bicuculline), anthraquinones, steroids (e.g. stigmasterol, â-sitosterol, daucosterol).	anti-pyretic, anti-dotary; anti-bacterial activities, anti-inflammatory, anti-oxidation, cardiovascular diseases, promotes blood circulation, central nervous system, analgesic effects, and anti-injury for hepatocyte

Contd.

Botanical name	Family	Habit	Chemical constituents	Uses/effects
Cynanchum L. / 283 spp. (Type specimen-LT: Cynanchum acutum L.)	Apocynaceae	climber or liana	benzene derivatives (acetophenones, e.g. cynandione A, cynanchone), C21 steroids (e.g. four-ring pregnane type, 14,15-secopregnanetype, 13,14:14,15-diseco-pregnane type, aberrant 14,15-seco-pregnane type, 12,13-seco-14,18-nor-pregnane type), steroidal saponins, alkaloids (e.g. phenanthroindolizidine, antofine, vincetene, tylophorine), flavonoids (e.g. glycosides, kaempferol, quercetin, hyperin and terpene (e.g. monoterpene diglycosides neohancoside A &B, taraxasterol, â-amyrin, lupeol, oleanolic acid).	anti-pyretic, promotes blood circulation, anti rheumatic, anti-tumor, neuroprotective and anti-fungal effects, anti-cancer, anti-inflammatory, anti-virus, appetite suppressing.
Delphinium L. / 457 spp. (Type specimen- CT: Delphinium peregrinum L.)	Ranunculaceae	herb	alkaloids (e.g. barbinine, barbinidine, delavaine A/B, deoxylycoctonine, delcosine, grandiflorine, deltatsine, 14-acetyldictyocarpine, methyllycaconitine), flavanoids (e.g. kaempferol).	anodyne, anti-rheumatic, anti-inflammatory.
Dendrobium Sw./ 1523 spp. (Type specimen- CT: Dendrobium moniliforme (L.) Sw. typ. cons. Epidendrum moniliforme L.	Orchidaceae	epiphyte	phenols (e.g. gallic acid), alkaloids, coumarins, terpenes (e.g. ursolic acid, â-sitosterol, lupeol), flavonoids.	used as tonic, for febrile diseases with thirst and dry mouth, dry cough and chronic tidal fever, cytotoxic effect, anti-oxidant

Contd.

Taxon	Family	Habit	Phytochemicals	Uses/Activity
Fritillaria Tourn. ex L. / 141 spp. (Type specimen-LT: *Fritillaria meleagris* L.)	Liliaceae	herb	alkaloids (e.g. cevanine, jervine, solanidine, puqiedine, lichuanisinine, pengbeimine), saponins (e.g. steroidal aglycone, pallidifloside A-H) and terpenoids (e.g. kaurane diterpene), and various other phytochemicals.	anti-tussive, expectorant, treating cancers, used for treating lumps beneath skin.
Gentiana L. / 359 spp. (Type specimen-LT: *Gentiana lutea* L.)	Gentianaceae	herb	gentiolutelin, gentioluteol 5-carboxyl-3,4-dihydrogen-1 H-2-benzopyran-1-one, roburic acid, erythrocentauric acid, oleanolic acid, gentiopicroside	anti-pyretic, anti-oxidant activity, anti-dotary, heptoprotective, anti-inflammatory, anti-rheumatic, diuretic, stomachic, anti- tumor immunomodulatory,
Hypericum L. / 458 spp. (Type specimen-LT: *Hypericum perforatum* L.)	Hypericaceae	herb	naphtodiantrones (hypericin, pseudohypericin), flavonoids (hyperoside, rutin, quercetin quercitrin kaempferol 3 -rutinoside, rutin-acetyl), tannins, caftaric acid, chlorogenic acid, acylphloroglucinols (hyperforin, adhyperforin), flavonol glycoside (hyperoside), essential oil constituents.	anti-microbial, anti-dotary, anti-rheumatic, diuretic,diuretic, stomachic, anti-oxidant, wound healing, analgesic, anti-inflammatory activity.
Ilex L. /475 spp. (Type specimen-LT: *Ilex aquifolium* L.)	Aquifoliaceae	shrub or tree	anthocyanins, flavonoids, terpenoids, sterols, amino acids, alkaloids (e.g. theobromine, caffeine), fatty acids, alcohols,	anti-microbial, anti-inflammatory, anti-fungal, pyretic, antidotary for the treatment of

Contd.

			carbohydrates, carotenoids, cyanogenic glucoside, phenols and phenolic acids.	cardiovascular illness.
Isodon (Schrad. ex Benth.) Spach / 107 spp. (Type specimen-LT: *Isodon rugosus* (Wall. ex Benth.) Codd., basionym: *Plectranthus* sect. *Isodon* Schrad. ex Benth.)	Lamiaceae	herb or shrub	glycosides, alkaloids, phenolic compounds, flavonoids, saponins, essential oils, tannins and terpenoids ; 7,20-epoxy -ent-kaurene diterpenoids, phyllostachysins D–H, rabdoloxins A–B, rabdoinflexin B, amethystoidin A, rabdokunmin D, macrocalyxin E, daucosterol, 5,7-dihydroxy-42 -hydroxylflavone, oleanolic acid.	anti-microbial, anti-cancer, anti-pyretic, anti-rhematic, anti-inflammatory, anti-diabetic, cytotoxic, phytotoxic, anti-oxidant, anti-spasmodic, anthelmintic.
Lysimachia L. /193 spp. (Type specimen-LT: *Lysimachia vulgaris* L.)	Primulaceae	herb	flavonol and flavone (e.g. myricetin, quercetin, kaempferol, isorhamnetin and flavonoid mono-, di- and tri-glycosides).	anti-rheumatic, blood circulation, anti-inflammatory emmenagogue, anti-oxidant activities.
Persicaria Mill. / 66 spp. (Type specimen-LT: *Persicaria maculosa* Gray.)	Polygonaceae	herb	Flavonoids (e.g. catechin, (–)-epicatechin, hyperin, isoquercitrin, isorhamnetin, kaempferol, quercetin, quercitrin, rhamnazin, rutin), sesquiterpenes (e.g. 3-β-angeloyloxy-7-epifutronolide, 7 -ketoisodrimenin, changweikangic acid, dendocarbin L, (+)-fuegin, futronolide, polygonumate, (+)-winterin), phenylpropanoid esters	anti-dysentric, haemostatic, anti-pyretic, anti-oxidant, anti-microbial, anti-helminth, anti-feedant, cytotoxicity, anti-inflammatory, oestrogenicity, anti-fertility, anti-adipogenicity, anti-cholinesterase, neuroprotection.

Contd.

Polygala L. / 623 spp. (Type specimen-LT: *Polygala vulgaris* L.)	Polygalaceae	herb	(e.g. hydropiperosides A & B, vanicosides A,B&E, phenolic acids–caffeic acid, chlorogenic acid, ρ-coumaric acid). xanthones (e.g. sibiricoxanthone, methylmangiferin, lancerin), triterpenic saponin (e.g. polygalasaponins I–XIX), oligosaccharides (e.g. senegoses A–O, glomeratoses A–G), aryltetrahydronaphthalene lignans,	anti-inflammatory, anti-cancer, tonic, relieves weakness, tranquilizer; folk medicine as anesthetics, for treatment of disturbances of bowel, kidney, and central nervous system (CNS).
Polygonum L. / 217 spp. (Type specimen-LT: *Polygonum aviculare* L.)	Polygonaceae	herb	quinones (e.g. physcion, emodin, fallacinol, anthraglycoside, rhein, polyganin, cuspidatumin), stilbenes (e.g. polydatin, resveratrol-42 -O-glucoside, resveratrol 4-O-D-(62 -galloyl)- glucopyranoside, sodium and potassium trans -resveratrol-3- O-β-D-glu copyranoside-43 -sulfate), flavonoids (e.g. rutin, quercetin, hyperoside, (+)-catechin), coumarins, lignans, and others phytochemicals such as Protocatechuic acid, oleanolic acid, β-sitosterol, tartaric acid; volatile oils (e.g. 2-hexenal, 3-hexen-1-ol, n-hexanal, 1-penten-3-ol, 2-penten-1-ol ethy vinyl ketone.	anti-dysentric, haemostatic, anti-pyretic, anti-oxidant, anti-microbial, anti-helminth, anti-feedant, cytotoxicity, anti-inflammatory, anti-nociceptive, oestrogenicity, anti-fertility, anti-adipogenicity, anti-cholinesterase, neuroprotection; Alzheimer×s disease, Parkinson×s disease, hyperlipidaemia, and cancer.

Contd.

Rubus L. / 1494 spp. (Type specimen-LT: *Rubus fruticosus* L.)	Rosaceae	scandent or busy shrub	anthocyanins, phenolics and ascorbic acids; 3,5-dihydroxy benzoic acid, gallic acid, ethyl galactoside oleanolic acid, β-sitosterol, 3-O-[β-D-galactopyranosyl-(12)-D-glucopyranoside.	astringent, anti-cancer, heart disease (e.g. antioxidant, anti inflammatory and cell regulatory effects), anti-inflammatory
Salvia L. / 986 spp. (Type specimen-LT: *Salvia officinalis* L.)	Lamiaceae	herb	polyphenols (e.g. caffeic acid, carnosic acid, kaempferol, oleanolic acid, rosmarinic acid, ursolic acid), volatile constituents (e.g. monoterpene hydrocarbons, oxygen -containing monoterpenes and oxygen- containing sesquiterpenes; alpha-pinene, 1,8-cineole, linalool, limonene, myrcene, beta-caryophyllene, spathulenol. viridiflorol, beta- caryophyllene oxide, carene, alpha-bisabolol)	treat coronary heart diseases, anti-pyretic, anti-dotary, microbial infections, anti-plasmodial, cancer, malaria, anti-inflammation, loss of memory, promotes blood circulation.
Scutellaria L. / 468 spp. (Type specimen -LT: *Scutellaria galericulata* L.)	Lamiaceae	herb	flavones (e.g. 4-hydroxy wogonin, apigenin, scutellarin, luteolin, quercetin, rivulari, apigenin 5- O-β-glucopyranoside, baicalein); flavonoid glycosides (e.g. scutellarin, baicalein, 5,8-dimethoxyflavone-7-O-D-glucuronopyranoside, kaempferol-3-O-β-D-rutinoside), flavanones (e.g. carthamidin, isocarthamidin, eriodictyol, 7-hydroxy-5,8,2'-trimethoxy flavanone), alkaloids (e.g.	anti-pyretic, anti-dotary, promotes blood circulation, haemostatic, anti-inflammatory, anti-viral, sedative, anti-oxidant, anti-convulsant, hepatoprotective, memory improvement.

Contd.

Swertia L. / 105 spp. (Type specimen-LT: *Swertia perennis* L.)	Gentianaceae	herb	scutebarbatine A,), diterpenes (e.g. scutebarbatine F), triterpenoids (e.g. scutellaric acid), polysaccharide, essential oils (e.g. hexahydrofarnesyl acetone, 3,7,11,15- tetramethyl-2-hexadeceen-1-ol, menthol, 1-octen-3-ol). xanthones (e.g. mangiferin, neolancerin, swertipunicoside, swertiabisxanthone-I), flavonoids (e.g. isoorientin, isovitexin, luteoli, quercetin, apigenin, swertisin, isoswertisin), iridoid glycosides (e.g. swertiamarin, gentiopicroside, amarogrentin, amaroswerin) and triterpenoids (e.g. oleanane, ursane, taraxerane, lupane, hopene, isohopane, gammacerane, swertane, chiratane, lanostane skeletons, sweriyunnanosides A-C), oleanolic acid, hederagenin, and several others	tonic, anti-pyretic, stomachic, anti-oxidant, hepatoprotective, anti-hepatotoxic, anti-microbial, mutagenicity, anti-carcinogenic, anti-leprosy, hypoglycemic, anti-malarial, anti-cholinergic, CNS depressant, anti-inflammatory.
Thalictrum Tourn. ex L. / 157 spp. (Type specimen-LT: *Thalictrum foetidum* L.)	Ranunculaceae	herb	benzylisoquinoline-derived alkaloids, aporphine, pavine, phenanthrene, protoberberine groups, imers of the bisbenzylisoquinoline, secobisbenzylisoquinoline, and aporphine-benzylisoquinoline groups.	anti-pyretic, anti-dysentric, anti-microbial.

Note: Authority: L.= Carl von Linnaeus (1707-1778, Sweden); Sw.= Olof Swartz (1760-1818, Sweden, West Indies), DC.= Augustin Pyramus de Candolle (1778-1841, Switzerland), Wall. = Nathaniel Wallich 1786-1854, London).

The data presented are gathered from the online sources such as The Plant List (http://www.theplantlist.org), Tropicos (http://www.tropicos.org), published revisionary works, using google search and consultation of books on drug discovery pigeoned in libraries. The aim of this tabulation is to provide one place information of different important plant genus that occurs in India which will help in characterization of new molecules and preparation of formulation for human diseases in near future.

It is estimated that approximately 9,500 registered herbal industries and a multitude of unregistered cottage level herbal units set-up in India depends on the continuous supply of medicinal plants for manufacturing of herbal medical formulations and extracts which is mostly based on traditional system of medicines such as Ayurveda, Unani and Amchi system (Ved and Goraya 2007). International trade of medicinal plants and plant extracts has become a major force in the global economy. Demands of medicine and nutraceutical plants have been increased both in under-developing and developed countries. Now a days, the supply of medicine and food raw material faces challenges in the form of fullfiling demands of people in their daily needs, in relation to cultivation, conservation, maintaining standard, sustainability, world marketing and other similar issues related to plants and human survival. Cultivation can increase the production and yield, but requires proper understanding of plant-environment interactions that could alter the environmental metabolic pathways (Craker and Simon 2004). World trade of medicinal plants can make raw materials available, but this requires a regulatory system to impose uniform standards to maintain the quanlity standards and their control parameters. Table 3 provide highly traded species of Indian Himalaya that has commercial demand in International markets for product development and for pharmaceutical sectors in drug formulations.

Further, illustration relates to interaction of human and environment is evident from the fact that all living organisms depends on or look for their multiple use of distinct ecological environment in search of wild resources (Ladio and Lozada 2004, Singh 2016). As already discussed above, that the plant diversity of the Indian Himalaya were utilized by the native communities for medicine, food, fodder, fuel, timber and tools, and it can be concluded that wild plants of Himalaya can be used as commercial food items at the time of food scarcity and absence of cultivated crops (Samant and Dhar 1997). Samant and Dhar (1997) pointed out that 675 plant species belonging to 384 genera

Table 3: Highly traded species of Indian Himalaya for food and medicine (Modification after Singh et al. 2016, Singh and Bedi 2017)

Botanical name / Family	Part traded	Trade category		Market price [Rate/kg (INR)]	
		Food	Medicine	Village market	National level
Achillea millefolium L. / Asteraceae	dried aerial parts	NA	available	NA	70-100
Agaricus campestris L. /Agaricaceae	fresh aerial parts	available	NA	10-20	40-60
Alisma plantago-aquatica L./ Alismataceae	fresh rhizomes	available	NA	10-15	NA
Allium carolinianum DC./ Amaryllidaceae	fresh tubers	available	NA	15-20	NA
Allium consanguineum Kunth/ Amaryllidaceae	fresh tubers	available	NA	20-30	NA
Allium humile Kunth/ Amaryllidaceae	fresh tubers	available	NA	10-20	NA
Amaranthus caudatus L./ Amaranthaceae	fresh aerial parts	available	NA	5-10	NA
Angelica archangelica L./ Apiaceae	fresh leaves	available	NA	5-10	NA
Angelica glauca Edgew./ Apiaceae	dried roots	NA	available	NA	70-100
Artemisia dracunculus L./ Asteraceae	dried leaves	NA	available	NA	75-100
Cousinia lappacea Schrenk/ Asteraceae	dried roots	NA	available	NA	50-60
Asparagus racemosus Willd./ Asparagaceae	dried tubers	NA	available	NA	220-240
Berberis lycium Royle/ Berberidaceae	fresh berries	available	NA	20-25	NA
Persicaria amplexicaulis (D.Don) Ronse Dec/ Polygonaceae	dried roots	NA	available	NA	50-60
Bunium persicum (Boiss.) B. Fedtsch/ Apiaceae	dried fruits	NA	available	NA	500-600
Cannabis sativa L./ Cannabaceae	dried leaves	NA	available	NA	200-250
Carum carvi L./ Apiaceae	dried seeds	NA	available	NA	300-350
Centella asiatica (L.) Urb./ Apiaceae	dried leaves	NA	available	NA	90-120
Chenopodium album L./ Amaranthaceae	fresh twigs	available	NA	10-15	NA
Cichorium intybus L./ Asteraceae	dried aerial parts	NA	available	NA	60-120
Codonopsis ovata Benth./ Campanulaceae	dried roots	NA	available	NA	70-80
Cyperus rotundus L./ Cyperaceae	dried tuber	NA	available	NA	75-100
Dioscorea bulbifera L./ Dioscoreaceae	fresh tubers	available	NA	15-20	NA

Contd.

Botanical name/ Family	Part used	Availability	Availability	Market price	Market price
Diplazium esculantum (Retz.) Sw./Athyriaceae	fresh twigs	available	NA	10-15	NA
Eriobotrya japonica (Thunb.) Lindl./ Rosaceae	fresh fruits	available	NA	12-15	NA
Fagopyrum esculentum Moench/ Polygonaceae	fresh twigs	available	NA	8-10	NA
Foeniculum vulgare Mill./ Apiaceae	dried seeds	NA	available	NA	180-200
Fragaria nubicola (Lindl. ex Hook.f.) Lacaita/ Rosaceae	dried roots	NA	available	NA	150-160
Fragaria vesca L./ Rosaceae	fresh fruits	available	NA	10-20	40-50
Elaeagnus rhamnoides (L.) A.Nelson	fresh fruits	available	NA	10-15	50-70
Juglans regia L./ Juglandaceae	dried fruits	available	NA	20-25	180-350
Lactuca sativa L./ Asteraceae	fresh twigs	available	NA	10-20	NA
Maianthemum purpureum (Wall.) LaFrankie/ Asparagaceae	fresh twigs	available	NA	10-20	NA
Malus domestica Borkh./ Rosaceae	fresh fruits	available	NA	20-40	80-200
Mentha arvensis L./ Lamiaceae	fresh leaves	available	NA	35-50	NA
Morchella esculenta Dill. ex Pers. / Morchellaceae	whole plant	available	available	3000-5000	20000-30000
Morus alba L./ Moraceae	fresh fruits	available	NA	30-40	NA
Oxyria digyna (L.) Hill./ Polygonaceae	fresh leaves	available	NA	10-15	NA
Prunus cornuta (Wall. ex Royle) Steud.	fresh fruits	available	NA	10-20	NA
Rheum emodi Wall./ Polygonaceae	dried roots	NA	available	NA	110-120
Rubia cordifolia L./ Rubiaceae	dried tubers	NA	available	NA	150-170
Rubus niveus Thunb./ Rosaceae	fresh berries	available	NA	10-20	NA
Salvia moorcroftiana Wall. ex Benth./ Lamiaceae	dried roots	NA	available	NA	120-140
Sinopodophyllum hexandrum (Royle) T.S.Ying / Berberidaceae	dried rhizomes	NA	available	60-70	130-150
Thymus serpyllum L./ Lamiaceae	dried aerial parts	NA	available	20-30	110-130
Trillium govanianum Wall. ex D.Don	dried rhizomes	NA	available	40-50	150-170
Zingiber officinale Roscae / Zingiberaceae	fresh rhizomes	available	available	20 25	70-90

*NA indicate not available in the local market as food or as medicine**Market price values varies from place to place, and depends on seasonal availability

and 145 families are used as wild edible food in Indian Himalayas, and parts of these plants are either consumed as raw, roasted, boiled, fried, cooked or consumed as oil, spice, seasoning materials, jam and pickle. Table 4 presented provide a list of several wild edible plants used by tribal communities residing in Indian Himalaya.

Table 4: Wild edible plants used by tribal communities of Western Himalaya (modification after Singh et al. 2016, Samant and Dhar 1997)

Scientific name/Family	Local Usages
Abies pindrow (Royle ex D.Don) Royle/ Pinaceae	dried bark used for making herbal teas.
Achillea millefolium L./Asteraceae	aerial parts used for making alcoholic local drinks.
Agaricus campestris L./ Agaricaceae	aerial parts fried and cooked as vegetables.
Agrimonia pilosa Ledeb./Rosaceae	aerial parts used for making herbal teas.
Alisma plantago-aquatica L. / Alismataceae	rhizome boiled, and stewed with meat as vegetables.
Allium carolinianum DC./ Amarylidaceae	dried powdered tubers added to soups.
Allium consanguineum Kunth / Amarylidaceae	fresh tubers cooked as vegetable.
Allium humile Kunth / Amarylidaceae	whole plants cooked as vegetable.
Allium victorialis L./Amarylidaceae	whole plants can be cooked as food along with meat and vegetables.
Amaranthus caudatus L. / Amaranthaceae	aerial parts cooked as vegetable
Anaphalis triplinervis (Sims) Sims ex C.B. Clarke/Asteraceae	flower buds eaten as salad by shepherds.
Angelica archangelica L./ Apiaceae	leaves dried as well as fresh leaves added as flavoring agent while cooking vegetables.
Angelica glauca Edgew./Apiaceae	root dried and powdered, and use as flavoring agent in vegetables.
Artemisia dracunculus L. /Asteraceae	dried leaves used in preparation of local drinks.
Asparagus racemosus Willd./ Liliaceae	fresh tubers consumed as raw food
Berberis lycium Royle /Berberidaceae	ripe blue berries consumed as raw foods.
Berberis pachyacantha Bien. ex Koehne/ Berberidaceae	ripe pink berries consumed raw as food.
Bunium persicum (Boiss.) Fedtsch./Apiaceae	matured fruits dried and used as spices.
Cannabis sativa L./ Cannabiaceae	dried leaves used in preparation of local soup.

Contd.

Capsella bursa-pastoris (L.) Medk./ Brassicaceae — aerial parts fried or cooked as vegetables.

Carum carvi L./ Apiaceae — seeds poured in salad for taste and quality improvement.

Centella asiatica (L.) Urb./ Apiaceae — fresh leaves consumed as salads, and also added while cooking meat and fish.

Cerastium davuricum Fisch. ex Spreng/ Caryophyllaceae — young twigs cooked as vegetables.

Chenopodium album L./ Amaranthaceae — aerial parts fried or cooked as vegetable.

Cichorium intybus L./ Asteraceae — young twigs stewed with meat and eaten as source of tonic for human

Codonopsis ovata Benth./ Campanulaceae — fresh roots consumed raw as tonic

Codonopsis rotundifolia Benth./ Campanulaceae — fresh roots edible as wild ready-made food.

Colchicum luteum Baker/ Liliaceae — dried powdered seeds taken after heavy food.

Crataegus rhipidophylla Gand./ Rosaceae — fresh red ripe fruits eaten as fruits

Cyperus rotundus L./ Cyperaceae — fresh tubers eaten raw.

Dioscorea bulbifera L./ Dioscoreaceae — tubers boiled or fried in oil as vegetables or meat.

Dioscorea deltoidea Wall. ex Grseb./ Dioscoreaceae — young twigs boiled, and consumed as vegetables.

Diplazium esculantum (Retz.) Sw. / Dryopteridaceae — fresh young fronds cooked as vegetables.

Dipsacus inermis Wall./ Dipsacaceae — Young twigs cooked as vegetables.

Duchesnea indica (Jacks.) Focke/ Rosaceae — fresh fruits edible as wild readymade energy yielding food

Elsholtzia densa Benth./Lamiaceae — fresh leaves used for making local chutney.

Elsholtzia eriostachya (Benth.) Benth./ Lamiaceae — fresh leaves used for making local chutney.

Epilobium parviflorum Schreb./ Onagraceae — flowers used in preparation of local herbal tea and wines.

Eremurus himalaicus Baker/ Xanthorrhoeaceae — aerial parts cooked as vegetables.

Eriobotrya japonica (Thunb.) Lindl./Rosaceae — fruits eaten fresh.

Fagopyrum esculentum Moench/Polygonaceae — young twigs cooked as vegetable

Ficus auriculata Lour./ Moraceae — pinkish ripe fruits eaten raw.

Filipendula vestita (Wall. ex G.Don) Maxim./ Rosaceae — fresh flowers used in preparation of herbal teas.

Foeniculum vulgare Mill/Apiaceae — dried seeds used as spices

Fragaria nubicola (Lindl. ex Hook.f.) Lacaita/ Rosaceae — fresh as well as dried roots used in making herbal teas.

Fragaria vesca L./Rosaceae — red ripe fruits consumed as raw food.

Gagea lutea (L.) KerGawl./Liliaceae — dried tubers used as spices

Galium odoratum (L.) Scop./Rubiaceae — young twigs cooked as vegetables.

Gentiana tianschanica Rupr. ex Kusn./ — fresh parts consumed as salad or

Contd.

Gentianaceae

Gentianella moorcroftiana (Wall. ex Griseb) Airy Shaw/ Gentianaceae — roots used in local alcoholic drinks fresh roots chewed as tonic or as herbal teas.

Geranium wallichianum D.Don ex Sweet/ Geraniaceae — dried powdered roots used in preparation of vegetables.

Heracleum candicans Wall. ex DC./Apiaceae — fresh twigs eaten as salad

Hippophae rhamnoides L./ Elaeagnaceae — fruits eaten fresh or used to make alcoholic juice.

Hypericum perforatum L./Hypericaceae — flowers used for making alcoholic drinks.

Impatiens glandulifera Royle/Balsaminaceae — fresh seeds edible.

Juglans regia L./Juglandaceae — kernal of seeds eaten fresh or stir-fried, and used for making vegetable oil.

Juniperus communis L./ Cupressaceae — berries used in making herbal teas.

Lactuca lessertiana (Wall. ex DC.) Wall. ex C.B. Clarke/ Asteraceae — leaves boiled, fried and used as vegetables.

Lactuca sativa L./ Asteraceae — fresh young twigs eaten raw as salad.

Lathyrus humilis (Ser.) Spreng./ Fabaceae — fresh seeds edible.

Maianthemum purpureum (Wall.) LaFrankie/ Liliaceae — young stems boiled, stir-fried or used in making local soup.

Malus domestica Borkh./ Rosaceae — ripe fruits eaten fresh raw.

Malva neglecta Wall./ Malvaceae — leaves boiled, fried and consumed as vegetables.

Medicago minima (L.) L./ Fabaceae — leaves boiled or fried in oil as vegetable or fresh leaves eaten

Mentha arvensis L./ Lamiaceae — fresh leaves chewed as mouth freshner.

Mentha longifolia (L.) L./ Lamiaceae — fresh leaves used as chutney.

Morus alba L./ Moraceae — ripe fruits eaten as readymade food and home made chutney prepared from unripe fruits.

Oxalis acetosella L./ Oxalidaceae — fresh tubers consumed to alleviate thirst.

Oxyria digyna (L.) Hill/ Polygonaceae — leaves eaten as chutney.

Persicaria alpina (All.) H.Gross/ Polygonaceae — fresh stem chewed tonic, and also used for making local home made chutney.

Persicaria amplexicaulis (D.Don) Ronse Decr/ Polygonaceae — dried powdered roots used for preparation of local herbal teas.

Phlomoides bracteosa (Royle ex Benth.) Kamelin & Makhm./ Lamiaceae — leaves dried powered used in preparation of herbal teas.

Phytolacca acinosa Roxb./ Phytolaccaceae — tender young stems cooked as vegetables.

Pinus wallichiana A.B.Jacks/ Pinaceae — seeds eaten fresh by children

Plantago depressa Willd./ Plantaginaceae — whole plants boiled or fried in oil as vegetables.

Contd.

Plantago lanceolata L./ Plantaginaceae	leaves boiled, and cooked as vegetables.
Polygonum aviculare L./ Polygonaceae	fresh young leaves boiled, and cooked as vegetables.
Potentilla atrosanguinea G.Lodd. ex D.Don/ Rosaceae	fresh leaves cooked as vegetables
Prunus armeniaca L./ Rosaceae	kernel of ripe fruits eaten raw.
Prunus cornuta (Wall. ex Royle) Steud./ Rosaceae	ripe fruits eaten fresh.
Rheum emodi Wall./Polygonaceae	young twigs consumed as seasonal vegetables.
Rheum webbianum Royle/ Polygonaceae	fresh young stem consumed as salad, and also used in making local chutney.
Ribes alpestre Wall. ex Decne./ Grossulariaceae	ripe fruits eaten fresh or used to prepare alcoholic drinks.
Ribes orientale Desf./ Grossulariaceae	ripe fruits eaten fresh.
Rosa macrophylla Lindl./ Rosaceae	dried berries used in preparation of herbal teas.
Rosa webbiana Wall. ex Royle/ Rosaceae	ripe fruits eaten fresh.
Rubia cordifolia L./ Rubiaceae	fresh aerial twigs boiled, and used in local alcholic drinks
Rubus alceifolius Poir./ Rosaceae	ripe fruits consumed fresh.
Rubus caesius L./ Rosaceae	ripe fruits consumed fresh.
Rubus idaeus L./ Rosaceae	ripe fruits consumed fresh.
Rubus niveus Thunb./ Rosaceae	ripe fruits consumed fresh.
Rubus saxatilis L./ Rosaceae	ripe fruits consumed fresh.
Rumex acetosa L./ Polgonaceae	young leaves cooked as vegetables.
Rumex nepalensis Spreng./ Polygonaceae	young leaves cooked as vegetable
Rumex patientia L./ Polygonaceae	fresh leaves used for making local chutney.
Salvia moorcroftiana Wall. ex Benth./ Lamiaceae	dried roots mixed with vegetables. and consumed as foofitem.
Sambucus wightiana Wall. ex Wight & Arn./ Adoxaceae	fresh fruits edible.
Saussurea costus (Palc.) Lipsch./ Asteraceae	fresh thick roots boiled, and used in vegetables.
Silene baccifera (L.) Roth./ Caryophyllaceae	leaves and young twigs cooked as vegetables.
Sinopodophyllum hexandrum (Royle) T.S.Ying/ Berberidaceae,	fresh ripe fruits consumed as wild food.
Solanum americanum Mill./ Solanaceae	black ripe fruits eaten fresh.
Sonchus oleraceus (L.) L./ Asteraceae	fresh leaves used as salad and consumed as source of tonic.
Taraxacum campylodes G.E. Haglund / Asteraceae	herbal tea prepared from fresh leaves and used as tonic.
Thymus serpyllum L./Lamiaceae	aerial parts used for preparation of herbal teas.

Contd.

Trifolium pratense L./ Fabaceae	whole plant or fresh parts consumed as tonic, and as salad
Trifolium repens L./ Fabaceae	fresh parts consumed as source of body tonic, and used as salad
Trillium govanianum Wall. ex D.Don / Liliaceae	fresh rhizomes used for making herbal teas.
Urtica dioica L./ Urticaceae	leaves boiled, fried and cooked as vegetable along with potatoes and meat.
Verbascum thapsus L./ Scrophulariaceae	fresh flowers used for making herbal teas.
Viola canescens Wall./Violaceae	fresh flowers used for making herbal teas.
Zingiber officinale Roscoe/Zingiberaceae	rhizomes used for preparation of herbal teas.

The content of the book "Plants for Human Survival and Medicine" presents twenty-five chapters written by different scientists, professors and research scholars working in different research organizations. A brief synopsis with reference to key-points and content of particular chapter is summarized below:

Chapter 1 'Himalaya is a repository of wild food and medicine, a close look on plants for human survival' is described by Singh. This chapter deals with the importance of wild edible plant used as food and medicine by humans. Illustration to these words, this section focused on how the plants plays significant role in human survival and how they are contributing to sustainable development of human society. It is of the record that plants are the major source of different bioactive compounds that can be used directly as drugs and medicines. This compilation reports that Indian Himalaya is repository of 8000 plant species and most of these plants used in Indian traditional system of medicines, *viz.* ayurveda, unani, siddha or amchi since ancient time. Ethnobotanical knowledge contributes greatly to the discovery of several drugs, and, therefore, this chapter also focuses on the economic plants of Himalayan origin and their application as ethnomedicine, wild fruits and leafy vegetables, herbal teas, alcoholic drinks, and other important human usages. According to this section, 122 compounds have been obtained from ethnomedicinal plants that are globally used as direct drugs. This indicate that tribal knowledge on plants and drug discovery cannot be separated, and new drugs for deadly diseases can only be discovered from ethnobotanical leads. This chapter is then followed by a chapter on Astavarga plant by Singamanenia et al. where eight endangered Himalayan herbs are described. These includes four orchids *Habenaria intermedia* D.Don, *Platathera edgeworthii* (Hook. f.ex collett) R.K. Gupta, *Malaxis acuminata* D.Don, *Malaxis muscifera* Lindl., three liliaceae members (*Lilium polyphyllum* D.Don, *Polygonatum verticillatum* L., *Polygonatum cirrhifolium* Wall.) and one zingiber (*Roscoea purpurea* Sm.). Chemistry and pharmacology of these plants are described in detail which indicates that these plants are popular in various Ayurvedic formulations, as for instance, Chyawanprash which is well recognized for strengthening vital force to the body, help in cell regeneration, and help in building immune system. Bioactive ingredients isolated from these plant such as diosgenin, β-sitosterol, catechin, and p-coumaric acid shows positive biological functions and can be used as anti-microbial, anti-oxidant, and anti-inflammatory agents in drug discovery programmes.

Chapter 3 deals with *Colebrookea oppositifolia* Sm. plants as an important hepato-protective phytopharma plant that are used in drug discovery and this section deals with important findings of Katare et al. This chapter discusses the phytochemical and the pharmacological aspects of this species leading to new discovery as hepato-protective plant. This is evident from its role as anti-microbial, anti-fungal and anti-oxidant properties which is due to the presence of the bioactive molecules such as n-triacontane, acetyl alcohol, 32-hydroxydotriacontyl ferulate, β-sitosterol and 5,6,7,42 -tetramethoxy flavones. An attempt has been made to provide data on biology, natural products, constituent of volatile oil, salient markers, medicinal chemistry, result showing hepatoprotective parameters, and future direction for research on *C. oppositifolia*. This chapter was then followed by chapters on two important Himalayan plants *Dysoxylum binectariferum* Hook.f., and *Zanthoxylum armatum* DC. which is presented in chapter 4 and chapter 5, respectively. As per Jain, rohitukine has been emerged as potential preclinical cyclin-dependent kinase (Cdk) inhibitor in drug discovery. Camptothecine and 9-methoxy-camptothecine were isolated from *Dysoxylum binectariferum* for the first time by bioassay-guided fractionation. Presented chapter provides original work which was carried out at CSIR-Indian Institute of Integrative Medicine, Jammu, (India) on this plant. Phytochemistry of *D. binectariferum*-dysoline, camptothecin and 9-methoxy-camptothecin, medicinal chemistry and preclinical evaluation of IIIM-290 as a semi-synthetic derivative of rohitukine is also discussed in this chapter. Thakur and Singh explained the *Zanthoxylum armatum* DC. as an important plant used in traditional ayurvedic system of medicine. Biology and chemistry are explained briefly along with future conservation perspectives of this species and its application in development of value added pharmaceutical products. Volatile oils of this species indicate active constituents as linalool, limonene and ligan. *Z. armatum* has been declared as an endangered species as per International Union for Conservation of Nature and Natural Resources (IUCN) Red List Category.

Chapter 6 deals with *Grewia asiatica* L., locally called 'Phalsa' which yields berries rich in calories, proteins, carbohydrates, dietary fibers, vitamins, macro-and micro-nutrients such as calcium, phosphorus and sodium. As per Gochar et al., CSIR-Indian Institute of Integrative Medicine (IIM) Jammu has developed and released a new variety of this plant for rural prosperity and for development of value-added products such as health drinks and nutraceuticals. The chemical constituents of this species show anti-oxidant, anti-diabetic, anti-microbial, hepatoprotective, anti-fertility, anti-fungal, analgesic and anti-viral activities. Nongbri et al. described *Panax* L. of Northeast India in chapter 7 and pointed out that this genus has high commercial demand for its ginsenoside contents. Taxonomy, ecology and phenology of five species of *Panax* found in north-eastern India *viz., P. assamicus* R.N.Banerjee, *P. bipinnatifidus* Seem., *P. pseudoginseng* Wall., *P. variabilis* J.Wen and *P. sokpayensis* is discussed in this chapter. This is then followed by chapter 8 on *Butea monosperma* (Lam.) Taub. which is known as 'Flame of the Forest'. This plant species is used in Ayurveda, and Unani system of medicine due to presence of important constitutent which causes medicinal properties. This species is used in treatment of diabetes, diarrhoea, and sore throat. As per Bajpai et al., the recent studies revealed that the phytochemicals responsible for osteogenic activity is present in the bark of *B. monosperma*, and therefore, for the first time, Ultra-Performance Liquid Chromatography Tandem Mass Spectrometry (UPLC-MS/MS) method was developed for identification and simultaneous quantitation of bioactive compounds from different parts of *B. monosperma*. Chapter 9 by Mir et al. discussed the diversity, distribution and commercial importance of Indian Magnolias, and reports that 25 species of Magnolia occurs in India, out of which *Magnolia gustavii* and *M. pleiocarpa* are critically endangered, *M. pealiana* is endangered, *M. manii* and *M. nilagirica* is vulnerable, nine are Least Concern and ten Data Deficient at Global level. All species of Magnolia have huge economic potential and are used for a number of purposes including ornamental, medicinal, culinary, timber and joinery works.

Bhatia et al. describes chapter 10 which deals with traditional medicinal plants used to cure dermatological disorders in Udhampur district of J&K State. Ethnomedicinal data for this study were collected by interviewing 91 infomants between the age group 26-89 years and this investigation revealed that 64 plant species belonging to 59 genera and 43 families were used by the tribals and natives of district Udhampur for the treatment of 25 ailments of skin disorders. Another important Himalayan plant, *Picrorhiza kurroa* Royle ex Benth. is presented in chapter 11 by Sharma, which by deals with the chemistry and biological activities associated with *P. kurroa* chemical constituent, and this is helpful in development of new drug as medicine. Basumatary and Das describes chapter 12 which deals with plant resources used by riparian communities of Dhir and Diplai Wetlands of Assam. According to this study, total 48 plant species recorded were used as vegetables, fruits, fodder, fuel, wood, medicine, roofing materials, craft making materials, bio-fertilizer and fish poisoning agent.

The 'Indian Frankincense' botanically known as *Boswellia serrata* Roxb. ex Colebr. known for boswelic acid as major chemical constituent is presented in chapter 13 by Nagar et al. The plant shows pharmacological activities such as anti-rheumatic, anti-pyretic, anti-cancerous, anti-inflammatory, anti-hyperlipidemic, anti-coronary, analgesic and hepato-protective properties which is due to the presence of several loaded bioactive molecules such as α-amyrins, β-boswellic acid, acetyl-β-boswellic acid, 11-keto-β-boswellic acid, acetyl-11-keto-β-boswellic acid and tetracyclic triterpenoic acids. Besides, this species also yield essential oils whose major constituents are monoterpenoids such as α-pinene (2.05-64.7%), cis-verbenol (1.97%), trans-pinocarveol (1.80%), borneol (1.78%), myrcene (1.71%), verbenone (1.71%), limonene (1.42%), thuja-2,4(10)-diene (1.18%) and p-cymene (1.0%) and all has pharmaceutical applications in one way or the other.

Phosphate solubilizing bacteria, *Withania somnifera* (L.) Dunal as multiferous plants, volatile profiling of *Cymbopogon khasianus* [IIIM (J) CK-10 Himrosa] and *in vitro* investigation of anti-diabetic plant *Dendrophthoe falcata* (L.f.) Ettingsh is presented by Saikas and Baishya, Das et al., Paudwal and Gupta, and Bhanwaria et al. which is described detail in chapter 14, 15, 16, 17 and 18, respectively. Chapter 19 deals with crop-weather interactions, phytochemistry, pharmacology and evaluation of the phenological models for *Echinacea purpurea* (L.) Moench. by Koul et al. and describes that this species is a repository of several bioactive moleculer loaded in the form of phenylpropanoids, flavonoids, terpenoids, lipids, nitrogenous compounds, carbohydrates and some others such as ascorbic acid, sitosterol, linoleic acid, cyanidin glycosides, and sesquiterpene esters in different quantities. Chapter 20 of Sharma and Singh describes commercial valuable medicinal wealth of Hamirpur district of Himachal Pradesh. Today, cosmetic and pharmaceutical industries are the fast growing sectors that uses the botanical extracts to maintain the health and support integrity of the skin and other parts of human body. Cosmeceutical additives of market made from plants as herbal botanicals are of high demand as these causes fewer or no side effects, and therefore, the concept of ethnomedicine is gaining immence importance for the cosmetic and the value added products development of natural origin.

Human and animal skin is the largest covering of the body as it provide barrier to the internal tissues, and protect the body from infections and toxic chemicals. Keeping in mind the importance of skin, investigations and literatures works were consulted on plants to review the status of skin protecting plants. The potential high altitude plants of Himalaya were screened out to be used in skin care and treatment of various skin pathogens and all these are preented in chapter 21 which entitled revisiting Himalayan High altitude plants for skin care and disease. In this chapter, total 81 species of plants under 72 genera belonging to 40 families were presented that have ethnobotanical application in skin care and usages in curing skin diseases. This chapter provide the base-line information for cosmetic and pharmaceutical companies for development of various new value-added products for human and animal health care. Similarly, describes forest resources of Sikkim Himalaya with special reference to Pangolakha Wildlife Sanctuary. Forty nine economical valued plant species belonging to 46 genera and 35 families were documented and their importances for human needs is described in this chapter. Singh and Sneha presented chapter 23 of this book on *Saussurea costus* (Falc.) Lipsch. that describes the traditional uses, therapeutic potential and conservation aspects. According to this chapter, *S. costus* is traditionally used for the treatment of more than forty diseases happen to occurs in hilly and mountaineous regions, and this is due to the presence of various bioactive compounds such as costunolide, germacrene, lappadilactone, isodihydro-costunolide, cynaropicrin, linoleic acids, cyclocostunolide, alantolactone, isoalantolactone and sesquiterpene-saussureamines in plant parts. Chapter 24 deals with herbarium and importance of digitization of plant vouchers for studying biodiversity. In this chapter, the development of the original Janaki Ammal

Herbarium (JAH) database that contains a comprehensive and searchable data of voucher snaps, historical background and statistical analysis of 23, 225 voucher specimens is described and highlighted the importance of taxonomy and authentic certification of plant sample. Chapter 25 deals with medicinal genus *Aconitum* which is in the limelight of current scenerio of Pharmacopeia. A large number of species under this genus is placed under/IUCN Red Data category. Chemistry indicate that this genus is a repository of aconitine, mesaconitine and pyroaconine. Tubers are used as remedy for pain, neuronal disorders, inflammation and rheumatism. Therefore, a short review on diversity, pharmacology and conservation aspects is presented in this chapter. It can be concluded that all chapters in two book are of worth importance and published information in the form of book **"Plants for Human Survival and Medicine"** will provide one place several aspects which will be useful in drug discovery and for generation of come.

ACKNOWLEDGEMENT

Editor would like to thank all authors to of this book for their research contribution in publishing this book.

REFERENCES CITED

Chopra RN, Nayar SL, Chopra IC, Asolkar LV, Kakkar KK. 1956-1992. Glossary of Indian Medicinal Plants. Council of Scientific & Industrial Research, New Delhi, India.

Chopra RN. 1933. Indigenous drugs of India. Art Print, Calcutta, India.

Chopra RN, Nayar SL, Asolkar LV, Kakkar KK, Chakre OJ, Verma BS. 1956-1992. Glosssary of Indian Medicinal Plants, 3 Vols. Council of Scientific and Industrial Research, New Delhi, India.

Chopra VL, Vishwakarma RA. 2018. Plants for Wellness and Vigour. New India Publishing Agency, New Delhi, India.

Cracker LE, Simon JE. 2004. XXVI International Horticultural congress: The future for medicinal and aromatic plants. *ISHS Acta Horticulture*, ISBN 978-90-66055-07-0.

Fabricant DS, Farnsworth NR. 2001. The value of plants used in traditional medicine for drug discovery. *Environmental Health Perspectives* 109(1):69-75.

Ladio AH, Lozado M. 2004. Patterns of use and knowledge of wild edible plants in distinct ecological environments: a case study of a Mapoche community from North Western Patagonia. *Biodiversity and Conservation* 13(6): 1153-1173.

Meulenbeld GJ. 1999-2002. History of Indian medical literature. Published by Egbert Forsten, Groningen.

Samant SS, Dhar U. 1997. Diversity, endemism and economic potential of wild edible plants of Indian Himalaya. *The International Journal of Sustainable Development and World Ecology* 4(3): 179-191.

Singh B, Bedi YS. 2017. Eating from Raw Wild Plants in Himalaya: traditional knowledge documentary on Sheena tribes along LoC Border in Kashmir. *Indian Journal of Natural Products and Resources* 8(3): 269-275.

Singh B, Sultan P, Hassan QP, Gairola S, Bedi YS. 2016. Ethnobotany, Traditional Knowledge, and Diversity of Wild Edible Plants and Fungi: A Case Study in the Bandipora District of Kashmir Himalaya, India. *Journal of Herbs, Spices & Medicinal Plants* 22(3): 247-278.

Singh B. 2019. Plants of Commercial Values. New India Publishing Agency, New Delhi, India. (ISBN: 978-93-87973-503).

Taylor L. 2000. The healing power of rainforest herbs. Published in plants based drugs and medicine (www.rain-tree.com).

Ved DK, Goraya GS. 2007. Demand and supply of medicinal plants in India. NMPB, New developmental FRZHT, Bangalore, India.

Wujastyk D. 2003. The Roots of Ayurveda. Published by Penguin, London.

1

Himalaya is a Repository of Wild Food and Medicine: A Close Look on Plants for Human Survival

Bikarma Singh

[1]Plant Sciences (Biodiversity and Applied Botany), CSIR-Indian Institute of Integrative Medicine, Jammu-180001, Jammu and Kashmir, INDIA
[2]Academy of Scientific and Innovative Research, New Delhi-110001 INDIA
*Corresponding email: drbikarma@iiim.res.in, drbikarma@iiim.ac.in

ABSTRACT

Plants play an important role in human survival and contribute to sustainable development as they are the major source of different bioactive compounds such as taxol, digoxin, reserpine and vinblastin used directly as drug and medicines for human diseases. As plants are basic need and requirement for living, they are used as herbal remedy, and known to produce lead compounds for semi synthesis of higher activity or lower toxicity. Indian Himalaya regions are repository of more than 8,000 plants species and out of which 50% are endemic and unique due to the presence of morphological and chemical structure of peculiar characters. Most of the plant used in Indian traditional system of medicines such as Ayurveda, Unani, Siddha or Amchi described for different formulations and ethnomedicinal applications had distributional range restricted to the hills and the valleys of Himalayas. Hence, in this communication, an attempt has been made to present economic plants of Himalayan origin used as ethnomedicine, wild food plants as vegetables, herbal teas, home-made soups, wild salads, chutney, underground snacks, alcoholic drinks, nutrients as burnt powder and other human usages. Besides, this unique Himalayan plant has potential to be used as wild source for the development of new drugs and medicines which can sustain future to human survival and had the ability to produce medicine against deadly diseases. It is estimated that 80% of the world population uses traditional medicine made from herbal botanicals to cure disease and illness. It is of the record that 122 compounds obtained from ethnomedicinal plants is globally used as drugs. This indicates that tribal knowledge on plants and drug discovery cannot be separated, and new drugs for unique deadly diseases can be discovered only from ethnobotanical leads. More than 5,000 species reported from Indian Himalayas are of ethnobotanical values and they are used as medicine or wild food by different tribal communities. It can be concluded that Himalayan tribals are endowed with strong culture associated with botanicals and have age-old tradition with respect to the use of wild plants as food and medicine. Plants provide a real substitute for the primary health care systems in hills and valleys, and definitely going to serve as jewel for mankind in near future to come.

Keywords: Himalayan plants, Wild food, Medicine, Human survival, Drug discovery.

INTRODUCTION

Himalayan mountain of South-Central Asia extends to approximately 2400 km and includes Indian States such as Arunachal Pradesh, Assam, Himachal Pradesh, Jammu and Kashmir, Meghalaya, Uttarakhand and Sikkim. These hilly belts of India encompassed with nine of the world's ten highest mountain peak ranges (Anonymous 2018). Hills and Valleys of this unique hotspot represented with variation in the elevation ranges from 305 m to 8849 m above mean sea level (ASL), precipitation and humidity varies as elevation preceeds from low altitude towards high, higher diversity in species composition (lower or higher group), and even patterns of human livelihoods differs (Singh et al. 2018). It is estimated that this mountaineous belts provide ecosystem services to 52.7 million people including India and neighbouring countries, such as Bhutan, China, India, Nepal and Pakistan (Singh and Thadani 2015).

Schild (2008) called Himalaya mountains as 'Water Tower of Asia' as this region provide a suitable environment for the origin of several perennial rivers and support the highest biodiversity and referred as endemism centre for unique plants (Xu et al. 2009). It is of record that Cherrapunji in Meghalaya of eastern Himalaya averages annual rainfall of 12,000 mm, while Leh-Ladakh division in western Himalaya receives only 100 mm rainfall annually (Singh and Thadani 2015), which indicate diverse variation in climatic variables. According to Bolch et al. (2011), the glaciers velocity show great variability with many glaciers in Karakoram ranges of northern Himalaya, whereas in other parts, glaciers diversity is limited. From the ecological point of views, the forest productivity in Himalayan regions are very high as it serves as an important repository of carbon pool stock (Singh and Thadani 2013). It has been recorded that as Himalaya regions extend from the east to the west arc, biodiversity composition, rainfall intensity and living culture of local inhabitants varies accordingly (Bhattacharjee et al. 2017). As far as forest types and vegetation composition is concern, its diversity ranges from tropical to arctic, and from very moist to almost xeric environment (Singh and Thadani 2015).

Few of the world's major river system such as the Indus, the Ganges and the Brahmaputra originates from the Himalayas, and these high ranges causes a profound effect on the climate of the region, and helps to keep the monsoon rains on the Indian soils and limiting rainfall on the tibetan plateaus. The climate ranges from tropical at the base of the mountains to permanent ice and snow at the highest elevations as indicated above, and estimated that the total number of higher plants (angiosperms and gymnosperms) on this planet is 250,000, and out of these 18,664 species recorded from India, and placed India as the tenth richest country in terms of biodiversity. Himalayas alone supports more than or approximately 8,000 plant species, out of which about 50% species are endemic and has localized distribution.

As plants are the main sources for food and medicine, only about 6% species of the world screened for their biological activities, and of these again only 15% has been evaluated for phytochemicals. Illustrating further, publication of Fabricant and Farnsworth (2001) reported that in the United States of America, for every 10,000 pure compounds that are biologically evaluated, only 20 used to get tested in animal models, 10 of these gets clinically evaluated, and only one compound reaches to the United States Food and Drug Administration for getting approvals in drug formlation and for marketing, and the total time taken under this process estimated to be about 10 years at a cost of $231 million (Vagelos 1991, Falsrieant and Farnswoah 2001).

Since plants play significant role in human survival and are one of the major source of therapeutic agents, several bioactive compounds gets isolated for direct use as drugs such as morphine, reserpine and taxol. Besides, researchers can produce bioactive compounds of novel or known structures as lead compounds for the semi-synthesis to produce patentable entities of higher activity and/or lower toxicity, and can be used as for the biological testing, and more importantly various herbal remedy for treatment of local seasonal diseases. Literatures reveals that extensive ethnobotanical works on Himalayan tribes have been undertaken to study plant Human interactions with nature and some of them includes Gujjar, Kashmiri, Pahari, Bakarwal, Garo, Sheena, Himachali, Arunachali, Khasis and Boto (Rashid et al. 2008, Azad 2013, Singh et al. 2012, 2014; Sajem and Gosai 2010, Lokho and Narasimham 2013, Singh et al. 2016, Singh and Bedi 2017, Singh et al. 2018). In this context an attempt also have been made to give the information regarding ethnobotanical plants used by the hilly people of Himalayas as wild food and medicine, and future perspectives of Himalayan origin plants in drug development and new value added products formulation.

Ethnobotany and Drug Discovery

At least 25% of the prescribed drugs issued in USA and Canada markets contain bioactive compounds that are derived from plants as the natural products and fine ingradients(Farnsworth 1966). Many of these drugs were discovered by following leads provided from indigenous knowledge systems and herbal formulation prepared by different tribes for curing their day-to-day diseases (Gairola et al. 2016). It is scientifically validated that traditional knowledge leads have proven effective in drug discovery. Several reviews have been published pertaining to approaches for selecting plant as a candidate for drug discovery programmes (Phillipson and 1989, Kinghorn 1994, Farnsworth 1966, Harvey 2000, Spjut and Perdue 1976). Table 1 deals with reviewed ethnobotanical information published on some plants by different ethnobotanists as wild food and medicine which can be used as potential plants for future drug discovery and formulation of new products from plants as needed for human survival (Dar and Khuroo 2013, Kachroo et al. 1997, Koul 1997, Kuchroo and Nahni 2006, Malik et al. 2013, Sharma and Kuchroo 1983).

Table 1: Traditional usages of Himalayan plants for curing diseases and consumed as wild food

Plant name/ Population status	Documented studies on ethnobotany in India
Anaphalis triplinervis (Sims) Sims ex C.B.Clarke/ Endemic to Asia; common in Kashmir Himalaya	Fresh roots and leaves are used in stomach pain, and dried leaves used in fever (Gorsi and Miraj 2002). Flower buds are consumed raw as salad (Ballabh et al. 2007).
Asparagus racemosus Willd./ Endemic to Asia; sparsely distributed in Himalaya belts	Tubers consumed raw as fruits by Garo tribe of Meghalaya (Singh et al. 2012). It is also used in treatment of diabetes, jaundice, and urinary disorder by people in Sikkim (Das et al. 2012). Roots and tubers are crushed, mixed with hot water and extract drunk in case of pneumonia (Singh et al. 2014). Half cup tuberous root's decoction is diluted with equal amount of milk and taken once a day for three month as remedy for treatment of epilepsy by Tripuri and Reang tribes in Assam (Das et al. 2009). Decoction made from tubers of *Asparagus racemosus* and bark of *Azadirachta indica* is given twice a day for treatment of diabetes and also to check blood sugar level (Swarnkar and Katewa 2008). Decoction obtained from root used to cure blood diseases, diarrhoea, dysentery, cough, bronchitis and mental debility; also root is boiled with cow milk and used for increasing milk secretion (Sankaranarayanan et al. 2010). Leaves are dried, powdered and taken orally to cure stomach disorders by Lushai tribe of Assam (Sajem and Gosai 2010).
Berberis lycium Royle/ Endemic to Asia; common in Himalaya belts	Root used as tonic, good for cough and throat troubles; leaves used to cure jaundice (Gorsi and Miraj 2002). Boiled water of root is used to cure bone fracture (Mir 2014). Roots and young apical shoots are dried, in shade, boiled in water, and dried, and then decoction prepared called Rasaunt, which is used to cure eye infection (Uniyal et al. 2006). Ripe fruits are consumed raw in Suru and Zanskar region of Ladakh (Ballabh et al. 2007). Fruit extract taken to cure stomach pain, diarrhoea, jaundice and liver diseases; roots extract called *Rasaunt* used as cooling agent and eye lotion (Gupta et al. 2013). Ripe fruits are eaten raw in Himachal Pradesh by tribal communities (Sharma et al. 2009). Fruits are eaten by local people in Kishtawar area of Himalaya (Kumar and Hamal 2009).

Contd.

Berberis pachyacantha Koehne ssp. *zabeliana* Jafri/ Rare and endemic to Kashmir Himalaya

Fruits used as medicine by people in Cold Desert of Himalaya (Kala 2006).

Centella asiatica (L.) Urban/ Common throughout Asia,

Whole plant is boiled with water, fried with mustard oil and consumed as medicine for treatment of dysentery (Singh et al. 2014). Whole plant parts are crushed, and used to cure leprosy, tuberculosis and asthma (Sajem and Gosai 2010). Roots and leaves are eaten raw or cooked as vegetable (Kayang 2007). Aerial parts used as vegetable with smalled fishes, and also prefered as salad and chutney in Poga area of Assam (Pegu et al. 2013). Juice of leaves applied on cuts, wounds and boils, and grind paste from whole plant prepared is used for treatment of syphilis (Begum and Nath 2014), asthma, urinary discharges and improving brain memory (Das et al. 2012).

Codonopsis ovata Benth./ Rare and endemic to Kashmir Himalaya

Fresh roots are consumed raw by Ladakh people of Himalaya (Ballabh et al. 2007). Powdered roots are used in treatment of ulcers and wounds (Kapahi et al. 1993).

Codonopsis rotundifolia Benth./ Rare and endemic to Kashmir Himalaya

Fruits eaten as vegetable, and roots considered as aphrodisiac (Gupta et al. 2013). Extract prepared from aerial parts used for treatment of asthma and body weakness in livestock (Mir 2014). Leaves and roots are used in treatment of wounds and cuts.

Crataegus rhipidophylla Gand./ Scarce in Himalaya belts

Fruits eaten as wild food

Cyperus rotundus L./ Common in Himalaya belts

Tuber paste is used as appetizer. Decoction made after crushing with root of *Solanum torvum* and stem of *Tinospora cordifolia* is used in treatment of puerperal diseases, and tuber paste mixed with honey given in dyspepsia (Saharia and Sharma 2011). Tubers are crushed and taken with honey in case of diarrhoea and indigestion (Naidu and Kharim 2010).

Elsholtzia densa Benth./ Common in Himalaya belts

Leaves are eaten raw as chutney and salad by people in Ladakh (Ballabh et al. 2007). Flowers grinded and paste used on skin rashes (Begum and Nath 2014). Juice extract is used in dysentery and stomach pain; tender shoots kept at the angle of earlobe, believe to ward off evil spirits (Lokho and Narasimhan 2013).

Elsholtzia eriostachya (Benth.) Benth./ Common in Himalaya belts

Leaves are eaten raw as chutney as well as salad (Ballabh et al. 2007).

Contd.

Ficus auriculata Lour./ Common in Himalaya belts	Ripe fruits are eaten by Garo tribe in Meghalaya (Singh et al. 2012). Khasi tribe uses the sprouts as vegetable (Jain and Dam 1997).
Fragaria nubicola Lindl. ex Lacaita/ Common in Himalaya belts	Fresh rhizomes grinded to a fine powder and mixed with sugar, taken approximately 2-5 mg daily continue for a month to cure tonsillitis (Khan et al. 2004). Decoctions made from the aerial parts are consumed twice a day for 5-6 days in morning and evening to get relief from fever (Uniyal et al. 2006). Ripe fruits are edible, and roots used as tea substitute (Kumar and Hamal 2009). Fruits are laxative and purgative (Qureshi et al. 2007).
Fragaria vesca L./ Rare in Kashmir Himalaya belts	Herbal tea prepared from roots in winter. season by local people of Kashmir Himalaya (Gorsi and Miraj 2002). Ripe fruits are consumed raw in Himachal Pradesh by local people (Sharma et al. 2009).
Gentiana tianschanica Rupr. ex Kusn./ Common in Kashmir and Ladakh Himalaya belts	Whole plant is used as salads by tribal people in Cold Desert (Ballabh et al. 2007).
Heracleum candicans Wall./ Common in Kashmir Himalaya belts	Fruit powder is considered as aphrodisiac (Kapahi et al. 1993).
Hippophae rhamnoides L./ Very common in Kashmir and Ladakh Himalaya belts	Ripe fruits are used in preparation of local juice and leaves are eaten raw as chutney and salad (Ballabh et al. 2007). It is used in as improvement of digestion, anti-oxidants, tumours, liver ailments, eye ailments, bronchial asthma, skin wrinkles and high cholesterol (Singh and Bedi 2017). Fruit jelly consumed to cure hepatic enlargement (Kapahi et al. 1993). Seeds used for treatment of cancer, and fruit juice used as a cough syrup (Gorsi and Miraj 2002).
Juglans regia L./ Very common in Kashmir and Ladakh Himalaya belts	Bark and green pericarp of fruit is used to make gum stronger and also to clean teeth (Kapahi et al. 1993). Bark is used as dye, and also acts as detergent (Das et al. 2012). Leaves are used as tonic by people in Khanabad village (Gorsi and Miraj 2002). Nuts used to cure hypertension, and barks used to clean teeth commonly called as Dandasa (Mir 2014). Extracted oil is used in case of headache (Das et al. 2012). Fruit kernels are edible, and catkins cooked as vegetable (Kumar and Hamal 2009).
Lactuca sativa L./ Cultivated in Himalaya belts of Asia	Young leaves are eaten raw as salad and dried them for future use in winter season by local people of Cold Desert in Himalaya (Ballabh et al. 2007). *Contd.*

Lathyrus humilis (Ser.) Fisher ex Spreng./ Common in Kashmir and Ladakh Himalaya belts

Seeds and pods are consumed raw by tribal communities at Kashmir Himalaya (Ballabh et al. 2007, Gorsi and Miraj 2002).

Malus domestica Borkh./ Cultivated in Kashmir Himalaya belts

Cultivated in the North India especially in Kashmir for fruits.

Mentha longifolia L. / Commonly occurs in Kashmir and Ladakh Himalaya belts

Herbal tea is prepared from fresh leaves and taken to cure abdominal pain (Gorsi and Miraj 2002). It is also used as carminative in case of diarrhea and dysentery (Qureshi et al. 2007). Stem shoot is used for stomach pain and gas problems, and juice of leaves consumed to expel worms from stomach (Mir 2014). Leaves are used in the preparation of local chutney, and dried leaves are used to flavor local dishes during winter season in Cold Desert of India (Ballabh et al. 2007).

Morus alba L./ Common in Asian countries

Ripe fruits are consumed raw by tribal people in Ladakh (Ballabh et al. 2007). Fruits are used to cure sore throats (Gorsi and Miraj 2002). Leaves are used in rearing of silkworm (Mir 2014).

Oxyria digyna (L.) Hill/ Sparsely occurs in high altitude areas of Kashmir and Ladakh regions

Whole plant is a rich source of Vitamin 'C' and eaten raw as salad by local people in Ladakh (Ballabh et al. 2007). Leaves serves as the source of refrigerant, cooling and catarrh (Gorsi and Miraj 2002).

Oxalis acetosella L./ Common in Himalaya belts

Whole Plants used to make local chutney (Singh and Bedi 2017).

Persicaria alpina (All.) H.Gross/ Common in Kashmir and Arunachal Himalaya belts

Local syrup 'Sher' is prepared from roots and flowers (Gorsi and Miraj 2002). Herbal tea is prepared and taken as remedy for flue fever and joint pains (Gorsi and Miraj 2002).

Prunus armeniaca L./ Common in Himalaya belts

Fruits are consumed fresh and after dehydration stored for winter use as local badam in Northern Himalaya (Ballabh et al. 2007, Kumar and Hamal 2009). Gum obtained from stem known for anticancer activity (Gorsi and Miraj 2002). Fruits are edible (Kumar and Hamal 2009).

Prunus cornuta (Wall. ex Royle) Steud. / Common in Himalaya belts

Rheum webbianum Royle/ Common in Kashmir and Ladakh Himalaya belts

Fresh tubers used in treatment of dysentery (Gorsi and Miraj 2002). Leaf stalk are rich source of vitamin 'C' and used as wild salad and chutney by people of Ladakh (Ballabh et al. 2007).

Ribes alpestre Wall. ex Decne./ Common in Kashmir Himalaya belts

Shepherds eats ripe fruits in Ladakh (Ballabh et al. 2007).

Contd.

Ribes orientale Desf./ Common in Kashmir and Ladakh Himalaya belts	Children in Ladakh eats ripe fruits (Ballabh et al. 2007).
Rosa webbiana Wall ex Royle/ Common throughout Himalaya belts	Shepherds consumed ripe fruits in Cold Desert in India (Ballabh et al. 2007). Tender shoots and leaves used as vegetable. Petals used in making local drinks (Kumar and Hamal 2009).
Rubus alceifolius Poir./ Common throughout Himalaya belts	Ripe fruits are eaten by tribal communities in Meghalaya (Singh et al. 2012). It is also edible by Mao-Naga tribes of Manipur (Lokho and Narasimhan 2013).
Rubus caesius L./ Common throughout Himalaya belts	Matured berries are consumed as raw.
Rubus idaeus L./ Sparsely occurs in Himalaya belts	Leaves contain flavonoid and tannin (Madhuri and Pandey 2009). Fruits are rich source of vitamins (A,B,C) and ellagic acid.
Rubus niveus Thunb./ Common throughout Himalaya belts	Ripe fruits are consumed by tribal communities in Meghalaya (Sharma et al. 2009, Kayang 2007). Fruit juice extracted consumed by tribal community of Himachal Pradesh.
Rubus saxatilis L./ Rare in Himalaya belts	Berries are eaten raw Bernus one. species.
Rumex patientia L. ssp. *orientalis* (Bernh. ex Schult. & Schult.f.) Danser/ Common throughout Kashmir and Ladakh Himalaya belts	Leaves are used in making chutney in some areas of Ladakh (Ballabh et al. 2007).
Sinopodophyllum hexandrum (Royle) T.S.Ying/ Common throughout Northern Himalaya belts	Rhizomes and roots reported for curing cancer (Gorsi and Miraj 2002). Rhizome paste applied on tumor (Begum and Nath 2014). Roots crushed, mixed with warm water, filtered, and consumed in treatment of acidity and heart abnormalities (Mir 2014). Powdered roots are also used in case of hepatic enlargement (Kapahi et al. 1993). Ripe fruits are eaten by children in Zanskar area of Ladakh (Ballabh et al. 2007).
Solanum americanum Mill./ Common throughout Himalaya belts	Tibetan children consumed ripe fruits (Ballabh et al. 2007) as vegetable and laxative. Fruits extract mixed with grains are fed to birds to prevent bird flu and associated diseases (Lokho and Narasimhan 2013).
Sonchus oleraceus (L.) L. / Common throughout Kashmir and Ladakh Himalaya belts	Green leaves are taken as salad in Cold Desert area of Himalaya (Ballabh et al. 2007). Fruits are eaten raw by tribal communities in Meghalaya (Kayang 2007).
Trifolium repens L./ Common throught Himalaya belts	Leaves are poisonous to horses and cow (Gorsi and Miraj 2002). Whole plant crushed and paste applied on head to remove dandruff. Dried powdered plant used as herbal tea and act as an expectorant (Kapahi et al. 1993).

Himalayan Plants as Leafy Vegetables

Wild plant species growing in Himalaya are eaten and consumed during the food scarcity are or lean period (Singh et al. 2016). Plant parts such as young and succulent stems, petioles, young twigs, flowers and fruits were commonly used as leafy vegetables (Singh and Borthakur 2014). Vegetables prepared from a single species or from a combination of different species varies based on traditional and tribal use from place to place. In temperate and alpine Himalayas, several wild leafy species grows as seasonal and highly perishable. Therefore, to preserve such vegetables for winter season, sun-drying in growing period of fresh parts and local method of storage is preferred. Plants growing in nitrogen-rich ecosystem such as *Chenopodium album, Eremus himalaicus, Fagopyrum esculentum, Lactuca lessertiana, Plantago lanceolata, Polygonum aviculare, Rheum emodi, Rumex acetosa, Rumex nepalensis* and *Urtica dioica* is particularly attractive as leafy vegetables in Himalayas. *Alisma plantago-aquatica, Cichorium intybus,* and *Dioscorea bulbifera* is cooked along with meat by tribal communities of Himalaya. Tubers and roots of *Alisma plantago-aquatica, Allium consanguineum, Dioscorea bulbifera* and *Articum lappa* were boiled, and used them as vegetable. Plant such as *Diplazium esculentum, Chenopodium album, Malva neglecta* and *Medicago minima* were common vegetable consumed after boiling and frying in cooking oil by tribal population of Himalaya.

Herbal Teas from Himalayan Plants

Tribal communities of Himalaya has a long tradition with regard to the preparation of aqueous mixture from wild plants such as herbal teas total drinks, beverages or any other home-made remedies with potential health benefits (Singh et al. 2016). The most commonly used species as herbal drinks or teas in Western Himalayas are *Abies pindrow, Bistorta amplexicaulis, Juniperus communis, Thymus serpyllum, Verbascum thapsus* and *Zingiber officinalis*. Tribal people believe that herbal teas or drinks prepared from *Zingiber officinalis* and *Thymus serpyllum* are good for nausea and motion sickness. Drinks prepared from plants as for example *Abies pindrow, Juniperus communis* and *Bistorta amplexicaulis* are used to treat the pain in head, joint-pains and other similar disorders associated with humans. Leaves of *Phlomoides bracteosa* and *Taraxacum officinale*, flowers of *Epilobium parviflorum, Filipendula vestita, Verbascum thapsus* and *Viola canescens*, berries of *Rosa macrophylla*, roots of *Bistorta amplexicaulis, Fragaria nubicola* and

Gentianella moorcroftiana, and rhizomes of *Trillium govanianum* are also used as herbal teas by different Sheenas and Gujjar tribes of Himalayas (Singh et al. 2016).

Himalayan Plants as Home-Made Soups

Tribal populations in many parts of Himalaya collect herbs in growing season to be used as home made soups for winter season and for local parties. For the utilization of the local soups, *Gagea lutea, Bunium persicum, Foeniculum vulgare* and *Epilobium parviflorum* are the most commonly used plants added to enhance taste and curative properties of vegetables and soup in Himalaya.

Wild Fruits Plants from Himalayan

Himalayan plants produces fruits which are delicious and unique in nature. Fruits contain several essential or non-essential bioactive compounds such as anthocyanins, alkaloids, flavonoids, polyphenols and tannin. Besides, these fruits are rich sources of nutrients such as amino acids, minerals, proteins and vitamins which modulate the activity of a wide range of enzymes and cell receptors. Consumption of such plants and their parts are useful for preventing various diseases associated with oxidative stress such as cancer, cardiovascular and neurodegenerative diseases (Singh et al. 2016). Anorexia, asthma, diarrhea, fevers, cough, indigestion, thirst, toxemia, spermatorrhoea, stomatitis, tuberculosis and sexual disorders (Pallavi et al. 2011) can be cured by eating Himalayan wild fruits. *Berberis lyceum, Duchesnea indica, Fragaria vesca, Morus alba, Sambus wightiana, Solanum americanum, Rubus saxatilis* and *Rubus niveus* are eaten by the shepherds and the herdsmen in forest field when they feel hungry or thirsty (Singh et al. 2018). Some fruits are eaten raw by local people just after collection and peeling the outer skin owing to its good taste such as of *Crataaegus rhipidophylla, Eriobotrya japonica, Ficus auriculata* and *Ribes orientale*. Seeds of plant species such as of *Pinus wallichiana, Impatiens glandulifera* and *Rosa webbiana* consumed raw by children playing in the valleys of Himalayan tribal communities (Singh et al. 2016).

Plants Consumed as Salads and Chutneys

Traditional preparations such as salads, chutneys and pickles were consumed along with rice, chapathi, and various snack foods, as side dishes in India, which increase appetite (Singh et al. 2016). The plants growing in nitrogen-rich farmlands or pastures are eaten as salads or local chutneys. These foods investigated as one of the most common food items included during lunch or dinner, and suitable as a ready-to-eat food for immediate consumption to the people (Singh and Bedi 2017). The most commonly used plant species for making the local chutneys by local peoples are *Elsholtzia densa, Mentha longifolia, Oxyria digyna, Rumex patientia* and, however, these plants were also investigated to be used as salads. *Centella asiatica, Gentiana tianschanica, Heracleum candicans, Lactuca sativa, Rheum webbianum, Sonchus oleraceus* and *Trifolium pratense* are the most investigated plant species used as salads. Seeds of *Carum carvi*, a member of apiaceae family and leaves of *Mentha arvensis*, a member of lamiaceae group, added to salads to improve the taste (Singh et al. 2016). Fresh buds of *Anaphalis triplinervis* consumed by shepherds in salads prepared at workplace, and believe that it adds more taste to food.

Himalayan Food from Underground Plant Parts

Tuberous plant species play a vital role in supplementing food necessity of tribal people in remote locations through stored foodstuffs during the time of food shortage and lean agricultural seasons (Singh et al. 2016). The nutritional value of indigenous uses of tubers or root vegetables is higher than several known common cultivated species. Most of the tubers or roots have potential for income generation, easy propagation, making varieties of food items, storage and preservation ability, however, on the other hand, they failed to compete with exotic or cultivated taxa due to lack of awareness (Singh et al. 2017). Investigation proven that the practice of use of wild tuberous plants is still alive among the tribal people in Himalayan belts. The hilly undulating terrain with limited agricultural lands, non-availability of sufficient food, poor accessibility and low income from yield are the main reasons for the use of wild tubers or roots as food items in the investigated area. Tubers of *Asparagus racemosus, Cyperus rotundus*, and *Oxalis acetosella* and also roots *Codonopsis ovata, Codonopsis rotundifolia* are eaten as fresh snacks just after collection and peeling the outer skin by the *Gujjars, Himanchalis, Sheenas* and *Arunachalis* when they feel hungry. Fresh tubers of *Allium carolinianum* added to soup, and dried powdered tubers of *Gagea lutea* is used as spices. The tubers of *Dioscorea bulbifera* and roots of *Saussurea costus* can be cooked for consumption after a process of slicing, soaking in running water and prolonged boiling to eliminate the bitter taste (Ju et al. 2013).

Alcoholic Drinks from Himalayan Plants

Consumption of home-made alcoholic drinks has been decreased in the 20th century, and one of the reason is introduction of license legislation for home production of alcoholic drink in most parts of hilly regions in the country (Singh et al. 2016). The measures taken by the government to protect the people from alcoholism probably contributed to reduce the consumption of traditional home-made drinks. Liqueurs can be prepared from various wild plants of Himalayas as indicated and used by various tribal communiies of Himalayas. *Hippophae rhamnoides* is one of the most common endemic plants of western Himalaya, used as food and medicine. Its fruits are eaten fresh or used to make beverage and local wines, and also used to treat cough and invigorate the circulation of blood (Singh and Bedi 2017). *Ribes alpestre* fruits are eaten fresh and also used to prepare local alcoholic drinks. The roots of *Gentiana tianschanica* seemed to be an age old tradition. Aerial portions of *Achillea millefolium*, *Artemisia dracunculus*, *Persicaria alpina* and *Rubia cordifolia*, and flowers of *Hypericum perforatum* used in making local alcoholic drinks. Dry stem of *Taraxaxum officinale* are also used as alcoholic drink by Sheenas of LoC Kashmir Himalaya.

Nutrients from Himalayan Plants after Burnt and Dry Powder

Tuberous parts of wild plants are to be eaten burnt as such after removing the skin or washing with running water. Underground parts such as roots, rhizomes, bulbs and tubers of several wild plant species were burnt in the fire and consumed as snacks, and is the most preferred food of children and shepherds of Sheenas residing along LoC (Singh et al. 2016). Tubers (*Asparagus racemosus*, *Dioscorea bulbifera*) and roots (*Codonopsis ovata*, *C. rotundifolia*, *Saussurea costus*,) were mostly recorded to be used as wild food by Sheena tribe of Kashmir valleys (Singh et al. 2016). Sometimes, the underground parts of wild plants such as *Asparagus racemosus*, *Rubia cordifolia* and *Zingiber officinale* are dried, grinded to powdered and taken raw in pudding or cake. The powdered roots of *R. cordifolia* are added with local herbal tea or alcoholic drinks to enhance taste and potency by the Gujjars and the Bakarwals in Western Himalaya (Singh et al. 2016).

Himalayan Plants in Primary Health Care

Plants of medicinal and therapeutic values can be defined as plants or plant parts such as roots, barks, stems, leaves, flowers, fruits and seeds or chemical substances derived from these parts, used in tradiional systems of medicine such as amchi, ayurveda, siddha and unani globally known for their curative properties (Singh 2019, Bannerman et al. 1983). Table 2 is provided with common Himalayan medicinal plants available for local people in treatment of their day to day diseases.

Table 2: Medicinal plants with potential applications in primary health care

Botanical name/Family	Constituent(s)	Activity/indications
Aegle marmelos (L.) Correa /Rutaceae	Aeglemarmelosine	Laxative, anti-inflammatory
Ageratum conyzoides (L.) L./ Asteraceae	Ageratochromone	Wound healing
Azadirachta indica A.Juss. /Meliaceae	Nortriterpenoids	Antimalarial, antipyretic, seed insecticidal
Balanites aegyptica (L.) Delile/ Zygophyllaceae	Steroidal glycosides, furanocoumarines	Laxative, anti-inflammatory, molluscicidal
Bridelia ferruginea Benth. Phyllanthaceae	Coumestans, flavonoids	Antifungal, mouth infections
Cajanus cajan (L.) Millsp./ Fabaceae	Amino glycosides, phenylalanine	Management of sickle-cell anaemia
Carica papaya L. Caricaceae	Proteolytic enzymes (volatile oils in leaves)	For fevers, antidiabetic
Curcuma longa L. Zingiberaceae	Curcumin	Tonic, choleretic
Cymbopogon citratus (DC.) Stapf./Poaceae	Volatile oils	Diuretic, tonic
Dracaena mannii Baker Asparagaceae	Saponins	Local antifungal, anti-protozoan
Eucalyptus globulus Labill. /Myrtaceae	Essential oil	Local antiseptic, cold colds, rubefacient
Garcinia cowa Roxb. ex Choisy/Clusiaceae	Biflavonoids	Antihepatotoxic, antiviral, adaptogen, plaque inhibitor
Morinda lucida Benth. / Rubiaceae	Anthraquinones	Antimalarial, jaundice
Murraya koenigii (L.) Spreng. /Rutaceae	Indole alkaloids	Antimalarial, broad spectrum antiprotozoan
Ocimum gratissimum L./ Lamiaceae	Terpenes, xanthones	Antiseptic, coughs, fevers
Piper betle L./Piperaceae	Lignans, alkaloids	Antimicrobial, insecticidal, tonic, anti-inflammatory
Psidium guajava L. Myrtaceae	Essential oils, vitamins	Carminative
Senna obtusifolia (L.) H.S. Irwin & Barneby / Fabacaeae	Anthraquinone, glycosides	Laxative
Tamarindus indica L. /	Ascorbic acid, citrates	Laxative, nausea

Contd.

Caesalpiniaceae		
Vernonia amygdalina Delile /Asteraceae	Sesquiterpenes, saponins	Tonic, antidiabetic
Zanthoxylum armatum DC. /Rutaceae	Aromatic acids	Management of sickle-cell anaemia
Zingiber officinale Roscoe / Zingiberaceae	Terpenes	Antihypertensive, anti-histamine

Drugs and Medicines Discovered from Ethnobotanical Leads

New value added products prepared from natural herbal botanicals are replacing existing market products on a regular basis due to the best quality improvement from safety point of view (Singh 2019). Table 3 is provided with different drugs discovered from ethnobotanical leads of Himalayan plants or plants recorded growing in geographic regions of India.

Table 3: Drugs discovered from ethnobotanical leads of Himalayan plants or plants recorded growing in Himalayan regions of India, their correlation and sources (Modified after Fabricant and Farnsworth 2001).

Botanical name/Family	Drug	Action/Clinical Usages
Aesculus hippocastanum L. / Sapinadaceae	Aescin	Anti-inflammatory
Ammi majus L./ Apiaceae	Xanthotoxin	For vitilago and leukoderma
Anamirta cocculus (L.) Wight & Arn./ Menispermaceae	Picrotoxin	Barbiturate antidote and analeptic
Ananas comosus (L.) Merr. /Bromeliaceae	Bromelain	Anti-inflammatory, proteolytic agent
Andrographis paniculata (Burm.f.) Nees/Acanthaceae	Andrographolide	Bacillary dysentery
Ardisia japonica (Thunb.) Blume / Primulaceae	Bergenin	Antitussive
Areca catechu L./Areaceae	Arecoline	Anthelmintic
Artemisia maritima L./ Asteraceae	Santonin	Ascaricide
Atropa belladonna L./ Solanaceae	Atropine	Pupil dilator and anticholinergic
Berberis vulgaris L./ Berberidaceae	Berberine	Bacillary dysentery
Brassica nigra (L.) K.Koch/ Brassicaceae	Allyl isothiocyanate	Rubefacient
Camellia sinensis (L.) Kuntze/ Theaceae	Caffeine, Theophylline	Stimulant, diuretic, antiasthmatic and bronchodilator
Cannabis sativa L./	Tetrahydrocannabinol	Antiemetic

Contd.

Cannabaceae

Carica papaya L./Caricaeae	Papain	Attenuator of mucus, proteolytic and mucolytic
Catharanthus roseus (L.) G.Don	Vinblastine, Vincristine	For Hodgkin disease, and paediatric leukaemia
Centella asiaca (L.) Urban/Apiaceae	Asiaticoside	Vulnerary
Cinchona pubescens Vahl/Rubiaceae	Quinine, Quinidine	For malaria prophylaxis and cardiac arrhythmia
Cinnamomum camphora (L.) J.Presl./Lauraceae	Camphor	For rheumatic pain
Colchicum autumnale L. / Colchicaceae	Colchicine	For gout
Crotolaria sessiliflora L./Fabaceae	Monocrotaline	Antitumer agent
Cullen corylifollium (L.) Medik. /Caesalpinaceae	Psoralen	For vitilago
Curcuma longa L./Zingiberaceae	Curcumin	Cholreretic
Datura stramonium L./ Solanaceae	Scopalamine	For motion sickness
Digitalis purpurea L./ Plantaginaceae	Digoxin, Emetine	For atrial fibrillation, and atrial fibrillation
Ephedra sinica Stapf Ephedraceae	Ephedrine, Pseudoephedrine	Bronchodilator and rhinitis
Filipendula ulmaria (L.) Maxim./Rosaceae	Aspirin	Analgesic and antiinflammatory
Frangula purshiana Cooper/ Rhamnaceae	Cascara	purgative
Glycyrrhiza glabra L./Fabaceae	Glycyrrhizin	Sweetener
Hamamelis virginia L. / Ham amelidaceae	Gallotanins	Haemorrhoid suppository
Hydrangea macrophylla (Thunb.) DC./Hydrangaceae	Phyllodulcin	Sweetner
Hyoscyamus niger L./ Solanaceae	Hyoscyamin, Ipratropium	Anticholinergic and Bronchodilator
Melilotus officinalis (L.) Pall. / Fabaceae	Dicoumarol	Antithrombotic
Papaver somniferum L./ Papaveraceae	Codeine, Morphine, Noscapine, Papaverine	Analgesic, antitussive, antitussive, and antispasmodic
Physostigma venenosum Balf./ Caesalpinaceae	Physostigmine, Sigmasterol	For glaucoma, Steroidal precursor
Rauvolfia serpentina (L.) Benth. ex Kurz /Apocynaceae	Rescinnamine, Reserpine	Antihypertensive
Rauvolfia tetraphylla L. / Apocynaceae	Deserpidine	Antihypertensive
Rhododendron molle G.Don/	Rhomitoxin	Antihypertensive

Contd.

Ericaceae		
Rorippa indica (L.) Hochr./	Rorifone	Antitusssive
Brassicaceae		
Salix alba L./ Salicaceae	Salicin	Analgesic
Senna alexandrina Mill./	Sennoside A&B	Laxative
Caesalpinaceae		
Silybum marianum (L.)	Silymarin	Antihepatotoxic
Gaertn./Asteraceae		
Sinopodophyllum hexandrum	Podophyllotoxin	For condyloma
(Royle) T.S. Ying /		acuminatum
Berberidaceae		
Stevia rebaudiana Bertoni/	Stevioside	Sweetener
Lamiaceae		
Strychnos guianensis (Aubl.)	Toxiferine	Relaxant in surgery
Mart./ Loganiaceae		
Syzygium aromaticum (L.)	Eugenol	For toothache
Merr. & L.M.Perry /		
Myrtaceae		
Thymus vulgaris L./Lamiaceae	Trichosanthin	Abortifacient
Valeriana officinalis L./	Valepotriates	Sedative
Valerianaceae		

Himalayan Plant Species Used as Medicine to Cure Animal and Bird Diseases

Herbal botanicals have always been a form of therapy for livestock among poor farmers. The indiginous knowledge of ethnoveterinary medicine from plants and its significance has been identified by the traditional communities through the process of experience over hundred and hundreds of years, because livestock provide a wide range of services and products including animal power, wool and supplementary nutritions (Phondani et al. 2010). It is evident from published literatures that there are lots of plant abode in Himalayas, which has traditional applications in tribal medicines with reference to their domesticated livestocks and birds. *Lyonia ovalifolia, Curcuma domestica, Bombax ceiba, Vigna mungo, Dendrocalamus strictus, Zanthoxylum armatum, Piper nigrum, Terminalia chebula, Acacia catechu, Picrorhiza kurroa, Stephania glabra, Berberis arista, Dalbergia sisso, Pinus roxburghii, Aconitum heterophyllum, Artemisia nilagerica, Boswellia serrata, Adansonia digitata, Aloe vera, Cyperus articulatus, Allium sativum* and various other plants are used as medicine for livestocks (Table 4).

Table 4: Himalayan Medicinal Plant Used to Cure Animal and Bird Diseases

Botanical name/Family	Parts used
Acacia catechu (L.f.) Willd./ Mimosaceae	Leaves/bark
Aconitum heterophyllum Wall. ex Royle /Ranunculaceae	Whole plant
Acorus calamus L./ Acoraceae	Roots
Allium sativum L./Amaryllidaceae	Whole plants/tuber
Aloe vera (L.) Burm.f./Xanthorrhoeaceae	Leaves
Artemisia nilagirica (C.B.Clarke) Pamp. /Asteraceae	Aerial plarts
Berberis arista DC./ Berberidaceae	Stem
Bombax ceiba L./ Bombacaceae	Stem
Boswellia serrata Roxb. ex Colebr./Burseraceae	Gum
Cassia fistula L./ Ceaselpinaceae	Whole plant
Cissampelos praira DC./ Menispermaceae	Whole plant
Curcuma longa L./ Zingiberaceae	Rhizome
Cyperus articulatus L./Cyperaceae	Root
Dalbergia sissoo DC./ Caesalpiniaceae	Stem/leaves
Dendrocalamus strictus (Roxb.) Nees/ Poaceae	Stem/young parts
Desmodium adscandens DC./ Fabaceae	Whole plant
Geophila gracilis DC./ Rubiaceae	Whole plant
Lyonia ovalifolia (Wall.) Drude/Ericaceae	Leaves/stem
Oxalis corniculata L./ Oxalidaceae	Whole plant
Picrorhiza kurroa Royle ex Benth/ Plantaginaceae	Roots
Pinus roxburghii Sarg./Pinaceae	Leaves
Piper nigrum L./ Piperaceae	Fruits
Sapium baccatum Hemsl./ Euphorbiaceae	Whole plant
Spiranthera odoratissima St.Hil./ Rutaceae	Leaf
Stephania glabra (Roxb.) Miers./ Menispermaceae	Tuber
Terminalia arjuna (Roxb. ex DC.) Wight & Arn./ Combretaceae	Whole plant
Terminalia chebula Retz./Combretaceae	Fruits
Vigna mungo (L.) Hepper/Fabaceae	Whole plant/seeds
Zanthoxylum armatum DC./Rutaceae	Whole plants

Himalayan Plants for Aroma Therapy and Product Development

Himalayan regions are repository of wild aroma bearing plants. More than 3000 species reported world wide are used for production of essential oils. The volatile chemical constituents of aromatic plants of India as well as few exotic plants are cultivated for essential oils and for industries having application in product development. The analyses of volatile oils are usually performed in a combined gas chromatography and mass spectro-photometry system coupled with a computer programmed for automatic identification of the essential oil's chemical constituent. Computer programs based on real and simulated Kovats Chromatographic Retention Index are developed to optimize the identification work. The following are some examples showing the potential application of volatile oils in the fields of pharmacy, cosmetics and other industries growing in Himalayas.

Thymol Oil

This oil is obtained from *Monarda citriodora* L. (Lamiaceae) annual plant and is very similar to thymus oil, because of the presence of thymol in concentrations of 55-70%. It can be obtained from the aerial parts or flowering twigs of this plants and its yield ranges from 0.4-0.7%. There are new developments to promote its use as a substitute for chamomile oil in the cosmetic industry.

Spearmint Oil

This oil is obtained in low yield from leaves of *Mentha spicata* L., an aromatic herb which is cultivated throughout India for essential oils. The major chemical constituent of this oil is piperitenone oxide (21%), which is supposed to be the active principle of this medicinal plant. The plant is a active herb used against the intestinal parasites *Amoeba hystolitica* and *Giardia llamblia* (Faloricant and Farnsworth 2001).

Lemongrass Oil

This oil is obtained from *Cymbopogan* species of Poaceae family. This species is a grass rich in β-citral (80-85%), cis-ocimene (15%) and other component in minor quantities. Oil recovery varies from 0.5-0.6% and has application in perfumery, flavour and fragrance industry.

Lavender Oil

This oil is obtained from *Lavandula angustibolia* Mill. (Lamiaceae) rich in linalool (35-45%), linaliyl acetate (10-12%), borneol and lavandulyl acetate. It has application in cosmatics and perfumery industries growing mainly in Kashmir Himalayas of temperate environmental and snow-bound areas.

Geranial Oil

This oil is obtained from *Pelargonium graveolens* plants and rich in citronellol (25-26%), geraniol (20-21%), linalool (10-11%), citronellyl formate (7-8%) and isomenthore (7-8%). The plant best grows in subtropical and temperate environment of Himalayas. The chemical constitutents has wide application in pharmaceutical, flavour and fragrance industry.

Linking Between Medicinal Plants, Drug Development and Conservation

It is presumed that the discovery of a new plant-derived drug will ultimately be of value to preservation and conservation efforts, especially in Himalayan forest regions. This idea is based on the profit potential and economic impact, as well as of the opinion that the government and people will somehow place a greater value on a resource if it or its derivatives can produce a product for international markets. The potential of Himalayan medicinal plants to support conservation efforts could have three perspectives: local herbal traditional medicine, national herbal botanical industry and international pharmaceutical industry. In traditional medical systems, they accrue to collectors who sell the plant to traditional practicenor or to the healer themselves. The local and national herbal industries benefit a broad range of people and institutions, including collectors, whole-salers, brokers and companies that produce and sell herbal formulations. In the international pharmaceutical industry, at the corporate level, also involved in wholesale and retail sales. It is estimated that the international herbal, pharmaceutical companies are 10 times the size of US Herbal Industry; whose value is about £1300 million annual (Fabricant and Farnsworth 2001). Hence it can be estimated that the market value of traditional medical products, used by thousands of people across the world earn thousands and millions of dollars each year.

Current Challenges

National meetings, seminars and policy making assumed that all the local members of a Himalayan tribal community in a forested area must equally be benefited, if the community is to develop effective resource management institutions. Despite the provision of benefit sharing, the sharing provisions are questioned on various counts. Again, the benefits derived from the establishment of local village level reformed committees are doubtful in terms of its sustainability in long term scenerio (Singh et al. 2011.). Tribal people of low caste/poor tribal families collect small timber, firewood, fodder and various NTFPs from the nearby forest area to sustain their livelihood. But, these families were least empowered, neglected and increasingly alienated from participation in decision-meeting forums. Therefore, the lack of participatory process still remains same in the planning, implementation, monitoring and evaluation of management programmes in Himalayan regions of India and elsewhere in the world.

The concept of co-management or collective action by local institutions where both the state and the local communities have rights and responsibilities over the resources, have been widely accepted worldwide. Eco-tourism in Himalayan regions has the potential to generate substantial revenue, but that is not enough to meet the requirement of conservation and preservation of gene pool of rare, endemic and threatened plants. Therefore, there in need of proper planning and managements

Limitations of the Ethnomedical Research in Drug Development

For the most part, taxonomists and ecologists who conduct field work in areas where use of medicinal plants is a way of life are not trained or do not fully understand the disease state relates to loss of biodiversity. The information gathered is generally inadequate for the laboratory researchers to evaluate in terms of selecting the plants for expensive biological investigations. Illustrating further, the literatures in enthnobotany, ethnomedicine and ethnopharmacology usually documents informations on the latin binomials common or local names of the plant used, plant parts used, geographical area of collection and list of ethnobotanical applications. Data that are required for the assessment to the value of plant as medicine is usually missing from ethnobotanical writings includes method of preparation of the medicine, amount and frequency of use, source of informations, traditional healer, person who knows someone who uses the plant for formulation, route of administration (oral or external) and the specific medical use and/or symptomatology of the disease.

There are numerous examples that can be cited relates to these deficiencies. An informant states that a plant is used to allay thirst. Since thirst is one of the many symptoms of the diabetes, follow-up questions should be made to see if the user urinates frequently, is susceptible to fungal infections or has any other characteristic symptoms of diabetes. In the case of information concerning use of plant as a 'contraceptive', rarely does the literature indicate whether the plant is taken by men, women or both. If taken by a woman about the fertility of the woman, (vaginal infections are not uncommon among women living in Himalayas, which result in sterility), age, sexual activity, normal or abnormal menstrual patterns and other. The point is that if ethnomedical information is to be of value in drug discovery, it must be collected in more detail. Otherwise, this approach is no better than random selection followed by targeted biological screening. In order to remedy this deficiency in the type of ethnomedical information found in the literature, scientists trained in botany, who intend to pursue studies in which the information they collect will be of predictive value for drug discovery, should be encouraged to include in their educational programme as curriculum courses in medical terminology or in pharmacology. Michael Balick (1994) has pointed out that there is an urgent need to collect good ethnomedical information before the current healers pass on and their information is lost forever. This will require more well-trained ethnobotanists.

Future Perspectives

The rich and the unique biodiversity of Himalayan regions are threatened by anthropogenic induced multiple factors followed by the natural factors operating at different spatial scales (Bhattacharjee et al. 2017). Deforestation, biological invasion, pollution, over-exploitation, fire, and mining in fragile regions degrade natural ecosystems, while poaching is responsible for the decline of several keystone Himalayan species (Singh et al. 2011, Singh 2019). The challenges commonly faced in Himalayan biodiversity conservation are the lack of appropriate interaction between the local communities and the Forest Departments, absence of a follow-up to the management plans. A local forester mentioned that there was an increase in tourists from urban areas which was a menace to them and politicians disturbed the functioning of the same. There was a lack of facilitators to solve issues within and among stakeholders. Opportunities and incentives (like ecotourism, value addition of NTFPs) need to be created for the benefit of local communities so that they will actively engage in the convention and management plan. Conservation of biodiversity in these hotspot regions requires a broader approach, as for example, identification and protection of the most sensitive areas outside the present core area of the reserves, maintaining the remaining forest matrix under the sustainable management regimes, involving the local communities in the process of management, and encouraging the uses of synergetic with the conservation plan. As the plant and the animal species of the region represents a significant biodiversity hotspot component that requires again and again research attention and protection, given the observed climatic trends and alarming future projections (Bhattacharjee et al. 2017). Data on seasonal research and publications on physiological thresholds and population dynamics of peculier rare and endemic plant species in these belts are must to established fundamental baseline information regarding the biodiversity of Himalayas and the South Asian subcontinent as a whole for saving future of human in time to come.

ACKNOWLEDGEMENT

Author would like to thank Director IIIM for necessary facilities. This publication bears institutional publication number IIIM/2238/2018.

REFERENCES CITED

Anonymous. 2018. Himalaya mountains. www.yourdictionary.

Azad SA. 2013. Plants used against Gynaecological diseases by the Gujjar, Bakerwal and Pahari tribes of district Rajouri (Jammu and Kashmir). *Indian Journal of Scientific Research* 4(11): 135-136.

Ballabh B, Chaurasia OP, Pande PC, Ahmed Z. 2007. Raw edible plants of Cold desert Ladakh, India. *Indian Journal of Traditional Knowledge* 6(1): 182-184.

Ballabh B. 2002. Ethnobotany of Boto tribe of Ladakh Himalaya, (Kumaon University, Nainital, India) (Unpublished).

Bannerman RHO, Burton J, Ch'en WC. 1983. Traditional Medicine and Health Care Coverage: A Reader for Health Administrators and Practitioners. Geneva: World Health Organization.

Begum D, Nath NC. 2014. Ethnobotanical review of medicinal plants used for skin diseases and related problems in North Eastern India. *Journal of Herbs, Spices and Medicinal Plants* 7(3): 55-93.

Bhattacharjee A, Anadón JD, Lohman DJ, Doleck T, Lakhankar T, Shrestha BB, Thapa P, Devkota D, Tiwari S, Jha A, Siwakoti M, Devkota NR, Jha PK, Krakauer NY. 2017. The Impact of Climate Change on Biodiversity in Nepal: Current Knowledge, Lacunae, and Opportunities. *Climate* 5: 80; doi:10.3390/cli5040080.

Bolch T, Pieczonka T, Benn DI. 2011. Multi-decades mass loss of glaciers in the Everest area (Nepal, Himalaya) derived from Stereo imagery. *Cryosphere*, 5: 349-358.

Dar GH, Khuroo AA. 2013. Floristic diversity in the Kashmir Himalaya: Progress, Problems and Perspects. *Sains Malaysiana* 42(10): 1377-1386.

Das BH, Majumdar K, Datta BK, Ray D. 2009. Ethnobotanical uses of some plants by Tripuri and Reang tribes of Tripura. *Natural Product and Radiance* 8(2): 172-180.

Das T, Mishra SB, Saha D, Agarwal S. 2012. Ethnobotanical survey of medicinal plants used by ethnic and rural people in eastern Sikkim Himalayan regions. *African Journal of Basic and Applied Sciences* 4(1): 16-20.

Fabricant DS, Farnsworth NR. 2001. The value of plants used in traditional medicine for drug discovery. *Environmental Health Perspectives* 109(1):69-75.

Farnsworth NR. 1966. Biological and phytochemical screening of plants. *Journal of Pharmacological Sciences* 55: 225-276.

Gairola S, Sharma J, Bedi YS. 2016. A cross-cultural analysis of Jammu, Kashmir & Ladakh (India) medicinal plant use. *Journal of Ethnophamcology* 155(2): 925-986.

Gorsi MS, Miraj S. 2002. Ethnomedicinal survey of plants of Khanabad village and its allied areas, district Gilgit. *Asian Journal of Plants Sciences* 1(5): 604-6015.

Gupta SK, Sharma OMP, Raina NS, Sehgal S. 2013. Ethnobotanical study of medicinal plants of Paddar Valley of Jammu and Kashmir, India. African *Journal of Traditional, Complementary and Alternative Medicine* 10(4): 59-65.

Harvey A. 2000. Strategies for discovering drugs from previously unexplored natural products. *Drug Discovery Today* 5: 294-300.

Jain SK, Dam N. 1997. Some ethnobotanical notes from North Eastern India. *Economic Botany* 33(1): 52-56.

Ju Y, Zhuo J, Liu B, Long C. 2013. Eating from the wild: diversity of wild edible plants used by Tibetans in Shangri-La region, Yunnam, China. *Journal of Ethnobiology and Ethnomedicine* 1:28.

Kachroo P, Sapru BL, Dhar U. 1997. Flora of Ladakh. Bishen Singh Mahendra Pal Singh, Dehra Dun, India.

Kala CP. 2006. Medicinal Plants of the high altitude Cold deserts in India: diversity, distribution and traditional uses. *International Journal of Biodiversity Science and Management* 2(1): 43-56.

Kapahi BK, Srivastava TN, Sarin YK. 1993. Traditional medicinal plants of Gurez, Kashmir-An ethnobotanical study. *Science of Life* 1: 119-124.

Kaul MK. 1997. Medicinal Plants of Kashmir & Ladakh-Temperate & Cold Himalaya. Indus Publishing Company, New Delhi, India.

Kayang H. 2007. Tribal knowledge on wild edible plants of Meghalaya, Northeast India. *Indian Journal of Traditional Knowledge* 6(1): 177-181.

Khan ZS, Khuroo AA, Dar GH. 2004. Ethnomedicinal survey of Uri, Kashmir Himalaya. *Indian Journal of Traditional Knowledge* 3(4): 351-357.

Kinghorn AD. 1994. The discovery of drugs from higher plants. *Biotechnology* 26: 81-108.

Kuchroo P, Nahni IM. 2006. Ethnobotany of Kashmiris forest flora of Srinagar and plants of neighbourhood, Dehra Dun India, 239-263.

Kumar S, Hamal A. 2009. Wild edibles of Kishtwar high altitude National Park of Northwest Himalaya, Jammu & Kashmir, India. *Ethnobotanical Leaflets* 13: 195-202.

Lokho K, Narasimhan D. 2013. Ethnobotany of Mao-Naga tribe of Manipur, India. *Pleione* 7(2) (2013): 314-324.

Madhuri S, Pandey G. 2009. Some anticancer medicinal plants of foreign origin. *Current Science* 96(6): 779-783.

Malik AK, Khuroo AA, Dar GH, Khan ZS. 2013. Ethnomedicinal uses of some plants in the Kashmir Himalaya. *Indian Journal of Traditional Knowledge* 10(2): 362-366.

Mir MY. 2014. Ethnobotanical survey of plants from Kupwara, Jammu & Kashmir, India. *International Journal of Advance Research* 2(1): 846-857.

Naidu KA, Kharim SM. 2010. Contribution to the floristic diversity of ethnobotany of Eastern Ghats in Andhra Pradesh, India. *Ethnobotanical Leaflets* 14: 920-941.

Pallavi, K. J., R. Singh, S. Singh, K. Singh, M. Farswan and V. Singh. 2011. Aphrodisiac agents from medicinal plants: A review. *Journal of Chemical and Pharmaceutical Research* 3: 911–921.

Pegu R, Gogai J, Tamuli AK, Teron R. 2013. Ethnobotanical study of wild edible plants in Poba Reserve forest, Assam, India: multiple functions and implications for conservation. *Research Journal of Agriculture and Forestry Science* 1(3): 1-10.

Phillipson JD, Anderson LA. 1989. Ethnopharmacology and Western medicine. *Journal of Ethnopharmacology* 25:61-72.

Qureshi RA, Ghufran MA, Gilani SA, Sultana K, Ashrat M. 2007. Ethnobotanical studies of selected medicinal plants of Sudhan Gali and Ganga Chotti hills district ban, Azad Kashmir. *Pakistan Journal of Botany* 39(7): 2275-2283.

Rashid A, Anand VK, Serwar J. 2008. Less known wild plants used by the Gujjar tribe of district Rajouri, Jammu and Kashmir state-India. *International Journal of Botany* 4(2): 219-224.

Saharia S, Sharma CM. 2011. Ethnomedicinal studies on indigenous wetland plants in the tea garden tribes of Darrang and Udalguri district, Assam, India. *NeBIO* 2 (1): 27-33.

Sajem A, Gosai K. 2010. Ethnobotanical investigations among the Lushai tribes in North Cachar Hills district of Assam, Northeast India. *Indian Journal of Traditional Knowledge* 9(1): 108-113.

Sankaranarayanan S, Bama P, Ramachandran J, Kalaichandran PT, Deicaraman M, Vijayalakshimi M, Dhamotharan R, Dananjeyan B, Bama SS. 2010. Ethnobotanical study of medicinal plants used by traditional users in Villupuram district of Tamil Nadu, India. *Journal of Medicinal Plant Research* 4(12): 1089-1101.

Schild A. 2008. ICIMOD'S position on climate change and mountain systems. *Mountain Research and Development* 28 (3): 328-331.

Sharma BM, Kachroo P. 1981-1983. Flora of Jammu and plants of neighbourhood. Bishen Singh Mahendra Pal Singh, Dehra Dun, India.

Sharma SS, Gautam AK, Bhadauria R. 2009. Some important supplementary food plants and wild edible fungi of upper hilly regions of district Shimla, Himachal Pradesh, India. *Ethnobotanical Leaflets* 13: 1020-1028.

Singh B, Bedi YS. 2017. Eating from Raw Wild Plants in Himalaya: traditional knowledge documentary on Sheena tribes along LoC Border in Kashmir. *Indian Journal of Natural Products and Resources* 8(3): 269-275.

Singh B, Borthakur SK, Phukan SJ. 2014. A survey of ethnomedicinal plants utilized by the indigenous people of Garo Hills with special reference to the Kashmir Biosphere Reserve (Meghalaya), India. *Journal of Herbs Spices and Medicinal Plants* 20(1): 1-30.

Singh B, Borthakur SK. 2014. Phenology and Geographic Extension of Lycophyta and Fern flora in Nokrek Biosphere Reserve of Eastern Himalaya. *Proceedings of the National Academy of Sciences, India Section B: Biological Sciences* 85(1): 291-301.

Singh B, Phukan SJ, Sinha BK, Singh VN, Borthakur SK. 2011. Conservation strategies for *Nepenthes khasiana* in the Nokrek Biosphere Reserve of Garo Hills, Northeast, India. *International Journal of Conservation Sciences* 2(1): 55-64.

Singh B, Singh S, Singh B, Kitchlu S, Babu V. 2018. Assessing ethmic traditional knowledge, biology and chemistry of *Lepidium didymum* L., Lasser known wild plants of western Himalaya. *Proceeding National Academy of Sciences, India Section B: Biological Sciences* DOI: https://doi.org/10.1007/S40011-018-1027-4.

Singh B, Sinha BK, Phukan SJ, Borthakur SK, Singh VN. 2012. Wild edible plants used by Garo tribes of Kashmir Biosphere Reserve in Meghalaya, India. *Indian Journal of Traditional Knowledge* 11(1): 166-171.

Singh B, Sultan P, Hassan QP, Gairola S, Bedi YS. 2016. Ethnobotany, Traditional Knowledge, and Diversity of Wild Edible Plants and Fungi: A case study in the Bandipora district of Kashmir Himalaya, India. *Journal of Herbs, Spices & Medicinal Plants* 22(3): 247-278.

Singh B. 2019. Plants of Commercial Values. New India Publishing Agency, New Delhi, India. ISBN: 978-93-87973-503.

Singh SP, Thadani R. 2013. Valuing ecosystem services flowing from the Indian Himalayan states for incorporation into national accounting: IN; Lowmann M, Devy S, Ganesh J, editors. Treetops at risk: challengesd of global canopy, ecology and conservation. New York, Springer, 423-434.

Singh SP, Thadani R. 2015. Complexities and controversies in Himalayan Research: a call for collaboration and rigor for betles data. *Mountain Research and Development* 35 (4): 401-405.

Spjut RW, Perdue RE. 1976. Plant folklore: a tool for predicting sources of antitumor activity? *Cancer Treatment Reports* 60: 979-985.

Swarnkar S, Katewa SS. 2008. Ethnobotanical observation on tuberous plants from tribal areas of Rajasthan (India). *Ethnobotanical Leaflets* 12: 647-666.

Uniyal SK, Singh KN, Jamwal P, Lal B. 2006. Traditional use of medicinal plants among the tribal communities of Chhota Bhangal, Western Himalaya. *Journal of Ethnobiology and Ethnomedicine* 2(14): 1-8.

Vagelos P.R. 1991. Arc prescription drug prices high. *Science* 252: 1080-1084.

Xu J, Grumbine E, Shreshta A, Eriksson M, Yang X, Wang U, Wilkes A. 2009. The melting Himalayas: Cascading effects of climate change on water, biodiversity and livelihoods. *Conservation Biology* 23(3): 520-530.

2

Endangered Ayurvedic Himalayan Herb: A Review on Chemistry and Pharmacology of Ashtavarga Plants for Scientific Intervention

Venugopal Singamaneni,[1,3], Upasana Sharma[2] and Prasoon Gupta[1,3]*

[1]*Natural Product Chemistry Division, Indian Institute of Integrative Medicine Canal Road, Jammu-180001, Jammu and Kashmir, INDIA*
[2]*Department of Applied Chemistry, Mahant Bachittar Singh College of Engineering and Technology, Babliana, Jammu-181101, Jammu and Kashmir INDIA*
[3]*Academy of Scientific & Innovative Research (AcSIR), New Delhi-110025, INDIA*
*Corresponding email: guptap@iiim.ac.in

ABSTRACT

In Ayurvedic system of medicine, the word Astavarga represents eight endangered Himalayan herbs, which includes four orchids *Habenaria intermedia* D.Don, *Platanthera edgeworthii* Hook.f., *Crepidium acuminatum* D.Don, *Malaxis muscifera* Lindl., three liliaceae members (*Lilium polyphyllum* D.Don, *Polygonatum verticillatum* L., *Polygonatum cirrhifolium* Wall.) and one zingiber (*Roscoea purpurea* Sm.). These herbs are popular in various ayurvedic formulations, as for instance, Chyawanprash which is well recognized for strengthening vital force to the body, help in cell regeneration, and build immune system. These plants are repository of various bioactive ingredients such as diosgenin, β-sitosterol, catechin, p-coumaric acid and many more which shows positive biological functions such as anti-microbial, anti-oxidant, anti-inflammatory and other associated health promoting activities. In traditional system, different parts of these plants are used for different purpose, and mainly rhizomes are used as herbal formulations. In terms of ecology, these plants are growing wild in Himalaya belts between the elevation ranges of 1200-4000m above mean sea level.

Subtropical, temperate and alpine environment provide condusive adaptation and habitat for these species to grow and florish. Due to lack of proper documentation and certification, the identification of most of the species became difficult and illusory. Considering the medicinal importance and endangerment status, attempts have been made in the present communication to provide a comprehensive account on phytochemistry and pharmacological activities associated with astavarga species. The data presented in this communication will put door open for various research activities related to discovery of new drugs and nutraceutical products for human welfare in days to come.

Keywords: Ashtavarga Plants, Medicine, Endangered, Chemistry, Pharmacology, Indian Himalaya, Drug Discovery

INTRODUCTION

Ayurveda is the science of life and its traditions are everlasting. In ancient India, great discoveries were made by Dhanvantri, Aswin kumars, Susruta and Charaka. Their contributions were investigations into the properties of medicinal plants for Jievaniya (increasing the power of life), Vayastapana (antiageing), Swastya vardhaka (increasing health), Roga pratirodhaka (disease resistance) and Kshamta vadhaka (increasing capacity). In those days, traditional knowledge about the use of plants was transmitted verbally among the people. With the disappearance of Gurukul system of ancient education, which had more practical and less theoretical approach, the knowledge of medicinal plants started fading away coupled with no written details most of the medicinal plants over the several centuries had a great confusion about their actual identity had taken its deep roots. This was the case with Astavarga which included eight plant species with their sanskrit names as Ridhi, Vridhi, Jeevak, Rishbhak, Kakoli, Kshirkakoli, Meda and Mahameda (Sharma and Balakrishna 2005). All these plants have their natural occurrence habitats in Himalaya especially the North-Western Himalaya. Their natural habitats are specific in ecological requirements and hence these occur only in small restricted habitats and ecosystems patches.

The World Health Organisation (WHO) estimated that 80% of the population of developing countries relies on traditional medicines, mostly plants, for their primary health care needs (WHO 2005). Modern pharmacopoeia still contain at least 25% of drugs derived from plants, as well as synthetic analogues built on prototype compounds isolated from plants (Frans-worth 1994, Mukherjee and Wahil 2006). India holds the highest proportion of medicinal plants known for their medicinal value in the world (Kala et al. 2006). The Indians have been using plants to cure diseases since ancient history and according to ancient documentation on traditional medical systems based on plants: the Rigveda (4500-1600 BC), Charak Samhita (1000-800 BC) and Sushrut Samhita (800-700 BC) (Dhyani et al. 2010).

Over the centuries of Ayurvedic history, much confusion about identity of plants were further added. In Nighantus by various authorities, and commentators it was made clear that Astavarga is even rare to kings and therefore suggested use of substituents. This suggestion put forth full stop on further efforts to explore these plants in their habitats. After independence and revival of interest in Ayurveda provided the necessary enthusiasm and also the taxonomic system of plant classification facilitated the task of correct identification of Astavarga plants.

HISTORICAL EVIDENCE

In Dhanwantri Nighantu, there is mention of Astavarga plants, *viz.*, Ridhi (*Malaxis acuminata* D.Don), Vriddhi (*H. edgeworthii* Hook.f.), Jeevak (*Crepidium acuminatum* D.Don), Rishbak (*M. muscifera* Lindl.), Kakoli (*Roscoea purpurea* Sm.), Kshirakakoli (*Lilium polyphyllum* D.Don), Meda (*Polygonatum verticillatum* L.), and Mahameda (*P. cirrhifolium* Wall.). The great Ayurvedic sages Charak and Sushrut made a mention of 5 Astavarga plants for use in Chayavanprash. These were Rdhi, Jeevak, Rishbhak, Meda and Kakoli. For the first time the introduction of Ashtavarga in literature comes in *Jivaniya Mahakasaya* and *Jivaniya Pancamula* in the classical ayurvedic text of CHARAKA SAMHITA where ten medicinal plants were described, but these medicinal plants were not mentioned as Ashtavarga. Here Kakoli, Kshirakakoli, Jivaka, Rishbhaka, Meda, Mahameda, Mudgaparni, Masaparani, Jivanti, Madhuka were mentioned in the Jivaniya Mahakasaya of Charaka. During this early phase of Ayurvedic development, Ashwani kumars, who has a vast reputation as Ayurvedic wonder healers, saw the old and frail, emaciated body of Rishi Chayavan, decided to rejuvenate his body through medication. For this, they invented Astavarga – a group of eight medicinal plants and did the miracle of rejuvenating the body of Rishi Chayavan as youthful. Since then after the name of Rishi Chayavan the preparation was called as Chayavanprash and has been a favourite and the most demanded medicine for kings and rich people.

From the eternal times of Ayurveda, the Jeivaniya drugs of Astavarga (group of eight specific medicinal plants) have been a secret. The world famous Chyawanprash drug and health tonic has the eight plants of ashtavarga as an important ingredient. But the problem remains of mystery and uncertainity in the botanical identification. More recently, studies on 'Chyavanprash' suggest that it neutralizes the free radicals and acts as an immune modulator and blood purifier and possesses anti-ageing properties (Velpandian et al. 1998).

Due to the lack of proper identification and availability of astavarga plants classical preparation of Chyavanprash had been full of uncertainty and worries. At present in best samples of Chyavanprash, only four plant species namely, Vridhi, Meda, Kakoli and Jeevak are being utilized. As mentioned in different Ayurvedic books these plants have been found highly efficacious in a range of serious health/disease problems, which even in modern system of treatment still poses a challenge.

The Astavarga plants are known as (i) Jeevaniya (drugs strengthening vitality, immunity system etc) which includes Jeevak, Rishbhak, Meda, Mahameda, Kakoli and Kshirkakoli (ii) Brhnayiya (increase flesh in the body by activating cell regeneration even in old age) which includes Kakoli and Kshirkakoli, and (iii) Vayasthapan (metabolic processes especially anabolism become active and leads to youthful body complexion) which include Meda, Mahameda, Ridhi and Vridhi. The combination is sheetvirya, tasty, mans dhatu vardhak, sukrajanak, vipak guru, quickly cures bone fracture, tonic, increases kapha and is useful in vata-pitta, rakta, thirst, burning sensation in the body, fever, urinary problems including diabetes. These plants had so far been excluded from the modern researches because of confusion on their identity and difficulty in availability of sufficient drug materials.

Ashtvarga' medicinal plants distributed in Northwestern Himalaya region in small patches with restricted geographic location has distributional altitudinal range of 1600-4000 m above mean sea level. This plant species are well known for (i) rejuvenating health promoting activity, (ii) strengthening vital force of the body, (iii) enhancing cell regeneration capacity, and (iv) improving immunity system (Mathur 2003, Pandey 2005, Singh and Duggal 2009, Balkrishna et al. 2012). Plants of this group are considered useful in curing the healing fractures, seminal weakness, fever, diabetes, etc. Ashtvarga plants are considered threatened in its natural habitat. For instance, *Lilium polyphyllum* and *H. intermedia* are critically endangered, *Malaxis muscifera* is Vulnerable, *Platanthera edgeworthii* is rare (Saha et al. 2015a,b; Chinmay et al. 2009) and other four species i.e. *Polygonatum verticillatum, P. cirrhifolium, Roscoea purpurea* and *Crepidium acuminatum* are scarcely found in the wild.

ADULTERANTS AND SUBSTITUTES

The companies mostly used substitutes of the Ashtavarga species i.e. *Withania somnifera* and *Curculigo orchioides* in place of *Roscoea purpurea* Sm., *Withania somnifera* (L.) Dunal, *Chlorophytum arundinaceum* Baker, *Fritillaria roylei* D.Don and *F. oxypetala* D.Don in place of *Lilium polyphyllum* D.Don, *Pueraria tuberosea* Willd., *Centaurea behen* Lam. and *Tinospora cordifolia* Willd. in place of *Crepidium acuminatum* D.Don; *Centaurium roxburghii* D.Don in place of *M. muscifera* Lindl., *Asparagus racemosus* Willd. and *Eulophia campestris* Wall. in place of *Polygonatum verticillatum* L., *A. racemosus* Willd., *Sida caefolia* Lam. and *P. multiflorum* L. in place of *P. cirrhifolium* Wall.; *Tacca integrifolia* Ker-Gawl., *Sida cordifolia* L. and *Asparagus filicinus* Buch.-Ham. in place of *Habenaria intermedia* D.Don; *Tacca integrifolia* Ker-Gawl., *Dactylorhiza hatagirea* D.Don and *Sida acuta* Burm.f. in place of *Platanthera edgeworthii* Hook.f. (Balkrishna et al. 2012).

None of the Astavarga drugs are true or authentic materials in the market. Various Ayurvedic Nighantu books describe the medicinal properties of all the Astavarga plants either as a group or alone. Till date none has tried their chemical evaluation in the modern laboratories. Certainly, as is evident, they need further investigation on scientific lines to add value to their excellent text described qualities.

Astavarga is cooling, tasty, nutritious tonic, aphrodisiac, nourishes body and increases kapha. It is beneficial in seminal weakness, increases fat in the body, heals bone fracture, and cures vata, pitta, and rakta doshas, abnormal thirst, burning sensation in the body, fever and diabetic condition. It is one of the excellent combinations of herbal drugs which restores human health immediately, strengthens immunity system and rectifies defects in anabolism or body growth processes and works as anti-oxidant in the body. There is a need to grow these species on large scale to take up clinical trials and chemical investigations. The tissue culture programme may be advantageous to save these species and achieve fast multiplication.

ASHTAVARGA PLANTS BASED ON ANCIENT TRADITIONAL LITERATURE

Until the development of modern botanical nomenclature and classification system for the identification of plants, there is a great confusion and contradiction in the identification of Ashtavarga plants mentioned in classic Ayurvedic literature. The traditional Nighantu texts mentioned many synonyms for each plant and each synonym attributed to the property, morphology and habitat of that plant or its meaning. Along with it, the names given to the plants were linked to other natural forms, to which they resembled. This format was considered appropriate during that era because people were very close to nature at that time. During vedic era, there was not much mystery and confusion regarding medicinal plants, because education was imparted in the forests, which was very much practical and close to nature. The dissemination of knowledge was through verbal communication. After this period the gradual development of knowledge of Ayurveda and medicinal plants got restricted. The rich and great traditions of Ayurveda became suspected and complicated, this created difficulty in correct identification and use of medicinal plants. Still we have collected all the existing available literature regarding Ashtavarga and have put efforts to compile all possible information with full evidences and support for the first time related to history of Ashtavarga.

Now a days none of the medicines advertised as containing Ashtavarga in the market, had right plant ingredient. Presently in the market, only four herbs are available in the name of Ashtavarga, (1) Meda/ Mahameda, (2) Kakoli/Kshirkakoli, (3) Jeevak/Rishbhak and (4) Ridhi/ Vridhi. But these are also adulterated or unavailable. On the basis of our research experience and literature search, the botanical description of species, substituents, geographical distribution, chemical constituents and pharmacological activities of each species included in this review.

DIVERSITY AND DISTRIBUTION OF ASTAVARGA PLANTS

Polygonatum Mill. (Liliaceae) include about 50 species occur in the world and out which 15 species are found in India. Two species *P. verticillatum* and *P. cirrhifolium* are included in the Astavarga of Chayavanprash which are used as aphrodisiac or tonic for weakness. These grow in shrubberies open grassy slopes, and as under growth in forests on altitudes varying between 1500 to 3700 m in temperate

Himalaya. Roots show sclerenchymatous central zone in *P. verticillatum* which absent in *P. cirrifolium* (Pandey et al. 2006).

Malaxis Soland ex Sw. (Orchidaceae) genus contains about 300 species distributed all over the world, out of which 17 species occur in India. These are mainly terrestrial, rarely epiphytic or lithophytic herbs. The colour of the plant bears a striking relation to the kind of habitat under which it is grown. Two species namely *M. acuminata* and *M. muscifera* occur in temperate Himalaya. These grow as undergrowth in forests over the altitudes varying from 600 to 4000 m ASL. They prefers shruby habitat grassy slopes, moist and humus rich soil and shady rocks.

Roscoea Sm (Zingiberaceae) includes about 15 species found in different parts of the world out of only 4 species occur in India in which *R. Purpurea* belongs to Astavarga group. These are perennial herbs with thick fleshy roots occurs mostly in forest edges, open slopes on altitudes ranging from 1500 to 3000 m ASL.

Lilium L. (Liliaceae) contains about 80 species distributed throughout world of which 10 species are found in India. These are perennial, erect, leafy stemed herbs which bear flowers in axillary or terminal racemes. One species *L. polyphyllum* belongs to Astavarga group found in the temperate Himalaya. Generally this species found in open slopes, grassy lands in forests altitude ranging between 1800 to 3800 m ASL.

Habenaria Willd (Orchidaceae) has 600 species in the world of which 100 species occur in India. These are terrestrial leafy herbs bearing flowers in spikes of racemes. Two species *H. intermedia* and *H. edgeworthii* are found in temperate Himalaya on grassy slopes between 1800 to 2900 m ASL elevations.

The detail studies undertaken by different authors are provided below in subheads.

(1) MEDHA PLANT AS ASTAVARGA

Botanical Name / Family / Common English Name
Polygonatum verticillatum (L.) All./Liliaceae/ Whorled Solomon's Seal

Distribution
Found around the world in Europe, Turkey, North and Central Asia, Afghanistan, Tibet and Pakistan up to an elevation of around 4500 m. In India this species occurs in the Himalayas from Kashmir (at an altitude of around 2000-3600 m ASL), Himachal Pradesh, Uttarakhand

(altitude around 1600-3500 m ASL) to Sikkim (at an altitude around 2600-4000 m ASL).

Chemical Constituents

Rhizomes of *P. verticillatum* (Figure 1) contain lysine (1), serine (2), aspartic acid (3), threonine (4), diosgenin (5), β-sitosterol (6) (Khan et al. 2013). Several compounds have been isolated from the rhizomes including lectins (Antoniuk 1993), 5- hydroxymethyl-2-furaldehyde (7) and diosgenin (5) (Khan et al. 2011). New steroidal glycoside (25S)-spirost-5-en-3β-ol 3-O-β-D-glucopyranosyl-(1,3)-[β-D-fucopyranosyl-(1,2)]-β-D-glucopyranosyl-(1,4)-β-D-galactopyranoside isolated from rhizomes of *P. verticillatum* (Gvazava and Skhirtladze 2016). *P. verticillatum* can be a good source of diosgenin for pharmaceutical industry as it contains about 2.2% diosgenin (Motohasi et al. 2003). Bioactivity-guided fractionation led to the isolation of 2-hydroxybenzoic acid (8), santonin (9), propyl pentadecanoate (10), 2,3,-dihydroxypropyl pentadecanoate (11), Quinine (12) and β-sitosterol (6) (Khan et al. 2013). Ethanol extract showed maximum TPC (0.126 mg/g, gallic acid equivalent in dry sample), TFC (0.094 mg/g, rutin equivalent in dry sample) and TTC (29.32 mg/g, catechin equivalent in dry sample). GCMS spectrometry demonstrated that the oily components of the aerial parts were α-bulnesene (1.5648%), Linalyl acetate (0.4535%), Eicosadienoic (0.3702%), Pentacosane (0.3319%), Piperitone (0.3091%), Docasane (0.1720%), and Calarene (0.1321%) (Khan et al. 2013). GC/MS analysis of different rhizome extracts of *P. verticillatum* revealed the presence of 90 different phytoconstituents among the extracts (Patra and Singh 2018).

Pharmacological Studies

This plant species is diuretic and can be used as a pain reliever, for pyrexia, for burning sensations, phthisis and for treating general weakness. Aphrodisiac, emollient, appetizer, tonic, and galactagogue. The extract of the aerial parts showed leishmanicidal activity against *Leishmania major* which causes Kala Azar. The crude methanolic extract of its rhizomes showed antinociceptive activity in rats. As a complex administration, *P. verticillatum* together with *P. sibiricum* improved chronic hepatitis B (Motohasi et al. 2003). *P. verticillatum* has also been studied for its analgesic (Khan et al. 2010, Khan et al. 2011), antimalarial and antioxidant (Khan et al. 2011, 2013), metal accumulant and cytotoxic (Saeed et al. 2010), insecticidal (Saeed et al. 2010), antibacterial (Khan et al. 2012), tracheorelaxant and anti-inflammatory (Khan et al.

2013), bronchodilatory (Khan et al. 2013) and antipyretic and anti-convulsant (Khan et al. 2013) activities. Chloroform extract exhibits the strongest cytotoxicity against the human breast cancer cell line, MCF-7 (Patra and Singh 2018). Leaf aqueous extract possess aphrodisiac activity (Kazmi et al. 2012).

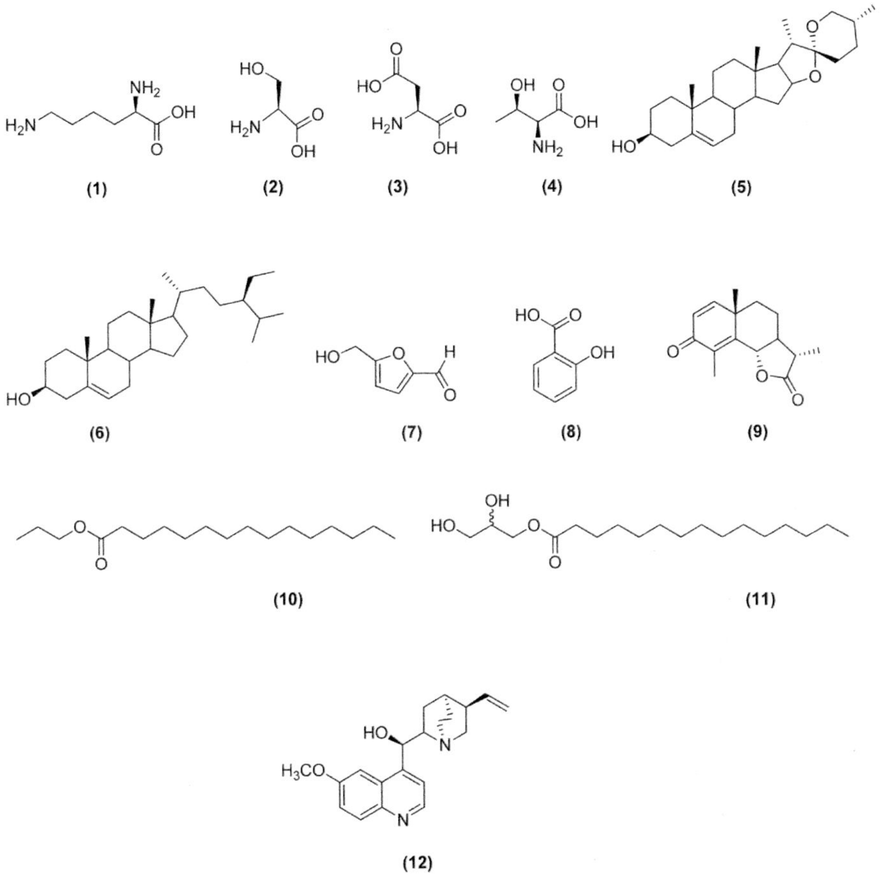

Fig. 1: List of chemical constituents reported from *P. verticillatum*

Part Used

Rhizome

Substitutes

Shatavari (*Asparagus racemosus*), Salam misri (*Eulophia campestris* Wall.)

(2) MAHAMEDHA PLANT AS ASTAVARGA

Botanical Name / Family / Common English Name

Polygonatum cirrhifolium (Wall.) Royle/ Liliaceae/ Kings Solomons Seal.

Distribution

P. cirrhifolium found in Northern Himalayas, China, Pakistan, Nepal and Bhutan between an altitude of 2000-4000 m. ASL. In India it is found in the temperate Himalayas from Himachal Pradesh eastwards to Sikkim at an altitude of around 1500-3300 m and in other regions of Uttarakhand up to 2000-3000 m ASL.

Chemical Composition

Rhizome of *Polygonatum cirrhifolium* (Figure 2) is reported to contain glucose, sucrose and two new steroidal saponins sibricoside A (1) and sibricoside B (2).

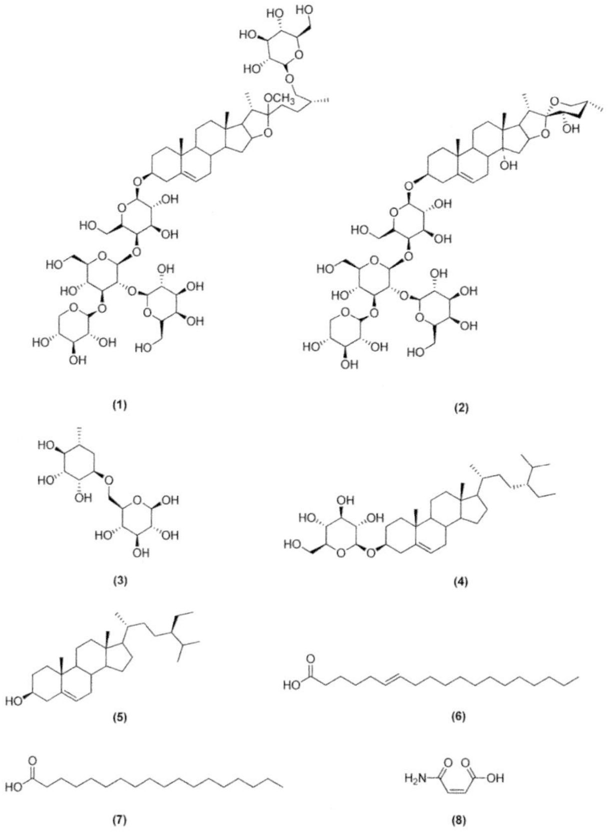

Fig. 2: Chemical constituents of *Polygonatum cirrhifolium*

The main ingredients of n-butanol extract of rhizomes are steroids, terpenoids, polysaccharides, phenol and tannin. Recent spectroscopic studies identified following compounds α-L-rhamnopyranosyl, β-D-glucopyranoside (3), daucosterol (4), β-sitosterol (5), 6-nonadecenoic acid (6), stearic acid (7), maleamic acid (8) (Balakrishna et al. 2012, Virk et al. 2017).

Pharmacological Studies

Cooling, mild laxative, galactagogue, aphrodisiac, depurative, wound healer, febrifuge, expectorant and tonic are some of the pharmacological properties of *P. cirrhifolium* compounds extracted from its rhizomes showed fungicidal activities.

Part Used

Rhizomes

Substitutes

Shatavari (*Asparagus racemosus* Willd.), Nagbala (*Sida veronicaefolia* L.), Shakaakul mishri (*Polygonatum multiforum* L.)

3. JEEVAK PLANT AS ASTAVARGA

Botanical Name / Family / Common English Name

Crepidium acuminatum D.Don)/ Orchidaceae/ Jeevaka

Distribution

Wide distribution reported from China, Cambodia, and South-East Asia at an elevation of 1400 m. ASL. In India this species occcur in temperate and subtropical Himalayas at an altitude of around 1200-2100 m ASL. from Himachal Pradesh, Uttarakhand to Arunachal Pradesh, Assam, Nagaland, Manipur, Mizoram and Tripura.

Chemical composition

Pseudobulb of *Crepidium acuminatum* (Figure 3) contains alkaloid, glycoside, flavonoids and β-sitosterol (1). Also contains piperitone (2), O-methyl batatasin (3), 1,8-cineole (4), citronellal (5), eugenol (6), choline (7), limonene (8), p-cymene (9) and ceryl alcohol (10) ((Balakrishna et al., 2012). Isorhamnetin-O-glycoside (11), bulbophythrin (12), gigantol (13), batatasin III (14), 3'-O-methyl batatasin (3), lusianthrine (15), 2,3-dimethoxy-9,10-dihydrophenanthrene-4,7-diol (16), liparacid C (17), stigmasterol (18), stigmasterol-3-O-glucoside (19) identified from ethanolic extract of pseudobulbs by the rapid de-replication using HPLC-ESI-QTOF-MS/MS (Singh et al. 2017).

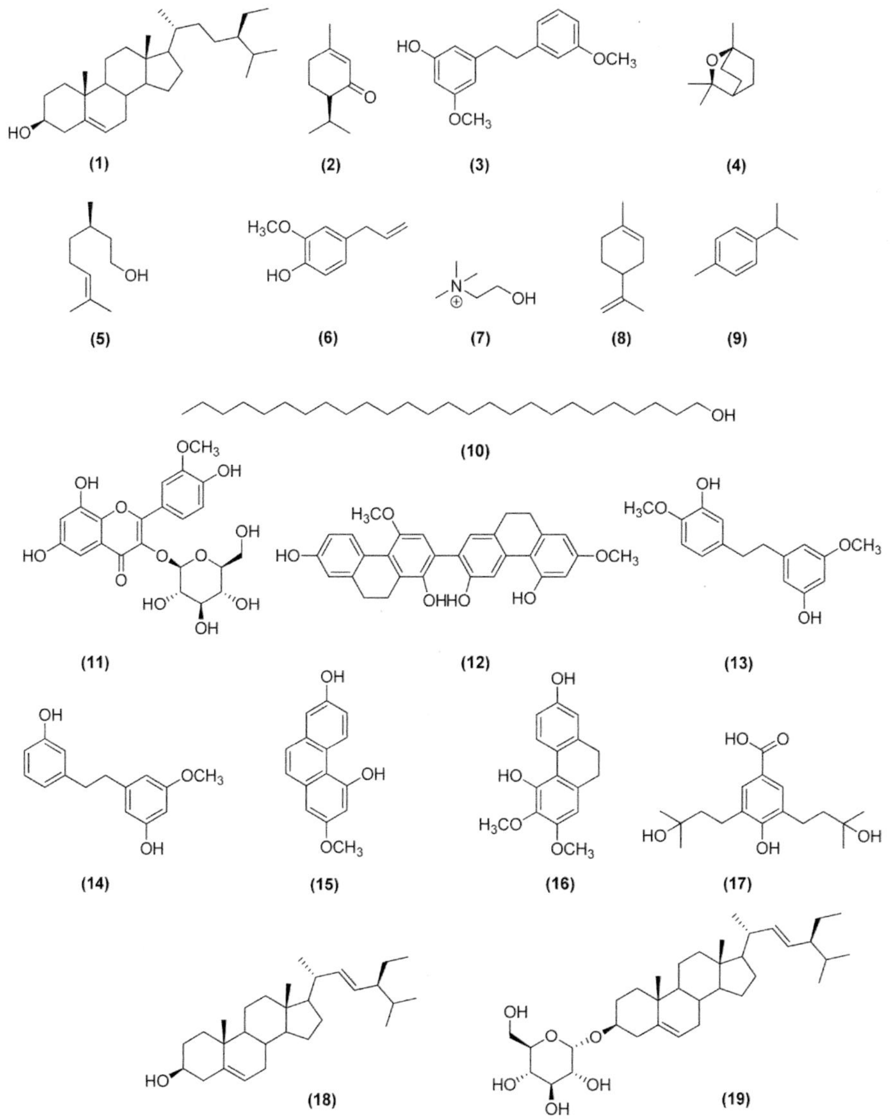

Fig. 3: Chemical constituents of *Crepidium acuminatum*

Pharmacological Studies

The pseudobulbs are sweet, refrigerant, aphrodisiac, febrifuge and tonic. This species is useful in haematemesis, fever, semen related weakness, burning sensation, dipsia, emaciation, tuberculosis and general debility.

The ethanolic extract of its tuber showed analgesic and anti-inflammatory activity in experimental animals.

Part Used

Pseudobulbs

Substitutes

Vidari Kand, Bahaman safed (*Centaurea behen*), Guduchi (*Tinospora cordifolia*).

4. RISHBHAK PLANT AS ASTAVARGA

Botanical Name / Family / Common English Name

Malaxis muscifera (Lindl.) Kuntze. /Orchidaceae/ Rishbhaka

Distribution

Malaxis muscifera widely distributed in the Himalayan mountains across Afghanistan, Pakistan, China, Nepal and Bhutan between an elevation of 2100-4100 m ASL. In India found in the temperate Himalayas around at an altitude of 2400-3600 m ASL. eastwards from Jammu& Kashmir, Himachal Pradesh, Uttarakhand and Sikkim.

Chemical Constituents

Bulb contains a bitter principle.

Pharmacological Studies

Sweet, refrigerant, aphrodisiac, haemostatic, antidiarrheal, styptic, antidysentric, febrifuge, cooling and tonic. It is useful in sterility, vitiated conditions of pitta and vata, semen related weakness, internal and external haemorrhages, fever, emaciation, burning sensation, and general debility. Tubers are efficacious cure for pitta, rakta, vata doshas. It is useful in phthisis, burning sensation in the body, fever and increases kapha and shukra.

Part Used

Pseudobulbs

Substitutes

Vidari kanda (*Pueraria tuberosa* DC.), Bahaman lal (*Centaurium roxburghii* (G.Don) Druce)

5. KAKOLI PLANT AS ASTAVARGA

Botanical Name / Family / Common English Name

Roscoea purpurea Sm./Zingiberaceae/ Roscoe's Lily

Distribution

Roscoea purpurea widely reported from the Himalaya regions of Pakistan, Bhutan and Tibet around an elevation of 1500-3100 m ASL. In India this species occurs in Central and Eastern Himalayan region from Uttarakhand to Sikkim and Assam around an altitude of 3300 m ASL. in alpine grass land, steppes, grassy hillslides, damp gullies and stony slopes.

Chemical composition

The rhizome of *Roscoea purpurea* (Figure 4) contains flavonoids, alkaloid, tannins, saponin, glycosides and phenolic compounds. Protocatechuic acid (1), syringic acid (2), ferulic acid (3), rutin (4), apigenin (5) and kaempferol (6) were identified by Srivastava et al. through Simultaneous reversed-phase high-performance liquid chromatography- ultraviolet (RP-HPLC) photodiode array detector and quantified as 0.774%, 0.064%, 0.265%, 1.125%, 0.128% and 0.528%, respectively. Total flavonoid content (TFC) and total phenolic content (TPC) were significantly ($P < 0.01$) rich in whole extract that is 26.78 mg/g QE and 3.03 g/g GAE, respectively. After fractionation, flavonoid content decreases in order of aqueous fraction and then Ethanol, acetone, petroleum ether, and chloroform fractions having 9.15, 1.71, 1.39, 0.27, and 0.216 mg/g QE, respectively. Phenolic content was also highest in aqueous (0.565 g/g GAE) fraction, followed by acetone (0.135 g/g GAE), ethanol (0.106 g/g GAE), chloroform (0.100 g/g GAE), and petroleum ether (0.040 g/g GAE). Miyazaki et al. isolated Kaempferide (7), Kaempferide 3-O-β-D-glucuronopyranoside (8), Kaempferol 3-O-β-D-glucuronopyranoside (9) and (Z) -3-hexen-1-ol-β-D-glucopyranoside (10) from aerial parts and Kaempferide (7), Kaempferol-3-O-methyl ether (11) and Adenosine (12) from rhizomes. Bag et al. (2014) reported the presence of the quercetin (13), kaempferol (6), pinocembrin (14), galangin (15), quinic acid (16) or their analogues by mass spectral analysis of the rhizome extract. Tannin content was found to be (p < 0.01) 4.75 mg tannic acid equivalent /g dry weight. Total flavonoids content was found to be 7.64 mg quercetin equivalent/ g dry weight. Total phenolic content was found to be 2.92 mg gallic acid equivalent /g dry weight.

Riboflavin (17) and thiamine (18) were found to be 3.80 and 4.75 mg/ g dry weight respectively. Gallic acid (19) was found to be 92.84 mg/ 100 g dry weight. Catechin (20) and p-coumaric acid (21) were found to be 12.05 and 0.14 mg/100 g dry weight respectively (Rawat et al. 2014). Quantification of secondary metabolites reveals that kaempferol (6) (0.30%) was the major metabolite among the other identified markers and then follows the order, vanillic acid (22) (0.27%), protocatechuic acid (1) (0.14), syringic acid (2) (0.08%) and ferulic acid (3) (0.05%). Total phenolic and flavonoid content in methanol extract was 7.1 mg/g GAE and 6.1 mg/g QE as estimated by regression analysis of Gallic acid and quercetin (0.1 mg/ml) as standard (Misra et al. 2015).

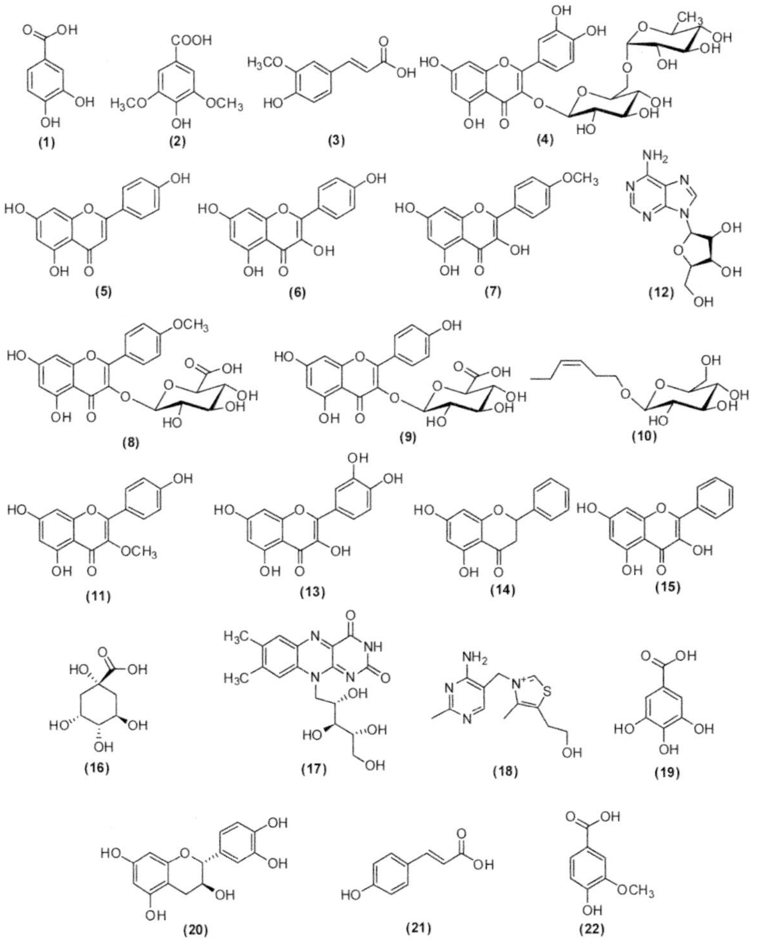

Fig. 4: Chemical constituents of *Roscoea purpurea*.

Pharmacological Studies

Rhizomes are reported to have antirheumatic, febrifuge, galactagogue, haemostatic, expectorant, sexual stimulant, diuretic, tonic, sweet, bitter and cooling properties.

Anti-oxidant Property

Radical Scavenging Activity by DPPH Assay

The scavenging effect of DPPH radical was concentration dependant and potentially varied for ascorbic acid, quercetin, rutin and butylated hydroxy toluene (BHT) and methanol extract. Ascorbic acid exhibits maximum inhibition of 77.57% which is followed by quercetin, rutin, methanol extract and BHT having inhibition of 72.43, 71.48, 69.57, 62.10% respectively, although the IC_{50} decreases in order of RPE > rutin > quercetin > ascorbic acid > BHT and hence indicating that BHT is potent inhibitor of free radical in all (Misra et al. 2015). Antioxidant activity varied considerably among the whole extract and fractions of tubers when compared to standard reference viz., ascorbic acid, quercetin, and rutin. In standards, maximum inhibition of free radicals was observed in ascorbic acid (77.57%, IC_{50}: 3.86 ± 0.057 μg/mL), followed by quercetin (72.43%, IC_{50}: 5.93 ± 0.115 μg/mL), and rutin (71.48%, IC_{50}: 6.80 ± 0.173 μg/mL). Whole extract exhibits IC_{50} at 0.925 ± 0.005 mg/mL. Among the *Roscoea* fractions, inhibition of radicals varies from 5.2% to 83.21%. Chloroform fraction possesses significant IC_{50} value at 25 mg/mL (Srivastava et al. 2015).

Reducing Ability by FRAP Assay

Ferric reducing antioxidant power assay was performed by adding FRAP reagent to methanolic extract and kept at 37° C for 8 min. Absorbance was taken at 593 nm by using UV-VIS spectrophotometer (Rawat et al. 2014). The reducing power assay of methanol extract served as significant indicator of its potentiality as reducing agent, which in turns signifies its anti oxidant activity and data reveals that reducing power of methanol extract increases linearly (r^2 = 0.946) with increase in concentration, similar to standards i.e. ascorbic acid, quercetin, rutin and BHT respectively.

β-Carotene Bleaching Method

Antioxidant activity estimated by bleaching of β-carotene for standard viz ascorbic acid, quercetin, rutin and BHT were estimated. BHT exhibited IC_{50} at 1.22 mg/ ml, thus act as potential anti-oxidant which is followed by quercetin, rutin and methanol extract. Ascorbic acid did not respond to this assay. IC_{50} of methanol extract, however was significantly different to that of standards when compared at 5% and 1% level of significance ($p < 0.01$) in both DPPH radical scavenging assay and β-carotene bleaching method. There is positive correlation between IC_{50} value of DPPH and β-carotene beaching method with phenolic and flavonoids content i.e. the antioxidant activity increases linearly with increase in content of phenolics and flavonoids, moreover the values were also significantly ($p < 0.01$) correlated with each other. Data depicted that correlation is more to flavonoid content (higher r^2 value) than to phenolic content (Misra et al. 2015).

Acute Oral Toxicity Study

There was no considerable signs of toxicity shown when administered the animals with doses of *Roscoea purpurea* ethanolic extract from 50 mg/kg body weight up to the dose 5000 mg/ kg body weight.

Immunomodulatory Activity

Ethanolic extract of *Roscoea purpurea* rhizomes administered orally at doses 300 mg/kg and 600 mg/kg, p.o. to mice. Result revealed that, the foot pat thickness of ethanolic extract group ($P<0.05$) significantly enhanced the production of circulating antibody titre in response to Sheep red blood cells (SRBC) and phagocytic functions of mononuclear macrophages and non-specific immunity. There was ($P<0.05$) significant decrease in foot pad thickness, WBC and total platelet counts in cyclophosphamide group as compared to control group. While extract-treated group animals showed ($P<0.05$) significant increase (Sahu et al. 2013).

Antidiabetic and Hypolipidemic Activity

The methanolic rhizome extract of *Roscoea purpurea* at the dose of 400, 600 and 800 mg kg^{-1} body weight was administered orally once a day to the groups for 30 days. The fasting blood glucose, cholesterol, HDL cholesterol and serum triglyceride content were estimated in both normal and alloxan induced diabetic rats. The fasting blood glucose, cholesterol and serum triglyceride content were found to be significantly reduced ($p<0.05$) in treated rats whereas the extract also showed the potent elevation in the level of serum HDL cholesterol. Study reveals that *R.purpurea* has significant anti-diabetic activity and a hypolipidemic activity in alloxan induced and normal fasting rats (Bairwa et al. 2012).

Cytotoxic Activity

Potential activity was observed in the ethanolic extract against the tested cell lines exhibiting IC_{50} value of <10, 49.5, 83.8, and 141.6 µg/mL against A549, SiHa, C-6, and CHOK1, respectively. All the tested fractions exhibit dose-dependent cytotoxicity against the cell lines. Among them Petroleum ether fraction exhibit the highest activity (69.1 ± 0.6 and 60.5 ± 1.5) on SiHa and CHOK-1 cells at concentration of 100 and 150 µg/mL, respectively, and Chloroform fraction showed the highest activity (62.4 ± 1.0) on A549 cells at a concentration of 100 µg/mL. However, the activity against C-6 cells is similar in Petroleum ether and Chloroform fractions. In a nutshell, from above preliminary cytotoxic activity, it was observed that whole extract of *Roscoea* exhibits promising effect and after fractionation, the potentiation of action reduces and variable response was observed (Srivastava et al. 2015).

Part Used

Rhizome

Substitutes

Aswagandha (*Withania somnifera*) and Kali Musali (*Curculigo orchioides Gaertn.*)

6. KSHIRKAKOLI PLANT AS ASTAVARGA

Botanical Name / Family / Common English Name

Lilium polyphyllum D.Don/ Liliaceae/ White Himalayan lily

Distribution

L. polyphyllum distributed in the Himalayan region across Afghanistan, West China, Pakistan, Nepal, and Tibet around an elevation of 1800-3700 m ASL. In India it occurs in the Western temperate Himalayas at an altitude of 2000-4000 m ASL Jammu and Kashmir, Punjab, Himachal Pradesh and Uttarakhand are its centre of occurrence.

Chemical Composition

Raval et al. separated the phytochemicals from methanol extract of dried roots using Sepbox and identified by GC-MS analysis.

Pharmacological Studies

The bulbs are sweet, bitter, refrigerant, galactagogue, expectorant, aphrodisiac, diuretic, antipyretic and tonic. Shows soothing, astringent and anti-inflammatory properties. Useful in agalactia, cough, bronchitis, vitiated conditions of pitta, semen related weakness, burning sensation, hyperdipsia, fever, haematemesis, rheumaltagia and general debility.

Part Used

Bulbs

Substitutes

Aswagandha (*Withania somnifera*) and Safed Musali (*Chlorophytum arundinaceum*)

7. RDHI PLANT AS ASTAVARGA

Botanical Name / Family / Common English Name

Habenaria intermedia D.Don / (Orchidaceae)/ white wild orchid

Distribution

H. intermedia found in the Himalayan Mountains of Pakistan, Nepal and Bhutan at an altitude of 2000-3300 m ASL. In India found in the temperate Himalayas at an altitude of 1500-2400 m ASL from Kashmir and Himachal Pradesh to Uttarakhand and Sikkim.

Chemical composition

The tubers of the *H. intermedia* (Figure 5) contains steroids, flavonoids, coumarin glycosides, tannins (Khandelwal 2008, Kokate 1994) and also reported to possess taxol (1).

Scopoletin (2) and gallic acid (3) were isolated from ethyl acetate and ethanolic fractions respectively (Habbu et al. 2012).

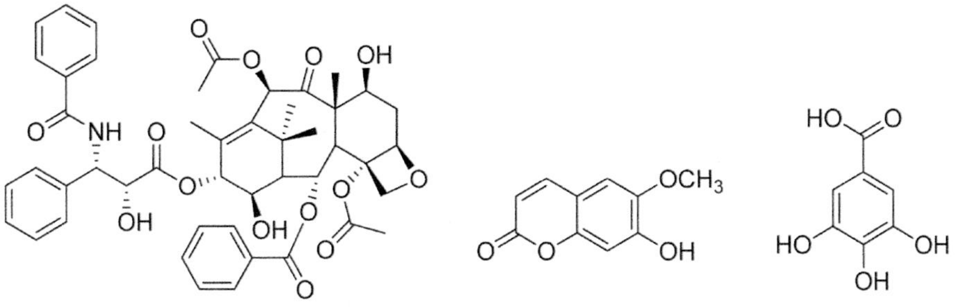

Fig. 5: Chemical constituents of *Habenaria intermedia*

Pharmacological Studies

The edible tubers are sweet, emollient, and used as intellect promoting, aphrodisiac, depurative, anthelmintic, rejuvenating and tonic. Tubers are useful in asthma, leprosy and skin diseases. This plant is an important ingredient of *Chyavanprasha* a well known polyherbal rejuvenator (Warrier et al. 1994, Kirtikar & Basu 1994). Antioxidant activity of polyherbal formulation containing tubers of *Habenaria intermedia* was investigated in nitric oxide scavenging activity (Jagetia et al., 2004). The tested doses of ethanolic extract (100 and 200 mg/kg, *p.o.*) and higher dose of ethyl acetate extract (200 mg/kg, *p.o.*) normalized altered serum biochemical parameters and the severity of ulcers in both acute and chronic stress. Ethanolic extract of *H. intermedia* showed DPPH scavenging activity with IC_{50} value of 35.46 µg/mL as compared with ascorbic acid (2.94 µg/mL), where as ethyl acetate extract showed DPPH scavenging activity with IC_{50} value of 32.88 µg/mL. The ability of the ethanolic and ethyl acetate extracts to scavenge the hydroxyl radical were found to be 52.38 µg/mL and 11.28 µg/mL respectively and IC_{50} value of standard mannitol was found to be 4.99 µg/mL. The inhibition of LPO by ethanolic extract was found to be 122.62 µg/mL where as ethyl acetate extract inhibit LPO with IC_{50} 42.75 µg/mL Anti-stress activity of *H. intermedia* may be attributed to the presence of scopoletin and gallic acid or their synergistic properties, which may be mediated through anti-oxidant mechanism (Habbu et al. 2012). Hepatoprotective potential against CCl_4 induced toxicity in albino rat liver was controlled significantly by restoration of the levels of serum bilirubin, proteins and enzymes as compared to the normal and standard drug silymarin treated groups (Goudar et al. 2014 and 2015). Significant anti-anxiety activity was observed in methanol extract of fruits with respect to control diazepam (Kumar et al. 2017). The effect of ethanolic extract of *H. intermedia* tubers on the foot pad thickness and hematological data such as WBC and Total Platelet counts of antigenically challenged mice, the ($P<0.05$) significant decrease in the foot pad thickness, WBC and Total Platelet counts in cyclophosphamide group as compared to control group. While in extract-treated group animals showed ($P<0.05$) significant increase. Ethanolic extract of *H. intermedia* tubers was found to have a promising immunostimulant potential (Sahu et al. 2013).

Part Used

Tubers

Substitutes

Varahi kand (*Tacca aspera* Roxb.), Utangan seeds (*Blepharis edulis* Pers.), Bala (*Sida cordifolia*) and Chiriya Musali (*Asparagus filicinus* Buch.- Ham.)

8. VRDDHI PLANT AS ASTAVARGA

Botanical Name / Family / Common English Name

Platanthera edgeworthii (Hoo.f. ex Collett) RK Gupta (Synonym: *Habenaria edgeworthii* Hook.f. ex. Collect)/ Orchidaceae/ Edgeworths Habenaria

Distribution

P. edgeworthii found across the Himalayas in the North Western parts of Pakistan and India. Himachal Pradesh, Uttarakhand and Nepal are its main centre where at distributed at an elevation of 2500-3000 m ASL. on grassy pastures.

Chemical Composition

The tubers of the *Platanthera edgeworthii* (Figure 6) contain bitter substances, minerals and starch. Gallic acid (1) and Hydroxybenzoic acid (2) have been reported from this species (Giri et al. 2011).

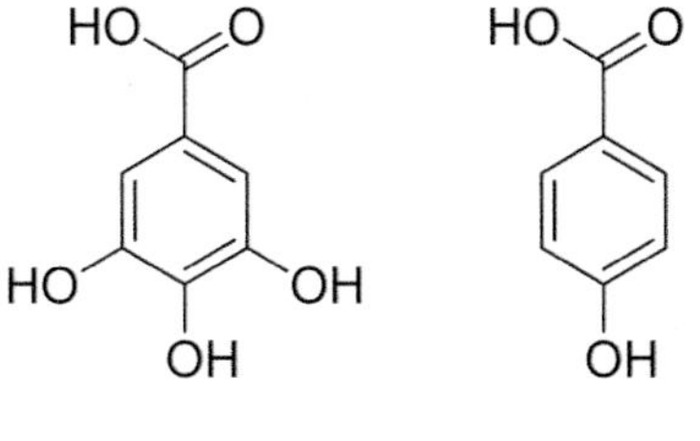

(1)   **(2)**

Fig. 6: Chemical constituents of *Platanthera edgeworthii*

Pharmacological Studies

Cooling, emollient, aphrodisiac, rejuvenating, brain tonic, blood purifier, appetizer, and is beneficial in aggravated vata and pitta, burning sensation in the body, excessive thirst, hyperdipsia, fever, cough, asthma, leprosy, epilepsy, emanciation and general debility. Tuber extracts showed significantly ($p < 0.05$) higher total phenolics and flavonoids among the populations. Antioxidant activity determined by 2, 22 - azinobis (3-ethylbenzothiazoline-6-sulfonic acid) (ABTS) radical scavenging, 1,1-diphenyl-2- picrylhydrazyl (DPPH) radical scavenging and ferric reducing antioxidant power (FRAP) assays exhibited considerable antioxidant potential (Giri et al. 2016).

Part Used

Tubers

Substitutes

Varahi Kand (*Tacca aspera* Roxb.), Salam panja (*Dactylorhiza hatagirea*), Mahabala (*Sida acuta*), seeds of *Sida cordifolia.*

ASHTAVARGA PLANTS IN AYURVEDIC FORMULATIONS

- Satavari ghritam
- Chaglangham ghritam
- Amritprasha ghritam
- Kumkumadi ghritam
- Bhallatkam ghritam
- Mahapaisaachik ghritam
- Shiva ghritam
- Nakulangh ghritam
- Vrchyamadi ghritam
- Ashok ghritam
- Kumarkalpdrum ghritam
- Piplayangh ghritam
- Trisatiprasaarni tailam
- Saptsateekprasarni tailam
- Ekadashsateekprasarni tailam
- Maashtailam dwitiyam
- Mahamaashtailam
- Balarishta
- Mahachandanadi tailam

- Balatailam
- Vishnu tailam
- Sudhakar tailam
- Narayan tailam
- Kamdev ghrit
- Chyawanprash
- Sudarshan churna

FUTURE PERSPECTIVES

Ashtavarga plants are placed in Ayushya group, which mitigate the diseases of the body, and increase the longevity of life or slow down the process of ageing. The active ingredients of the Ashtavarga plants develop the resistance to the human body and saves the body from disease. For the identification of Ashtavarga, traditional knowledge, deep exploration and research, and self realization and science came into use for just proving that these were the ashtavargas as per texts. By reading of this review reveals that there are several issues to be solved in future.The requirement of the day is the scientific investigation of the described properties by the ancient rishis; to prove its importance through chemical analysis, as per modern pharmacological sciences. These plants need immediate attention of conservations as some of them are rare and many many face extinction in times to come. The chemical analytical studies are another important area to add value to these species. Extensive and intensive survey on these species must be undertaken in order to assess their availability in their natural habitats. Another important study is their ecological life behavior and conditions to initiate work on their multiplication either under field conditions or taking the help of tissue culture. These plant species may prove excellent antioxidants as is apparent from their qualities described in literature. As all the eight species have been substituted, adulterant species requires chemical evaluation whether they can be exploited in drug use or not. Similarly the drug substitute species should be subjected to chemical evaluation and comparative study to determine worth of these species in comparison to Astavarga plants. Moreover, herbal companies procured these plant materials mostly by non botanists which need detailed pharmacognostic studies to identify actual species from adulterants or substitutes. Another aspect that needs to be done is to standardize the formulation to check whether correct amount of individual ingredients added or not.

REFERENCES CITED

Antoniuk VO. 1993. Purification and properties of lectins of *Polygonatum multiflorum* [L.] All. and *Polygonatum verticillatum* (L.) All. *The Ukrainian Biochemical Journal* 65: 41–48.

Bag BG, Dash SS, Roy A, 2014. Study of Antioxidant Property of the Rhizome Extract of *Roscoea purpurea* Sm. *(Kakoli)* and its Use in Green Synthesis of Gold nanoparticles. *International Journal of Research in Chemistry and Environment* 4: 174-180.

Balakrishna A, Srivastava A, Mishra RK, Patel SP, Vashistha RK, Singh A, Jadon V, Saxena P. 2012. Astavarga Plants–threatened medicinal herbs of the North-West Himalaya. *International Journal of Medicinal and Aromatic Plants* 2: 661-676.

Bairwa R, Basyal D, Srivastava B. 2012. Study of antidiabetic and hypolipidemic activity of *Roscoea purpurea* (Zingiberaceae). *International Journal of Institutional Pharmacy and Life Sciences* 2: 130-137.

Chinmay R, Kumari S, Bishnupriya D, Mohanty RC, Dixit R, Padhi MM, Babu R. 2011. Pharmacognostical and phytochemical studies of *Roscoea procera* (kakoli) and *Lilium polyphyllum* (ksheerkakoli) in comparison with market samples. *Pharmacognosy Journal* 3: 32-38.

Chinmay R, Kumari S, Dhar B, Mohanty RC, Singh A. 2009. Pharmacognostical and phytochemical evaluation of rare and endangered *Habenaria* spp. (Riddhi and Vriddhi). *Pharmacognosy Journal* 1: 94–102.

Dhyani A, Bahuguna YM, Semwal DP, Nautiyal BP, Nautiyal MC. 2009. Anatomical features of *Lilium polyphyllum* D. Don ex Royle (Liliaceae). *Journal of American Science* 5: 85-90.

Dhyani A, Nautiyal BP, Nautiyal MC. 2010. Importance of Astavarga plants in traditional systems of medicine in Garhwal, Indian Himalaya. *International Journal of Biodiversity Science, Ecosystem Services & Management* 6: 13-19.

Dhyani A, Sharma G, Nautiyal BP, Nautiyal MC. 2014. Propagation and conservation of *Lilium polyphyllum* D. Don ex Royle. *Journal of Applied Research on Medicinal and Aromatic Plants* 1: 144-147.

Fransworth NR. 1994. Ethnopharmacology and drug development. In: Chadwick DJ, Marsh J, editors. Ethnobotany and the search for new drugs. Ciba Foundation Symposium (185). Chichester (England): John Wiley & Sons. 42–51.

Giri L, Jugran A, Rawat S, Dhyani P, Andola H, Bhatt ID, Rawal RS, Dhar U. 2012. In vitro propagation, genetic and phytochemical assessment of *Habenaria edgeworthii*: an important astavarga plant. *Acta Physiologiae Plantarum* 34: 869-875.

Giri L, Jugran AK, Bahukhandi A, Dhyani P, Bhatt ID, Rawal RS, Nandi SK, Dhar, U. 2016. Population genetic structure and marker trait associations using morphological, phytochemical and molecular parameters in *Habenaria edgeworthii*- a threatened medicinal orchid of west Himalaya, India. *Applied Biochemistry and Biotechnology* 181: 267-282.

Goudar MA, Jayadevappa H, Mahadevan KM, Shastry RA, Habbu PV, Sayeswara HA. 2015. Isolation and characterization of secondary metabolite from *Habenaria intermedia* D. Don for screening of hepatoprotective potential against carbon tetrachloride induced toxicity in Albino rat liver. *International Journal of Current Pharmaceutical Research* 7: 57-61.

Goudar MA, Jayadevappa H, Mahadevan KM, Shastry RA, Sayeswara HA. 2014. Hepatoprotective potential of tubers of *Habenaria intermedia* D.Don against carbon tetrachloride induced hepatic damage in rats. *International Journal of Current Research* 6: 10090-10097.

Gvazava LN, Skhirtladze AV. 2016. Steroidal saponin from *Polygonatum verticillatum*. *Chemistry of Natural Compounds* 52: 1052-1055.

Habbu PV, Smita DM, Mahadevan, KM, Shastry, RA, Biradar, SM. 2012. Protective effect of *Habenaria intermedia* tubers against acute and chronic physical and psychological stress paradigs in rats. *Brazilian Journal of Pharmacognosy* 22: 568-579.

Jagetia GC, Rao SK, Bakiga MS, Babu KS. 2004. The evaluation of nitric oxide scavenging activity of certain herbal formulations *in vitro*: a preliminary study. *Phytotherapy Research* 18: 561-565.

Kala CP, Dhyani PP, Sajwan BS. 2006. Developing the medicinal plant sector in northern India: challenges and opportunities. *Journal of Ethnobiology and Ethnomedicine* 2: 32-46.

Kazmi I, Afzal M, Rahman M, Gupta G, Anwar F. 2012. Aphrodisiac properties of *Polygonatum verticillatum* leaf extract. *Asian Pacific Journal of Tropical Disease* 2: S841-S845.

Khan H, Saeed M, Gilani AH, Ikram-ul H, Ashraf N, Najeeb-ur R, Haleemi A. 2013. Antipyretic and anticonvulsant activity of *Polygonatum verticillatum*: comparison of rhizomes and aerial parts. *Phytotherapy Research* 27: 468–471.

Khan H, Saeed M, Gilani AH, Khan MA, Dar A, Khan I. 2010. The antinociceptive activity of *Polygonatum verticillatum* rhizomes in pain models. *Journal of Ethnopharmacology* 127: 521–527.

Khan H, Saeed M, Gilani AH, Khan MA, Khan I, Ashraf N. 2011. Anti-nociceptive activity of aerial parts of *Polygonatum verticillatum*: attenuation of both peripheral and central pain mediators. *Phytotherapy Research* 25: 1024–1030

Khan H, Saeed M, Gilani AH, Mehmood MH, Rehman NU, Muhammad N, 2013. Bronchodilator activity of aerial parts of *Polygonatum verticillatum* augmented by anti-inflammatory activity: attenuation of Ca^{2+} channels and lipoxygenase. *Phytotherapy Research* 27: 1288–1292.

Khan H, Saeed M, Khan MA, Izhar-ul H, Muhammad N, Ghaffar R. 2013. Isolation of long chain esters from the rhizome of *Polygonatum verticillatum* with potent tyrosinase inhibition. *Medicinal Chemistry Research* 22: 2088–2092.

Khan H, Saeed M, Khan MA, Khan I, Ahmad M, Muhammad N, Khan A. 2011. Antimalarial and free radical scavenging activities of rhizomes of *Polygonatum verticillatum* supported by isolated metabolites. *Medicinal Chemistry Research* 21: 1278–1282.

Khan H, Saeed M, Mehmood MH, Rehman NU, Muhammad N, Haq IU, 2013. Studies on tracheorelaxant and anti inflammatory activities of rhizomes of *Polygonatum verticillatum*. *BMC Complementary and Alternative Medicine* 13: 197-204.

Khan H, Saeed M, Muhammad N, Ghaffar R, Khan SA, Hassan S. 2012. Antimicrobial activities of rhizomes of *Polygonatum verticillatum*: attributed to its total flavonoidal and phenolic contents. *Pakistan Journal of Pharmaceutical Sciences* 25: 463–467.

Khan H, Saeed M, Muhammad N, Tariq S A, Ghaffar R, Gul F. 2013. Antimalarial and free radical scavenging activities of aerial parts of *Polygonatum verticillatum* and identification of chemical constituents by GC-MS. *Pakistan Journal of Botany* 45: 497-500.

Khandelwal KR. 2008. *Practical Pharmacognosy: Technoques and Experiments*. Nirali Prakashan Pune, India. 19th Edition, 2: 149.

Kirtikar KR, Basu BD. 1994. Indian Medicinal plants. 4: 2413.

Kokate, C.K. 1994. Practical Pharmacognosy. Vallabh Prakashan. New Delhi, 2: 107.

Kumar P, Madaan R, Sidhu S. 2017. Screening of antianxiety activity of *Habenaria intermedia* D. Don fruits. *Journal of Pharmaceutical Technology, Research and Management* 5: 71-75.

Chunekar KC. 1969. Vanaspatika anusandhan darshika, vidhya bhavan.Varanasi.

Mathur DR, 2003. Yogtarangini. Chaukhamba Amarabharati Prakashan, Varanasi (India).

Misra A, Srivastava S, Verma S, Rawat AKS. 2015. Nutritional evaluation, antioxidant studies and quantification of poly phenolics, in *Roscoea purpurea* tubers. *BMC Research Notes* 8: 324-330.

Miyazaki S, Devkota HP, Joshi KR, Watanabe T, Yahara S. 2014. Chemical constituents from the Aerial parts and rhizomes of *Roscoea purpurea*. *Japanese Journal of Pharmacognosy* 68: 99-100.

Motohasi N, Zhang GW, Shirataki Y. 2003. *Oriental Pharmacy and Experimental Medicine* 3: 163-179.

Mukherjee PK, Wahil A. 2006. Integrated approaches towards drug development from Ayurveda and other systems of medicine. *Journal of Ethnopharmacology* 103: 25–35.

Pandey MM, Govindarajan R, Khatoon S, Rawat AKS, Mehrotra S. 2006. Pharmacognostical studies of *Polygonatum cirrifolium* and *Polygonatum verticillatum*. *Journal of Herbs, Spices & Medicinal Plants* 12: 37-48.

Pandey D. 2005. Sarangadhara Samhita. Chaukhamba Amarabharati Prakashan,Vara- nasi (India).

Patra A, Singh SK. 2018. Evaluation of phenolic composition, antioxidant, anti-inflammatory and anticancer activities of *Polygonatum verticillatum* (L.). *Journal of Integrative Medicine* 16: 273-282.

Prajapati NS, Purohit SS, Sharma AK, Kumar T. 2003. A Handbook of Medicinal Plants: A Complete Source Book; Agrobios: Jodhpur, India.

Raval SS, Mandavia MK, Mahatma MK, Golakiya BA. 2015. Separation and identification of phytochemicals from *Lilium polyphyllum* D.Don (*Kshirkakoli*), an ingredient of ashtavarga. *Journal of Cell and Tissue Research* 15: 5247-5254

Rawat S, Andola H, Giri L, Dhyani P, Jugran A, Bhatt ID, Rawal RS. 2014. Assessment of nutritional and antioxidant potential of selected vitality strengthening himalayan medicinal plants. *International Journal of Food Properties* 17: 703–712.

Saboon, Bibi Y, Muhammad A, Sabir S, Amjad MS, Ahmed E, et al. 2016. Pharmacology and biochemistry of *Polygonatum verticillatum*: a review. *Journal of Coastal Life Medicine* 4: 406–415.

Saeed M, Khan H, Khan MA, Khan F, Khan SA, Muhammad N. 2010. Quantification of various metals and cytotoxic profile of aerial parts of *Polygonatum verticillatum*. *Pakistan Journal of Botany.* 42: 3995–4002.

Saeed M, Khan H, Khan MA, Simjee SU, Muhammad N, Khan SA. 2010. Phytotoxic, insecticidal and leishmanicidal activities of aerial parts of *Polygonatum verticillatum*. *African Journal of Biotechnology* 9: 1241–1244.

Saha D, Ved D, Ravikumar K, Haridasan K. 2015a. *Malaxis Muscifera*. The IUCN Red List of Threatened Species. e.T50126625A50131390.

Saha D, Ved D, Ravikumar K, Haridasan K, Dhyani A. 2015b. *LiliumPolyphyllum*. The IUCN Red List of Threatened Species . e.T50126623A79918170.

Sahu MS, Sahu RA, Verma A. 2013. Immunomodulatory activity of alcoholic extract of *Habenaria intermedia* in mice. *International Journal of Pharmacy and Pharmaceutical Sciences* 5: 406-409.

Singh D, Kumar S, Pandey R, Hasanain M, Sarkar J, Kumar B. 2017. Bioguided chemical characterization of the antiproliferative fraction of edible pseudobulbs of *Malaxis acuminata* D. Don by HPLC-ESI-QTOF-MS. *Medicinal Chemistry Research* 26: 3307-3314.

Singh A, Duggal S. 2009. Medicinal orchids-an overview. *Ethnobotanical Leaflets* 13: 399-412.

Sharma BD, Balakrishna A. 2005. Vitality strengthening Astavarga plants (Jeevaniya and Vayasthapan Paudhe), Uttaranchal (India), Divya Publishers, Divya yog mandir.

Srivastava S, Ankita M, Kumar D, Srivastava A, Sood A, Rawat A. 2015. Reversed-phase high-performance liquid chromatography-ultraviolet photodiode array detector validated simultaneous quantification of six bioactive phenolic acids in *Roscoea purpurea* tubers and their *In vitro* cytotoxic potential against various cell lines. *Pharmacognosy Magazine* 11: 488-495.

Virk JK, Bansal P, Gupta V, Kumar S, Singh R, Rawal RK. 2017. First report of isolation of maleamic acid from natural source *Polygonatum cirrhifolium*-A potential chemical marker for identification. *Journal of Liquid Chromatography & Related Technologies* 40: 1031-1036.

Velpandian T, Mathur P, Sengupta S, Gupta SK. 1998. Preventive effect of Chyavanprash against steroid induced cataract in the developing chick embryo. *Phytotherapy Research* 12: 320–323.

Warrier PK, NambiarVPK, Thakur RS, Ramankutty C 1994. Indian medicinal plants-a Compendium of 500 species. Orient Longman, Delhi, Vol I, p 191.

Watanabe T, Rajbhandari KR, Malla KJ, Devkota HP, Yahara S. 2013. A handbook of Medicinal plants of Nepal Supplement I, Ayurseed Life Environmental Institute, Kanagawa. pp. 254-55.

WHO., 2005. National policy on traditional medicine and regulation of herbal medicines: Report of a WHO global survey. World Health Organization, Geneva, Switzerland.

3

Colebrookea oppositifolia Sm., An Important Hepato-protective Phytopharma Plant in Drug Discovery

Anil Kumar Katare[1]*, Inshad Ali Khan[2]*, Durga Prasad Mindala[1], Naresh Kumar Satti[3], Bal Krishan Chandan[4] Bikarma Singh[5], Gurdarshan Singh[6], Mowkashi Khullar[4] and Neelam Sharma[4]

[1]cGMP Division, [2]Clinical Microbiology Division, [3]NPC Division, [4]Inflammation Pharmacology Division, [5]Plant Sciences (Biodiversity and Applied Botany Division), [6]PK-PD Toxicology and Formulation Division CSIR-Indian Institute of Integrative Medicine, Jammu-180001, Jammu & Kashmir, INDIA
*Email: iakhan@iiim.ac.in, akkatare@iiim.ac.in, dpmindala@iiim.ac.in nksatti@iiim.ac.in, bkrishan@iiim.ac.in, singh_gd@iiim.ac.in drbikarma@iiim.ac.in

ABSTRACT

Drug discovery is the process to identify the active ingredients from traditional remedies as new candidate for medications. It involves the sources, identification of screening hits, medicinal chemistry and optimization of those hits to increase the affinity, potency, reduce the potential of side effects, increase the half life and oral bioavailability followed by clinical trials. While studying and screening the plant resourses of Himalaya, *Colebrookea oppositifolia* Sm. is identified as an important Heptoprotective Phytopharma plant. This is an important Indian medicinal plant of the family lamiaceae endemic to the Southeast Asia. The active ingredients of this species possess antimicrobial, antifungal and antioxidant properties as evident from experimentation which may be due to the high content of flavonoids and polyphenols as a major chemical constituents. In Ayurvedic traditional system of medicine, leaves of *C. oppositifolia* is used for the treatment of wounds and fractures, where as roots helps in curing epilepsy. Chemical investigations on the petroleum ether extract of the leaves of *C. oppositifolia*, some active molecules such as n-triacontane, acetyl alcohol, 32-hydroxydotriacontyl ferulate, β-sitosterol and 5,6,7,42 -tetramethoxy flavones have been identified which has hepatoprotective properties as evident from biological investigation. Structures of these compounds were interpreted on the basis of spectral (IR, NMR, MS) data. While studying acute toxicity of these bioactive ingredients, the species can be classified as category V as per GHS of OECD test guidelines and principles of GLP. In the present communication, an attempt has been made to provide scientific data on biology, natural products, constituents of volatile oils, salient markers, medicinal chemistry, results showing hepatoprotective activity and future direction for research to be carried out on this plant species. It also provide data on lyophilized extract or at least one bioactive fraction obtained from leaves of *C. oppositifolia* and treatment of liver toxicity caused by the conditions hepatoprotective against alcohol induced and tetrachloride (CCl_4) models. The presented data will be helpful in future for drug discovery.

Keywords: Himalaya plant, *Colebrookea oppositifolia*, Hepatoprotective, Acetoside, Lyophilized extract Drug discovery.

INTRODUCTION

Historically drug were discovered through identifying the active ingredients from traditional remedies or by serendipitous discovery (Warron 2011). This was the case of aspirin, which was derived from the bark of willow tree (Takenaka 2011). Plants and microbes are natural repository of various bioactive ingredients and fine chemicals. The application of this group of organisms provides a baseline data for research to be carried in the field of drug discovery. Liver has a pivotal role in regulation of physiological processes and hepatocyte alteration results in acute and chronic dysfunction (Rosenack et al. 1980). One of the major causes of morbidity and mortality in human and animal community is the liver diseases which affects humans and animal health of all ages. Hepatitis, a high incidence ailment around the world, is induced by toxic chemicals, viruses, alcohol, lipid peroxidative products and various drugs. Hepatic fibrosis is a common process of chronic liver injuries, characterized by increased deposition and altered composition of extracellular matrix leading to cirrhosis. The World Health Organization reports that approximately two billion alcohol consumers are there in worldwide and approximately 76.3 million people are with diagnosable alcohol use-disorders (WHO 2004).

Globally, alcohol causes 3.2% deaths for all causes (1.8 million deaths annually) and accounts for 4.0% of disease burden (Schermer 2006). The association between ethanol intake and alcoholic liver disease has been well documented, though liver cirrhosis develops only in a small proportion to frequent drinkers. Mortality studies have demonstrated that heavy drinkers dies from cirrhosis at a much higher rate than the general population (Jiang et al. 2014). The progression of Alcoholic Liver Disease (ALD) continues according to the amount of alcohol consumed resulting in a considerable liver fibrosis tissue with few functional hepatocytes, and chronic inflammation effecting structure and function of the liver. Corticosteroids are widely used in selected patients, and the treatment with pentoxifylline appears to be a promising anti-inflammatory therapy. Considering the importance of liver, studies on Himalayan medicinal plants were carried out to screen them as for hepatoprotective plant, and *Colebrookea oppositifolia* Sm. is one of them, which shows hepatoprotective properties.

The genus *Colebrookea* belongs to the flowering plants placed under the family Lamiaceae, which was first described by James Edward Smith in 1806 (Figure 1). *Colebrookea oppositifolia* Sm. is a monotypic species under this genus known by four synonymous names, *viz.*, *Buchanania oppositifolia* Sm., *Colebrookea ternifolia* Roxb., *Elsholtzia oppositifolia* (Sm.) Poir. and *Sussodia oppositifolia* (Sm.) Buch.-Ham. This particular species known for antimicrobial, antifungal and antioxidant properties (Ishtiaq et al. 2016), and traditionally used for treating sore eyes, corneal opacity or conjunctivities due to its anti-inflammatory effects (Torri 2012). This species is used to cure the diseases such as epilepsy, fever, headache, urinary problems and shows active properties of hepatoprotective, cardioprotective and anti-inflammatory attributes (Ishtiaq et al. 2016). Essential oils can be extracted from leaves and young flowering twigs and these reported to have fungitoxic properties (Holley and Cherla 1998). Venkateshappa and Sreenath (2013) reported anthelmintic properties in *C. oppositifolia* which is used in the management of ringworms, and employed in the treatment of dermatitis, nose bleeds, coughs and dysentery.

Fig. 1: Habit morphology and pharmagonostic characters of
Colebrookea oppositifolia

BOTANICAL RAW MATERIAL

Authentication and Certification

Plants are widely used as raw ingredients for many preparations in conventional medicine system. To confirm the authencity of the raw ingredients and to detect the presence of adulterant stains, there is need to carry out both taxonomical and pharmacognostic evaluations for the plant species. Usually these raw drugs were collected by the traditional workers who usually get engaged in herbal, ayurvedic or any other complementary system of medicine. Their identification is commonly based on macroscopic structural features or other unique visible characteristics. Therefore, in such manual practices there is a chance of accidental collection of improper or wrong plant samples. Anatomical, physico-chemical and phytochemical screening is helpful to avoid any ambiguity and chances of adulteration or mixing minimizes. For *Colebrookea oppositifolia*, taxonomical as well as pharmacogonostic studies were carried out for proper authentication and certification of the plant species.

Colebrookea oppositifolia in the form of leaves were shade dried at 35°C in laboratory room, till moisture content not more than 10%w/w and powdered using sieve mesh. The powdered raw material packed in low density polyethylene bags and kept in air tight container, and was analysed for its quality prior taken up for further process.

Taxonomical Enumeration of Plant Used

Colebrookea oppositifolia shows habit as large shrubs to small trees, 3-5 m tall. Stems and branches were densely tomentose. Leaves elliptic 5-20 cm long, 2-9.5 cm broad, crenulate, acute to acuminate; nerves upto 10 pairs, oblique; petioles hairy. Inflorescence terminal, spikelets, 5-10 cm long, 0.2-0.5 cm diam., panicled, densely hairy. Flowers white, in close whorls; sepals 1.5 mm long, tube short, densely white-hairy, lobes subulate; petals upto 2 mm long, equally 4-lobed; stamens 4 in number, exserted in male flowers, included in female flowers; filaments glabrous; style bi-fid at tip. Nutlets hairy at apex, fruiting calyx large.

PHARMACOGNOSTIC FEATURES

The transverse section of leaf and stem of *Colebrookea oppositifolia* shows the basic profile of botanical anatomy with a little contrast in having lunar shaped vascular bundles in the leaf, while a continuous vascular bundle with primary and secondary xylem and phloem with presence of sclerenchyma all around. Fibres and trachieds are with annular and spiral thickening in the medullary region. Parenchyma and collenchyma in papillose with oil globules. Phloroglucinol shows the presence of lignins in leaf and stem, while iodine solution indicate negative result for starch. Ferric chloride solution specified the presence of tannins.

Midrib

Outer layer of midrib wavy on both the upper and the lower sides. Glandular and non-glandular trichomes were present on the surface of midrib. Non-glandular hairs observed to be multicellular and uniseriate. The glandular hairs were sessile and filled with cell contents. Epidermal layer were followed by 1-2 layered collenchymatous hypodermis. The cells of this region are variable in size. Major region of the midrib is made up of parenchymatous tissue. Vascular bundle centrally located, and were semilunar in shape. Sclerenchymatous cells were in group of 3-20 and present outside the vascular bundle. These cells are conjoint, collateral and endarch. Phloem is made up of thin walled and irregular shaped cells and also few of them were filled with light brown substance (Figure 2).

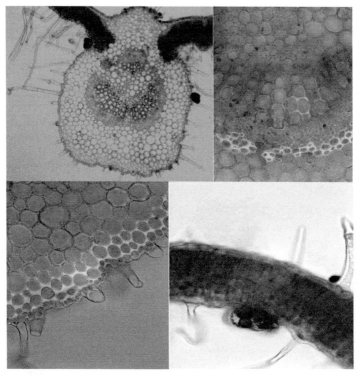

Fig. 2: Transverse section of *Colebrookea oppositifolia* (Upper two figures: Midrib and magnified view through vascular bundle; Lower two figures: magnified view through outer region, and lamina showing multicellular hairs and sessile glands).

Lamina

Upper and lower epidermis observed to be made up of oval to columnar cells, covered with thin cuticles. Mesophyll differentiated into palisade parenchyma and spongy parenchyma. Palisade parenchyma consists of 1 or -layers of elongated, compactly arranged cells. Below the palisade cells, 4-6 layered loosely arranged spongy parenchyma cells were present and were of irregular in shape. Vascular bundles were surrounded by sclerenchymatous bundle sheath. Both glandular and non-glandular trichomes were present on the surfaces. Non-glandular trichomes were unicellular and multicellular where as glandular trichomes were sessile. Druse crystals were observed in parenchyma region.

Powder Microscopy

Microscopic studies of the powdered leaves of *Colebrookea oppositifolia* shows multicellular hairs of variable size. Druse crystals were observed in parenchymatous cells. Spongy parenchymatous cells were present (Figure 3).

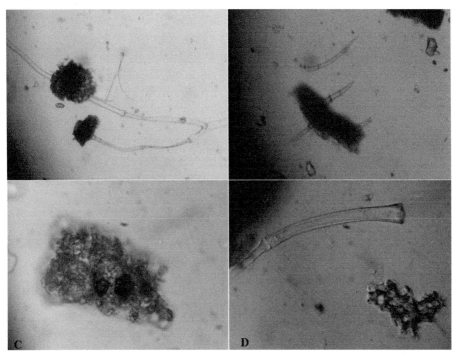

Fig. 3: Powdered analysis of *Colebrookea oppositifolia* showing multicellular hairs and sessile glands.

ESSENTIAL OIL CONSTITUENTS IN COLEBROOKEA OPPOSITIFOLIA

Chemical examination of *Colebrookea oppositifolia* revealed the presence of essential oil in leaves. GC-MS analysis were carried out using gas chromatography–high resolution mass spectrophotometer by using 2 µl of essential oil sample in Allegiant Hp 7880 with column 25m and Helium gas as carrier gas at constant flow rate. NIST (National Institute of Standard of Technology) online library proves that the leaves contain phytol (41.28%), n-hexadecanoic acid (27.52%), octanoic acid tridec-2-eny ester (5.1%), 9,12,15 octadecatrienoic acid (9.9%), 2-dodecen-1-ny, succinic anhydride(4.4%). The 9,12,15 octadecatrienoic acid also called α-linolenic acid is an essential fatty acid and necessary for human health because they are not produced within the human body, and must be acquired through diet (Burr et al. 1930). Octanoic acid tridec-2-eny ester also called caprylic acid is used commercially in the production of perfumes and used in manufacturing of dyes (Papamandjaris et al. 1998). The n-hexadecanoic acid or palmitic acid reported to be used in manufacturing of soaps, cosmetics and release agents (Benoit et al. 2009). Hydrogenation of palmitic acid yields acetyl alcohol, which is used to produce detergents and cosmetics. Sodium palmitate is permitted as a natural additive in organic products (Kingsbury et al. 1961).

Reference Chemical Compounds

Standard marker compounds, acetoside, and flavanoids (CO_1, CO_3 and CO_4) were isolated from the *Colebrookea oppositifolia* with optimized protocol. The reference standards, reagents and chemicals were used for further studies and isolation of compoundes.

Apparatus

Agilent triple quad LC-MS/MS,HPLC (Shimadzu L), columns (YMCODS-A,3 mm, 150x4.6 mm; YMCODS-A, 5 mm, 250x4.6 mm; Inertsil ODS-3,3 mm, 150x4.6 mm; BDS Hypersil C18, 5 mm, 100x4.6 mm), Chromoliths high resolution RP-18 end capped, 50x4.6 mm, LC-MS (Water), Oasis HLB1cc cartridge (Water), sonicator and micropipettes were used for the study.

Extraction, Isolation and Qualitative Chemical Analysis of Markers

Dried leaves of *Colebrookea oppositifolia* (6.0 kg) were placed in a metal still with ethanol (70 L) and extracted under reflux conditions using steam in the still jacket for 5 hrs. The extracts were drained. The marc was extracted three times more under similar conditions. Combined extract was concentrated under reduced pressure to obtain a gummy residue. The solvent free residue was suspended in distilled water (5 L) and mechanically stirred for 2 hrs, and then allowed to stand for sometime. Water soluble portions were decanted. This process of fractionation was repeated four times more with water insoluble residue to extract the maximum water soluble fraction. The combined water soluble fractions were centrifuged. The supernatants were freeze dried to obtain acteoside enriched fraction (AEF) a greenish power (1. 20 Kg). Standardization of the AEF was carried out on the bases of four chemical markers isolated from the bioactive fraction. The structure of the compounds were established as acteoside/verbascoside, CO-1 (5,7,4'-Trihydroxy flavone-3-O-glucuronide), CO-3 (5,6,7-Trimethoxy flavone) and CO-4 (5,6,7,4'-Tetramethoxy flavone) (Andary et al. 1982)

Standardization and Quality Control of Extract

Standardization and quality control of extracts were carried out in terms of CMC information. The extracts were evaluated for following parameters, *viz.*, heavy metals (Pb, Cd, Hg and As), aflatoxins contents (B1, B2, G1 and G2), acid insoluble ash, LOD at 105°C, total ash, water insoluble extractive and alcohol soluble extractive.

For determination of heavy metals approximately 0.25-1.0 gm of test samples were digested with 5-8 mL of nitric acid (concentrated) and then the vessel was placed in the microwave oven. After completing digestion, vessel was removed from microwave and cooled, and then the solution was transferred in 50 mL centrifuge tube and made up to 50mL with elemental water. 100 µL of 10 ppm internal standard was added to the sample in volume 50 mg (e.g. 20 ppb). The solution was injected directly into ICP-MS for determination of metals (AOAC 2002).

Analysis of aflatoxins were performed as follows: To 50 g of extract, 200 mL of methanol and 50 mL of 0.1N HCl was added. The resulting mixture was shaken at high speed for 5 min and filtered through Whatman Filter Paper No. 1 to 50mL of the filtrate, 50 mL of 10% NaCl solution and hexane were added and then the mixture was shaken for 30 seconds. Hexane layer was discarded and aqueous layer was again partitioned using 25z dichloromethane. The lower layers of dichloromethane were collected and anhydrous sodium sulphate was added to remove water, if any. The partitioning were performed twice with dichloromethane as mentioned above. The collected elutes were concentrated on steam bath and loaded on to silica gel column for separation of aflatoxins. Aflatoxins were eluted with 100 mL of mixture of dichloromethane and acetone (9:1). Elutes were evaporated on steam bath up to 6 mL and was divided into 3 parts for further analysis. After transferring, 2 mL of elute onto vial, and it was then evaporated to dryness using nitrogen.

Derivatization of sample were carried out by adding 200 mL of hexane and 50 mL TFA. To this solution, 2 mL of ACN-water mixture (1:9) was added and vortexed for 30 seconds, 25 mL of lower aqueous layer was injected into LC system. For this, Shimadzu LC-10ATVP with auto sampler SIL-10 ADVP or equivalent; Fluorescence Detector Shimadzu RF-10 AXL EX360 nm, Em 450 nm fitted with PC windows 2000 was used. LC column-25 cm X 4.6 mm id, 5 mm RP-18 or equivalent with 20 cm X 1 cm id were used. The LC elution solvent was Water : Acetonitrile:Ethanol (700:200:200), individual aflatoxins concentration was calculated from standard calibration curve (AOAC 1990).

Microbial testing of extract were performed to find out total aerobic bacterial count; total yeast, mold count; enterobatecriace count; *Escherichia coli, Salmonella* spp., *Staphylococcus aureus* and *Pseudomonas aeruginosa*.

ANALYSIS AND QUANTIFICATION OF STANDARD MARKERS

Reagents and Solvents

Acetic acid (AR grade); Water (Milli-Q); Methanol (HPLC grade)

Chromatographic conditions

- Instrument: Shimadzu (Nexera): LC-30 AD; Autosampler SIL-30 AC; Detector SPD-M20A; Communication bus module CBM-20 A; Column Oven CTO-20AC; Software –Labsolutions.

- Column : Merck Purospher STAR RP-18e, 5∝m, 250 x 4.0 mm

- Mobile Phase: (A) Buffer (1.5% Acetic Acid in Water), (B) Organic (Methanol)

- Column oven temperature: 30 °C

- Detector wavelength: 335 nm

- Injection volume : 10 µL

- Run time: 75 min

Table 1: Gradient system indicating percentage of buffer and organic

Mobile phase composition	Time (min)	Buffer %	Organic %
	0.01	92	8
	5.0	92	8
	25.0	50	50
	60.0	25	75
	65.0	25	75
	70.0	92	8
	75.0	92	8

Preparation of Buffer (1.5% AcOH in water)

Take 985 ml milli-Q water and to this add 15 ml AcOH, shake well and sonicate for fifteen minutes. Filter through 0.45∝m membrane filter and degas it.

Preparation of Standard Stock Solutions

Weigh accurately and transfer 1.1 mg of acteoside in a 5 ml volumetric flask and add about 5 ml of diluent (MeOH), vortex and sonicate it to dissolve. Make up the volume to 5 ml with MeOH. Weigh accurately and transfer 1.0 mg of CO-1 in a 5 ml volumetric flask and add about 5 ml of diluent (water), vortex and sonicate it to dissolve. Make up the volume to 5 ml with water. Weigh accurately and transfer 1.8 mg of CO-3 in a 5 ml volumetric flask and add about 5 ml of diluent (MeOH), vortex and sonicate it to dissolve. Make up the volume to 5 ml with MeOH. Weigh accurately and transfer 1.4 mg of CO-4 in a 5 ml volumetric flask and add 5 ml of diluent (MeOH), vortex and sonicate it to dissolve. Make up the volume to 5 ml with MeOH.

Preparation of Working Solution of Standards

All the four Standards – acteoside, CO-1, CO-3 and CO-4 were mixed together in equal ratio (200 µl each). The mixture was injected six times in the volumes 1 µl, 2 µl, 3 µl, 4 µl, 5 µl and 6 µl.

Preparation of Extract Solution

Weigh accurately 2.6 mg of extract in a culture tube and add 1 ml of methanol and sonicate for fifteen minutes. Filter it through 0.45 ∝m membrane filter. Procedure: Inject blank, mixed standards and all set of samples into the HPLC and record the chromatogram. The retention times of acteoside, CO-1, CO-3 and CO-4 are 29.964 mins, 32.267 mins, 55.605 mins and 58.155 mins respectively.

PHARMACOGNOSTIC INFORMATION

Pharmacognostic description of the botanical comprises the plant name and its microscopic (midrib and powder characteristics) and macroscopic characteristics as mentioned below:

Pharmacognostic Protocols for BRM

Botanical name: *Colebrookea oppositifolia* Sm.

Family: Lamiaceae

Source: CSIR-IIIM Farm Chatha, Cultivated

GPS Coordinates: Latitude-32.664715, Longitude-74.816203

Collection time: May-June

IUCN Status: Not available

Growth condition: Harvested during flowering/fruiting stage

Collection: Fresh leaves were collected and dried in shade at 35°C.

Moisture content: Below 10%

Storage condition: cGMP facility at CSIR-IIIM Jammu

Drug part: Leaves

Phytochemicals constituents

Broad range of medicinal properties such as antimicrobial, antifungal, and antioxidant properties have been awarded to the high content of flavonoids and polyphenols present in *Colebrookea oppositifolia*. Chemical investigations on the petroleum ether extract of the leaves of this species identified five major compounds namely, n-triacontane (1), acetyl alcohol (2), 32-hydroxydotriacontyl ferulate (3), β-sitosterol (4) and 5,6,7,42 -tetramethoxyflavone (5). Structures of the compounds were interpreted on the basis of spectral (IR, NMR, MS) data (Verma et al. 2012). Some of the other chemical constituents of *Colebrookea oppositifolia* includes: flavone glycosides *viz.* chrysin (6), negletein (7) and landenein (8); leaves contain 5,6,7-trimethoxyflavone (9), 5,6,7, 4'-tetramethoxyflavone, acteoside (10) and quercetin (11) in the bark (Yang et al. 1996, Aziz et al. 1974, Patwardhan et al. 1981, Madhavan et al. 2011). Root contains stearic (12), palmitic (13), oleic acids (14), triacontanol (15), flavone glycoside echioidin (16), 5,6,7-trimethoxyflavone and 4',5,6,7- tetramethoxyflavone (Ansari et al. 1982, Mukherjee et al. 2001, Reddy et al. 2009). Sugars and vitamins have been isolated from this plant species (Tyagi et al. 1995) (Fig. 4 and 5).

1. Triacontane

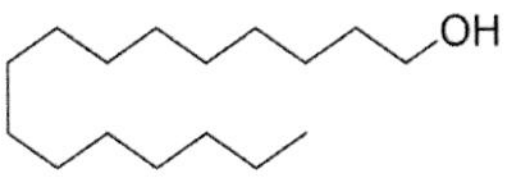

2. Cetyl alcohol

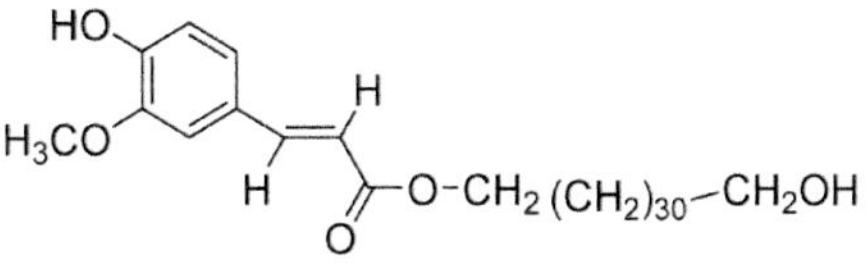

3. 32-Hydroxydotriacontyl ferulate I

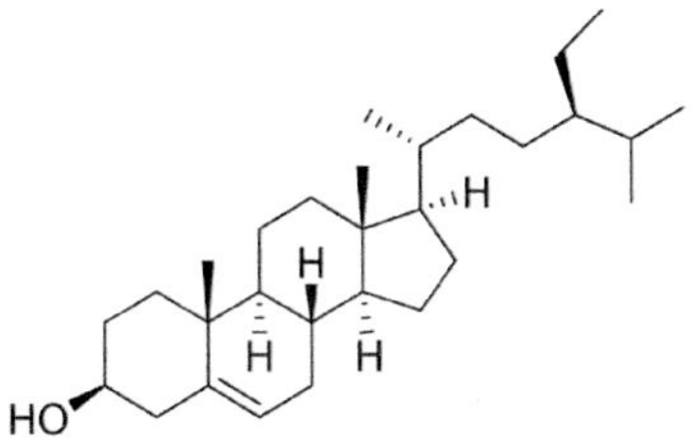

4. β-sitosterol

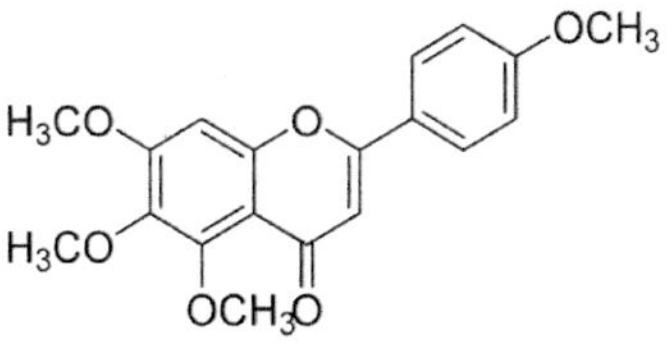

5. 4', 5,6,7-tetramethoxyflavone

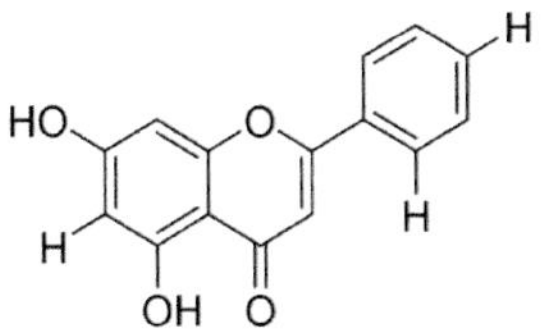

6. Chrysin

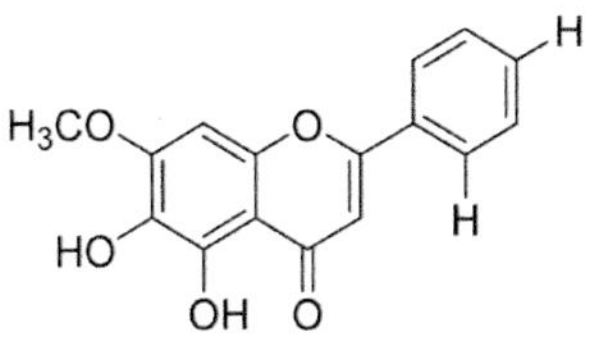

7. Negletein

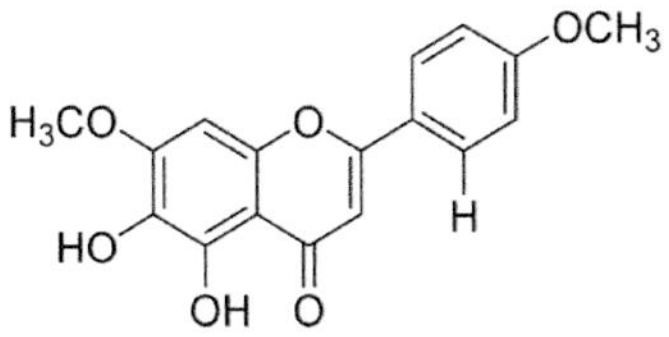

8. Ladanein

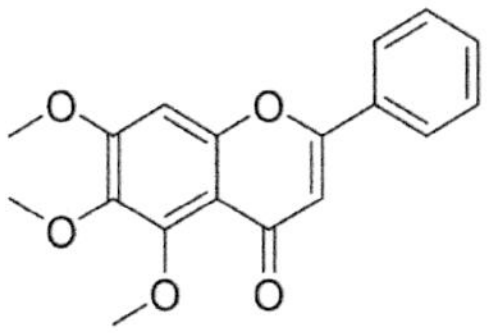

9. 5,6,7-trimethoxy flavone

Fig. 4: Phytochemical constituents of *Colebrookea oppositifolia.*

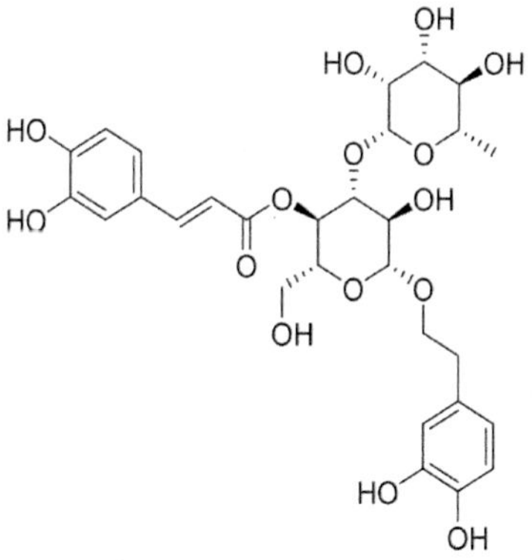

10. Acteoside

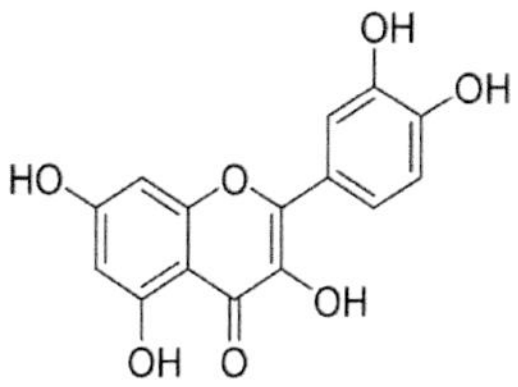

11. Quercetin

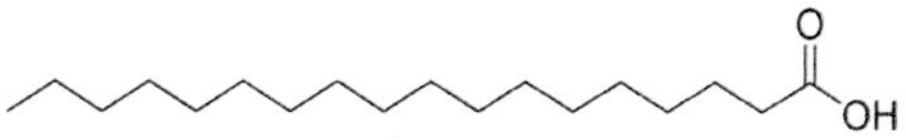

12. Stearic acid

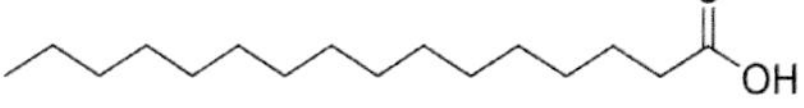

13. Palmitic acid

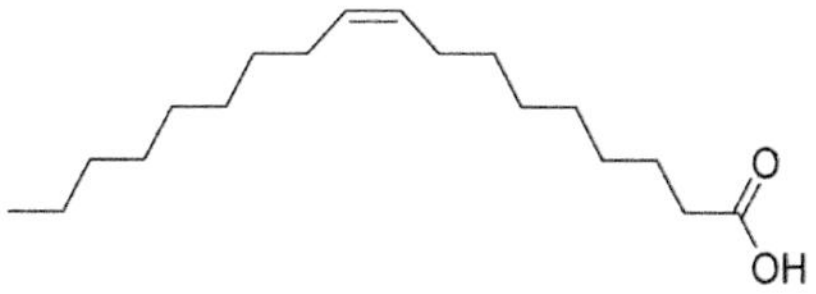

14. Oleic acid

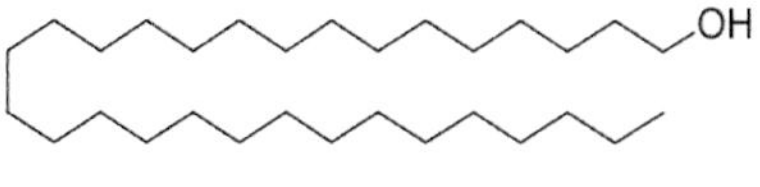

15. Tricontanol

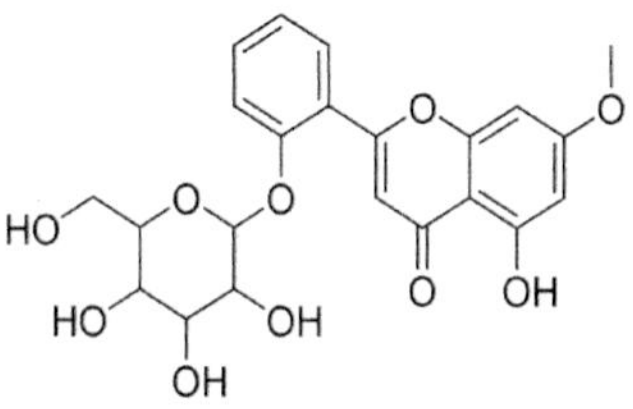

16. Echioidin

Fig. 5: Phytochemical constituents of *Colebrookea oppositifolia*.

Structure Elucidation of Bioactive Markers

1. *Acetoside*

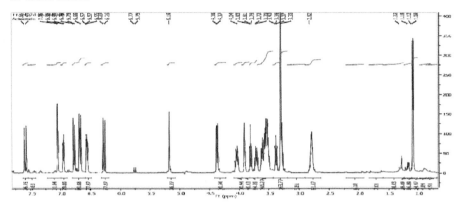

Fig. 6: ¹H-NMR Spectra of acteoside in *Colebrookea oppositifolia*

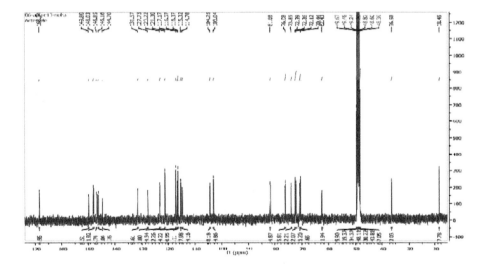

Fig. 7. ¹³C-NMR of acteoside in *Colebrookea oppositifolia*

NPC Division IIIM Jammu

Reported by User: System

Project Name   NP

SAMPLE INFORMATION

Sample Name: Acteoside	Acquired By: System
	Date Acquired: 12/18/2015 7.31.09 PM
Sample Type: Standard	Acq. Method Set: Belerica 3
Vial. 1	Date Processed. 12/21/2015 10.51.42 AM
Injection #: 1	Processing Method: RJM0662
Injection Volume: 15.00 ul	Channel Name: W2996 335.0nm-1.2
Run Time: 75.0 Minutes	Proc. Chnl. Descr.: W2996 PDA 335.0 nm at 1.2
Sample Set Name: Colebrookea	Flow rate: 0.5 ml/min
column_name RP-18,5um	Sample conc: 2.5 mg/10 mL MeOH
	Mobile phase: MeOH 1.5% AcOH in water(gradient)

Auto-Scaled Chromatogram

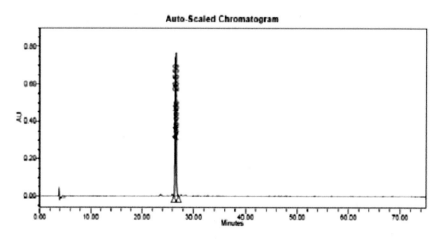

	Peak Name	RT	Area (µV*sec)	% Area	Height (µV)	Amount	Units
1	Acteoside	26.559	9279014	100.00	743932	3750.000	ng

Fig. 8: HPLC profile of actoside in *Colebrookea oppositifolia*

2. 5,6,7-Trimethoxy flavone (CO-3)

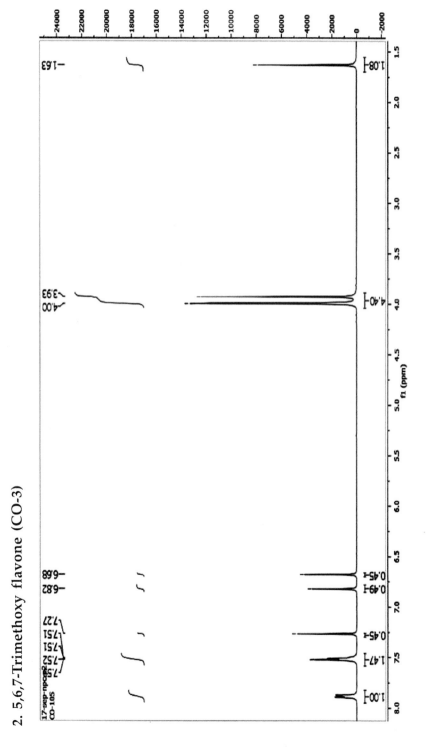

Fig. 9: ¹H-NMR spectra of CO-3 in *Colebrookea oppositifolia*

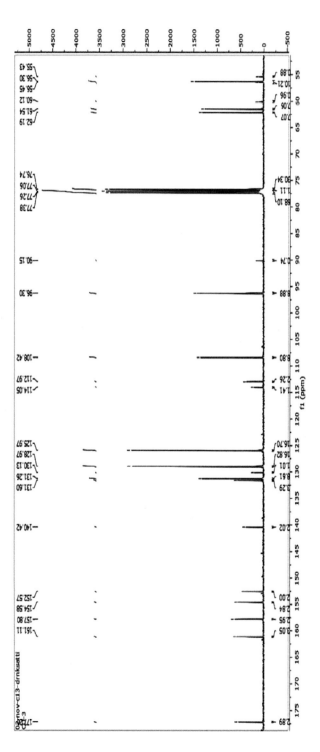

Fig. 10: ^{13}C-NMR spectra of 5,6,7-Trimethoxy flavone *Colebrookea oppositifolia*

NPC Division IIIM Jammu

Reported by User: System   Project Name: NP

SAMPLE   INFORMATION

Sample Name:	CO-3	Acquired By:	System
		Date Acquired:	6/29/2015 11:11:20 PM
Sample Type:	Unknown	Acq. Method Set:	Belerica 3
Vial:	7	Date Processed:	7/1/2015 11:23:55 AM
		Processing Method:	CO Frs
Injection #:	1	Channel Name:	W2996 335.0nm-1.2
Injection Volume:	5.00 ul	Proc. Chnl. Descr.:	W2996 PDA 335.0 nm at 1.2
Run Time:	75.0 Minutes	Flow rate: 0.5 ml/min	
Sample Set Name:	Colebrookea	Sample conc:	
column_name	RP-18,5um	Mobile phase: MeOH:1.5%AcOH in water(gradient)	

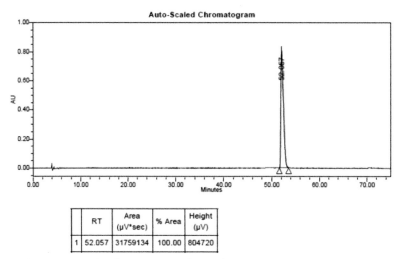

Auto-Scaled Chromatogram

	RT	Area (µV*sec)	% Area	Height (µV)
1	52.057	31759134	100.00	804720

Fig. 11: HPLC of 5,6,7-Trimethoxy flavone *Colebrookea oppositifolia*

3. 5,6,7,4'-Tetramethoxy flavone (CO-4)

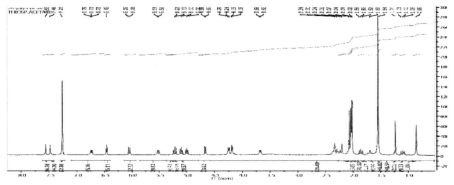

Figu. 12: ¹H-NMR Spectra of 5,6,7,4'-Tetramethoxy flavone in *Colebrookea oppositifolia*

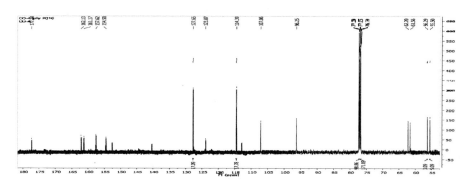

Fig. 13: ^{13}C-NMR Spectra of 5,6,7,4'-Tetramethoxy flavone in *Colebrookea oppositifolia*

NPC Division IIIM Jammu

Reported by User: System

Project Name: NP

SAMPLE INFORMATION

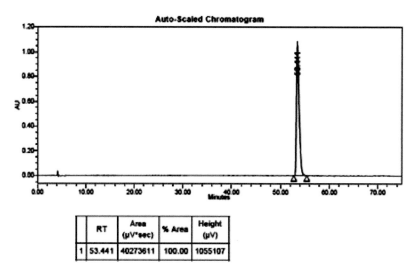

	RT	Area (µV*sec)	% Area	Height (µV)
1	53.441	40273611	100.00	1055107

Fig. 14: HPLC of 5,6,7,4'-Tetramethoxy flavone in *Colebrookea oppositifolia*

4. 5,7,4'-Trihydroxy flavone -3-O-Glucuronide (CO-1)

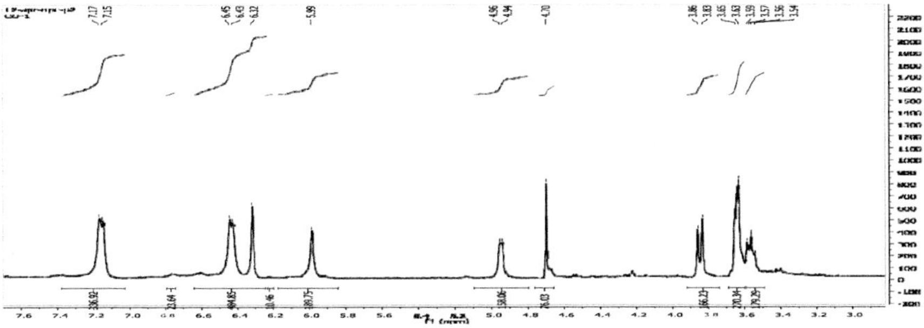

Fig. 15: ^{1}H-NMR spectra of 5,7,4'-Trihydroxy flavone -3-O-Glucuronide (CO-1) in *Colebrookea oppositifolia*

Hepatoprotective Activity Against Alcohol Induced Hepatitis in rats

Alcohol induced hepatitis occurs in waster rats when they are fed increasing doses of ethanol (5 ml to 12 ml) for 24 days. Alcohol administration results in significant elevation of ALT, AST, and the damaged cell liberates these enzymes into systematic circulation (Leelavinothan and Arumugam 2008). Triglycerides, one of the most useful clues to the alcohol induced fatty liver was also raised significantly (Pornpen et al. 2007). Albumin is known to decrease in alcoholic hepatic damage was also observed in this study (Lieber 1997). Treatment with acteoside enriched extracts (AEE) from day 8 to day 24, 30 minutes after alcohol administration significantly and dose dependently reversed the increased levels of ALT, AST, TG and decreased levels of albumin depicting its protective effect.

Ethanol consumption caused elevation of endotoxin. In the liver endotoxins interacts primarily with the kupffer cells and lead to secretion of TNF-α which then interacts with receptors on both kupffer cells and hepatocytes leading to the production of other inflammatory cytokines such as IL-1, IL-6 and IL -8. Enhanced TNF-α production by alcohol can also increase the production of reactive oxygen species by macrophages (Goossens et al. 1995). This can stimulate lipid peroxidation, and reduce GSH contents which are thought to have an important pathogenic role in alcohol-induced liver damage (Lieber 1997, Subir et al. 2005). Treatment with AEE significantly and dose dependently reversed the increased levels of Lipid peroxidation and restore depleted GSH depicting its protective and antioxidant effect. Estimation of cytokines like TGF-βl (Transforming growth factor β1) TNF-α (Tumor necrosis factor-alpha) and IL-6 (Interleukin 6) was carried out in the serum to test the protective efficacy of AEE against alcoholic hepatitis in rats. Treatment of animals with AEE normalized the increased levels of these parameters depicting its significant hepatoprotective potential is provided in table-2. Values represent the mean $\perp$ SE and within parentheses hepatoprotective activity percent of five animals in each group; Compared to vehicle; [b]:Compared to vehicle + alcohol.; [c]:compared to silymarin; ns: not significant, *: p<0.05; **:p<0.01; ***:p<0.001Units: Each unit is m mole pyruvate/min /L; LP: lipid peroxidation (n mole MDA/g liver; GSH: glutathione (m mole GSH/g liver).

Hepato Protective Activity Against CCl_4 Model in Rats

Hepato toxicity induced by CCl_4 results in marked elevation of ALT and AST. The damaged cells liberates these enzymes into systematic circulation. Hyperbilerubineaemia, one of the most useful clues to the severity of necrosis and hepatic damage was observed by CCl_4 administration. Treatment with AEE significantly and dose dependently reversed the increased levels of ALT, AST and bilirubin depicting its protective effect. Reactive metabolite formed by number of drugs and chemicals e.g. CCl_4 are detoxified by conjugation with glutathione. Administration of toxicants increases lipid peroxidation (Seki et al., 2000). Lipid peroxidation leads to membrane damage and subsequent alteration in calcium homeostasis, which are essential components of cell death (Jacquier et al. 1994). An accelerated lipid peroxidation and drastic fall in hepatic GSH content by toxicants have been observed with CCl_4 (Brien et al. 2000; Chandan et al. 2008). Treatment with AEE significantly and dose dependently reversed the increased levels of Lipid peroxidation and restore depleted GSH depicting its protective effect against free radicals.

Table 2: Hepatoprotective potential of *Colebrookea oppositifolia* AEE against alcohol induced hepatitis

Treatment	Dose mg/kg	Serum parameters				Hepatic parameters		Cytokines			Overall
		ALT (Units)	AST (Units)	Albumin (g/dl)	TG (g/dl)	LP	GSH	IL-6 (pg/ml)	TGF β1 (ng/ml)	TNF-α (pg/ml)	% protection
Vehicle	-	50.73 ±2.22	64.50 ±3.62	4.30 ±0.17	63.86 ±5.36	96.08 ±10.37	4.88 ±0.35	58.77 ±3.71	75.48 ±8.18	4.99 ±0.67	-
Vehicle + Alcohol	-	90.62 ±7.83a**	103.94 ±7.09a***	2.94 ±0.09a***	250.80 ±29.51a***	187.07 ±7.07a***	2.21 ±0.15a***	93.74 ±7.90a**	118.33 ±8.12a**	10.53± 0.97 a***	-
AEE+Alcohol	50	66.40 ±3.46b* (60.71)	81.00 ±4.29b* (58.16)	3.74 ±0.16b* (58.82)	157.82 ±20.37b* (49.73)	132.07 ±7.79b*** (60.44)	3.85± 0.13b*** (61.42)	71.42± 4.46b* (63.82)	83.62± 5.96 b** (81.00)	6.91± 0.73 b* (65.34)	62.16± 2.77cns
AEE+Alcohol	25	69.96± 2.02b* (51.79)	84.20± 3.82b* (50.05)	3.51± 0.13b* (41.91)	171.06± 12.90b* (42.65)	136.52± 8.09b*** (55.55)	3.42± 0.10b*** (45.31)	77.95± 6.22bns (45.15)	87.26± 7.99 b* (72.50)	7.83± 0.51 b* (48.73)	50.40± 3.13cns
AEE+Alcohol	12.50	73.17± 3.59b* (43.74)	88.36± 3.27bNS (37.91)	3.39± 0.16b* (33.08)	195.18± 8.72bns (29.75)	150.21± 10.44b* (40.50)	2.89± 0.24b* (25.46)	86.30± 6.68bns (21.27)	94.68± 5.56b* (55.19)	8.95± 0.70 bns (28.51)	35.04± 3.48c***
Silymarin + Alcohol	50	65.33± 3.68b** (63.39)	79.68± 4.90b* (61.51)	3.54± 0.16b* (44.11)	149.14± 18.38b* (54.84)	134.22± 6.74b*** (58.08)	3.65± 0.11b*** (53.93)	69.93± 5.48b* (68.08)	81.60± 4.49b** (85.71)	7.37± 0.62 b* (57.03)	60.74± 3.84
ED 50		21.3 27.31	27.31	33.43	47.62	21.5	39.42	30.98	8.74	27.43	28.63± 3.74

Values represent the mean ± SE and within parentheses hepatoprotective activity percent of five animals in each group;
Compared to vehicle; b:Compared to vehicle + alcohol.; c :compared to silymarin; ns: not significant, *: p<0.05; **:p<0.01; ***:p<0.001
Units: Each unit is mole pyruvate/min /L; LP: lipid peroxidation (n mole MDA/g liver; GSH: glutathione (mole GSH/g liver).

Table 3: Hepatoprotective potential of *Colebrookea oppositifolia* AEE against CCl$_4$ (1ml/kg.p.o.) induced hepatotoxicity in rats.

Treatment	Dose mg/kg	Serum parameters			Hepatic parameters		Overall
		ALT (Units)	AST (Units)	Bilirubin mg%	LP	GSH	% protection
Vehicle	-	68.37±6.54	87.69±5.58	0.25±0.03	54.48±3.77	5.79±0.13	-
Vehicle + CCl4	-	1430.35± 176.30a***	962.90± 101.99a***	0.92±0.06a***	112.17±5.26a***	3.36±0.10a***	-
AEE + CCl4	50	563.53±60.82b*** (63.64)	449.58±44.21b*** (58.65)	0.45±0.02b*** (70.14)	77.23±3.95b*** (60.30)	4.93±0.13b*** (64.60)	63.46± 1.98c**
AEE + CCl4	25	691.46±69.62b** (54.25)	555.40±44.87b** (46.56)	0.54±0.03b*** (56.71)	83.64±5.95b** (49.45)	4.58±0.09b*** (50.20)	51.43± 1.80cns
AEE + CCl4	12.50	900.99±63.29b* (38.86)	650.11±53.45b* (35.73)	0.66±0.03b** (38.80)	91.97±4.42b* (35.01)	4.22±0.08b*** (35.39)	36.75± 0.85c***
AEE + CCl4	6.25	1056.49±61.61bns (27.44)	753.01±37.34bns (23.98)	0.75±0.04b* (25.37)	100.31±6.19bns (20.55)	3.87±0.17b* (20.98)	23.66± 1.30c***
AEE + CCl4	3.12	1224.12±66.97bns (15.14)	837.78±60.71bns (14.29)	0.85±0.04bns (10.44)	105.11±8.19bns (12.23)	3.67±0.17bns (12.75)	12.97± 0.82c***
Silymarin + CCl4	50	693.66±40.33b** (54.08)	485.83±46.77b** (54.50)	0.53±0.03b*** (58.20)	84.28±4.85b** (48.34)	4.51±0.14b*** (47.32)	52.48± 2.03
ED50		22.04	30.13	19.50	27.91	24.89	24.89± 1.92

Values represent the mean ± SE of 6 animals in each group and within parentheses hepatoprotective activity percent. a :Compared to vehicle; b:Compared to vehicle + CCl$_4$; c :compared to silymarin; *: p<0.05; **:p<0.01; ***:p<0.001, ns: not significantUnits: Each unit is m mole pyruvate/min /L; LP: lipid peroxidation (n mole MDA/g liver) GSH: glutathione (m mole GSH/g liver).

CONCLUSION

In summary, *Colebrookea oppositifolia* is one of the most important medicinal plant of himalayan origin which has high demand in pharmcuetical sector for development of medicines and drug. Standardization, characterization and quality control of *Colebrookea oppositifolia* extracts were carried out at Quality Control and Quality Assurance Department, IIIM Jammu, based on the available markers such as Acetoside, CO-1, CO-3 and CO-4. Acetoside is the most bioactive marker for this species. The present new findings is advantageous over the earlier reported researches wherein only the hepatoprotective property against CCl_4 induced hepatotoxicity in rat was reported and involved cumbersome process and use of toxic solvents for fractionation, followed by column chromatography for the isolation of principle bioactive constituent namely acteoside. The present method of preparation is viable, where in no use of toxic solvents and column chromatography is required. Histopathological studies revealed that among all the treated groups, animals treated with AEE in doses of 50 and 25 mg/kg exhibited significant liver protection against ethanol induced hepatotoxicity as evident by the normal hepatic tissue architecture, absence of fatty infiltration, lack of necrotic cells, presence of normal hepatic cords with normal lobular pattern, and well preserved cytoplasm, decreased lymphocyte infiltration which was almost comparable to the silymarin treated groups. Therefore, the extract and the product to be prepared from *Colebrookea oppositifolia* will be considered to be a safe botanical for the human use. The current finding will provide base data for further research to be carried out in the field of drug discovery and medicine formulation.

Abbreviations

AEE   : Acetoside enriched extract

ALD   : Alcoholic Liver Disease

ALT   : Alanine Transaminase

AST   : Aspartate Transaminase

BRM   : Botanical Raw Material

CSIR   : Council of Scientific and Industrial Research

CMC   : Chemistry, manufactruing and control

GLP   : Good Laboratory practices

GSH   : Glutathione

ICP-MS : Inductively coupled plasma mass spectrometry

IUCN   : International Union for Conservation of Nature

HPLC   : High Performance Liquid Chromatography

LOD   : Loss on Drying

LP   : Lipid Peroxidase

NIST   : National Institute of Standard of Technology

OECD   : Organization for Economic Co-operation and Development

RRLH   : Regional Research Laboratory Herbarium

TG   : Triglycerides

TGF-â1 : Transforming growth factor beta 1

TNF-á   : Tumor necrosis factor-Alpha

WHO   : World Health Organization

ACKNOWLEDGEMENTS

Authors are thankful to Director IIIM, Dr. Ram A Vishwakarma for necessary facilities and encouragement. QC/QA Department of CSIR-IIIM Jammu is duly acknowledged for the analysis of extract of *Colebrookea oppositifolia,* and also would like to thank different scholars and staff members involved in findings associated with *Colebrookea oppositifolia* as medicinal plants. It's a part of BSC-0205, GAP-1112 and GAP-113 project undertaken at CSIR-IIIM Jammu and Bears Institutional Publication number IIIM/2246/2018.

REFERENCES CITED

Andary C, Wylde R, Laffite C, Privat G, Winternitz F. 1982. Structures of verbascoside and orobanchoside, caffeic acid sugar esters from Orobanche rapum-genistae. *Phytochemistry* 21: 1123–1127.

Ansari S, Dobhal MP, Tyagi RP, Joshi BC, Barar FSK. 1982. Chemical investigation and pharmacological screening of the roots of *Colebrookia oppositifolia* Smith. *Pharmazie* 37: 70.

Aziz SA, Siddiqui SA, Zaman A. 1974. Flavones from *Colebrookea oppositifolia*. *Indian Journal of Chemistry* 12: 1327-1328.

Benoit SC, Kemp CJ, Elias CF, Abplanalp W, Herman JP, Migrenne S, Lefevre AL, Cruciani-Guglielmacci C. 2009. Palmitic acid mediates hypothalamic insulin resistance by altering PKC-è subcellular localization in rodents. *Journal of Clinical Investigation* 119(9): 2577–2587.

Burr GO, Burr MM, Miller E. 1930. On the nature and role of fatty acid essential in nutrition. *Journal Biology and Chemistry* 86(5): 1-9.

Goossens V, Grooten J, De Vos K, Fiers W. 1995. Direct evidence for tumour necrosis factor-induced mitochondrial reactive oxygen intermediates and their involvement in cytotoxicity. *Proceedings of National Academy of Sciences of the USA* 92: 8115–8119

Holley J, Cherla K. 1998. Medicinal plants sector in India- A review of medicinal and aromatic plants program in Asia (MAPPA). SARO/ IDRC, New Delhi, India.

Houghton PJ, Hikino H. 1989. Anti-Hepatotoxic Activity of Extracts and Constituents of Buddleja Species. *Planta Medica* 55: 123–126.

Ishtiaq S, Meo MB, Afridi MSK, Akbar S, Rashool S. 2016. Pharmacognostic studies of aerial parts of *Colebrookea oppositifolia* Sm. *Annals of Phytomedicine* 5(2): 161-167.

Jacquier S, Fuller K, Richard MJ, Polla BS (1994). Protective effects of hsp70 in inflammation. *Experientia* 50: 1031-1038.

Jiang H, Livingston M, Room R, Dietze P, Norström T, Kerr WC. 2014. Alcohol Consumption and Liver Disease in Australia: A Time Series Analysis of the Period 1935–2006. *Alcohol* 49: 363–368.

Kingsbury KJ, Paul S, Morgan DM. 1961. The Fatty acid composition of humen deposition fat. *Botanical Journal* 78: 541-550.

Krajian AA (1963). Tissue cutting and staining. In: Gradwohl's Clinical Laboratory Methods and Diagnosis. Sam Frankel and Stanley Reitman (eds.) 6th edn. The CV Mosbay Co. Saint Louis (USA). Pp 1639.

Leelavinothan P, Arumugam S. 2008. Effect of grapes (*Vitis vinifera* L.) leaf extract on alcohol induced oxidative stress in rats. *Food and Chemical Toxicology* 46: 1627-1634.

Lieber CS. 1997. Role of oxidative stress and antioxidant therapy in alcoholic and nonalcoholic liver diseases. *Advances in Pharmacology* 38: 601–628.

Madhavan V, Yadav DK, Murali A, Yoganarasimhan SN. 2009. Wound healing activity of aqueous and alcohol extracts of leaves of *Colebrookea oppositifolia* Smith. *Indian Drugs* 46(3): 209-213.

Mukherjee PK, Mukherjee K, Hermans-Lokkerbol ACJ, Verpoorte R, Suresh B. 2001. Flavonoid content of *Eupatorium glandulosum* and *Coolebroke oppositifolia*. *Journal of Natural Remedies* 1: 21–24.

Papamandjaris AA, MacDougall DE, Jones PJ. 1998. Medium chain fatty acid metabolism and energy expenditure: obesity treatment implications. *Life Sciences* 62 (14): 1203–15.

Patwardhan SA, Gupta AS. 1981. Two New flavones from *Colebrookea oppositifolia*. *Indian Journal of Chemistry* 20 B: 627.

Pornpen P, Chanon N, Somlak P, Chaiyo C. 2007. Hepatoprotective activity of *Phyllantus amarus* Schum. ex. Thonn. extract in ethanol treated rats: in vitro and in vivo studies. *Journal of Ethnopharmacology* 114: 169-173.

Rasenack J, Koch HK, Nowack J, Lesch R, Decker K. 1980. Hepatotoxicity of d-Galactosamine in the isolated perfused rat liver. *Experimentaal and Molecular Pathology* 32: 264–275.

Reddy RVN, Reddy BAK, Gunasekaran D. 2009. A new acylated flavone glycoside from *Colebrookea oppositifolia*. *Journal of Asian Natural Products and Resources* 11: 183– 186.

Schermer CR. 2006. Alcohol and injury prevention. *Journal of Trauma Injury Infection and Critical Care* 60: 447–451.

Seki M, Kasma K, Imai K. 2000. Effect of food restriction on hepatotoxicity of carbon tetrachloride in rats. *The Journal of Toxicological Sciences* 25: 33-40.

Subir KD, Vasudevan DM. 2005. Effect of ethanol on liver antioxidant defence system: A dose dependent study. *Indian Journal of Clinical Biochemistry* 20: 80-84.

Takenaka T. 2011. Classical vs. reverse pharmacology in drug discovery. *BJU International* 88. (Suppl) 2:3-10.

Torri MC. 2012. Mainstreaming local health through herbal gardens in India: A tool to enhance women active agency and primary healthcare. *Environment, Development and Sustainability* 14: 389-406.

Tyagi S, Saraf S, Ojha AC, Rawat GS. 1995. Chemical investigation of some medicinal plants of Shiwalik. *Asian Journal of Chemistry* 7(1): 165-170.

Venkateshappa SM, Sreenath KP. 2013. Potential medicinal plants of Lamiaceae. *AIJRFANS* 3(1):82-87.

Warren JB 2011. Drug discovery: lessons from evolution. *British Journal of Clinical Pharmacology* 74(4): 497-503.

WHO. 2004. WHO Global Status Report on Alcohol 2004: Barbados 3: 1–7.

Xiong Q, Hase K, Tezuka Y, Tani T, Namba T, Kadota S. 1998. Hepatoprotective activity of phenylethanoids from *Cistanche deserticola*. *Planta Medica* 64: 120–125.

Yang F, Li XC, Wang HQ, Yang CR. 1996. Flavonoid glycosides from *Colebrookea oppositifolia*. *Phytochemistry* 42: 867–869.

4

Dysoxylum binectariferum Hook.f., An Indian Medicinal Plant as a Source for Anti-Cancer Agents

Shreyans Kumar Jain

Department of Pharmaceutical Engineering and Technology
Indian Institute of Technology (BHU), Varanasi -221005
Uttar Pradesh, INDIA
Email: sjain.phe@iitbhu.ac.in

ABSTRACT

Dysoxylum binectariferum Hook.f. (Meliaceae) is phylogenetically related to the Ayurvedic plant *D. malabaricum*. *D. binectariferum* has been identified as an enriched source of rohitukine, an alkaloid. Two synthetic flavones, flavopiridol (Alvocidib; L868275; HMR-1275; NSC 649890 of Sanofi-Aventis + NCI) and P-276-00 (Piramal) are advanced in clinical trials for the treatment of cáncer. These two molecules are considered as potent Cdk inhibitors. Cdk has been discovered as a key regulator of cell cycle. Therefore, Cdk inhibitors or modulators are of great interest to explore as novel therapeutic agents. These two molecules can be considered as rohitukine inspired drugs since their stuructural novelty originated from a compaign to develop a route of total synthesis of rohitukine. Recently a semisynthetic derivative of rohitukine; IIIM-290 has been emerged as potential preclinical cyclin-dependent kinase (Cdk) inhibitor. Moreover, phytochemical investigation of *D. binectariferum* yielded dysoline a new region-isomer of rohitukine. Dysoline seems to be a potent inflammatory and cytotoxic compound. Similarly, camptothecine and 9-methoxy-camptothecine were isolated from *Dysoxylum binectariferum* for the first time by bioassay-guided fractionation. Present chapter is reviewing some original work which were carried out at CSIR-Indian Institute of Integrative Medicine, Jammu, India. The present studies focused on the phytochemistry of *Dysoxylum binectariferum*; dysoline, camptothecin and 9-methoxycamptothecin and medcinal chemistry and preclinical evaluation of IIIM-290 a semi-sythestic derivative of rohitukine.

Keywords: Anti-cancer plant, *Dysoxylum binectariferum*, Rohitukine, Cdk inhibitors, Flavopiridol, IIIM-290

INTRODUCTION

There are around eighty different species in genus *Dysoxylum* Blume, which belongs to the family Meliaceae. These species are growing widely across the regions of South and South-East Asia, the Western Pacific Ocean, Australia, and distributed on the tropics between the Pacific and Indian oceans. *Dysoxylum binecteriferum* Hook.f. occurs in Sri Lanka and India. In India, it is distributed in South India and Western, Ghats (Elamalai, Anaimalai in South Sahyadri). (Sasidharan 2004) Many species are tree in habit and have commercial values in wood and timber industries for building construction, boxes, turnery and plyboard making. In Southern India, *D. binecteriferum* is considered as sacred tree and is used instead of sandal tree. Decoction of wood is useful in arthritis, anorexia, cardiac debility, expelling intestinal worms, inflammation, leprosy and rheumatism. (Kumar 2009). However there are more than 120 different kind of metabolites has been isolated and investigated for different biological activities from the genus *Dysoxylum*, but very less phytochemical investigation has been carried out on *D. binectariferum* and only few compounds have been reported from this species. Since *D. binectariferum* belongs to family Meliaceae which is known for biological sources of alkaloids. Rohitukine (**1**) is a chromone alkaloid, which was first isolated from leaves and stem of *Amoora rohituka* and then from the stem barks of *D. binectariferum*, (Mohanakumara et al. 2009) both belonging to the family Meliaceae (Yang et al. 2004).

Rohitukine (**1**) was isolated at pharmaceutical company Hoechst India Ltd. in 1990s as an anti-inflammatory and immunomodulatory agent. Later company started synthesis and medicinal chemistry around this nature-derived flavone alkaloid (Naik et al. 1988). After total synthesis of rohitukine, a large number of molecues were synthesized for strutural-activity relationship. A chloro-derivative Flavopiridol (2, Alvocidib; L868275; HMR-1275; NSC 649890) was found to be a potent anti-cancer after preliminary screening. This discovery influenced the research around the natural product rohitukine. Later Indian pharmaceutical company Piramal Life Science Ltd, also identified a a synthetic flavone alkaloid P-276-00 as potent anti-cancer. Currently both flavopiridol and P-276-00 are in clinical trial by Sanofi-Aventis + NCI and Piramal Life Science India respectively (Jain et al. 2012b).

The biological targets of these scaffolds is cyclin-dependent kinases (Cdk). (Galons et al. 2010, Huwe et al. 2003, Kelland 2000, Senderowicz 2001). The cyclin-dependent kinase is a key regulator of cell cycle. There are thirteen members of human Cdk family; all belonging to serine/threonine protein kinases (Huwe et al. 2003) Cdk 1, 2, 3, 4, and 6 directly interfere in the cell cycle (Morgan 1997) Cdk7 play role as a activator and have indirect role. Furthermore, Cdk 7, 8, and 9 plays crucial role as regulator of transcription (Huwe et al. 2003). Subsequent discoveries have revealed a pathogenic link between cancer and altered Cdk activity. A number of cancers are associated with hyper-activation of Cdks as a result of mutation of the Cdk genes or Cdkinhibitor genes (Michalides et al. 1995). Therefore, Cdk inhibitors or modulators are of great interest to explore as novel therapeutic agents for the treatment of cancer, and has led to the discovery of flavopiridol as a first Cdk inhibitor get entered in clinical studies (Bible and Kaufmann 1997; Blagosklonny 2004, Bradbury 2010, Brusselbach et al. 1998, Burdette-Radoux et al. 2004, Byrd et al. 1998, Carlson et al. 1996, Christian et al. 2007, Dasmahapatra et al. 2006, Drees et al. 1997, El-Rayes et al. 2006, Filgueira de Azevedo Jr et al. 2002; Galons et al. 2010, Getman and Bible 2004, Gojo et al. 2002, Grendys Jr et al. 2005, Jain et al. 2012a, Kelland 2000, Losiewicz et al. 1994, Melillo et al. 1999, Motwani et al. 2003; Rapella et al. 2002, Sedlacek et al. 1996, Senderowicz 1999, Thomas et al. 2002, Vesely et al. 1994, Wang and Ren 2010, Wittmann et al. 2003, Wu et al. 2002) and several other candidates are under advanced clinical studies such as P-276-00 (Lapenna and Giordano 2009, Raje et al. 2009).

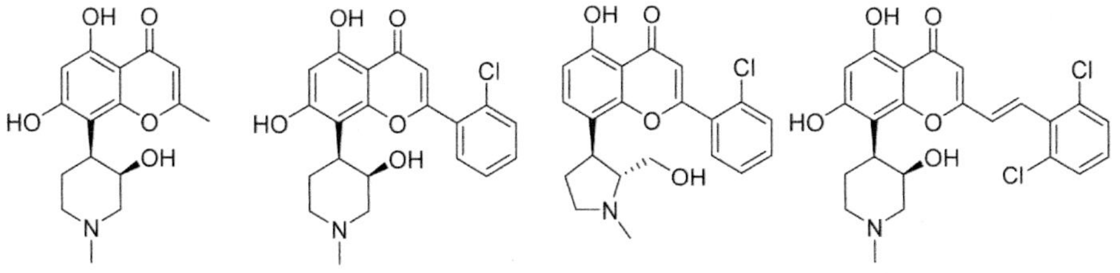

Rohitukine (**1**) Flavopiridol (**2**) P-276-00 (**3**) IIIM-290 (**4**)

Fig. 1: Chemical structure of rohitukine and related Cdk inhibitors which are in advanced clinical and preclinical stage.

Later, the *D. binecteriferum* bark was identified as rich source of rohitukine (Mohanakumara et al. 2009). Later leaves has been utilized for the production of rohitukine in sufficient amount to accomplish semi-synthestic studies, without a chromatography purification. (Kumar et al. 2016). Extensive amount of medicinal chemistry efforts around rohitukine (**1**) have been reported (Jain et al. 2012b). On the basis of reported work on rohitukine inspired drug flavopiridol, we designed new semisynthetic derivatives of rohitukine and for that *D. binectariferum* was also used for phytochemical investigations.

PHYTOCHEMISTRY

As discussed that *D. binectariferum* bark and leaves has been identified as rich source of rohitukine. Rohitukine can be easily extracted in alcoholic solvents and can be enriched through chromatographic free procedure (Kumar et al. 2016). Another alkaloid; rohitukine –N-oxide (**5**) was also isolated and known compounds were characterized by comparson of its spectroscopic data with literature values (Domg-Hui Yang et al. 2004). During isolation of rohitukine, some of the fractions showed the presence of another closely associated Dragendorff-positive spotin the TLC ($CHCl_3$: MeOH, 85:15, triple run).

To characterize the newly identified alkaloid, LC-HRMS analysis of the fraction was carried out, which showed presence of two peaks appearing at t_R 17.52 min and 19.59 min, both showing similar masses (*m/z* 306, M+H$^+$) (Figure 3). After performing a LC-HRMS analysis of pure rohitukine standard, it was found that the peak appearing at t_R 17.52 min belongs to the rohitukine (**1**). Interestingly, the LC-HRMS analysis indicated that the MS (*m/z* 306.1311; M+H$^+$) and molecular formula ($C_{16}H_{19}NO_5$)$^+$ of both peaks is exactly similar. Thus, it was speculated that the peak appearing at t_R 19.59 min could be a new isomer of rohitukine. The crude fraction, upon further purification using Sephadex LH20 column chromatography led to the isolation of the newly identified compound **5** (5 mg, 0.0003%), which was then characterized by extensive 2D-NMR data, and named as dysoline. Flow chart indicated isolation procedure for rohitukine, rohitukine-N-oxide and dysoline (Figure 2).

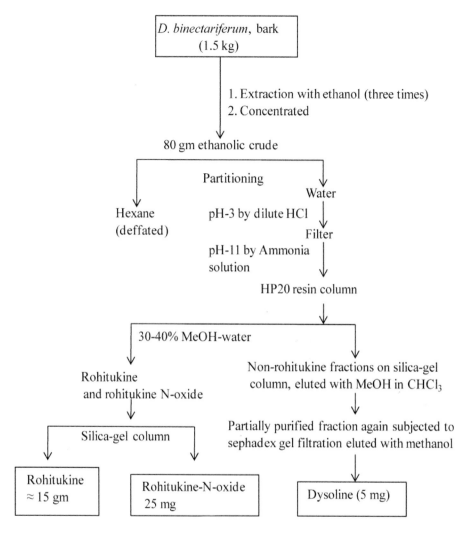

Fig. 2: Isolation scheme of rohitukine, rohitukine-N-oxide and dysoline.

Dysoline showed identical ^{1}H, ^{13}C, DEPT NMR and same molecular formula $C_{16}H_{19}NO_5$ (*m/z* 306, M+H$^+$ in MS spectrum) to that of rohitukine (**1**), however the melting point, TLC R_f and HPLC t_R were different. As shown in Figure 2, LCMS study indicated two different peak with same mass. Major peak was identified as rohitukine while minor as dysoline. Dysoline (**5**) was screened for cytotoxicity in a panel of cancer cell lines Colo205 (Colon), HCT116 (Colon), HT1080 (Fibrosarcoma), NCIH322 (Lung), A549 (Lung), Molt-4 (Leukemia) and HL60 (Leukemia). It displayed potent cytotoxicity in HT1080 cells

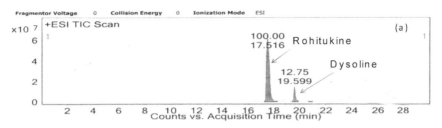

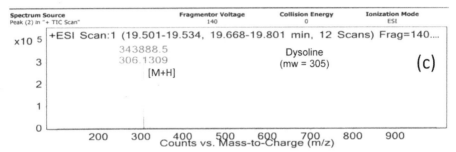

Fig. 3: LCMS of fraction showing presence of rohitukine and dysoline, (b) MS spectrum of peak at t_R 17.52 min, (c) MS spectrum of peak at t_R 19.59 min.

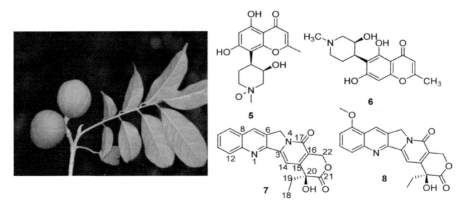

Fig. 4: Image of fruit bearing part of plant (Image copied from Rufford foundation) and structure of Rohitukine-N-Oxide (5), dysoline (6), camptothecin (7) and 9-methoxycamptothecin (8).

with IC_{50} value of 0.21 µM, whereas, it was less active against other cell lines (IC_{50} > 10 µM). Furthermore, dysoline displayed significant anti-inflammatory activity, as indicated by 47% and 83% inhibition of proinflammatory cytokines TNF-α and IL-6, respectively at 0.1 µM.

In another bioassay guided isolation studies, *Dysoxylum binectariferum* bark was identified as a new source of camptothecin. The ethanolic extract of *D. binectariferum* displayed good cytotoxicity in HL-60 cell (65% GI at 10 µg/mL). The extract was fractionated using HP20 resin and all fractions were tested for cytotoxicity. As shown in Figure 5, Fraction 4 was most active (95% GI at 10 10 µg/mL) showing IC_{50} = 1.5 µg/mL. This fractionwas Dragendorff-positive and showed a characteristic fluorescent band at higher wavelength (366 nm) on TLC. After repeated purifications on alumina column, two additional alkaloids were isolated. These compounds were characterized as CPT

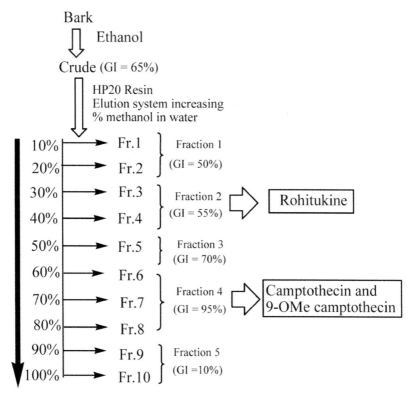

Growth inhibition (GI) in cell line, % inhibition at 10 µg/mL

Fig. 5: Protocol of bioassay-guided isolation of camptothecin and 9-methoxy camptothecin

and 9-methoxy CPT by spectroscopic data including UV, ^{1}H, ^{13}C, 2D NMR, MS and IR data and further their comparison with literature values (Ezell and Smith 1991, Wall et al. 1966) The CPT (7) was further confirmed via comparison of TLC and HPLC profile with commercially available reference standard.

The HP-20 resin was investigated for removal of rohitukine from the extract. Dianion HP-20 resin was used and the volume of resin used was 15% v/v of the material to be loaded. The extract was suspended in deionised water and was loaded on HP-20 column. The column was eluted with increasing amounts of MeOH in water (Figure 5). Rohitukine was eluted at ~35% MeOH in water whereas the CPT was eluted at ~65% MeOH in water. Highly pure CPT was obtained (~ 1 g) by repeating the HP-20 protocol. This protocol has several advantages such as it is environmentally-friendly, work process involves semi-aqueous medium, it is faster and reproducible at large scale; and it do not involve any heat or acid-base treatment. Next, our aim was to devise a practical and economical chromatography-free protocol for enrichment of crude extract for CPT content. For this, the obvious aim was to remove rohitukine from the extract as much as possible. The LCMS (MRM)-based quantification was performed to quantify CPT in plant. The concentrated EtOH extract was then suspended in minimum cold MeOH and subjected to repeated precipitation of rohitukine by adding cold EtOAc which resulted in isolation of rohitukine. The filtrate was concentrated, re-suspended in water and extracted with EtOAc. The LCMS analysis of EtOAc fraction in TIC mode (Figure 6a) showed three major peaks with mass m/z 305 (at t_R 6.6 min), m/z 349 (at t_R 7.9 min) and m/z 379.6 (at t_R 8.4 min) as depicted in Figure 6b-d. After running a LCMS chromatogram with pure reference standards and performing a spiking experiment (Figure 6e), these peaks were identified as rohitukine (1), CPT (7) and 9-methoxy CPT (8), respectively. The protonated molecular ions [M+H]$^+$ of rohitukine (1) and CPT (9) which were predominantly generated were chosen for multiple reaction monitoring (MRM) analysis. The fragmentor and CE (collision energy, eV) were optimized in order to obtain the maximum sensitivity of analysis.

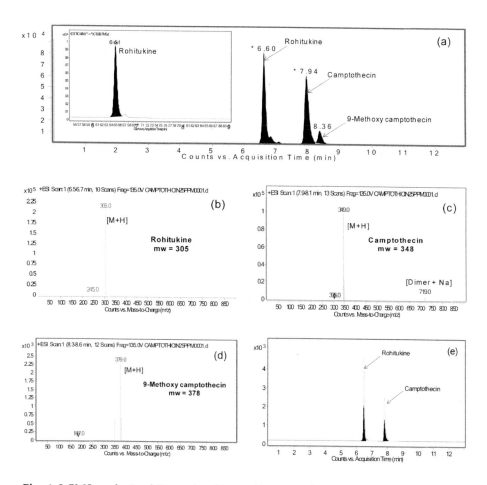

Fig. 6: LCMS analysis of *Dysoxylum binectariferum* crude extract for identification of CPT and rohitukine. (a) LCMS chromatogram of CPT-enriched EtOAc fraction obtained after repeated precipitation of rohitukine from EtOH extract (Inset: LCMS chromatogram of initial EtOH extract of *D. binectariferum* bark); (b) MS spectrum of peak eluted at t_R 6.6 min; (c) MS spectrum of peak eluted at t_R 7.9 min; (d) MS spectrum of peak eluted at t_R 8.36 min; (e) LCMS chromatogram showing spiking of rohitukine and CPT in crude extract. Thechromatographic on chromolith RP18$_e$ column (50 mm x 4.6 mm) protected by a chromolith guard column. The mobile phase consisted 0.1% v/v of formic acid in water (A), and acetonitrile (B). A gradient elution was used over 15 min: 0-2 min 5-10% B; 2-4 min 10-40% B; 4–9 min 40% B and 9-12 min 40-5% B; 12-15 min 5% B. The flow rate was 0.5 mL/min. The injection volume was 10 μL.

Both negative and positive ion modes were conducted, which showed that the positive ion mode to be more sensitive for both compounds. Based on these results, the precursor/product ion pair of transition mass m/z306/245 and 306/70 were taken as quantifier for rohitukine and pairs m/z 349/305.1 and 349/249 were taken as qualifier for CPT for the MRM scan. The peak areas obtained from the MRM of both the standards were utilized for quantification. The calibration equation of rohitukine and CPT were obtained by plotting LC-MS peak area (y) versus the concentration (x, mg/ml) of calibrators as y = 14.079943x + 253.704408 (r^2 = 0.997) and y = 14.552128x + 142.943875 (r^2 = 0.996), respectively. The equation showed very good linearity over the range 25-2500 ng/ml. The CPT content of EtOAc fraction was found to be 209 µg/mg of the EtOAc extract. The total CPT content in the dry plant material was found to be 0.105% (41.8 mg in 40 g dry plant material). Next, we also investigated leaves, seed, and seed coats of *D. binectariferum* for the presence of CPT alkaloids. The EtOH extracts of these parts were processed in similar fashion and were analyzed by LCMS (MRM) method. The LCMS results indicated that, rohitukine was the major component present in these parts; whereas they do not contain CPT alkaloids. The content of CPT in different sources is summarized in Table 4. The highest concentration of CPT was reported in *C. acuminata* young leaves (0.4 to 0.5% of dry powder) (Ramesha et al. 2008) and *N. nimmoniana* bark (0.3 to 0.7% of dry powder) (Ramesha et al. 2008). The CPT content in *D.binectariferum* bark (0.11%) was found to be similar to *Ervatamia heyneana* bark (0.13%) (Gunasekera et al. 1979) and *Ophiorrhiza pumila* roots (0.10%) (Arisawa et al. 1981) In general, the young leaves comprises the higher contents of secondary metabolites, thus bioprospecting of various parts of *D. binectariferum* at different life stages and their clonal multiplication via developing appropriate *in-vitro* production systems will further lead to identification of optimized high-yielding CPT source.

MEDICINAL CHEMISTRY

The co-crystal structure of flavopiridol and Cdk reveals several key interactions (De Azevedo et al. 1996). The Cdk-2 interactions with ATP are characterized by predominantly hydrophobic and Van-der waals interactions, The adenine ring is enclosed in a hydrophobic pocket formed by Ile10, Ala31, Val64, Phe80, Phe82, and Leu134 and forms two hydrogen bonds, between the N6 atom of the adenine and the carbonyl oxygen of Glu81, and between N1 and the backbone amide of Leu83. Key hydrogen bonds are formed to Lys33 and Wat384 with O(3) of the hydroxyl-piperidine moiety and to Asp145 with the N(11)H and O(3)H. Core hydrogen bonds are formed to Glu81 with O(5)H and to Leu83 with O(4). As shown in Figure 7a-b, compound essentially takes the same position as ATP. Cdk binding is mainly achieved by hydrophobic interactions, Hydrophobic side chains such as chlorophenyl produced potent inhibitory activity, since it points out of the ATP binding pocket of the enzymes and occupies an area that cannot be utilized by ATP.This additional interaction seems to be responsible for the high affinity and selectivity of inhibitor for Cdks (Huwe et al. 2003, Zhang et al. 2008).

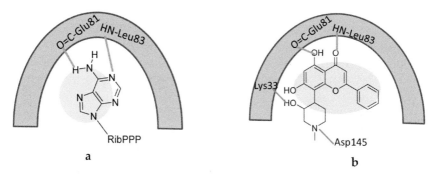

a   b

Fig. 7: (a). Schematic representation of the ATP binding pocket of Cdk-2; (b) Schematic representation of the ATP binding pocket of Cdk-2 in the complex with des-chloroflavopiridol. Source of a and b.(Huwe et al. 2003) hence based on above described reported findings and the co-crystal structure of compound with Cdk. Following four series of compounds were designedfor synthetic modifications on rohitukine

Series I: O-benzylation and benzoylation at 7-hydroxy position

Series II: Mannich adducts at position 6 with different secondary amines

Series III: Baylis Hillman adducts at position 3 by different aldehydes

Series IV: Styryl analogs by condensation of aldehydes at 2-methyl position.

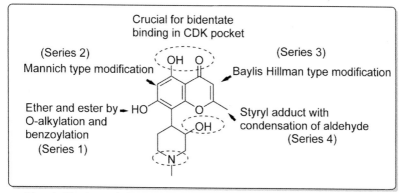

Fig. 8: Design of four different series of rohitukine derivatives.

Synthesis of derivatives

To synthesize series, I, various base catalyzed O-benzylation/ benzoylation methods were attempted.A mixture of mono, di and tri-O-benzylation and benzoylation productsformation was observed, under different basic conditions (K_2CO_3, NaH, NaOH, CH_3ONa etc.) and solvents (acetone, THF, Dioxane, DCM etc). Subsequent attempts discovered that KF/Alumina catalyzed regioselective method under solvent free grind-stone condition (Figure 9, reaction path a) was successful.(Jain et al. 2012c) to synthesize 7-O-benzylation/benzoylation regioselectively. In series II, Mannich products (b) at position 6 in rohitukine (1) were synthesized by conventional procedure (Figure 9, reaction path b). Mannich reaction is one of the oldest method for alkaloidal type compounds, and Shiff base. The reaction was performed with different secondary amines in presence of formalin solution, and DMSO was used as a solvent under the heating (60-70 °C, 5 h). for series III and IV. the Baylis-Hillman adducts (c) and styryl derivatives (d) were formed when rohitukine (1) was subjected to reaction with aldehydes under alkaline condition (Figure 9, c and d).

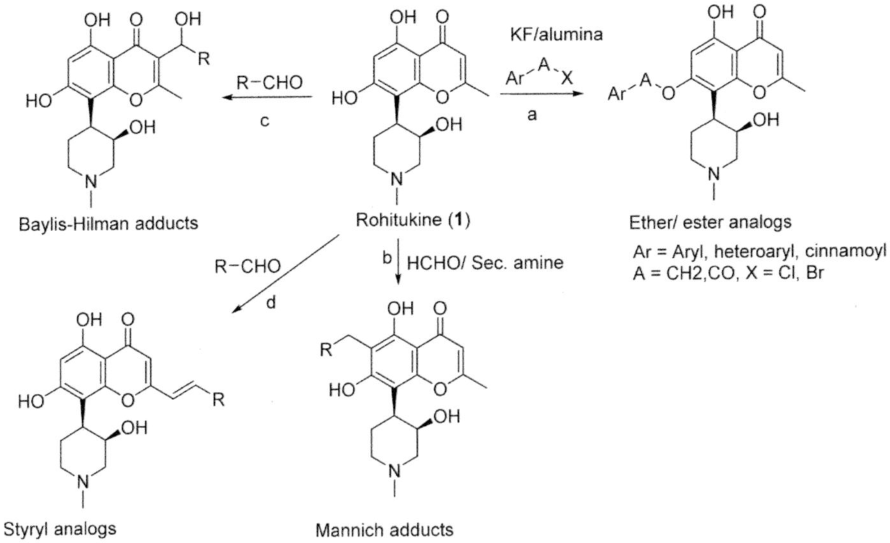

Fig. 9: Reagents and conditions for synthesis of rohitukine derivatives: (a). KF/ Alumina (5 mol%); grinding, substituted benzyl or benzoyl halide, 5-15 min, rt, 60-80%; (b). Formalin solution (1.7 eq), secondary amine, DMSO, 60 °C, 5 h, 70-90 %. (c) DABCO (1.2 eq.), substituted aldehydes (1 eq), MeOH, rt, 15 days, 30-65%. (d). KOH (12 eq. or 25% aq sol.), substituted aldehydes (2.5 eq), MeOH, 110 °C, 10 h, 40-65%.

The Baylis-Hillman type product (**c**)formation was observed at weakbasic condition (DABCO, MeOH at rt) and the Claisen-Schmidt Condensation styryl product (**d**) at high temperature with strong alkaline condition (KOH; 25% sol) / EtOH, 110 °C). By applying the optimized conditions, a series of many compounds has been synthesized. All synthesized products were purified on routine silicagel column and/or preparative TLC and isolated products were characterized based on their spectral properties.

Screening

First rohitukine was screened for Cdk inhibition (IC_{50}: 7.3 and 0.3 µM for Cdk-2/A and Cdk-9/T1 respectively), and cytotoxicity. These preliminary results suggested that rohitukine is a selective Cdk-9 inhibitor which also indicated role of Cdk-9 in cancer need to be addressed in future. All synthesized compounds were screened against Cdks (Cdk-2 and Cdk-9 at 500 nM) and tested for cytotoxicity against a panel of cancer cell line. (Bharate et al. 2018). Result indicated that series IV; styryl derivatives compounds were found to be potent Cdk inhibitors; however other series compounds were weak to moderately active.The most promising compounds wasIIIM-290 (Table 4. IC_{50}, Cdk-2/A: 15 ± 0.94 nM, Cdk-9/T1: 1.9 ± 0.61 nM).

Table 1: Cdk, cell line cytotoxicity and pharmacokinetics of rohitukine

Compound	Cdk	(IC_{50} µM) Cytotoxicity	
Rohitukine	Cdk-2/A = 7.3 Cdk-9/T1 = 0.3	HL60 = 7 MOLT-4 = 10 PC-3 = 26 MCF-7 = 28	Panc-1 = 24 A549 = 33 786-O = 30

Similarly, IIIM-290 showed cytotoxicity against wide panel of cancer cell line (Table 2). These results prompted us for further investigations of IIIM-290 for detailed preclinical studies.

Table 2. Cytotoxicity of IIIM-290 against a large panel of cancer cell lines

Cell line	IC_{50} (μM)	
	IIIM-290	Flavopiridol
MDAMB-231	4	n.d.
T47D	6	n.d.
BT549	5	n.d.
MDAMB468	4	0.4
A549	4	n.d.
MOLT-4	0.5	n.d.
MCF-7	4	n.d.
Colo-205	7	0.3
SW620	0.3	0.2
HOP-92	3	1
NCIH322	2	4
LOXIMVI	4	1
DU145	5	2
HCT116	5	1
Panc-1	4	1
NCIH522	5	0.8
HL-60	0.9	n.d.
PC-3	4	n.d.
A431	8	n.d.
MIAPaCa-2	0.6	n.d.
HGF (Normal)	18	n.d.
FR-2 (Normal)	19	n.d.

Table 3. Cdks profiling of IIIM-290

Cdks	IC_{50} (nM)	
	IIIM-290	Staurosporine
Cdk-1/cyclin A	4.93	2.5
Cdk-1/cyclin E	75.9	2.5
Cdk-2/cyclin O	138	1.7
Cdk-2/cyclin A	15.5	n.d.
Cdk-3/cyclin E	>1000	9.8
Cdk-4/cyclin D3	22.5	1.8
Cdk-5/p25	15.5	2.4
Cdk-5/p35	15.7	1.6
Cdk-6/cyclin D1	45	3.9
Cdk-6/cyclin D3	199	49.3
Cdk-7/cyclin H	711	54.0
Cdk-9/cyclin K	412	166
Cdk-9/cyclin T1	1.9	n.d.

Structure- Activity Relationship

All ether and ester derivatives at 7-hydroxy position of rohitukine (series I) were screened against Cdk-2 and Cdk-9 however none of them was found to be significant potent activity. These results indicated that substitution at 7-hydroxy position is not suitable with respect to biological activity. Esters were found to more active than ether perhaps because of hydrolysis of ester is easier than ether to get parent molecule. All compounds from series II, showed less activity against Cdk-2 and Cdk-9 at 500 nM, indicating that introduction of any hydrophilic functional group (nitrogen containing) at this position is not tolerable probably because of orientation of molecule get changed in enzyme cavity. However, few derivatives showed moderate cytotoxicity against a panel of cancer cell line. Here the biological results were not observed in a specific manner with respect to their chemical structure. Moreover, the substitution at C-3 position is not tolerable with respect to Cdk-2 and -9 activities. In series IV, the modification at 2-methyl position of rohitukine was most favorable and tolerable. Different halogen (especially Cl and F) substituted aldehydes were used to synthesize styryl series of products (Series IV).

Cdk screening results indicated that most of the compounds of this series showed >50% inhibition of Cdk-9 at 500 nM. It is also clear that increase in bond length provided more affinity toward Cdk-9 when results were compared with flovopiridol. Position of halogen substituent on ring D plays an important role in selectivity for Cdk-9 versus Cdk-2. In general activity of chlorosubstituted analoges have ten time higher against Cdk-9 than Cdk-2. Similarly, monoflouro substitution followed same activity pattern but the difference in selectivity of Cdk-2 and Cdk-9 was not significant as compared to monochloro substituted analoges. Other potent compound dichloro phenyl styryl analog displayed 7-fold selectivity toward Cdk-9 (IC_{50} = 1.9 nM) versus Cdk-2 (IC_{50}=15.5 nM). Monochloro substitution in phenyl ring at different position has different activity. The substitution at position 2 showed potent activity in both Cdk-2 and Cdk-9 while substitution at 3 and 4 position have less activity against Cdk-2 while retained activity for Cdk-9. SAR summarized in Figure 10.

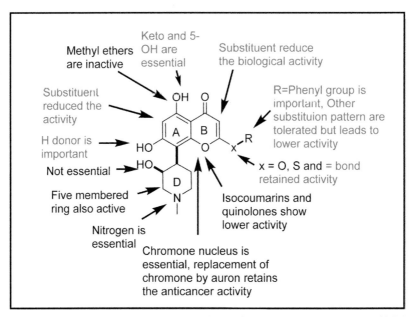

Fig. 10: Structure-activity relationship of chromone alkaloid scaffold.

Anti-cancer Mechanism

It has been reported that the selective Cdk-9 inhibitors targets survival proteins and induce apoptosis. Therefore, compound IIIM-290, was subjected to apoptosis related studies in MIAPaCa-2 (pancreatic cancer) cell line. In apoptosis study, there was an increased formation of apoptotic bodies and chromatin condensation in nuclei of cells treated with IIIM-290, which suggests that this compound inhibits the growth of MIAPaCa-2 cells by inducing apoptosis. The scanning electron microscopy of treated cells confirmed the vesicle formation inside the cells, compared to control cells. The untreated cells showed healthy and round nuclei with no DNA fragmentation. The compound IIIM-290 was also found to trigger mitochondrial membrane potential (MMP) loss in MIAPaCa-2 cells. We also studied the effect of compound IIIM-290 on apoptosis markers, viz. PARP, caspase-3, and caspase-7. Western-blot studies indicated the concentration-dependent cleavage of PARP, caspase-3, and caspase-7 by compound IIIM-290, confirming the apoptosis mediated cell death of MIAPaCa-2 cells.

Physico-chemical Characterization

Before preclinical animal studies compound IIIM-290 was subjected for physiochemical characterization. Compound IIIM-290 was found to possess moderately soluble in water (20 μg/ mL). The experimental log P, log D, and pKa of this compound were found to be 3.09, 1.65, and 5.4, respectively. Furthermore, the compound IIIM-290 also follows the Lipinski Rule of Five. The percent protein binding of IIIM-290 in human plasma was 96.7%, and in rat, mouse, and dog plasma, it was also in similar range (95"97.2% binding). Themetabolic profile results of these studies summarized in Table 5

Pharmacokinetics Studies

The pharmacokinetics of compound IIIM-290 was tested in BALB/c mice following a single 10 mg/kg dose administration by oral route and 1.0 mg/ kg dose administration by IV route. Following oral administration, elimination half-life (t1/2, ß) was found to be 4.65 h and $AUC_{0 t}$ was found to be 4503 nM·h. Following IV administration, elimination half-life ($t_{1/2}$,ß) was found to be 5.46 h, and clearance was <"55 mL/min/kg. The absolute oral bioavailability was 71%. The data presented in Table 6.

In-vivo Anti-cancer Activity

The preliminary anti-cancer activity of IIIM-290 was performed in Ehrlich solid tumors, a murine mice model to understand the eûective dose-range and máximum tolerable dose in mice in a 15-day study. Initial doseoptimization studies involved various doses ranging from 10 to 100 mg/kg. Compound III-290 at 70 and 100 mg/kg p.o. showed 31 and 39% tumor growth inhibition without causing any mortality (mortality: 0/7). The doses above 100 mg/kg resulted in mortality. The detailed results of anticancer activity in murine models are shown in Table 7.

Table 4: Physico-chemical characterization of IIIM-290.

Log P/pKa	Solubility(μg/ml ± SD)	PPB, % bound	Chemical stability	CYP inhibition IC_{50}(μM)
Log P: 3.09	Water 20.31 ±4.04	Mouse: 97.2	% drug remaining (after h)	1A2: >202
Lop D: 1.65p	PBS (pH 7.4): 2.29 ± 0.07	Rat: 95.4	pH 1.2: 100 (24)	D6: >202
Ka:5.4	SGF (pH 1.2): 8.81 ± 0.39	Dog: 95	pH 4.0: 100 (24)	C9: >202
	SIF (pH 6.8): 2.18 ± 0.16	Human: 96.7	pH 6.8: 100 (24)	C19: >20
		HSA: 96.7	pH 7.4: 100 (24)	3A4: >20
		AAG: 98.3	SGF: 100 (8)	
			SIF: 100 (8)	

Table 5: Metabolic profile of IIIM-290.

Species	Metabolic stability					
	Liver microsomes		Hepatocytes		S9 liver fraction	
	$t_{1/2}$ (min)	CL_{int} (μL/min/mg protein)	$t_{1/2}$ (min)	CL_{int} (μL/min/mg protein)	$t_{1/2}$ (min)	CL_{int} (μL/min/mg protein)
Mouse	>30	<8.8	186	14.9	86	16
Rat	>30	<13	376	9.2	31	45
Human	>30	<8.6	726	4.8	72	19
Dog	24	<29	281	12.3	117	12
Monkey	25	54.8	326	10.6	56	14

Table 6: Pharmacokinetic parameters of IIIM-290

Parameters	IIIM-290 BALB/c mice[2,a]	
	PO, 10 mg/kg[b]	IV, 1 mg/kg[c]
Half life $T_{1/2}$ (hr.)	4.65	5.46
C_{max} (ng/mL)	593	312
T_{max} (ng/mL)	0.25	-
C_0 (ng/mL)	-	396
AUC_{0-t} (ng'''h/mL)	2076	242
$AUC_{0-\infty}$ (ng'''h/mL)	2128	301
CL (mL/min/kg)	-	55.4
V_d (L/kg)	-	26.2
V_{dss} (L/kg)	-	14.1
Bioavailability	70.7 %	-

[a]$t_{1/2,\beta}$: terminal half life; AUC_{0-t}: the area under the plasma concentration-time curve from 0 to last measurable time point; $AUC_{0-\infty}$: area under the plasma concentration-time curve from time zero to infinity; C_{max}: maximum observed plasma concentration; C_0: extrapolated concentration at zero time point;CL: clearance; V_d: volume of distribution; V_{dss}: volume of distribution at steady state; T_{last}: time at which last concentration was found; F: bioavailability. [b]Time points considered for $t_{1/2,b}$ calculation 8 – 24 h, [c]Time points considered for $t_{1/2,\beta}$ calculation 2 – 8 h; C_0 calculated manually using initial 4 time points.

Table 7: Body weight of mice during *in-vivo* study of IIIM-290 in Ehrlich Ascites Carcinoma model.

	Dose(i/p)	Body weight (g)			
		Day 1	Day 5	Day 9	Day 12
Control	0.2 ml N.S.	22.8	26.2	27.7	28.4
IIIM-290	70 mg/kg	21.14	21.28	20.75	20.55
5-FU	20 mg/Kg	21.42	22.0	21.28	20.42

Further, anti-cancer activity of IIIM-290 was confirmed in three human xenograft models, viz. pancreatic, colon, and leukemia cancer at 25 and 50 mg/kg doses via a peroral route. The selection of these tumor types for xenograft study was based on cytocoxicity results. Compound IIIM-290 showed most prominent cytotoxicity in Molt-4 and MIAPaCa-2 cells. Results from xenograft activity showed up to 40"52% TGI at 50 mg/kg, p.o. dose, without causing any mortality. Results obtained from the in vivo study conûrmed the anticancer potential of the compound IIIM-290.

Table 8: *In-vivo* anti-cancer activity of IIIM-290 in Ehrlich Ascites Carcinoma

Drug	Dose(i/p)	Day 12					
		Av. Vol. of ascitic fluid (ml)	Av. Wt of ascitic fluid (g)	Av. No. of tumor cells	% Tumor cell growth	% Tumor growth inhibition	Mortality
Control	0.2 ml N.S.	10.02	9.90	183.14×10^7	100	-	0/10
IIIM-290	90 mg/kg	Intolerable		90.14	6/7		
IIIM-290	70 mg/kg	1.7	1.65	26.42×10^7	14.43	85.57	2/7
5-FU	20 mg/Kg	0.93	0.63	5.61×10^7	3.07	96.93	0/7

Table 8: *In-vivo* anti-cancer activity of IIIM-290 in Ehrlich solid tumor mice model

Route	Dose IIIM-290	Av. Body weights (g) of animals on days			Day 13	Av. Tumor wt. (mg)	%TGI	Mortality
		1	5	9	Av. Body wt (g)			
IP	50 mg/kg	21.21	21.92	22.3	22.46	989.12 ± 115.22*	33.11	0/7
	70 mg/kg	20.57	21.14	22.0	21.85	922±88.18**	37.65	0/7
	90 mg/kg	21.0	21.76	22.1	22.1	860.85±87.22***	41.75	0/7
	Saline, 0.2 ml	21.8	22.1	22.9	23.1	1478.9	–	0/10
Oral	70 mg/kg	25.85	26.71	26.14	26.71	1058.14±81.85*	30.42	0/7
	100mg/kg	25.85	26.14	25.14	25.71	978.85±75.06**	35.63	0/7
	Saline, 0.2ml	26.7	27.2	27.4	27.9	1520.85±140.39	0	0/10
IP	5FU, 22mg/kg	26.00	26.42	24.0	24.42	628.42±106.14***	58.67	0/7

Medicinal chemistry eforts on rohitukine have resulted in the identification of the first orally bioavailable Cdk inhibitor IIIM-290

EXPERIMENTAL SECTION

All chemicals were purchased from Sigma-Aldrich Company ^{1}H, ^{13}C and 2D NMR spectra were recorded on Brucker-Avance DPX FT-NMR 500 and 400 MHz instruments (for ^{13}CNMR 125 MHz or 100 MHz). Chemical data for protons and carbon are reported in parts per million (ppm) downfield from tetramethylsilane and are referenced to the residual ^{1}HNMR/^{13}CNMR in the NMR solvent (CDCl$_3$, 7.26/77.1 ppm; CD$_3$OD, 3.31/49.0 ppm, and DMSO-d$_6$ 2.50/39.5ppm). Melting points were recorded on digital melting point apparatus (Buchi). All chromatographic purifications were performed on silica gel (#60–120 or #100–200 from E. Merck, Germany). Thin layer chromatography (TLC) was performed on pre-coated silica gel 60 GF$_{254}$ aluminium sheets (Merck). ESIMS and HRMS spectra were recorded on Agilent 1100LC-QTOF and HRMS-6540-UHD machines. LC-ESI-MS/MS analysis was carried out on Agilent Triple-Quad LC"MS/MS system (model 6410). An Agilent 1260 liquid chromatography system (Agilent, USA) equipped with a quaternary solvent delivery system, an auto-sampler and a column compartment was used. A 6410 B triple quad LCMS system from Agilent was used for the detection, which is a hybrid triple quadrupole masss pectrometer equipped with Turbo V sources. The analyses were performed using an electrospray ionisation (ESI) source in positive and negative modes. The operation conditions wereas follows: scan rangeof 100–600 amu, V charging 2000 V, ion source temperature 325 °C, nebulizer 60 psi, gas flow 12 L/min, capillary voltage 4000, collision gas nitrogen, dwell time of 50 ms and a step size of 0.1 amu. Nitrogen was used in all cases. The dwell time was set at 60 ms in positive mode. Agilent Mass Hunter software (version B.04.00) was used for data acquisition and processing.

Plant Material

Stem barks of *D. binectariferum* were sampled from Kathagal located in the central Western Ghats region of Karnataka, India by Prof. R. Umashaanker group, from University of Agricultural Science (UAS), Karnataka, India. Voucher specimens (COF\DBT\WG-185-1-36) for each of the sample tree collected was deposited at the herbarium of the College of Forestry, Sirsi (UAS, Dharwad), India.

Extraction and isolation of compounds

The plant material (1.5 kg) was extracted with ethanol (3 L) and concentrated over rotary evaporator to get crude (110 g, extractive value 7.3%) material. Half of the crude (55 g) was again suspended water (150 ml) and pH was adjusted to 3-5 (color indicated by pH paper) by aq. HCl (5%) and placed over night at room temperature. Then it was filtered from celite and a clear solution was basified with liquid ammonia to get pH 11-13. The solution was adjusted to 2 L with water and passed through HP20 resin (100 ml, bed size 10% of final volume).

Isolation of rohitukine, rohitukine-N-oxide and dysoline

Adsorbed material was eluted with increasing proportion of methanol in water; rohitukine was isolated at 40% methanol. Purification was done by crystallization with water-acetone mixture (40: 60). Finally 17 g of rohitukine (1.13%) was isolated with > 98.5% purity. This method was found to be simple and fast method for purification of large amount of pure rohitukine without HPLC. However based on HPLC analysis it has been reported previously that content of rohitukine can be enriched up to 53% in the extract (Cui et al. 2008). Same results could be observed when this method was applied on small amount but this method was not found to be suitable for large scale isolation of rohitukine, here we developed a alternative procedure for the large scale purification of rohitukine with a green procedure without using organic solvents such as ethyl acetate and n-butanol and our method can be suitably for scaled up to gram and kg level. Another disadvantage observed in reported method; lose of rohitukine during alkaloid enrichment fractionation, in non alkaloidal butanol fraction. The compound was analysed by NMR and mass and it was matching with rohitukine alkaloid (Naik et al. 1988).

After isolation of rohitukine, the remaining fractions were pooled together and loaded again on basic alumina column and column was eluted with increasing proportion of methanol in chloroform. Fractions were monitored with TLC and Dragendorff-reagent. Dragendorff-positive fraction finally was purified with repeated sephadex LH20 column. The structure of known compound was confirmed based on earlier reports as rohitukine-N-oxide (15 mg, 0.001%).

Bioassay-guided isolation and quantification of camptothecin

Ethanolic extract (3 gm) was suspended in deionized water (500 mL) and loaded over HP20 resin (30 mL) to adsorb organic components on resin from water stream and fractionated with increasing methanol in water elution system. Fractions were pooled on the basis of TLC. Five fractions were prepared and submitted to cytotoxicity on HL60 cell line (% growth inhibition at 10 µg/mL. Crude 65%, fraction 1: 50%, fraction 2: 55%, fraction 3: 70% fraction 4: 95%, and fraction 5: 10%). Fraction 4 was found to be most potent (IC_{50} = 1.5 µg/mL). Repeated column chromatography of fraction 1 and 2 yielded rohitukine as major constituent while fraction 4 yielded camptothecin and 9-methoxy camptothecin. The isolated compounds were characterized by comparison of spectroscopic data with literature values.

LC-MS condition for quantification

Thechromatographic separation was performed on chromolith RP18$_e$ column (50 mm x 4.6 mm) protected by a chromolith guard column. The mobile phase consisted 0.1% v/v of formic acid in water (A), and acetonitrile (B). A gradient elution was used over 15 min: 0-2 min 5-10% B; 2-4 min 10-40% B; 4–9 min 40% B and 9-12 min 40-5% B; 12-15 min 5% B. The flow rate was 0.5 mL/min. The injection volume was 10 µL and the column temperature was 25 °C. Before making calibration curve, three blank injections were run to check the noise level of the system. The calibration equation of rohitukine and CPT were obtained by plotting LC-MS peak area (y) versus the concentration (x, mg/mL) of calibrators as $y = 14.079943x + 253.704408$ ($r^2 = 0.997$) and $y = 14.552128x + 142.943875$ ($r^2 = 0.996$), respectively. The equation showed very good linearity over the range. Precursor ion scan experiment of rohitukine and CPT also confirmed the origin of daughter ions from the same molecular ion; there by confirming their presence in the sample. All the quantification protocols were performed in triplicate.

Preparation of KF-Alumina

A mixture of potassium fluoride (45 g) and basic alumina (55 g, type T, Merck) in water (100 mL) was stirred at room temperature for 10 min. The resulting suspension was concentrated on rotary evaporator at 50 °C, and then dried in calcium chloride desiccator under vacuum for 15 h. Use of basic alumina in the solid support gave better results relative to the neutral one (Singh et al. 2010).

General procedure

Rohitukine (1 mmol), KF-Al$_2$O$_3$ (5 mol%), and benzyl/benzoyl halides (1.2 mmol) were mixed in a mortar and grinded intermittently using a pestle. The mixture changed to mushy state within a proper reaction time and then gets solidified itself. Formation of product was monitored by TLC. After filtration of the catalyst, the filtrate was washed with aqueous NaOH (10%) and the organic phase was evaporated under reduced pressure to furnish the desired product. The crude product was sufficiently pure.

Synthesis of Mannich derivatives of rohitukine

To the solution of rohitukine (61 mg, 2 mmol) in Methanol-water (10 ml, ratio 7:3) was slowly added to a solution of formaldehyde (1ml solution) and secondary amine (2 mmol) at 70 °C for 5 hr. Products were purified over sephadex gel filtration in methanol a gummy sticky mass solidified in acetone. The conditions used in the synthesis are mild and yields are excellent.

Synthesis of Baylis-Hillman derivatives of rohitukine

The conditions used in the synthesis are mild and yields are moderate. The solution of rohitukine (61 mg, 2 mmols) in methanol (5 mL) was stirred with substituted aromatic and aliphatic aldehydes (2 mmol) in presence of DABCO (1 mmol) as catalyst. The reaction mixture was continuously stirred for 10-15 days, however in none of the reaction starting material completely consumed. Desirable product was separated on preparative TLC as racemic mixture.

Synthesis of styryl derivatives of rohitukine

Synthetic scheme is depicted in Figure 23. The solution of rohitukine (61 mg, 2 mmol) in methanol or ethanol (5-10 mL) was stirred with substituted aromatic and aliphatic aldehydes (2 mmol) in presence of 15 % aqueous KOH (few drops) as catalyst. The reaction mixture was continuously stirred for 12-15 hr, however in these reactions, Baylis-Hillman product was observed as a side product. Yield was also varied in different reactions.An intense yellow colored band was separated using preparative TLC.

In-vitro cytotoxicity

The MTT assay was used to assess the effect of the test molecules on cell viability. Cell viability on eight cell lines *viz.* Colo205 (colon),

HCT116 (colon), HT1080 (fibrosarcoma), NCIH322 (lung), A549 (lung), Molt-4 (leukemia) and HL60 (leukemia) (Source: European Collection of Cell Cultures, ECACC UK; purchased through Sigma-Aldrich India.) were investigated. In each well of a 96-well plate, 3×10^3 cells were grown in 100 µL of medium. After 24 h, each test molecules were added to achieve a final concentration of 10 to 0.01 µmol/L, respectively. After 48 h of treatment, 20 µL of 2.5 mg/mL MTT (Organics Research, Inc.) solution in phosphate buffer saline was added to each well. After 48h, supernatant was removed and formazan crystals were dissolved in 200 µL of DMSO. Absorbance was then measured at 570 nm using an absorbance plate reader (Bio-Rad Microplate Reader). Data are expressed as the percentage of viable cells in treated relative to non-treated conditions. Each experiment was repeated three times and data was expressed as mean ± SD of three independent experiments (Mordant et al. 2010).

Effect of dysoline on production of pro-inflammatory cytokines TNF-± and IL-6

Splenocytes were seeded into three to four wells of a 96-well flat-bottom microtiter plate (Nunc) at 2×10^6 cells/ml. Cells were incubated with different concentrations of dysoline (0.1 µM -100 µM) along with Con A (2.5 µg/well) or LPS (1 µg/ml) for 72 h at 37 ^{0}C with 5% CO_2 in CO_2 incubator. The supernatants were harvested and the measurement of cytokines (TNF-α and IL-6) in the culture supernatants was carried out using commercial kits as per manufacturer's instructions by using ELISA kits (R&D, USA).(Llanos et al. 2008, Mosmann 1983)

General procedure for In-vitro cytotoxicity (MTT assay)

Inhibition of cell proliferation by different compounds was measured with the 3-(4,5-dimethylthiazole-2-yl)-2,5-diphenyltetrazolium bromide or MTT assay. Different cell line panel was purchased and cells were seeded in 96 well cell tissue culture plates (Source: European Collection of Cell Cultures, ECACC UK; purchased through Sigma-Aldrich India). Cells were treated with different concentrations of compounds for 48 h. MTT dye (2.5 mg/mL final concentration) was added to each well 4 hours prior to experiment termination followed by incubationat 37 °C. MTT-formazan crystals were dissolved in 140 µL DMSO and the OD measured at 570 nm with reference wavelength of 620 nm and IC_{50} value was determined.

General protocol for kinase assay

All assays were carried out using a radioactive (33P-ATP) filter-binding assay at International Centre for Kinase Profiling (ICKP), DSTT College of Life Sciences, University of Dundee, Dow Street DD1 5EH. http:// www.kinase-screen.mrc.ac.uk/kinase-assay. Cdk-2/cyclin A (5–20 mU diluted in 50 mM Hepes pH 7.5, 1 mM DTT, 0.02% Brij35, 100 mM NaCl) was assayed against Histone H1 in a final volume of 25.5 µl containing 50 mM Hepes pH7.5, 1 mM DTT, 0.02% Brij35, 100 mM NaCl, Histone H1 (1 mg/ml), 10 mM magnesium acetate and 0.02 mM [33P-g-ATP](500-1000 cpm/pmole) and incubated for 30 min at room temperature. Assays were stopped by addition of 5 µl of 0.5 M (3%) orthophosphoric acid and then harvested onto P81 Unifilter plates with a wash buffer of 50 mM orthophosphoric acid.Cdk-9/Cyclin T1 (5-20mU diluted in 50 mM Tris pH 7.5, 0.1 mM EGTA, 1 mg/ml BSA, 0.1% Mercaptoethanol) was assayed against a substrate peptide (YSPTSPSYSPTSPSYSPTSPKKK) in a final volume of 25.5 µl containing 50 mM Tris pH 7.5, 0.1mM EDTA, 10mM DTT, 1mg/ml BSA, 0.3 mM YSPTSPSYSPTSPSYSPTSPKKK, 10 mM magnesium acetate and 0.05 mM [33P-ã-ATP] (50-1000 cpm/pmole) and incubated for 30 min at room temperature. Assays were stopped by addition of 5 µl of 0.5 M (3%) orthophosphoric acid and then harvested onto P81 Unifilter plates with a wash buffer of 50 mM orthophosphoric acid.

Protocol for cell cycle analysis

Cells were incubated with IIIM-290 at indicated concentrations for 24 h. 400 µg of cells was collected, washed (PBS), and fixed with 70% ethanol for overnight at 4 °C. Than cells were incubated with RNase at a concentration of 0.2 mg/mL at 37 °C for 1 h and stained with propidium iodide (10 µg/mL) for 30 min in the dark. Cells were analyzed on flow-cytometer (FACS Calibur, Becton Dickinson), and data were collected in list mode on 10,000 events for FL2-A vs FL2-W. Resulting DNA distributions were analyzed by Modfit (Verity Software House Inc., Topsham, ME) for the proportions of cells inapoptosis, G1-phase, S-phase, and G2-M phase of the cell cycle as described (Bharate et al. 2018)

Scanning Electron Microscopic (SEM) Analysis

MIAPaCa-2 cells were incubated in a six-well tissue culture plate with differnt concentration of IIIM-290 for 24 h. The cells on a coverslip were fixed with 2.5% glutaraldehyde in 0.1 M phosphate buffer (pH

7.2) at 4 °C for 24 h, postfixed with 1% osmium tetraoxide in the same buffer for 4 h, dehydrated with graded ethanol solution, dried in a critical point drier using HMDS, and coated with gold using a sputter coater (Joel JEC-3000 FC). The specimens were examined with an electron microscope (JEOL-JSM-IT300) with ASID at 20 kV as described (Bharate et al. 2018)

In-vivo anticancer activity of IIIM-290 in Ehrlich Ascites Carcinoma (EAC)

Animals: Swiss, Sex: Female and Weight: 18-23 g, Ehrlich ascites carcinoma (EAC) cells were collected from the peritoneal cavity of the Swiss mice harbouring 8-10 days old ascitic tumor. 1×10^7 EAC cells were injected intraperitoneally in Swiss mice selected for the experiment on day 0. The next day, animals were randomized and divided into different groups. The treatment groups contained 7 animals each and control group contained 10 animals. Two treatment groups were treated with 70 mg/kg each of IIIM-290 intraperitoneally from day 1-9. One more of the treatment groups received 5-fluorouracil (20 mg/kg, i.p) and it served as positive control. The tumor bearing control group was similarly administered normal saline (0.2 ml, i.p.). On day 12, animals were sacrificed and ascitic fluid was collected from peritoneal cavity of each mouse for the evaluation of tumor growth. Percent tumor growth inhibition was calculated based on the total number of tumor cells present in the peritoneal cavity as on day 12 of the experiment using the following formula.(Geran et al. 1972) Percent tumor growth inhibition =[(Avg. no. of cell in control - Avg. no. of cell in treated)/ Avg. no. of cell in control] X100.

Anticancer activity of IIIM-290Ehrlich Tumor (Solid) animals

Swiss, Sex: Males and Weight: 18-23 g, Ehrlich ascites carcinoma (EAC) cells were collected from the peritoneal cavity of the swiss mice harbouring 8-10 days old ascitic tumor. 1×10^7 EAC cells were injected intramuscularly in right thigh of 24 swiss male mice selected for the experiment on day 0. The next day, animals were randomized and divided into three groups. Two treatment groups contained 7 animals each and one control group contained 10 animals.

Treatment was given as follows: Group I: IIIM-290(50 mg/kg i/p) from day 1-9. The second treatment group was treated with 5-fluorouracil (22 mg/kg, i.p) from day 1-9 and it served as positive control. The control group was similarly administered normal saline (0.2 ml, i.p.)

from day 1-9. On day 9 and 13, tumor bearing thigh of each animal was shaved and longest and shortest diameters of the tumor were measured with the help of vernier caliper. Tumor weight of each animal was calculated using the formula described in results and discussion. The percent tumor growth inhibition was calculated on day 13 by comparing the average values of treated groups with that of control group. Tumor growth in saline treated control animals was taken to be 100%. Tumor weight of each animal was calculated using the following formula.

Tumor weight (mg) = Length (mm) x [width(mm)]^2/ 2.

Solubility determination

The compound was dissolved in methanol to get stock solution of concentration 2000 µg/mL. the stock solution was introduced into 96-well plates and allowed to evaporate at room temperature to ensure that the compound (1, 2, 4, 8, 16, 25, 40, 80, 160 and 300 µg) is in solid form in the beginning of the experiment. Thereafter, 200µl of the dissolution medium (water/ PBS/ SGF/ SIF) was added to the wells and plates were shaken horizontally at 300 rpm (Eppendorf Thermoblock Adapter, North America) for 4 h at room temperature (25±1 °C). The plates were covered with aluminium foil in order to prevent degradation of compounds and were kept overnight for equilibration. Later, the plates were centrifuged at 3000 rpm for 15 min (Jouan centrifuge BR4i). Samples of 50 µl was withdrawn into UV 96-well plates (Corning® 96 Well Clear Flat Bottom UV-Transparent Microplate) for analyses with plate reader at corresponding ë$_{max}$ of the sample (SpectraMax Plus384). The analysis was performed in triplicate for each compound. The solubility curve of concentration (µg/mL) vs absorbance was plotted to find out saturation point and the corresponding concentration was noted.(Heikkila et al. 2011; Roy et al. 2001)

hrCYP P450 isoenzyme assay

 hrCYP P450 isoenzyme were aliquoted as per the total concentration required to conduct the study and stored at -70 °C until use. Total assay volume was adjusted to 200 µl and consists of three components: cofactors, inhibitor/vehicle and enzyme-substrate (ES) mix. 50µl of working cofactor stock solution was dispensed to all the specified wells in a black coloured nunc microtiter polypropylene plate. The 50µl of diluted working concentrations of NCE's/ positive control inhibitors/

vehicle were dispensed in triplicates to the specified wells as per the plate map design. Reaction plate with cofactor and test item was pre incubated at 37 °C ±1 °C shaking incubator for 10 minutes. Simultaneously, ES mix was prepared by mixing the hrCYP P450 isoenzyme. Remaining volume was made up with the buffer and pre incubated for 10 minutes at 37 °C ±1 °C. 100 µL of ES mix was dispensed per well as per the plate map design and incubated at 37 °C ±1 °C shaking incubator for predetermined time. A set of controls were incubated with hrCYP P450 isoenzymes and substrate without test or reference item. A set of blanks were incubated with substrate and test or reference item, in the absence of hrCYP P450 isoenzymes. Reaction was terminated by adding specific quenching solutions (For CYP1A2, CYP2C19 and CYP3A4 - 75µl of 100% acetonitrile; For CYP2C9 - 20µl of 0.25 M Tris in 60% methanol; For CYP2D6 -75µl of 0.25 M Tris in 60% methanol). The reaction is quenched by thoroughly mixing the final contents of the wells by repeated pippeting using multichannel pipette. The product fluorescence per well was measured using a spectrofluorimeter (Plate reader) at excitation and emission wavelength for respective hrCYP P450 isoenzyme flourogenic metabolites. Data was exported from the spectroflourimeter and analyzed using Excel spreadsheet and the % inhibition was calculated.(Crespi 1997)

Caco-2 permeability assay

Permeability studies are conducted with the Caco-2 monolayers cultured for 21days (TEER full form values >500 cm^2 in each well) and by adding an appropriate volume of buffer (HBSS buffer containing 10 mM HEPES) containing test compounds to apical chamber. Test Samples are taken from both apical and basolateral chambers at 0 and 90 min after incubation at 37 ^{0}C and analyzed by LC-MS/MS. Same experiment is repeated by adding an appropriate volume of buffer (HBSS buffer containing 10 mM HEPES) containing test compounds to basolateral chamber. The AUC defined the net influx and out flux of the test compound across the Caco-2 cell monolayer.

Pharmacokinetic studies

Oral and IV, pharmacokinetic (PK) studies of IIIM-290 was carried out in BALB/c male mice of age 4-6 weeks, For PO (10 mg/kg): 4.929 mg of IIIM-290 was wetted with ~30 µL of Tween-80 and triturated in a mortar and pestle, then slowly add 0.5% of methyl cellulose to make up the final volume up to 4.93 mL (in portions with the help of 1 mL pipette) with trituration. For IV (1.0 mg/Kg): 0.85 mg of IIIM-290 was dissolved in 425 μL of DMSO, to it add 425 μL of Solutol: absolute alcohol (1:1, v/v) and add normal saline to make up the final volume up to 8.50 mL (in portions with the help of 1 mL pipette) and mixed well. The formulations at dose of 10 mg/kg for oral and 1 mg/kg for IV administered. Plasma samples were collected at appropriate time points between the ranges of 0 hours to 24 hours and analyzed by LC-MS-MS. Mean plasma concentration calculated and data was further analysed for PK parameters evaluation using WinNonlin 5.3 software package.

CONCLUSION

From *D. binectariferum* bark, we have isolated rohitukine, rohitukine-N-oxide, dysoline, camptothecin and 9-methoxy camptothecin. Rohitukine was isolated in bulk quantity for semi-synthesis. Dysoline was a new regioisomer of rohitukine, characterized by extensive spectral studies. This compound has shown potent cytotoxicity in fibrosarcoma cell line (HT1080, IC_{50} = 210 nM) and significant anti-inflammatory properties as revealed by inhibition of pro-inflammatory cytokine (IL-6 and TNF-α) production. *D. binectariferum* was first time identified as an alternative source of an impotant anticancer drug camptothecin and 9-methoxy camptothecin. Furthermore, this is the first report of camptothecin isolation from Meliaceae family. A LC-MS based quantitative analysis revealed approximately 0.11% camptothecin in dry plant material (bark).

Rohitukine first time identified as Cdk inhibitor and medicinal chemistry effort on rohitukine led to identification of IIIM-290 as a promising candidate for preclinical development. Further, notable anticancer results were also obtained from *in-vivo* study in murine mice model, as well as good pharmacokinetic properties in BALB/c mice makes this molecule a suitable candidate for further development.

Abbreviations

ADME, absorption, distribution, metabolism, and excretion; AUC0t, the area under the plasma concentration"time curve from 0 to last measurable time point; AUC0 ", area under the plasma concentration" time curve from time zero to infinity; BALB/c mice, albino, laboratory-bred strain of the housemouse; Cmax, maximum observed plasma concentration; C0, extrapolated concentration at zero time point; CL, clearance; CLint, intrinsic clearance; CPT, camptothecin; CYP3A4, cytochrome P450 3A4; CYP2D6, cytochrome P450 2D6;CYP2C9, cytochrome P450 2C9; CYP2C19, cytochrome P4502C19; CPM, counts per minute; DABCO, 1,4-diazabicyclo[2.2.2]octane;DAPI, 42, 6-diamidino-2-phenylindole; DCM, dichloromethane; %F, percentage bioavailability; fR2, normalepithelial tissue; GST, glutathione S-transferase; hERG, humanether-a-go-go-related gene; HCT-116, human colon carcinomacell lines; HRMS, high resolution mass spectroscopy; HSA,human serum albumin; IR, infrared spectroscopy; MMC,mitomycin C; NAT, N-acetyltransferase; NADPH, nicotinamideadenine dinucleotide phosphate; PBS, phosphate buffersaline; PARP, poly(ADP-ribose) polymerase; PDB, ProteinData Bank; PRKCE, protein kinase C epsilon; PO, oral route;Ro5, rule of five; SD, standard deviation; SD rats, Sprague"Dawley rats; SGF, simulated gastric fluid; SIF, simulatedintestinal fluid; SULT, sulfotransferase; THF, tetrahydrofuran;TGI, tumor growth inhibition; Tlast, time at which lastconcentration was found; t1/2,β, terminal half-life; Vd, volume of distribution; Vdss, volume of distribution at steady state; XO,xanthine oxidase.

Competing Interest

No conflict of interest

LITERATURE CITED

Arisawa M, Gunasekera SP, Cordell GA, Farnsworth NR. 1981 Plant anticancer agents XXI. Constituents of Merrilliodendron megacarpum. *Planta Med.* 43: 404-407

Bharate SB, Kumar V, Jain SK, Mintoo MJ, Guru SK, Nuthakki VK, Sharma M, Bharate SS, Gandhi SG, Mondhe DM, Bhushan S, Vishwakarma RA. 2018 Discovery and preclinical development of IIIM-290, an orally active potent cyclin-dependent kinase inhibitor. *J. Med. Chem.* 61: 1664-1687.

Bible KC, Kaufmann SH. 1997 Cytotoxic synergy between flavopiridol (NSC 649890, L86-8275) and various antineoplastic agents: The importance of sequence of administration. *Cancer Res.* 57: 3375-3380.

Blagosklonny MV. 2004 Flavopiridol, an inhibitor of transcription. Implications, problems and solutions. *Cell Cycle* 3: 1537-1542.

Bradbury RH. 2010 Kinase inhibitors: Approved drugs and clinical candidates. In: Abraham DJ, Rotella DP (eds) Burger's medicinal chemistry, drug discovery, and development. John Wiley & Sons, Inc.

Brusselbach S, Nettelbeck DM, Sedlacek HH, Muller R. 1998 Cell cycle-independent induction of apoptosis by the anti-tumor drug flavopiridol in endothelial cells. *Int. J. Cancer* 77: 146-152.

Burdette-Radoux S, Tozer RG, Lohmann RC, Quirt I, Ernst DS, Walsh W. 2004 Phase II trial of flavopiridol, a cyclin dependent kinase inhibitor in untreated metastatic malignant melanoma. *Invest. New Drugs* 22: 315–322.

Byrd JC, Shinn C, Waselenko JK, Fuchs EJ, Lehman TA, Nguyen PL. 1998. Flavopiridol induces apoptosis in chronic lymphocytic leukemia cells via activation of caspase-3 without evidence of bcl-2 modulation or dependence on functional p53. *Blood* 92: 3804–3816.

Carlson BA, Dubay MM, Sausville EA, Brizuela L, Worland PJ. 1996 Flavopiridol induces G1 arrest with inhibition of cyclin-dependent kinase (CDK) 2 and CDK4 in human breast carcinoma cells. *Cancer Res.* 56: 2973 2978.

Christian BA, Grever MR, Byrd JC, Lin TS. 2007 Flavopiridol in the treatment of chronic lymphocytic leukemia. *Curr. Opin. Oncol.* 19: 573–578.

Crespi CL. 1997 Microplate plate assays for inhibition of human drug metabolizing cytochrome P450. *Anal. Biochem.* 248: 188-190.

Cui BS, Ma X-Q, Yang D-H, Hu X-Q, Cai S-Q. 2008 Purification method of rohitukine from the stem bark of *Dysoxylum binectariferum*. *J. Chin. Pharm. Sci.* 17: 153-157.

Dasmahapatra G, Almenara JA, Grant S. 2006 Flavopiridol and histone deacetylase inhibitors promote mitochondrial Injury and cell death in human leukemia cells that overexpress Bcl-2. *Mol. Pharmacol.* 69: 288-298.

De Azevedo WF, Jr., Mueller-Dieckmann HJ, Schulze-Gahmen U, Worland PJ, Sausville E, Kim SH. 1996 Structural basis for specificity and potency of a flavonoid inhibitor of human CDK2, a cell cycle kinase. *Proc. Natl. Acad. Sci. USA* 93: 2735-2740.

Domg-Hui Yang, Shao-Qing Cai, Zhao Y-Y, Liang H. 2004 A new alkaloid from *Dysoxylum binectariferum*. *J Asian Nat. Prod. Res.* 6.

Drees M, Dengler WA, Roth T, Labonte H, Mayo J, Malspeis L, Grever M, Sausville EA, Fiebig HH. 1997 Flavopiridol (L86-8275): selective antitumor activity in vitro and activity in vivo for prostate carcinoma cells. *Clin. Cancer Res.* 3: 273-279.

El-Rayes BF, Gadgeel S, Parchment R, Lorusso P, Philip PA. 2006 A phase I study of flavopiridol and docetaxel. Invest. *New Drugs* 24: 305–310.

Ezell E, Smith L. 1991 1H-and 13C-NMR spectra of camptothecin and derivatives. *J. Nat. Prod.* 54: 1645-1650.

Filgueira de Azevedo Jr W, Canduri F, Freitas da Silveira NJ. 2002 Structural basis for inhibition of cyclin-dependent kinase 9 by flavopiridol. *Biochem. Biophys. Res. Commun.* 293: 566-571.

Galons H, Oumata N, Meijer L. 2010 Cyclin-dependent kinase inhibitors: a survey of recent patent literature. Expert Opin. *Ther. Patents* 20: 377-404.

Geran RI, Greenberg NH, MacDonald MM, Schumacher AM, Abbott BJ. 1972 Protocols for screening chemical agents and natural products against animal tumors and other biological systems. *Cancer Chemother. Rep.* 3: 1-103.

Getman CR, Bible KC. 2004 Molecular targets of flavopiridol, a promising new antineoplastic agent. *Recent Res. Dev. Cancer* 6: 37-56.

Gojo I, Zhang B, Fenton RG. 2002 The cyclin-dependent kinase inhibitor flavopiridol induces apoptosis in multiple myeloma cells through transcriptional repression and down regulation of Mcl-1. *Clin. Cancer Res.* 8: 3527–3538.

Grendys Jr EC, Blessing JA, Burger R, Hoffman J. 2005 A phase II evaluation of flavopiridol as second-line chemotherapy of endometrial carcinoma: A gynecologic oncology group study. *Gynecol. Oncol.* 98: 249-253.

Gunasekera SP, Badawi MM, Cordell GA, Farnsworth NR, Chitnis M. 1979 Plant anticancer agents X. Isolation of camptothecin and 9-methoxycamptothecin from *Ervatamia heyneana*. *J. Nat. Prod* 42: 475–477.

Heikkila T, Karjalainen M, Ojala K, Partola K, Lammert F, Augustijns P, Urtti A, Yliperttula M, Peltonen L, Hirvonen J. 2011 Equilibrium drug solubility measurements in 96-well plates reveal similar drug solubilities in phosphate buffer pH 6.8 and human intestinal fluid. *Int. J. Pharm.* 405: 132-136.

Huwe A, Mazitschek R, Giannis A. 2003 Small molecules as inhibitors of cyclin-dependent kinases. *Angew. Chem. Int. Ed.* 42: 2122-2138.

Jain SK, Bharate SB, Vishwakarma RA. 2012a Cyclin-dependent kinase inhibition by flavoalkaloids. *Mini Rev. Med. Chem.* 12: 632-649.

Jain SK, Meena S, Singh B, Bharate JB, Joshi P, Singh VP, Vishwakarma RA, Bharate SB. 2012c KF/alumina catalyzed regioselective benzylation and benzoylation using solvent-free grind-stone chemistry. *RSC Adv.* 2: 8929-8933.

Kelland LR. 2000 Flavopiridol, the first cyclin-dependent kinase inhibitor to enter the clinic: current status. Expert Opin. Investig. *Drugs* 9: 2903-2911.

Kumar NA. 2009 Exploring the "Bio- Cultural" Heritage in Conservation of 5 Rare, Endemic & Threatened (RET) Tree Species of Western Ghats of Kerala. Saving Culture for Saving Biodiversity IUCN Organization, Gland, Switzerland.

Kumar V, Guru SK, Jain SK, Joshi P, Gandhi SG, Bharate SB, Bhushan S, Bharate SS, Vishwakarma RA. 2016 A chromatography-free isolation of rohitukine from leaves of *Dysoxylum binectariferum*: Evaluation for in vitro cytotoxicity, Cdk inhibition and physicochemical properties. *Bioorg. Med. Chem. Lett.* 26: 3457-3463.

Lapenna S, Giordano A. 2009 Cell cycle kinases as therapeutic targets for cancer. *Nat. Rev. Drug Discov.* 8: 547-566.

Llanos GG, Araujo LB, Jimenez IA, Moujir L, Bazzocchi IL. 2008 New cytotoxic withanolides from *Withania aristata*. *Planta Med.* 74: PB131.

Losiewicz MD, Carlson BA, Kaur G, Sausville EA, Worland PJ. 1994 Potent inhibition of CDC2 kinase activity by the flavonoid L86-8275. *Biochem. Biophys. Res. Commun.* 201: 589-595.

Melillo G, Sausville EA, Cloud K, Lahusen T, Varesio L, Senderowicz AM. 1999 Flavopiridol, a protein kinase inhibitor, down-regulates hypoxic induction of vascular endothelial growth factor expression in human monocytes. *Cancer Res.* 59: 5433–5437.

Michalides R, van Veelen N, Hart A, Loftus B, Wientjens E, Balm A. 1995 Overexpression of cyclin D1 correlates with recurrence in a group of forty-seven operable squamous cell carcinomas of the head and neck. *Cancer Res.* 55: 975-978.

Mohanakumara P, Sreejayan N, Priti V, Ramesha BT, Ravikanth G, Ganeshaiah KN, Vasudeva R, Mohan J, Santhoshkumar TR, Mishra PD, Ram V, Shaanker RU. 2009 *Dysoxylum binectariferum*Hook.f (Meliaceae), a rich source of rohitukine. *Fitoterapia* 81: 145-148.

Mordant P, Loriot Y, Leteur C, Calderaro J, Bourhis J, Wislez M, Soria J-C, Deutsch E. 2010 Dependence on phosphoinositide 3-kinase and RAS-RAF pathways drive the activity of RAF265, a novel RAF/VEGFR2 inhibitor, and RAD001 (everolimus) in combination. *Mol. Cancer Ther.* 9: 358-368.

Morgan DO. 1997 Cyclin-dependent kinases: engines, clocks, and microprocessors. Annu. Rev. Cell Dev. Biol. 13: 261-291.

Mosmann T. 1983 Rapid colorimetric assay for cellular growth and survival: Application to proliferation and cytotoxicity assays. *J. Immunol. Methods* 65: 55-63.

Motwani M, Rizzo C, Sirotnak F, She Y, Schwartz GK. 2003 Flavopiridol enhances the effects of docetaxel in vitro and in vivo in human gastric cancer cells. *Mol. Cancer Ther.* 2: 549–555

Naik RG, Kattige SL, Bhat SV, Alreja B, de Souza NJ, Rupp RH. 1988 An antiinflammatory cum immunomodulatory piperidinylbenzopyranone from *Dysoxylum binectariferum* : Isolation, structure and total synthesis. *Tetrahedron* 44: 2081-2086.

Raje N, Hideshima T, Mukherjee S, Raab M, Vallet S, Chhetri S, Cirstea D, Pozzi S, Mitsiades C, Rooney M, Kiziltepe T, Podar K, Okawa Y, Ikeda H, Carrasco R, Richardson PG, Chauhan D, Munshi NC, Sharma S, Parikh H, Chabner B, Scadden D, Anderson KC. 2009 Preclinical activity of P276-00, a novel small-molecule cyclin-dependent kinase inhibitor in the therapy of multiple myeloma. *Leukemia* 23: 961-970.

Ramesha BT, Amna T, Ravikanth G, Gunaga RP, Vasudeva, Ganeshaiah KN, Shaanker RU, Khajuria RK, Puri SC, Qazi GN. 2008 Prospecting for camptothecines from *Nothapodytes nimmoniana* in the western ghats, south india: identification of high-yielding sources of camptothecin and new families of camptothecines. *J. Chromat. Sci* 46: 362-368.

Rapella A, Negrioli A, Melillo G, Pastorino S, Varesio L, Bosco MC. 2002 Flavopiridol inhibits vascular endothelial growth factor production induced by hypoxia or picolinic acid in human neuroblastoma. *Int. J. Cancer* 99: 658–664.

Roy D, Ducher F, Laumain A, Legendre JY. 2001 Determination of the aqueous solubility of drugs using a convenient 96-well plate-based assay. *Drug Dev. Ind. Pharm.* 27: 107-109.

Sedlacek H, Czech J, Naik R, Kaur G, Worland P, Losiewicz M, Parker B, Carlson B, Smith A, Senderowicz A, Sausville E. 1996 Flavopiridol (L86 8275; NSC 649890), a new kinase inhibitor for tumor therapy. *Int. J. Oncol.* 9: 1143-1168.

Senderowicz AM. 1999 Flavopiridol: the first cyclin-dependent kinase inhibitor in human clinical trials. Invest. *New Drugs* 17: 313-320.

Senderowicz AM. 2001 Development of cyclin-dependent kinase modulators as novel therapeutic approaches for hematological malignancies. *Leukemia* 15: 1-9.

Singh IP, Jain SK, Kaur A, Singh S, Kumar R, Garg P, Sharma SS, Arora SK. 2010. Synthesis and antileishmanial activity of piperoyl-amino acid conjugates. *Eur. J. Med. Chem.* 45: 3439-3445.

Thomas JP, Tutsch KD, Cleary JF, Bailey HH, Arzoomanian R, Alberti D. 2002. Phase I clinical and pharmacokinetic trial of the cyclin-dependent kinase inhibitor flavopiridol. *Cancer Chemother. Pharmacol.* 50: 465–472.

Vesely J, Havlicek L, Strnad M, Blow JJ, Donella-Deana A, Pinna L, Letham DS, Kato JY, Deativaud L, Leclerc J, Meijer L. 1994. Inhibition of cyclin-dependent kinases by purine derivatives. *Eur. J. Biochem.* 224: 771-779.

Wall ME, Wani MC, Cook CE, Palmer KH, McPhail AT, Sim GA. 1966. Plant antitumor agents. . The isolation and structure of camptothecin, a novel alkaloidal leukemia and tumor inhibitor from *Camptotheca acuminata. J. Am. Chem. Soc.* 88: 3888-3890.

Wang LM, Ren DM. 2010. Flavopiridol, the first cyclin-dependent kinase inhibitor: recent advances in combination chemotherapy. *Mini-Rev. Med. Chem.* 10: 1058-1070.

Wittmann S, Bali P, Donapaty S, Nimmanapalli R, Guo F, Yamaguchi H, Huang M, Jove R, Wang HG, Bhalla K. 2003. Flavopiridol down-regulates antiapoptotic proteins and sensitizes human breast cancer cells to epothilone B-induced apoptosis. *Cancer Res.* 63: 93-99.

Wu K, Wang C, D'Amico M, Lee RJ, Albanese C, Pestell RG, Mani S. 2002. Flavopiridol and trastuzumab synergistically inhibit proliferation of breast cancer cells:

Association with selective cooperative inhibition of cyclin D1-dependent kinase and Akt signaling pathways. *Mol. Cancer Ther.* 1: 695-706.

Yang DH, Cai SQ, Zhao YY, Liang H. 2004. A new alkaloid from *Dysoxylum binectariferum*. *J. Asian Nat. Prod. Res.* 6: 233-236.

Zhang S, Ma J, Bao Y, Yang P, Zou L, Li K, Sun X. 2008. Nitrogen-containing flavonoid analogues as CDK1/cyclin B inhibitors: synthesis, SAR analysis, and biological activity. *Bioorg. Med. Chem.* 16: 7128-7133.

5

Zanthoxylum armatum DC., Perspectives of Biology and Chemistry in Medicine Discovery

Sushil Kumar[1] and Bikarma Singh[*]

Plant Sciences (Biodiversity and Applied Botany), CSIR-Indian Institute of Integrative Medicine, Jammu-180001, Jammu and Kashmir, INDIA
Academy of Scientific and Innovative Research, Anusandhan Bhawan New Delhi-110001, INDIA
[1]Department of Botany, Govt. Degree College, Ramnagar - 182122 Jammu and Kashmir INDIA
*Corresponding Email: drbikarma@iiim.res.in

ABSTRACT

Zanthoxylum armatum DC. is a plant species belonging to flowering family Rutaceae which has medicinal application in traditional Ayurvedic system of medicine. It is characterized by trifoliolate leaves with leaf-stalk winged, branches with straight prickles, and fruits are small reddish brown. In India, tribal community used this species as carminative, dyspepsia, tooth curing, treating malaria fever, stomach pain and expelling roundworms. Chemistry indicates that *Z. armatum* is a repository of many chemical compounds such as alkaloids, flavonoids, sterols and triterpenoids. Zanthonitrile, berberine, α-sitosterol, armatamide, limonene and linalool are major constituents. Leaves and young shoots yield essential oils which act as anti-bacterial, antifungal and anthelminthics, and has tremendous potential in development of value added products. Volatile oils are active constituents of linalool, limonene and ligan. *Z. armatum* has been declared as an endangered species as per International Union for Conservation of Nature and Natural Resources(IUCN) Red List category. Conservation practices needs to be adopted to conserve this valuable medicinal plant. There is a need to study on ecology, adaptive biology, chemistry and biotechnological intervention for future propagation and conservation of this medicinal plant species.

Keywords: Medicinal Plant, *Zanthoxylum armatum*, Traditional Knowledge, Product Development, Himalayas

INTRODUCTION

Zanthoxylum L. (Rutaceae) represented by 176 species (www. theplantlist. org) distributed across the globe in tropical, subtropical and temperate regions. Almost all the species within the genus has medicinal application in preparation and formulation of value-added products. *Zanthoxylum armatum* DC. is a medicinal bushy shrub, commonly known in Himalayan valleys as 'Timbru'. This species extends its distribution from Himachal Pradesh to Jammu and extending to hilly belts of Assam, Meghalaya and Arunachal Pradesh. This species occurs between the elevation range of 600 to 2000 m above mean sea level. Fruits, branches and thorns of this plant are used as carminative, in stomach pain and also act as remedy for tooth pain.

In Ayurveda, *Zanthoxulum armatum* DC. is used for the treatment of skin disease and abdominal pain. Himalayan tribes use leaves and fruits as mouth freshner and for curing tooth pain caused by cavity. Bark is used for intoxicating fishes, while the seeds of *Z. armatum* have a history of aplication in Indian system of medicine as disinfectant, antiseptic and for the treatment of fever, cholera and general debility (Annappana et al. 2015).

Chemically *Zanthoxylum armatum* consists of alkaloids, flavonoids, saponin, tannins, steroids, terpenes, glycosides, carbohydrates, phenolic, proteins and amino acids (Akhtar et al. 2009). Essential oil shows biological activities such as larvicidal, anti-inflammatory, anti-oxidant, anti-biotics, anthelminthic, anti-diabetic, anti-microbial, anti-spasmodic, anti-fungal, anti-viral and hepatoprotective activities (Dube et al. 1990, Annappan et al. 2015).

Indian Institute of Integrative Medicine, an establishment of Council of Scientific and Industrial Research (CSIR) has carried out extensive survey of *Z. armatum* covering far-flung areas of Kashmir Himalaya and Himachal Pradesh. Field studies were carried out in different seasons to study phenological phenophases. Ethnobotanical data were collected through different interview methods (participatory rural appraisal (PRA), direct observation, semi-structured interviews, individual discussions and questionaires. Field trips were undertaken for collection of plant samples along with the photographs and GPS coordinates. For review work, published literatures on *Z. armatum* across the globe were consulted and cited.

BIOLOGICAL ASPECTS

Systematic Position

- Kingdom: Plantae
- Division: Magnoliophyta (or Flowering plants)
- Class: Magnoliopsida (or Dicotyledons)
- Order: Sapindales
- Family: Rutaceae (or Rue family)
- Genus: *Zanthoxylum* L. (Pricklyash)
- Species: *Zanthoxylum armatum* DC.

Taxonomy

Large spiny bushy shrubs to small trees, deciduous or evergreen, grows upto 5 m tall. Leaves trifoliolate, rachis glabrous, pubescent, leaf-stalk winged. Leaflets stalkless, 2.5-7.5 cm long, 1.2-1.7 cm broad, elliptic to ovate-lanceolate, margin crenate, often revolute when dry, apex acute to acuminate. Inflorescences terminal on short lateral branchlets, occasionally axillary. Flowers 6-8; sepals acute; petals absent. Male flowers with 6-8 stamens, large anthers, yellowish. Female flowers 1-3 celled ovary, 3 mm in diameter, pale red, splitting into two when ripe. Seeds rounded, 3 mm in diameter, reddish brown.

Synonym

Fagara armata Thunb., *Zanthoxylum alatum* Roxb., *Zanthoxylum alatum* Hemsl., *Zanthoxylum alatum* var. *planispinum* (Siebold & Zucc.) Rehder & E.H. Wilson, *Zanthoxylum alatum* var. *subtrifoliolatum* Franch., *Zanthoxylum arenosum* Reeder & S.Y.Cheo, *Zanthoxylum armatum* var. *armatum*, *Zanthoxylum planispinum* Siebold & Zucc.

Common Names

- Andkaka Dhiva, Tejbatee or Tumburu in Sanskrit
- Chi-it, Sibit-paklauit in Philippines
- Faaghir, Kabaab-e-Khandaan in Unani
- Fagrieh in Arabic
- Gaira, Tejovati in Bengali
- Kabab-e-Khanda in Urdu

- Latin name-*Zanthoxylum armatum*
- Mak kak in Thailand
- Prumo, Prumu, Tebun, Tejbal, Tejphal, Timur and Yerma in Nepali
- Tejapatri in Kannada
- Tejbal in Gujrati
- Tejbal in Hindi
- Tejovati in Assamese
- Tejyovathi in Siddha
- Tezbal in Garhwal
- Tezbol in Oriya
- Timber' in Pahari.
- Timbru in Dogri (Jammu)
- Tirmira in Punjabi
- Toothache tree or 'Indian Prickly Ash' in English
- Tumburl in Telugu
- Tumpunal in Tamil

Diversity within Genus Zanthoxylum

Zanthoxylum L. is a genus under Rutaceae which has high medicinal, economical as well as ecological importance and reported to have wide distributon in tropical and temperate regions of the world (Negi et al. 2011). It is represented by 176 species (www. theplantlist. org). In India, 11 species of this genus reported distributed mainly in hilly tropical and subtropical regions, and out of these six species occur in Himalayan regions (Kala et al. 2005).

Occurrence and Distribution

Zanthoxylum is found in Jammu and Kashmir, Himachal Pradesh, Uttarakhand and Sikkim between an altitude range of 1000-2100 m above mean sea level. In Arunachal Pradesh and Khasi hills of Meghalaya, this genus is reported between an altitude range of 600-1800 m above mean sea level, whereas in Nagaland, Manipur, Eastern Ghats of Orissa and Andhra Pradesh at an altitude of 1200 m above mean sea level (Figure 1). It is also reported from Nilgiri of Tamil Nadu (ENVIS 2018). It has been reported that *Zanthoxylum armatum* is a native to North America, but globally extends its distribution to China,

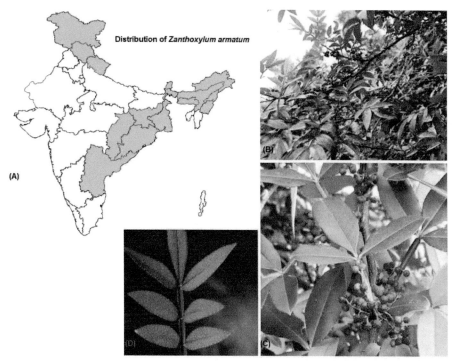

Fig. 1: Distribution of *Zanthoxylum armatum* in India

Japan, Bhutan, Bangladesh, Indonesia, Laos, Malaysia, Myanmar, Philippines, Thailand and Vietnam, where is grows between 1,300-1,500 m above mean sea level (https://www. bimbima.com/ayurveda).

Etymology

The name *Zanthoxylum* comes from the word 'Xanthoxylum' which is derived from the Greek word 'xanthon xylon' which means 'yellow wood'.

Phenology

Zanthoxylum armatum flowers in March-April, and fruits in June-July.

Cytology

Gametophytic count n=33 (Mehra and Khosla 1973).

Pollination and Propagation

Zanthoxylum armatum is insect pollinated plant species, and is usually propagated through seeds, although vegetative methods of propagation such as air layering and plant tissue culture have been developed (Purohit et al. 2014).

ETHNOBOTANICAL ASPECTS

Zanthoxylum armatum is known as an important medicinal plant as whole plant including stem, bark, fruit and seed possess medicinal properties and has been used as a medicine in both the traditional and Ayurvedic system of medicine since ancient time and for ages.

Traditional Usages of Fruits and Seeds

The bark, fruits and seeds of *Z.armatum* are extensively used in indigenous system of medicine as a carminative, stomachic and anthelmintic. The fruit and the seeds are employed as an aromatic tonic in fever and the dyspepsia. An extract of fruit is reported to be effective in expelling round worms. Because of their deodorant, disinfectant and antiseptic properties, fruits are used in dental troubles, and their lotion for scabies.

Seeds of timru are important ingredients to Zuroor-e-Qula (powdered polyherbal Unani formulation) which contain antimicrobial and anti-inflammatory activity. During winter, soup/decoction prepared from the dried fruits is consumed as it is believed to provide warmth to the body and relief from abdominal pain. Powdered fruit mixed with *Mentha longifolia* (Mints) and table salts are eaten with boiled egg is good for curing chest infection and digestive problems. Besides, pickles made from the fruits aids in relieving cold, cough, headache, tonsilitis, limbs numbness, and dizziness/vertigo. Due to their disinfectant and antiseptic properties, fruits are used to cure snake bites. Bacterial infections of the mouth, digestive system and urinary tract are successfully treated with timbru fruits.

In Ayurveda, the seeds of *Zanthoxylum armatum* are used in treatment of digestive problems, piles, heart diseases, hiccups, cough, throat disorders, asthma and dental diseases. In Unani system, fruits, seeds and bark are used as a carminative, stomachic and anthelmintic. (https://www.bimbima.com/ayurveda).

Bhotiya tribal people in Uttarakhand use timur fruit in the form of condiments, spices and medicines. The Bhotiya community brew liquor from timur, but the resulting liquor is palatable only to those highly addicted. Dried fruits (excluding seeds) are powdered and $1/4^{th}$ spoonful of powder is taken with a cup of boiled water to cure diarrhoea, dysentery, and stomachache. Powder of its dried fruit, *Mentha longifolia* dried leaves, *Trachyspermum ammi* seeds and black salts are taken with water during cholera and indigestion. Powder and decoction of fruits and seeds are used as piscicide, aromatic tonic in fever, dyspepsia and for expelling out roundworms from the body.

Usages of Stems

Young branches of *Z. armatum* are used as toothbrush (Miswak). Bark is utilized as traditional dye resource. An infusion in vinegar of stem is used to expel bugs or worms infecting ears.

Usages of Leaves

An infusion of the leaves is drunk to relieve stomach pains. A paste of leaves is applied externally to treat leucoderma.

Uses of Roots

Roots of *Z. armatum* are boiled in water for 20 minutes and the filtered water drunk as an anthelmintic.

Indiginous Usages of Resins

The resinous compounds contained in the bark, and especially in that of the roots, are powerfully stimulant and tonic substance. Significance and role of timbru in everyday life of local folks of Himalayas can be successfully illustrated in numerous works of folk poetry and songs. In Dogri (J&K State), there is a famous folk song dedicated to *Z. armatum* examples Asen kutti tez taraar chutney timbru di, naar-daana kane dhanya payaa, marchan paayiin char...chutney timbru di...

Ayurvedic Values and Action

Zanthoxylum armatum seeds are bitter and pungent in ras (taste) and dry, sharp and light in Guna (action). It taste after Vipak (digestion) is Katu or Pungent. It aggravates pitta and alleviates aggravated Vayu and Kapha. It cures krimi (parasitic infection), low appetite and durgandhya (foul smell coming out of the body). It mainly acts on excretory, circulatory, digestive and respiratory system. Some action reported for this species are Rasa (taste on tongue): Katu (Pungent) and Tikta (Bitter); Guna (Pharmacological Action): Laghu (Light), Tikshna (Sharp), Ruksha (Dry); Virya (Action): Ushna (Heating); Vipaka (transformed state after digestion): Katu (Pungent). Dipan (promote appetite but do not aid in digesting undigested food); Krimighna (destroys worms); Kapha–Vata har (Remover of the Humor of Kapha-Vata); Pachan (assist in digesting undigested food, but do not increase the appetite); Ruchikarak (improve taste) ((https://www.bimbima.com/ayurveda).

Phytochemistry

Phytochemical investigations have resulted in the isolation of a wide range of chemical compounds including alkaloids, tannins, triterpenes, tetranortriterpenoids, and steroids from different parts of this plant. Most of the natural products isolated were either from alcoholic or hydro-alcoholic extracts. An ample range of compounds have been reported from *Zanthoxylum armatum*. Chemical constituents in *Zanthoxylum* includes aramide, asarin, fargesin, α-amyrins, β-amyrins, lupeol, fargesin, armantamide, zanthonitrile, berberine, L-asarin, L-sesamin, L-planinin, β-sitosterol-β-D-glucoside, kaempferol, bergapten, umbelliferone, β-daucosterol, L-planinin, vanilic acid, 1-linoleo-2,3-diolein, α-amyrin acetate, armatonaphthyl arabinoside, limonene, terpineol, camphor, carvone, 1,8-cineole, cis-ocimene, limonene, linalool, myrcene, nerol, piperitone and many more chemical constituents of fine chemicals (Table 1).

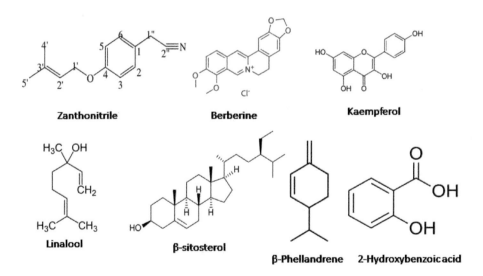

Fig. 2: Major chemical constituents of *Zanthoxylum armatum*

Table 1: Phytochemicals isolated from different parts of *Zanthoxylum armatum*

Part used	Compounds	References
Bark	Aramide, asarin, fargesin, α-amyrins, β-amyrins, lupeol, fargesin and armantamide	Kalia et al. (1999)
	Zanthonitrile	Read (1936)
	Berberine	Perry LM (1980)
	L-asarin, L-sesamin and L-planinin	Rao and Singh (1984)
	Berberine, β-Sitosterol-β-D-glucoside and asarinin	Ranawat et al. (2010)
	Zanthonitrile, kaempferol, bergapten and umbelliferone	Li et al. (2006).
	β-daucosterol, β-sitosterol L-asarinin, L-sesamin, L-planinin and vanilic acid	Li et al. (1996)
	1-linoleo-2,3-diolein, α-amyrin acetate and armatonaphthyl arabinoside	Agnihotri et al. (2017)
	β-sitosterol, armatamide, limonene and eudesmin, N-(3′,4′-methylene dioxy phenylethyl)-3,4-methylene dioxy cinnamoyl amide	Kumar et al. (2012)
Seed	α-Fenchol, α -thujene, α -terpineol, camphor, carvone, 1,8 cineole, cis-ocimene, ã-terpinene, limonene and linalool	Ahmad et al. (1993)
	α-Terpinene, α -thujone, α-pinene, α-terpineol, β-pinene, 1,8 cineole, geraniol, linalool oxide, limonene, linalool, myrcene, nerol, piperitone, sabinene, terpinen-4-ol, E-nerolidol, phellandral and oleic acid.	Tiwary et al. (2007)
	β-Phellandrene, limonene and linalool	Neetu et al. (2001)
	Citronellol, citronellal, linalool, methyl cinnamate, 8-methylnonanoic acid and 6-methylheptanoic	Yoshihito et al. (2000)
	Linalool, 1-α-phellandrene	Perry (1980)
	Stigmasta-5-en-3β-D-glucopyranoside, 1-Hydroxy -6,13-anthraquinone, 2-Hydroxybenzoic acid, 2-Hydroxy-4-methoxy, 3-Methoxy-11-hydroxy 6, 8-dimethylcarboxylate biphenyl, 3,5,6,7 Tetrahydroxy 3′, 4′-dimethoxyflavone-5-β-D-xylopyranoside, 5′-demethoxyepiexcelsin and epiaschantin	Akhtar et al. (2009)

Contd.

	(E)-methyl cinnamate, ethyl 9-hexadecenoate and ethyl hexadecanoate and linalool	Ramidi and Ali (1998)
	cis-10-Octadecenoic acid	Venkatachalam et al. (1996)
	Hydroxyalk-(4Z)-enoic acid, 8-hydroxy pentadec-(4Z)-enoic acid, 7-hydroxy-7-vinylhexadec(4Z)-enoic acid and hexadec-(4Z)-enoic acid	Ateeque et al. (1993)
Fruit	2α-Methyl-2β-ethylene-3β-isopropyl-cyclohexan-1β, 3α-diol; phenol-O-β-D-arabinopyranosyl-42 -(33 , 73 , 113, 153 -tetramethyl)-hexadecan-13 -oate; m-methoxy palmityloxy benzene, acetyl phenyl acetate, linoleiyl-O-α-D-xylopyranoside, m-hydroxyphenoxy benzene and palmitic acid	Nooreen et al. (2017)
	Eucalyptol, linalool, quercivorol, menthoglycol, trans-pipertiol, carveole, exo-2-hydroxy-1,8 cineole, theaspirane, piperitone, cuminol, artimisia alcohol, caryophyllene, α –asarone, β- asarone, 2-Hydroxy cylclopenta decanone, palmitic acid and 2-hydroxy cyclopentadecanone	Kayat et al. (2016)
Root and stem	L-Asarinin, sesamin, fargesin and eudesmin, sitosterol, syringaresinol, isodecaline and 4- hydroxybenzoic acid	Guo et al. (2012)
Leaves	Non-terpenic acyclic ketones and 2-undecanone	Singh et al. (2013)

PHARMACOLOGY ASPECTS

Anti-diabetic Activity

The stem has been shown to possess hypoglycemic activities. According to Karki et al. (2014) hydromethanolic extract of the bark of Z. armatum (HEBZA) was evaluated for its antidiabetic and antioxidant activity in streptozotocin-induced diabetic rats. Oral administration of HEBZA for 21 days (200 and 400 mg/kg) resulted in significant reduction in blood glucose, total cholesterol, triglycerides, low density lipoprotein, very low density lipoprotein and significant increase in high density lipoprotein and body weight of streptozotocin diabetic rats. In addition to that, significant decrease in lipid peroxidation and increase in catalase, superoxide dismutase and reduced glutathione were observed in streptozotocin diabetic rats.

Insecticidal and Anti-microbial Activity

Tiwary et al. (2001) have observed the larvicidal activity of *Zanthoxylum armatum* essential oil against three mosquito species, *Culex quinquefasciatus* was the most sensitive (LC50 = 49 ppm) followed by *Aedes aegypti* (LC50 = 54 ppm) and *Anopheles stephensi* (LC50 = 58 ppm). Of the four tested plants namely, *Artemisia vulgaris*, *Brugmansia arborea*, *Nicotiana tobacum* and *Zanthoxylum alatum* for their feeding deterrent and repellent activity against *Pieris brassiceae* larvae, *Z. alatum* was found to be the most potent (Paul and Sohkhlet 2012). The essential oil of fruits of *Z. armatum* have shown repellent activity to the insect *Allacophora foveicollis* and fungistatic to 24 fungi, including aflatoxin-producing strains of *Aspergillus flavus* and *A. parasiticus* at a minimum dose of $2 \cdot 0 \times 10^3$ µl (Dube et al. 1990). *Z. armatum* oil at 0.57 mg/cm^2 concentration gave significantly higher protection against mosquitoes both in mustard (445 min) as well as coconut oil (404 min) than the other repellents that were *Curcuma aromatica* (Jungli haldi) and *Azadirachta indica* (Neem) oils and Dimethyl Phthalate (DMP) (Das et al. 1999). *Z. armatum* oil was found at par with Citronyl and exhibited better results than dimethyl phthalate (DMP) and N-benzoyl piperidine (NBP) against leeches (Nath et al. 1993).

It has been found active against *Bacillus subtilis*, *E.oli* (Kalia et al. 1999, Bachwani et al. 2012). Two compounds N-(31,41-methylenedi-oxyphenylethyl)-3,4 methylenedioxy cinnamoyl amide and N-(31,41-dimethoxyphenylethyl)-3,4-methylenedioxy dihydro cinnamoyl amide exhibited moderate to potent activity against gram positive and gram negative bacteria when compared to their respective standards (Fluconazole and Benzyl penicillin) (Siddhanadham et al. 2017). The fruit extract of *Z. armatum* has been tested against *S. aureus*, *E. coli*, *Pseudomonas aeruginosa and Shigella boydii* for their antibacterial activity. The ethanolic extract was found to be inactive against *P. aeruginosa*, while it showed positive activity on the other three strains (Panthi and Chaudhary 2006). The essential oils of *Z. armatum* have been used widely for larvicidal and insecticidal activities. Further studies on other secondary metabolites mainly of alkaloids and individual terpenoids need to be examined, to explore further findings.

Anti-inflammatory Activity

Bose et al. (2009) have found that Bergapten, a coumarin extracted from *Z. armatum* exhibited significant inhibition to the production of pro-inflammatory cytokines, namely tumour necrotic factor-α (TNF-α) and interleukin-6 (IL-6) by PBMCs stimulated with lipopolysaccharide in a concentration dependent manner. Also, linalool and linalyl acetate are known to acquire inflammatory activity (Peana et al. 2002). The ethyl acetate fraction of the alcoholic extract of the root and stem of *Z. armatum* exhibit anti-inflammatory. The HPLC analysis of this fraction eight lignans eudesmin, fargesin, horsfieldin, kobusin, sesamin, asarinin, planispine A and pinoresinol-di-3,3-dimethylallyl were reported as the major components of this fraction (Guo et al. 2011).

Anti-cancerous Activity

Lupeol, a monoterpene found in timbru extracts, act as therapeutic and chemopreventive agent for the treatment of inflammation and cancer. The extract obtained from *Z. alatum* barks has been tested for anti-proliferative effect in keratinocytes (HaCaT cells) and it was found that the extract was highly active with an IC_{50} value of 11µg/mL (Kumar and Muller 1999).

Anti-oxidant Activity

The extracts obtained from leaves of *Z. armatum* exhibited antioxidant activities in a dose dependent manner. The IC_{50} values of methanol, chloroform, petroleum ether extracts and standard for DPPH were 3.63 ± 0.14 µg/mL, 9.33 ± 0.42 µg/mL, 7.70 ± 0.31 µg/mL and 7.87 ± 0.39 µg/mL; for Superoxide are 19.80 ± 0.96 µg/mL, 129.70 ± 2.57 µg/mL, 168.10 ± 2.26 µg/mL and 11.76 ± 0.72 µg/mL; for Nitric oxide are 105.0 ± 1.64 µg/mL, 157.60 ± 1.99 µg/mL, 185.30 ± 2.48 µg/mL and 28.48 ± 1.06 µg/mL; for Hydroxyl radical are 28.10 ± 0.75 µg/mL, 81.50 ± 1.88 µg/mL, 72.47 ± 1.76 µg/mL and 12.24 ± 0.82 µg/mL respectively (Karmakar et al. 2015).

COMMERCIAL ASPECTS

As a high value medicinal plant, various pharmaceutical companies use timbru fruit for making polyherbal unani and ayurvedic formulations (Tejowatyadya Grita, Tumbawardi Churna) and in different types of toothpaste. Due to its appealing aroma and valuable perfume, timbru is used in the manufacture of several health-care and cosmetic products. With its increasing commercial value, it has become a profitable non-timber forest product (NTFP). In the year 2000, the price in the local market was Rs 45 per kg, whereas the prices of timbru in the plains during the same year increased to Rs 150-200 per kg.

THREAT PERSPECTIVES

Population Status and Threat Perspectives

During the Biodiversity Conservation Prioritisation Project (BCPP), India undertook a prioritisation exercise for species, sites and strategies for conservation and management of biodiversity. The endangered species subgroup selected the conservation assessment and management plan process (CAMP) and the IUCN Red List Criteria (Revised 1994) for assessing the conservation status of species. According to recent population survey, Z. *armatum* is a nationally endangered species having restricted and fragmented range and there is decline in population of 70% in last 10 years. Climate change posses new threats to species loss, and there is urgent need to think of new approaches to measure the plant responses of Z. *armatum*, to detect potentially threatened status. Factors causing threats to this species is discussed below in subheads:

Habitat Loss and Fragmentation

Increasing human populations and modern development have brought about sudden, irreversible and often far-reaching disturbances of natural conditions essential for the survival of vast numbers of plant and animal species. For wild medicinal and aromatic plants like *Zanthoxylum* species, habitat loss and rapid habitat degradations is the primary cause of species loss at local, regional and global scales. The other treats such as urban development, water development, road building, recreation, fire-raising, fire-suppresion, agriculture and tree logging all destroy and degrade natural habitats of similar species. Habitat loss and degradation is harmful not only to a single species, but poses threats to the whole communities and related ecosystems because it fragments the remaining habitat, and the edges of habitats are strongly affected by their surrounding matrix.

Over-exploitation

The over-exploitation of species by human-being is the most significant cause of species disappearance. A certain amount of species use is sustainable, as plants populations will grow to replenish the stock taken. However, in many cases too much is taken, leaving the resource un-able to regenerate fast enough. Sometimes, the whole individuals are taken (as when logging for timber) but in other cases just parts of plants are used (when harvesting leaves for a herbal medicine). *Zanthoxylum armatum* has heavy demand in Ayurvedic system of medicine and applications in pharmaceutical industry for its medicinal and aromatic properties.

Exotic and Alien Species

Alien and exotic species are predators or pests that normally keep the species population in check. Biological trait that predisposes it to fast colonisation, such as a fast growth rate and the production of many seeds are common in himalayan climate. The rapid growth of floras like *Eupatorium adenophorum, Lantana camera, Artemisia species, Parthenium hysterophorum, Prosopis juliphera,* and *Argemone maxicana* in the natural Habitat of *Z. armatum* adversely affects the growth and population of this species and impose greater threat to their ecosystem. The problems caused by invasives can interact with other threats like climate change, species over-exploitation, habitat loss and pollution to cause catastrophic declines in biodiversity and ecosystem and all combine together contribute to the rapid decline and possible extinction of this species and other species elsewhere of the same habitat.

Climate Change

The distribution of different plant species, associations and vegetation types is controlled by a number of different climatic factors (such as annual and seasonal temperature, annual and seasonal precipitation, atmospheric CO_2 concentration) and their interactions. Therefore, the drastic variation in climatic factors responsible for the growth of *Zanthoxylum armatum* in the growing areas may affect the growth of species and cause species disappearance from their natural habitat.

CONSERVATION ASPECTS

Ex-Situ Conservation

Ex-situ conservation of *Z. armatum* means the conservation of population diversity outside their natural habitat to safeguard identified families or individual plant species from danger or loss. It includes collection, maintenance and conservation of sample organisms usually in the form of live whole plants, seeds, pollens, spores, vegetative propagules, tissue or cell cultures or other genetic material of growing or preserved individuals. Several options under *ex-situ* protection are available for conservation of *Zanthoxylum armatum,* such as botanical gardens, seed banks/ gene and DNA banks, captive cultivation in orchards/agricultural and forestry bodies, tissue culture and cryopreservation.

Depending upon the plant systems, a researcher has to select a cost-effective and appropriate plan. Ex-situ conservation procedures has an imperative role in the protection of rare plants and are source of fundamental research, education and publicity. Simultaneously, they offer to supply plant propagules to return to the wild as part of any future recovery endeavors. Ex-situ collections of *Zanthoxylum armatum* will be dependent on better characterization of existing collections and implementation of comprehensive sampling protocols. More integration of the activities undertaken by botanic gardens and gene banks is needed to ensure that shared priorities can be developed, and experiences, resources and technologies shared.

In-Situ Conservation

In-situ maintenance of *Zanthoxylum armatum* through the establishment of conservation and multiple-use areas offers distinct advantages over off-site methods in terms of coverage, viability of the resource, and the economic sustainability. It involves the natural habitat management to direct manipulation of the populations of *Zanthoxylum armatum*. The planting of this species in natural habitat could be done either by sowing of seeds or through vegetative propagules depending on the plant systems. It has to be kept in mind that the selected protected area will cover only a small portion of the total diversity of a rescued rare plant. Hence several areas will have to be conserved for a single species under *in situ* conservation.

Germplasm Enhancement

With *in-vitro* techniques, it is possible to provide a germplasm storage procedure which uniquely combines the possibilities of disease elimination and rapid clonal propagation. Further, the virus-tested cultures could provide an ideal material for international exchange and distribution of germplasm as they would be acceptable to plant quarantine authorities and comply with international quarantine regulations. Two basic approaches are followed to maintain germplasm collections in-vitro: (i) minimal growth, and (ii) cryopreservation. Minimal growth conditions for short to medium term storage can be followed in several ways - reduced temperature and/or light; incorporation of sub-lethal levels of growth retardants, induction of osmotic stress with sucrose or mannitol, and maintenance of cultures at a reduced nutritional status particularly reduced carbon, reduction of gas pressure over the cultures, desiccation and mineral oil overlay. The advantage of this approach is that cultures can be readily brought back to normal culture conditions to produce plants on demand.

However, the need for frequent subculturing may pose a great disadvantage including contamination of cultures as well as imposition of selection pressure with subsequent change in genetic make-up due to the somaclonal variation. Cryopreservation at the temperature of liquid nitrogen (-196°C) offers the possibility for long-term storage with the maximal phenotypic and genotypic stability. This method being relatively convenient and economical, and a large number of genotypes and variants could be conserved, thus maximize the potential for storage of genetically desirable material.

The development of new techniques for plant germplasm conservation offer new options and permit conservation of diversity in the form of seeds, pollens embryos and in-vitro cultures. In-vitro conservation using meristem and shoot tip cultures under minimal media, low temperatures and low light intensity provides adequate methods for germplasm conservation. Cryopreservation of seeds, pollens and in-vitro cultures put forward a unique method for preservation and use of germplasm. These can be especially rewarding for vegetatively propagated crops, plant species with recalcitrant seeds, wild and endangered species and also wild relatives of crop plants. The rapid strides made in the past few years with regard to these novel approaches have enhanced the value of gene banks and clonal repositories.

AYURVEDIC FORMULATION FOR COMMERCIAL VALUES

- Hingvadi Taila
- Madusnhi rasayan
- Dantmanjan

Seeds can be consumed in dose of 2-4 grams, 5-16 years: ½ adult usage levels, 1-5 years: ¼ adult usage levels, Stem bark 10-20 grams for preparation of decoction.

CONCLUSION

It is a matter of concern that *Zanthoxylum armatum* has been declared as an endangered species as per International Union for Conservation of Nature (IUCN) red list category. Therefore conservation practices needs to be adopted to conserve this valuable medicinal plant. There is a need of study on ecology, adaptive biology, chemistry and biotechnological intervention for conservation of this and other econominal plant species to some future of mankind.

ACKNOWLEDGEMENT

We thank the local people for their valuable information on indeginous uses presented in this research article.

LITERATURES CITED

Agnihotri S, Wakode S, Ali. 2017. Chemical constituents isolated from *Zanthoxylum armatum* stem bark. *Chemistry of Natural Compound* 53: 880-882.

Ahmad A, Misra LN, Gupta MM. 1993. Hydroxyalk-(4Z)-enoic acids and volatile components from the seeds of *Zanthoxylum armatum*. *Journal of Natural Product* 56: 456-460.

Akhtar N, Ali M, Alam MS. 2009. Chemical constituents from the seeds of *Zanthoxylum alatum*. *Journal of Asian Natural Product Research* 11: 91-95.

Annappan U, Rajkishore VB, Ramalingam R. 2015. A review of *Zanthoxylum alatum*: A review. *Research Journal of Pharmacognosy and Phytochemistry* 7(4): 223-226.

Ateeque A, Laxmi N, Massa, Madan MG. 1993. Hydroxyalk-(4z)-enoic acids and volatile components from the seeds of *Zanthoxylum armatum*. *Journal of Natural Products* 56(4): 456-460.

Bachwani M, Shrivastav B, Sharma V, Khandelwal R, Tomar L. 2012. An update review on Zanthoxylum armatum DC. *American Journal of Pharmaceutical Technical Research* 2(1): 274-286.

Bose SK, Dewanjee S, Sahu R, Dey SP. 2009. Effect of bergapten from *Heracleum nepalense* root on production of proinflammatory cytokines. *Natual Product Research* 1-6.

Brijwal L, Pandey A, Tamta S. 2013. An overview on phytomedicinal approaches of Zanthoxylum armatum DC.: An important magical medicinal plant. *Journal of Medicinal Plants Research* 7(8): 366-370.

Chinese Medicinal Plants of the Pen. 1936. Ts'ao Kang Mu, Peking, by BE Read, 3rd Edn, Peking Natural History Bulletin, China, p. 358.

Das NG, Nath DR, Baruah I, Talukdar PK, Das SC. 1999. Field evaluation of herbal mosquito repellents. *Phytotherapy Research* 13: 214-217.

Dube S, Kumar A, Tripathi SC. 1990. Antifungal and Insect-Repellent Activity of Essential Oil of *Zanthoxylum alatum*. *Annals of Botany* 65: 457-459.

Gaur RD. 2007. Traditional day yielding plants of Uttarakhand, India.

Guo T, Xie H, Deng YX, Pan SL. 2012. A new lignan and other constituents from *Zanthoxylum armatum* DC. *Natural Product Research* 26 (9): 859-864.

Gupta D, Sharma YP. 2017. Vanishing Timbru (http://www.dailyexcelsior.com/vanishing -timbru/).

Kala CP, Farooquee NA, Dhar U. 2005. Traditional Uses and Conservation of Timur (*Zanthoxylum armatm* DC.) through Social Institutions in Uttaranchal Himalaya, India. *Conservation and Society* 3: 224-230.

Kalia NK, Singh B, Sood RP. 1999. A new amide from *Zanthoxylum armatum*. *Journal of Natural Product* 62: 311-312.

Karki H, Upadhaya K, Pal H, Singh R (2014). Antidiabetic potential of *Zanthoxylum armatum* bark extract on streptozotocin induced diabetic rats. *International Journal of Green Pharmacy* 77-83. DOI: 10.4103/0973-8258.129568.

Karmakar I, Haldar S, Chakraborty M, Dewanjee S, Haldar PK. 2015. Antioxidant and Cytotoxic Activity of Different Extracts of *Zanthoxylum alatum*. *Free Radicals and Antioxidants* 5 (1): 21-28.

Kayat HP, Gautam SD, Jha RN. 2016. GC-MS analysis of hexane extract of *Zanthoxylum armatum* DC. *Fruits Journal of Pharmacognosy and Phytochemistry* 5(2): 58-62.

Kumar GVP, Divya R, Pratyusha V, Monica G. 2012. Phytochemical investigations on bark and invitro microbial assay of *Zanthoxylum armatum* DC. *Asian Journal of Research Chemistry* 5(6): 730-732.

Kumar S, Müller K. 1999. Inhibition of keratinocyte growth by different Nepalese *Zanthoxylum species*. *Phytotherapy Research* 13(3): 214–217.

Li H, Li P, Zhu L, Xie M, Wu Z. 2006. Studies on the Chemical Constituents of *Zanthoxylum armatum* DC. *Zhongguo Yaofang (Chinese Pharmacies)* 17: 1035-1037.

Li X, Li Z, Zheng Q, Cui T, Zhu W and Tu Z. 1996. Studies on the chemical constituents of Zanthoxylum armatum DC., Tianran Chanwu Yanjiu Yu Kaifa. *Natural Product Research and Development* 8: 24-27.

Manandhar NE. 2002. Plants and people of Nepal. Timber Press, Portland, Oregon. pp. 599.

Mehra PN, Khosla PK. 1973. Cytological studies of Himalayan Rubiaceae. *Silvae Genetics* 22: 182–188.

Nath DR, Das NG, Das SC. 1993. Persistence of leech repellent on cloth. *Indian Journal of Medical Research Section* 97: 128-131.

Neetu J, Srivastava SK, Aggarwal KK. 2001. Essential oil composition of Zanthoxylum alatum seeds from northern India. *Flavour and Fragrance Journal* 16: 408-410.

Negi JS, Bisht VK, Bhandari AK, Singh P, Sundriyal RC. 2001. Chemical constituents and biological activities of the genus Zanthoxylum: A review. *African Journal of Pure and Applied Chemistry* 5(12): 412-416.

Panthi MP, Chaudhary RP. 2006. Antibacterial Activity of Some Selected Folklore Medicinal Plants from West Nepal. *Scientific World* 4(4): 16-21.

Paridhavi M, Agrawal SS. 2007. Safety evaluation of a polyherbal formulation, Zuroor-e-Qula. *Natural Product and Radiance* 6(4):286-289.

Paul D, Sohkhlet M. 2012. Anti-feedant, repellent and growth regulatory effects of four plant extracts on Pieris brassicae larvae (Lepidoptera: Pieridae). *Journal of Entomological Research* 36(4): 287-293.

Peana AT, D'Aquila PS, Panin F, Serra G, Pippia P, Moretti MD. 2002. Anti-inflammatory activity of linalool and linalyl acetate constituents of essential oils. *Phytomedicine* 9: 721-726.

Perry LM. 1980. Medicinal plants of East and Southeast Asia, Massachusetts Institute of Technology, USA. Cambridge MIT Press.

Purohit S, Jugran A, Shyamal N, Bhatt I, Palni L, Bhatt A. 2014. Micropropagation of *Zanthoxylum armatum* DC-an Endangered Medicinal Plant of the Himalayan Region, and Assessment of Genetic Stability of in Vitro raised plants. *In Vitro Cellular and Developmental Biology* 50.

Ramidi R, Ali M. 1998. Chemical composition of the seed oil of *Zanthoxylum alatum* Roxb. *Journal of Essential Oil Research* 10: 127-130.

Ranawat J, Bhatt J, Patel J. 2010. Hepatoprotective activity of ethanolic extracts of bark of *Zanthoxylum armatum* DC. in CCl4 induced hepatic damage in rats. *Journal of Ethnopharmacology* 127: 777-780.

Rao GP, Singh SB. 1994. Efficacy of geraniol extracted from the essential oil of *Zanthoxylum alatum* as a fungitoxicant and insect repellent. *Sugarcane* 4: 16-20.

Siddhanadham AS, Prasad R, Prava R, Sama, JR and Koduru A. 2017. Isolation, characterization and biological evaluation of two new lignans from methanolic extract of bark of *Zanthoxylum armatum*. *International Journal of Pharmacognosy and Phytochemical Research* 9(3): 395-399.

Singh P, Mohan P, Gupta S. 2013. Volatile constituents from the leaves of *Zanthoxylum armatum* DC., a new source of 2-Undecanone. *Plant Sciences Research* 5: 1-3.

Singh TP, Singh OM. 2011. Phytochemical and pharmacological profile of *Zanthoxylum armatum* DC- An overview. *Indian Journal of Natural Products and Resources* 2 (3): 275-285.

Tiwary M, Naik SN, Tewary D, Mittal PK, Yadav S. 2007. Chemical composition and larvicidal activities of the essential oil of *Zanthoxylum armatum* DC (Rutaceae) against three mosquito vectors. *Journal of Vector Borne Diseases* 44: 198–204.

Useful tropical plants (http://tropical.theferns. info/viewtropical. php?id= Zanthoxylum +armatum accessed on 01/03/2018).

Venkatachalam SR, Hassrajani SA, Rane SS. 1996. cis-10-octadecenoic acid, component of *Zanthoxylum alatum* seed Oil. *Indian Journal of Chemistry Section B* 35: 514-517.

Yoshihito U, Yuriko N, Masayoshi H, Shuichi H and Seiji H. 2000. Essential Oil Constituents of Fuyu-sanshoo (*Zanthoxylum armatum* DC.) in Nepal. http:// jglobal. jst.go.jp/en/public.

Zulfa N, Kumar A, Bawankule DU, Tandon S, Ali M, Xuan TD, Ahmad A. 2017. New chemical constituents from the fruits of *Zanthoxylum armatum* and its *in vitro* anti-inflammatory profit. *Natural Product Research Letters*. DOI:https://doi.org/10.6084/ m9.figshare.5624560.v1.

6

Grewia asiatica L., an Important Plant of Shivalik Hills: Agrotechnology and Product Development

Rajendra Gochar[1], Bikarma Singh[2,3,], Mamta Gochar[1]*
Rajendra Bhanwaria[1,3] and Anil Kumar Katare[4]

[1]*Genetic Resources and Agrotechnology Division, CSIR-Indian Institute of Integrative Medicine, Jammu 180001, Jammu & Kashmir, INDIA*
[2]*Plant Sciences (Biodiversity and Applied Botany Division), CSIR-Indian Institute of Integrative Medicine, Jammu 180001, Jammu & Kashmir, INDIA*
[3]*Academy of Scientific and Innovative Research, Anusandhan Bhawan New Delhi - 110001, INDIA*
[4]*cGMP Division, CSIR-Indian Institute of Integrative Medicine, Jammu - 180001 Jammu & Kashmir, INDIA*
Corresponding author: drbikarma@iiim.ac.in; drbikarma@iiim.res.in

ABSTRACT

Grewia asiatica L. locally called 'Phalsa' (Family Tiliaceae) is a multipurpose gregarious plants species native to Asian belts, grows profusely in tropical and subtropical forests. Plant yield berries rich in calories, proteins, carbohydrates, dietary fibers, vitamins, macro- and micro-nutrients such as calcium, phosphorus and sodium. Their parts can be used as herbal medicine for treatment of various diseases. Recently, CSIR-Indian Institute of Integrative Medicine Jammu (IIIM) has developed and released a variety of this plant for rural upliftment and for development of value-added products such health drink and nutraceuticals. From its matured fruits of cultivated variety, a value added product called Shivalik Health Drink was developed by IIIM. For extension, agrotechnology of this plant has been transferred to various geographic regions of India such as Punjab, Rajasthan, Jammu & Kashmir, Uttar Pradesh, Maharashtra and Andhra Pradesh. This species prefers hardy and can be able to withstand in drought environment. For development of drugs and value added products, several therapeutic research works have been carried out which suggested that this plant species possess antioxidant, antidiabetic, antimicrobial, hepatoprotective, antifertility, antifungal, analgesic and antiviral activities. Further, there is a need to focus on biotechnological intervention with aim to develop seedless varieties and new nutraceutical products.

Keywords: Nutraceuticals, *Grewia asiatica*, Health drink, Product development, Shivalik Himalaya.

INTRODUCTION

Nutraceuticals provides medicine and health benefits to humans and animals. It shows advantage over medicine as they avoid side-effect, easily available, economically affordable and rich in composition of natural dietary supplements. Herbal nutraceuticals used as a powerful source for maintaining health and to act against nutritionally induced acute and chronic diseases, thereby promoting optimal health, longevity, and quality of human life. The food sources used as nutraceuticals are natural and can be dietary fibers, prebiotics, probiotics, polyunsaturated fatty acids, vitamins, polyphenols and spices.

Grewia asiatica L., locally called 'Phalsa' (family Tiliaceae) is a multipurpose gregarious plant species native to Asian belts, grows profusely in tropical and subtropical forests. Plant yield berries rich in calories, proteins, carbohydrates, dietary fibers, vitamins, macro- and micro-nutrients such as calcium, phosphorus and sodium elements.

Considering the importance of *G. asiatica* in Shivalik range, major initiative has been taken by CSIR-Indian Institute of Integrative Medicine, Jammu for conservation of this valuable plant. Lots of scientific data published on agrotechnology of this species have been established. Major extension activities has been undertaken for agrotechnology transfer and benefited lots of rural community under major projects such as J&K Aroma Arogya Gram Yojna (JAAG), end to end technology of Phalsa and CSIR-800 for rural prosperity. A value added product is developed by CSIR-IIIM in the form Shiwalik Health Drink. In this communication, we presented detail taxonomy for easy identification, experimentation trials undertaken for agrotechnology transfer, nutrient content in fruits for future development of value added products and nutraceuticals.

DIVERSITY AND DISTRIBUTION

Wide distribution of *Grewia asiatica* reported from tropical and subtropical vegetation, and extends its geographic locations throughout South-East Asia regions. In India, this species cultivated on a large scale in Uttar Pradesh, Punjab, Haryana, Rajasthan and Madhya Pradesh and to a limited scale in Jammu & Kashmir, Maharashtra, Gujarat, Andhra Pradesh, Bihar, West Bengal and Western Ghats along Malabar Coasts (Singh and Singh 2018, Tripathi 2009). Phalsa is one of the common wild edible fruit plants of the Shivalik Hills, whose distribution records from sub-Himalayan Mountain Range of the outer Himalayas. The hills have an average height of 900 to 1,200 m above sea level.

MEDICINAL USAGES

From medicinal point of view, the whole plant parts of *G. asiatica* have usage potential in different area. Stem used in marking baskets and small household agricultural containers. Bark used as medicine for gonorrhoea (Goyal 2012). Root used as a remedy for rheumatism and urinary tract problems (Zia-ul-Haq et al. 2012). There is a report that 50% ethanolic extract of aerial parts of *G. asiatica* shows hypotensive activity, while aqueous extract of stem bark shows anti-diabetic activity (Bhakuni et al. 1971). Infusion prepared from stem and root bark used as a demulcent, febrifuge and for treatment of diarrhea. It contains anthocyanin type cyanidin 3- glucoside, vitamin C, minerals and dietary fibers ((Nair et al. 2005, Gochar et al. 2017, Yadav et al. 2008, Yadav 1999). Leaves used to cure skin diseases especially pustular eruptions. Fruits reported to have astringent and cooling effect (Goyal 2012), and used in curing stomachache. Immatured fruits alleviates inflammation and has application in respiratory, cardiac, blood disorders and fever reduction (Mortan 1987). Seeds exhibit antifertility activity (Asolkar et al. 1992). There is a report that fruit extract of *G. asiatica* shows radioprotective effect in Swiss Albino Mice against Lethal Dose of α-irradiation (Ahaskar et al. 2007).

Experimental Agro-technology Trials

Before development of suitable variety for value added products, several field experimentation works have been carried out by CSIR-IIIM Jammu at different geographical locations in India. Table 1 shows experimentation works undertaken at different locations for phalsa plant.

Table 1: Location of experimental trials of Phalsa at different geographic zones

Experimental Sites	Chatha	IIIM Jammu	Dewamai	Varansi	Bikaner
State	J&K	J&K	J&K	UP	Rajasthan
Climate	Sub-tropical	Sub-tropical	Sub-tropical	Sub-tropical	Tropical
Annual rainfall (mm)	500–700	550-700	600-700	1000-1100	300-350
Altitude meter (mean sea level)	300	305	871	83	243
Latitude (N)	32°44′	32°73′	32°99′	25°36′	28°07′
Longitude(E)	75°55′	74°85′	74°93′	83°05′	73°34′

Propagation and Collection of Field Data

Phalsa can be propagated through conventional method by seeds. Besidesit, it also can be readily propagated by semi-hardwood cutting as well as by layering (Samson 1986). Seeds are soaked overnight in water to soft the seed coat. Seeds can be germinated in 15-20 days and one year old seedling can be used for Trans-planting. Growth promoting hormones Auxins (IAA, IBA, NAA) and Gibberellins improve rooting of phalsa plants (Singh et al. 2000).

Periodically field tours and data collection for the growth parameters such as number of branches, length of branches, number of leaves per branches, number of fruits clusters per branch and number of fruits per cluster at various experimental locations were collected and analyzed.

Taxonomy

Plants grow as large shrub or small sized tree, height varies from 1.5-8 m in length. Stems are greyish-white to brown bark covered, young seedling shoots have stellate hairy characters. Leaves broadly ovate to orbicular, 3.5-12 cm long, 2.5-5.2 cm broad, obliquely shallow, cordate at the base, serrate along the margin, hairy; stipules narrowly oblique-lanceolate or falcate, stellate hairy on both the sides. Inflorescence dischasial, usually 3-flowered, in axillary clusters, peduncles 2.2-3.0 cm long. Flowers orange-yellow, 1.7-2.2 cm across; pedicels 0.8-1.4 cm long; sepals oblong, 1-1.2 cm long, 0.2-0.3 cm broad, hairy outside, glabrous within; petals oblong, with a ring of hairs around white gland, irregularly lobed at the apex, orange-yellow; stamens numerous, filaments 0.4-0.6 cm long, orange-yellow, turns purple, anthers oblong; ovary globose, strigose; stigma 4-lobed. Fruits drupeceous, dark purple when matured, green when immature, globose, 0.5-1.2 cm diam., 2-lobed, hairy; mesocarp fibrous, acidic taste when eaten.

AGROTECHNOLOGY

Climate

Phalsa can be grown throughout the country except at higher altitudes and on slopy belts. It relish distinct winter and summer for best growth, yield and quality. In regions having no winter, this plant species doesn't shed leaves and produce flower more than once, thus yield poor quality fruits. Full-grown plants can tolerate freezing temperature for a short period. The plants can tolerate as high as 45°C.

Edaphic features

Phalsa can be grown on a wide variety of soils including moderate sodic soil. It prefers to grow in well-drained loamy and sandy loam soils. Usually, the plants are sensitive to water logging as this causes chlorotic. Therefore, while selecting sites for its cultivation, soils poor in sub surface drainage and water logged condition usually should not be suitable for its cultivation. The basic physico-chemical requirements of edaphic features suitable for *Grewia asiatica* is given in table 2.

Table 2: Physico-chemical properties of soil suitable for *Grewia asiatica* cultivation

Soil parameters	Analytical Value	Methods
Mechanical properties		
Sand (%)	66.4	Bouyouchos Hydrometer method (Bouyouchos 1962)
Silt (%)	18.0	Bouyouchos Hydrometer method (Bouyouchos 1962)
Clay (%)	15.6	Bouyouchos Hydrometer method (Bouyouchos 1962)
Textural Class	Sandy loam	
Physico-chemical properties		
EC(dsm^{-1})	0.24	Suspension of soil and water 1:2.5 with glass calomel electrode (Richard 1954)
pH$_2$	8.10	Salt bridge measurements from the suspension used for pH determination (Richard 1954)
Organic Carbon (%)	0.56	Dichromate oxidation of organic matter (Walkley and Black 1934)
Bulk Density (mg m^{-3})	1.41	Core method (Black and Hartge 1986)
Available Nitrogen (Kg ha^{-1})	211.0	Alkaline permanganate method (Subbiah and Asija 1956)
Available Phosphorus (Kg ha^{-1})	17.20	0.5 N Sodium bicarbonate (pH 8.5) (Olsen et al. 1954)
Available Potassium (Kg ha^{-1})	130.0	Ammonium acetate extraction method (pH 7.0) using flame photometer (Jackson 1973)

Propagation and Planting

Grewia asiatica can be propagated through seeds collected from matured fruits, and is the easiest and the most commonly used technique for propagation. Stem cutting (Samson 1986) and air layering propagation technique can be used with the help of growth regulators such as IAA, IBA and NAA which improves rooting (Mohammed and Chauhan 1970). Before plantation, land is prepared well before the plants are to be set in the field. Pits of size 50 cm x 50 cm x 50 cm are dug and re-filled with a mixture of top soil and well rotten FYM in the ratio of 1: 1. For transplantation, eight to twelve months old seedlings are better for planting. Usually, February-March season is suitable for plantation in the field as this period promote new growth for the plant and allow the seedling to get set in the field. Care should be taken while lifting the seedling from nursery that it should have bare roots. While planting, distance of 2.5-3.0 m from plant to plant and 3-4 m from row to row should be maintained for better yield and economics.

Nutrient Management with Respect to Manures and Fertilizers

Since *Grewia asiatica* is a hardy and stress tolerant plant, the use of fertilizer is not necessary. However, the application of some fertilizer encourages vegetative growth for new borne shoots. Use of 10-15 Kg well rotten FYM soon after planting promotes new growth and plants becomes more healthy. Nitrogenous fertilizers can be applied preferably in two split doses, first at the time of flowering, and second after the fruit setting or when flowers are in full bloom.

It has been experimented and observed that the response of fertilizers combinations on growth, yield and quality parameters in *Grewia asiatica* at Indian Institute of Integrative Medicine, Jammu observed that when nutrients and fertilizers were used in complex form (Nitrogen 100 gm + Phosphorus 50 gm + Potassium 50 gm) yields maximum number of branches per plant (192.33±6.81) as compared to control or non-fertilized plants which has less branches per plant (117.67±15.70) (Gochar et al. 2017, Saravanan et al. 2013). The detail of growth parameters of phalsa experimentation results is given in table 2. Application of nutrients gave positive difference in number of leaves per branch. Treatment N+P+K (Nitrogen 100 gm + Phosphorus 50 gm + Potassium 50 gm) produce maximum length of branch per plant, number of leaves per branch, number of fruits cluster per branch and number of fruits per cluster 124.44±8.22, 38.33±2.89, 19.78±1.68

Table 2: Growth parameter of *Grewia asiatica* under control and application of nutrients

Treatment	Control	Treatment of different nutrients and their combination effect on growth parameters						
		N	P	K	N+P	N+K	P+K	N+P+K
No. of branches / plants	117.67±15.70	130.00±10.0	123.00±9.64	140.00±10.0	150.00±5.00	158.33±7.64	153.33±10.41	192.33±6.81
Length of branches / plants	84.16±10.84	101.11±8.39	95.83±5.83	103.33±12.58	110.00±5.00	112.78±11.10	109.44±11.10	124.44±8.22
No. of leaves / branch	20.33±2.08	26.67±4.16	27.67±2.52	30.33±2.08	29.33±5.03	31.56±3.36	34.33±6.03	38.33±2.89
No. of fruits clusters/branch	11.56±3.10	13.00±2.33	13.11±2.34	14.11±3.47	17.11±1.02	15.22±2.41	15.44±0.77	19.78±1.68
No. of fruits/clusters	9.33±1.76	12.33±2.67	12.16±3.04	13.00±2.20	14.61±1.21	15.22±1.34	13.55±1.41	19.33±1.15

Note: C_o=Control, N= Nitrogen 100 g, P= Phosphorus 100 g, K= Potassium 50 g, N+P= Nitrogen 100g + Phosphorus 50 g, N+K= Nitrogen 100g + Potassium 50 g, P+K= Phosphorus 50 g+ Potassium 50 g, N+P+K= Nitrogen 100 g+ Phosphorus 50 g+ Potassium 50 g

and 19.33±1.15, respectively. These parameters recorded minimum in control plants which observed to haves 84.16±10.84, 20.33±2.08, 11.56±3.10 and 9.33±1.76 respectively (Gochar et al. 2017, Saravanan et al. 2013, Yadav et al. 2008). This plant is very sensitive to iron deficiency; however, it has been observed that *Grewia asiatica* showed beneficial effect due to usages of Zinc Sulphate (0.5%) and Ferrous Sulphate (0.4%) (Wali et al. 2005).

Irrigation

It has been experimentally proven that *Grewia asiatica* is a drought tolerant plant, but application of irrigation certainly increases yield and improves quality of fruits. At the time of transplantation, first irrigation is required. Even, after the application of fertilizers, probably in January February has good effect on growth parameters of the plant. Irrigation in summer, March-April at 2-3 weeks interval is considered as good for the plants.

Phenology

Flowering in *Grewia asiatica* starts in February and continues till May. The first flower opens at the base of the plant stem. Later, flowers starts appearing from the axils of the leaves. Flowers are small and yellowish, and are hermaphrodite in nature but cross pollination is crucial requirement for better fruit setting. Gill et al. (2001) reported that 61.60% fruit setting happens to be due to cross pollination than self-pollination which is observed to be only 23.00%.

Insect pollinators such as *Apis florea, A. mellifera, A. dorsata, Megachile bicolor* and *Chalicodoma cephalotes* were observed foraging both the nectar and the pollen of *Grewia asiatica* flowers. There are various other pollinators who foraged on plants for nectar and leaves throughout their growing seasons.

Harvesting, Yield and Quality

Fruits start appearing after second year of plantation of *Grewia asiatica*. Third years onwards, a good numbers of fruit yield can be recorded. In Shivalik range of Punjab and Haryana, the harvesting of phalsa fruit start by the end of May and can be harvested till the end of June, while in South India, March-April is the best season for fruit collection. Fruits harvested at the right stage of maturity has good nutrient contents and can be used fo product development. In matured plants, fruits become fully mature in 50-55 days after the fruit setting, and this maturity can be judged by observing fruit colour. It can be harvested when the colour changed to deep reddish brown, or when the pulp tastes sweeter. Fruits can be individually picked-up by manual/hand and can be stored in bamboo baskets cushioned with polythene sheet or newspapers. Several pickings are necessary because all fruits don't ripen at the same time, and therefore, fruit picking can be carried out on alternate days or after three days of picking.

It has been experimentally proved that appropriate usages of nutrients and manures in cultivated plants of *Grewia* increases length and breadth of fruits (table 3). It has been observed that maximum length and breadth of fruits were founded to be 1.15 cm and 1.52 cm respectively, treatment N+P+K (Nitrogen 100 gm + Phosphorus 50 gm + Potassium 50 gm) over other rest of treatments. The results regarding in yield attributing characters founded in all nutrients management treatment gave higher total fruits yield per plant in kg over control (1.55±0.22). Highest total fruits yield per plant reported in treatment N+P+K (Nitrogen 100 gm + Phosphorus 50 gm + Potassium 50 gm) 2.32±0.20. Similar finding reported by Saravanan et al (2013) and Yadav et al. (2009). Table-3 show the data of total soluble solid (%) most of the treatments were found to improve the quality of fruits by applying different doses of NPK. The maximum TSS content (%) was recorded in treatment of N+P+K (29.00±3.67) while minimum (18.78±2.52) were founded in control condition. The different doses of NPK significantly affected the weight of fruits and seeds (gm). Collected analysed data showed that weight of 50 phalsa fruits (40.77±10.61) was maximum in N+P+K treatment against the rest of all nutrient treatments. On an average, a matured plant can provide 2-4 kg of fruits per pant in a year.

Inter-culture and Inter-cropping

One or two ploughings after pruning the phalsa plant in orchard is desirable to control weeds and to incorporate FYM or compost. It can be perfered to grow green manure crop such as green gram, cowpea or black gram during the rainy season in early life of the orchard (first four years of planting). The green manure crop should be incorporated into the soil towards the end of rainy season or earlier for proper establishment of plants and also reduces weed population.

The phalsa fruit is borne in clusters at the axil of leaves on the new growing shoots produced during the current season. Annual pruning is therefore very essential to have new vigorous shoots to ensure regular and heavy fruiting. Phalsa plant is allowed to develop as a bush, hence,

Table 3: Yield attributing characters of *Grewia asiatica* in comparison to control and application of nutrients treatment

Fruit parameters	C_o	Treatment with nutrients and their combination on fruit yield						
		N	P	K	N+P	N+K	P+K	N+P+K
Length of fruits (cm)	0.98±0.04	0.99±0.03	1.01±0.05	1.01±10.06	1.04±0.04	1.05±0.03	1.00±0.04	1.15±0.10
Breadth of fruits (cm)	1.10±0.02	1.15±0.05	1.18±0.03	1.17±0.06	1.17±0.03	1.27±0.06	1.23±0.06	1.52±0.34
Total fruit production fruits (kg)	1.55±0.22	1.66±0.41	1.91±0.23	1.69±0.10	2.10±0.20	2.12±0.11	1.93±0.41	2.32±0.20
Total Soluble solid %	18.78±2.52	21.67±1.53	23.22±2.27	25.00±3.93	22.56±2.69	21.78±1.54	27.22±2.22	29.00±3.67
No. of 50 fruits wt. (gm)	30.21±2.91	31.80±3.30	33.30±3.76	33.40±2.20	37.17±2.08	34.23±4.15	35.47±3.61	40.77±10.61
No. of 50 seed wt. (gm)	2.57±0.15	2.33±0.06	2.30±0.30	2.53±0.21	2.27±0.12	2.23±0.15	2.50±0.10	2.13±0.15

Note: C_o=Control, N= Nitrogen 100 g, P= Phosphorus 100 g, K= Potassium 50 g, N+P= Nitrogen 100g + Phosphorus 50 g. N+K= Nitrogen 100g + Potassium 50 g, P+K= Phosphorus 50 g+ Potassium 50 g, N+P+K= Nitrogen 100 g+Phosphorus 50 g+ Potassium 50 g

no initial training is practiced. Both the severe and very light pruning affect the growth parameter of this species. The desirable height of pruning varies from fruit yield 50-100 cm from ground level. The best time for their pruning is December-January, when the plants shed their leaves and in all cases the operation should be finished well before the start of a new growth.

Nutrient Analysis of Value-Added Fruits

Ripe fruits of *Grewia asiatica* usually contain high amount of Vitamin A, Vitamin C, minerals (calcium, phosphorous) and fiber, however, low calorie and fat observed in this plant species. Threonine and methionine are found in fruit pulp and seeds. The attractive crimson red to dark purple colour of phalsa fruit is due to presence of anthocyanin pigment. The fruit shows very high antioxidant activity due to the presence of vitamin C, phenolics, flavonoids, tannins and anthocyanins (Tiwari et al. 2014).

Table 4: Comparison of Nutrient composition of phalsa fruits produced at Jammu (India) and Port Valley (Georgia).

Sr No.	Nutrient analyzed	Nutrient values/100 gram fruits	
		India (Jammu)	Georgia (Port Valley)
1	Calories (Kcal)	55.83	90.5
2.	Moisture (%)	-	76.3
3.	Fat (gm)	0.047	<0.1
4.	Proteins (gm)	0.433%	1.57
5.	Carbohydrates (gm)	13.44	21.1
6.	Dietary Fiber (gm)	-	5.53
7.	Ash (gm)	-	1.1
8.	Calcium (mg)	-	136
9.	Phosphorus (mg)	-	372
10.	Sodium (mg)	-	17.30
11.	Vitamin A (mg)	-	16.11
12.	Vitamin B1, Thiamin (mg)	15.0	0.02
13.	Vitamin B2, Riboflavin (mg)	0.20	0.264
14.	Vitamin B3, Niacin (mg)	-	0.825
15.	Vitamin C, Ascorbic acid (mg)	1230	4.385

Product Developed

CSIR-Indian Institute of Integrative Medicine Jammu has developed a value-added product in the form of tetrapack of phalsa health drinks and further R&D on this plants have been in progress.

Fig. 1: Habit of *Grewia asiatica* and product developed by CSIR-IIIM Jammu as health drink

CONCLUSION

It has been observed that *Grewia asiatica* is an underutilized wild fruit tree of India known for their nutritional, medicinal and valued added processed products. Fruits contain important growth promoting nutrients, vitamins, minerals and essential amino acids. Herbal medicine for the treatment of cancer, fever, rheumatism and diabetes can be manufactured from plant and plant parts of *G. asiatica*. Major drawback of *G. asiatica* fruits are its perishability, and cannot be stored for a long time. Future studies on storage mechanism needed for its proper utilization. For development of drugs and value added products, several therapeutic research works have been carried out which suggested that this plants possess antioxidant, antidiabetic, antimicrobial, hepatoprotective, antifertility, antifungal, analgesic and antiviral activities. There is a need to focus on biotechnological intervention with aim to develop seedless varieties and new nutraceutical products from fruits and medicines from plant parts of *G. asiatica*.

ACKNOWLEDGEMENT

Authors are thankful to Director IIIM, Dr Ram a Vishwakarma for necessary facilities and encouragement. This article bears institutional publicaton number IIIM/2236/2018.

REFERENCES CITED

Ahaskar M, Sharma KV, Singh S, Sisodia R. 2007. Radioprotective Effect of Fruit Extract of *Grewia asiatica* in Swiss Albino Mice against Lethal Dose of g-irradiation. *Asian Journal of Experimental Science* 21: 1-14.

Asolkar LV, Kakkar KK, Chakre OJ. 1992. Indigofera tinctoria. In: Second Supplement to Glossary of Indian Medicinal Plants with Active principles. New Delhi, Publication Information Directorate,

Bhakuni DS, Dhar ML, Dhar MM, Dhawan BN, Gupta B, Srimal RC. 1971. Screening of Indian plants for biological activity: Part III. *Indian Journal of Experimental Biology* 9: 91.

Bouyoucos HJ. 1962. A hydromater method for the determintion of textional classes of soils. *Technical Bulletin* 132.

Blake GR. Hartge KH. 1986. Bulk density. In: method of sol analysis, Part I (Klute A, Edition). American Society of Agronomy, Madison, Wiscons in USA, pp. 363-375.

Gill SS, Kaushik HD, Sharma SK. 2001 Effect of modes of pollination on fruit set and insect pollinators of phalsa (*Grewia subinaequalis* DC.). *Research on Crops* 2 (2): 193-196.

Goyal PK. 2012. Phytochemical and pharmacological properties of the genus grewia: a review. *International Journal of Pharmacy and Pharmaceutical Sciences* 4 (4): 72-78.

Gochar R, Koli B, Meena SR, Chandra S, Bhanwaria R, Gocher M. 2017. Response of nutrients management on growth, yield and quality of phalsa (*Grewia asiatica* L.) in Jammu region. *Trends in Biosciences* 10(24): 5195-5198.

Mohammed S, Chauhan KS. 1970. Vegetative propagation of phalsa (*Grewia asiatica* L.). *Indian Journal of Agricultural Science* 40: 581–586.

Morton JF. 1987. Phalsa, Fruits of warm climate. Julia Morton, Miami, F.L. 276

Nair MG, Wang H, Mody DK, Deueitt DL. 2005. Dietary food supplement containing natural cyclooxygenase inhibitors and methods for inhibiting pain and inflammation. United States Patent Application

Olsen SK, Cole VC, Wetanable FS, Dean LA. 1954. Estimation of available phosphorus in soil by extraction with sodium bicarbonate. USDA Cric. No. 939.

Richards LA. 1954. Diagnosis and improvement of Salinealkaline Soils. USDA Handbook No 60. US Developmnet of Agriculture, Washington DC, USA.

Samson JA. 1986. The minor tropical fruits. p. 316. In: Tropical fruits. Longman Inc., New York.

Sarvanana S, Chander P, Kumar R, Singh J. 2013. Influence of different level of NPK an growth, yield and quality of phalsa (*Grewis subinaequalis* L.). *Asian Journal of Horticulture* 81(2): 433-435

Singh KK, Singh SP. 2018. Cultivation and utilization in Phalsa (*Grrewia asiatica* L.) under Garhwal Himalayas region. *Journal of Medicinal Plants Studies* 6(1): 254-256

Singh Z, Grewal GPS, Singh L. 2000. Effects of gibberellin A4/A7 and blossom thinning on fruit set, retention, quality, shoot growth and return bloom of phalsa (*Grewia asiatica* L.) Acta Horticulturae, 525(Proceedings of the International Conference on Integrated Fruit Production

Subbaiah BV, Asija GL. 1956. A rapid procedure for the estimation of available nitrogen in soil. *Current Science* 25: 250-260.

Tiwari DK, Singh D, Barman K, Patel VB. 2014. Bioactive compounds and processed products of Phalsa (*Grewia subinaequalis* L.) Fruit. Popular Kheti

Tripathi P. 2009. Phalsa (Grawia subinaequalis) cultivation https://www. researchgate.net/publication/304765037.

Yadav DK, Pathak S, Yadav AL. 2008. Effect of integrated nutrient management on physico-chemical attribute of Phalsa (Grewia subinaequalis L.). *Plant Archiever* 8(1): 461-463.

Wali VK, Kaul R, Kher R. 2005. Effect of foliar sprays of nitrogen, potassium and zinc on yield and physico-chemical properties of Phalsa (*Grawia subinaequalis* DC) Cv. Purple round. *Haryana Journal of Horticultural Science* 34(1/2): 56-57.

Yadav AK. 1999. Phalsa: A Potential New Small Fruit for Georgia. J. anick. Pp 348

Walkley A, Black IA. 1934. Rapid titration method of organic carbon of soil. *Soil Science* 37: 29-33.

Zia-Ul-Haq M, Shahid SA, Muhammed S, Qayum M, Khan I, Ahmad S. 2012. Antimalarial, antiemetic and antidiabetic potential of *Grewia asiatica* L. leaves. *Journal of Medicinal Plants Research* 6: 3213–3216.

7

Taxonomy, Ecology and Phenology of *Panax* Species in North-Eastern India

Lucy B. Nongbri[1], Saroj Kanta Barik[1,2,]*
and Ladaplin Kharwanlang[1]

[1]*Centre for Advanced Studies in Botany, North-Eastern Hill University*
Shillong - 793022, Meghalaya, INDIA
[2]*CSIR-National Botanical Research Institute, Lucknow - 226 001*
Uttar Pradesh, INDIA
*Corresponding Email: sarojkbarik@gmail.com

ABSTRACT

Panax L. is an important medicinal plant species complex, which is commonly known as ginseng. Ginseng has high commercial value for its ginsenoside contents and is collected from the wild in the subtropical-temperate forests of north-eastern India. Taxonomy, ecology and phenology of five species of *Panax* found in north-eastern India *viz.*, *P. assamicus* R. N. Banerjee, *P. bipinnatifidus* Seem., *P. pseudoginseng* Wall., *P. variabilis* J. Wen and *P. sokpayensis* Sharma & Pandit, have been discussed in this paper. The distribution of these species is restricted to the states of Arunachal Pradesh, Manipur, Meghalaya, Nagaland, Sikkim and Darjeeling Hills of West Bengal. *P. pseudoginseng* has extended distribution range in Nepal. *P. variabilis* was reported from Nagaland, which is a new record for India. All the species of *Panax* prefer shady and moist habitat. The species grow in the soil texture that ranges from loamy sand to sandy. The pH ranges from 4.3-7.1. The species is perennial and pereniates through its underground rhizome. The phenophase starts with emergence of seedlings in the late March and senescence takes place in the middle of October after seed setting. Monitoring of the populations of all *Panax* species revealed that the population size have decreased by 40-50% over a period of 10 years due to over-harvesting from the wild by the local people for its medicinal value. Sustainable use of rhizome and its cultivation practice need to be standardized to address the conservation issues related to the plant.

Keywords: Ecology, North-eastern India, *Panax*, Phenology, Taxonomy

INTRODUCTION

For centuries, man has been using various plants as nutrients, beverages, cosmetics, dyes and medicines to maintain health and improve the quality of life. *Panax*, commonly and commercially known as ginseng belonging to the family Araliaceae is a medicinal herb, and has long been used in the far East, in particular Korea and China for maintaining physical vitality. *Panax* was first described by Carl Linnaeus in 1735 by naming two herbaceous species *viz.*, *P. quinquefolius* L. (American ginseng) and *P. trifolius* L. (dwarf ginseng) from eastern North America (Linnaeus 1735, 1753). The genus name *Panax* (Pan= all + axos=medicine) means 'cure all' in Greek. The herbal root is named as ginseng, because it is shaped like a man (Figure 1) and is believed to symbolise three essences (i.e. body, mind and spirit). It is known as the lord or the king of herbs (Hu 1976). Ginseng has been used mostly as a tonic to rejuvenate weak bodies, but rarely as a curative medicine. However, as per *Bancao Gangmu* (Encyclopedia of Herbs) written by Li Shizhen of China in 1596 A.D., *Panax* is an ingredient in several formulations for curing 23 diseases (Li 1596). Ginseng is frequently used in Asian countries as a traditional medicine and till today ginseng preparations are amongst the most popular and best-selling herbal medicines worldwide (Ernst 2002). The most commonly used *Panax* species are *P. ginseng* C.A.Mey. (Korean or Asian ginseng), *P. quinquefolius* (American ginseng), *P. notoginseng* (Burkill) F.H.Chen (Tienchi or Sanchi) and *P. japonicus* (T. Nees) C.A.Mey. (Japanese ginseng).

The ginseng genus has eastern Asian and eastern North American disjunct distribution (Wu 1983). *Panax* consists of approximately 18 species, of which 16 are from eastern Central Asia and two from eastern North America (Seemann 1868, Burkill 1902, Graham 1966, Proctor and Bailey 1987, Wen & Zimmer 1996, Wen 2001, Yoo et al. 2001, Lee & Wen 2004). Among the Asiatic species, several Himalayan taxa have been debatable due to sympatry of morphologically distinct taxa and the existence of occasional morphological intermediates (Wallich 1829, Wen & Zimmer 1996). In Asia, south-eastern China and eastern Himalayas are the centres of diversity for *Panax* (Burkill 1902, Hara 1970, Hu 1976, Wen & Zimmer 1996). In India, Araliaceae is represented by 15 genera distributed mostly in north and north-eastern region and *Panax* genus has five speciesviz., *P. assamicus* R. N. Banerjee, *P. bipinnatifidus* Seem., *P. pseudoginseng* Wall., *P. variabilis* J. Wen and *P. sokpayensis* Sharma & Pandit, all of which are found mostly in north-eastern part of India only.

Although taxonomic descriptions of some *Panax* species from North-east were published by earlier workers (Wallich 1829, Seeman 1868, Banerjee 1968, Pandey et al. 2007, Sharma & Pandit 2009), but there is no information on ecology and phenology of any *Panax* species except for *P. quinquefolius* from Illinois (Anderson et al. 1993).

The main objective of this paper is to present information on taxonomy, distribution, ecology and phenological events of *Panax* species found in North-east India, which could be useful in developing an effective management strategy for conservation of *Panax* species complex in India.

Fig. 1: *Panax ginseng* C.A.Mey. (Image source: www.Google.com)

INVENTORY OF PANAX IN NORTH-EASTERN INDIA

Secondary data were collected by consulting the earlier herbarium records at Central National Herbarium, Kolkata (CAL), Itanagar (ARUN), Shillong (ASSAM), and Chinese National Herbarium, Institute of Botany, Beijing (PE). The floras, published reports and research papers on *Panax* were extensively consulted. Primary information from local people and traditional healers on distribution were collected. Extensive field surveys were undertaken in the entire north-eastern region to validate the secondary data, and primary information along with the herbarium specimens were collected for *Panax*. The populations were present in Meghalaya, Manipur, Nagaland and Arunachal Pradesh.

Phenology

Phenological studies are important to understanding of species interactions, and community structure. Phenology was studied over a period of 3 years from 2010-2012 in Meghalaya, Manipur, Arunachal Pradesh and Nagaland. Phenological events, such as emergence and establishment of seedlings, flowering time, fruiting time and seed set were recorded at different places.

Soil texture and soil pH

Composite soil samples were collected from the natural habitats of *Panax*. Soil texture was determined by Buouyoucos hydrometer method (Allen et al. 1974) and a digital pH meter (Professional Meter, PP-20, Sartorius) was used to determine the pH of the soil.

Distribution of *Panax* species in North-east India

Panax species were reported from Arunachal Pradesh, Meghalaya, Manipur, Nagaland, Sikkim and also from Darjeeling Hills of West Bengal (Table 1).

TAXONOMY, ECOLOGY AND PHENOLOGY

Panax assamicus

Taxonomic description

Tall herb of 50-125 cm tall. Tuber horizontal, unbranched, elongated with thick internodes. Stem straw coloured, stout, erect, glabrous. Leaves whorled at the summit of the stem, exstipulate, digitately 5-7 foliate, 12-36 cm long, petiolate; petioles glabrous, stout, angular, 7-18 cm long; leaflets petiolulate, narrowly lanceolate, linear to broadly elliptic, 6-18 cm long, 1.2-3.0 cm wide, long acuminate, acumination up to 2 cm long, both surfaces setose on veins and midribs, base rounded, rarely attenuate, minutely uniformly serrate, midrib regular, rarely oblique, petiolules 0.3-1.8 cm long, lateral ones shorter than the rest. Inflorescence a terminal umbel, glabrous, 10-40 cm long, sometimes 2-7 umbels, pedunculated; peduncles upto 35 cm long, glabrous, stout, sometimes whorled at the top of the rachis; umbels 40-60 flowered, 2-4 cm in diameter; bracteoles linear, persistent, glabrous, pedicels pubescent up to 1.4 cm long. Flowers greenish-white in bud, bisexual, 1.2 cm long. Calyx green, cup-like obscurely toothed, teeth less than 1mm long, alternate to petals. Corolla polypetalous; petals 5, glabrous on both the surfaces, oblong one-nerved, inflexed-apiculate. Stamens 5, alternate with petals, filaments

Table: Distribution of genus *Panax* in North-eastern India

Species	Common Name	Distribution	Literature cited
Panax assamicus R. N. Banerjee	Shensheng (Khasi)	Meghalaya, Manipur and Darjeeling hills of West Bengal	Banerjee, (1968) Singh *et al.* (2000) Pandey *et al.* (2007) Pandey and Ali (2012)
Panax bipinnatifidus Seem.	Pearl ginseng, ge da qi and zhuzishen	Arunachal Pradesh, Sikkim and Darjeeling hills of West Bengal	Banerjee (1968) Pandey *et al.* (2007) Sharma and Pandit (2011)
Panax bipinnatifidus Seem. var. *angustifolius* (Burkill) J. Wen	-	Arunachal Pradesh, Sikkim, West Bengal (Darjeeling)	Banerjee (1968) Pandey *et al.* (2007)
Panax pseudoginseng Wall.	Jiarenshen, Himalayan, Nepal ginseng, Pseudoginseng	Sikkim, Nagaland, Arunachal Pradesh Nepal	Banerjee (1968) Pandey *et al.* (2007) Naithani (2013)
Panax sokpayensis Shiva K. Sharma & Pandit	Ginseng	Sikkim and Darjeeling hills of West Bengal	Sharma and Pandit (2009)
Panax variabilis J. Wen	-	Nagaland	Pandey *et al.* (2009)

filiform, 2 mm long; anthers oblong, 1.9 mm long, bilobed, dorsifixed; ovary inferior. Stigma 2-4, rounded. Fruit red, globose, up to 5 lobed, 0.5-1 cm long in diameter, with a black tip (Figure 2, 3).

Fig. 2: *Panax assamicus* (Meghalaya): Whole plant **(a)**; herbarium specimen **(b)**; inflorescence **(c)**; rhizome **(d)**; fruit **(e)**.

Fig. 3: *Panax assamicus* (Manipur): Whole plant **(a)**; herbarium specimen **(b)**; inflorescence **(c)**; rhizome **(d)**; fruit **(e)**.

Ecology

Grows in the elevation range of 1632-2251m asl in the subtropical broadleaved and temperate forests of Meghalaya and Manipur. The soil texture of the habitat ranges from loamy sand-sandy and soil pH ranges from 4.4-6.0. The species prefers shady and moist habitats. The associated species are ferns (Figure 2, 3) and other herbs such as *Hedera helix, Oplismenus burmanii, Impatiens* sp., *Tetrastigma serrulatum, Viola sikkimensis, Arisaema* sp., *Clintonia* sp., *Hydrocotyl asiatica, Hedychium coccineum, Ophiopogon* sp., *Crassocephalum* sp., *Pilea umbrosa* and *Elatostema* sp.

Phenology

The re-emergence of seedling takes place during early April up to late June. The vegetative growth occurs between April and early September. Flowering occurs during mid-June and continues up to late August. The fruiting period starts from late July and extends up to mid October. Seed setting starts during late August and continues till early-October. Most of the plants senesced during late-September to October. The phenophases vary between the populations from Meghalaya and Manipur by 15-30 days (Tables 2, 3).

Table 2: Phenology of *Panax assamicus* from Meghalaya

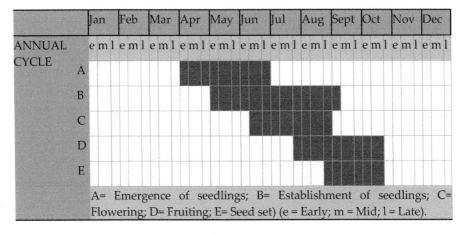

A= Emergence of seedlings; B= Establishment of seedlings; C= Flowering; D= Fruiting; E= Seed set) (e = Early; m = Mid; l = Late).

Table 3: Phenology of *Panax assamicus* from Manipur

	Jan	Feb	Mar	Apr	May	Jun	Jul	Aug	Sept	Oct	Nov	Dec
ANNUAL CYCLE	e ml	e ml	e ml	e ml	e ml	e ml	e ml	e ml	e ml	e ml	e ml	e ml
A				■	■							
B					■	■	■	■				
C						■	■	■				
D							■	■	■			
E									■	■		

(A= Emergence of seedlings; B= Establishment of seedlings; C= Flowering; D= Fruiting; E= Seed set) (e = Early; m = Mid; l = Late)

Panax bipinnatifidus

Perennial herb, 30-50 cm tall. Rhizome horizontal and elongated, with slender internodes and subglobose nodes. Stem straight, 30-45 cm long, slender, terete, glabrous, bearing whorl of 3-5 leaves at the top. Leaves palmate with 2-3 in a whorl, leaflets narrowly elliptic to broadly elliptic, sometimes slightly cleft to bipinnatifid, 12-19 cm long, upper surface sparsely hispid and ventral surface glabrous, petiolate, petioles glabrous, terete 3.5-7.0 cm long, lanceolate acuminate, lobes serrated, short petiolulate 0.4-1.1 cm long. Inflorescence terminal, solitary, simple, with a single umbel at the top or 2-5 clustered, 25-45 flowered, peduncles 8-14 cm long, glabrous, terete, articulated. Flowers green, bisexual, actinomorphic, bracteate, bracts caducous. Petals 5, triangular, Calyx cupular, sepals obscurely toothed; teeth 5, caducous, valvate. Stamens 5, anthers bilobed, dorsifixed, Ovary inferior, styles 2-3, erect. Fruit berries, globose, sometimes slightly compressed or triangular, red with a black tip, 1-2 seeded (Figure 4).

Fig. 4: *Panax bipinnatifidus* (Arunachal Pradesh): Whole plant **(a,c)**; herbarium specimen **(b)**; rhizome **(d)**.

Ecology

The species is found in the elevation range of 1500–3210 m asl in temperate forests of Arunachal Pradesh, Sikkim and Darjeeling Hills of West Bengal. The soil texture of its habitat is sandy and pH ranges from 4.3-4.8. Shady and moist habitats are preferred by the species. The associated species are, *Oplismenus compositus, Anaphalis margaritaceae, Elsholtzia strobilifera, Fragaria nubicola, Calanthe tricarinata* and *Pilea umbrosa*.

Phenology

The re-emergence of seedlings takes place during late March to late May. The vegetative growth occurs between late April and early August. Flowering starts during early June and continues upto late July. The fruiting period starts from early July and extends up to early September. Seed setting starts during early August and continues till late September. Most of the plants senesce during late September to October (Table 4).

Table 4: Phenology of *Panax bipinnatifidus* from Arunachal Pradesh

	Jan	Feb	Mar	Apr	May	Jun	Jul	Aug	Sept	Oct	Nov	Dec
ANNUAL CYCLE	e ml	e ml	E ml	e ml	e ml	e ml	e ml	e ml	e ml	e ml	e ml	e ml
A			▮	▮								
B				▮	▮	▮						
C					▮	▮	▮					
D						▮	▮	▮				
E							▮	▮	▮			

(A= Emergence of seedlings; B= Establishment of seedlings; C= Flowering; D= Fruiting; E= Seed set) (e = Early; m = Mid; l = Late).

Panax pseudoginseng

Tall herb of 30-60 cm height. Rhizome fusiform, short, elongated with thick internodes, unbranched, tubers single or fascicled. Stem, stout, erect, glabrous. Base of aerial stem with persistent glabrous and membranaceous scales. Leaves 3-4 verticillate at apex of stem, exstipulate, digitately 5-7 foliate, 14-21 cm long, upper surface with bristly hairs along veins and veinlets, lower surface less, petiolate; petioles glabrous, stout, angular, 7-11 cm long, leaflets petiolulate, lanceolate to narrowly elliptic, 6-10 cm long, 1.5-3.0 cm wide, long acuminate, acumination up to 1.2 cm long, base attenuate, margins biserrated, midrib regular, rarely oblique, petiolules 0.5-1.0 cm long, lateral ones shorter than the rest. Bases of petiole and petiolulesare with numerous lanceolate, stipule like appendages, obovate-elliptic to obovate-oblong. Inflorescence a terminal umbel with 1-3 umbel, each 40-60 flowered, glabrous, pedunculate, 7-11 cm long, glabrous, pedicel up to 1 cm long. Calyx green, cuplike obscurely toothed, teeth less than 1mm long, alternate to petals. Corolla polypetalous; petals 5, glabrous, oblong one-nerved, cuneate. Stamens 5, alternate with petals, anthers oblong, bilobed, dorsifixed; ovary inferior, 2-carpellate, styles 2. Fruit red, globose, 1-2 seeded (Figure 5).

Fig. 5: *Panax pseudoginseng* (Nagaland): Whole plant (a,b); rhizome types (c,d).

Ecology

Occurs in temperate forests of Arunachal Pradesh, Nagaland, Sikkim and Nepal in the elevation range of 1744–1883m asl. Soil texture and pH of the habitat range from loamy to sand-sandy and from 4.9-7.1, respectively. The preferred habitat of the species is shady and moist areas. The associated species are, *Begonia* sp., *Arisaema tortuosum*, *Eupatorium* sp. and *Pilea umbrosa*.

Phenology

The re-emergence of seedling takes place during mid April to early June. The vegetative growth occurs between mid May and mid July. Flowering occurs during late May and mid July. The fruiting period starts from late June and continues up to late August. Seed setting starts during mid July and continues till mid September. Most of the plants senesce during late September to October (Table 5).

Table 5: Phenology of *Panax pseudoginseng* from Nagaland

	Jan	Feb	Mar	Apr	May	Jun	Jul	Aug	Sept	Oct	Nov	Dec
ANNUAL CYCLE	e ml	e ml	E ml	e ml	e ml	e ml	e ml	e ml	e ml	e ml	e ml	e ml
A				▓								
B					▓	▓						
C					▓	▓	▓					
D							▓	▓	▓			
E								▓	▓			

(A= Emergence of seedlings; B= Establishment of seedlings; C= Flowering; D= Fruiting; E= Seed set) (e = Early; m = Mid; l = Late).

Panax variabilis

Herb attains 30–65 cm height. Horizontal tuber, elongated with thick internodes, unbranched. Stem green coloured, erect and glabrous. Leaves whorled at the summit of the stem, exstipulate, digitately 5–6 foliate, 15–30 cm long, petiolate; petioles glabrous, stout, angular, 7–14 cm long; leaflets petiolulate, obovate to broadly elliptic, 7–13 cm long, 1.5–2.6 cm wide, long acuminate, acumination up to 2 cm long, base rounded to cuneate, sometimes oblique, midrib regular, biserrated, both surfaces setose on veins and midribs; petiolules 0.5–1.7 cm long, lateral ones shorter than the rest. Inflorescence terminal, solitary, simple, bearing single umbel at the top or 2–3 clustered, 45–60 flowered; pedunculate, glabrous, stout, 13–24 cm long, bracteoles linear, persistent, glabrous. Flowers greenish-white in bud, bisexual, 1.2 cm long. Calyx green, cuplike obscurely toothed, alternate to petals. Corolla polypetalous; petals 5, glabrous, oblong one-nerved, inflexed apiculate.Stamens 5, alternate with petals, anthers oblong, bilobed, dorsifixed; ovary inferior. Fruits red with a black tip, globose (Figure 6).

Ecology

The species grows in temperate forest of Nagaland within the elevation range of 1581-1945m asl. The soil texture of the habitat is sandy and the species prefers acidic soil pH ranging from 4.8 to 5.9. The habitat preference is shady and moist areas along the streams. The associated species are *Hedera helix, Tetrastigma serrulatum, Oplismenus compositus, Impatiens* sp., *Pesicaria chinensis., Boehmeria* sp. and *Pilea umbrosa.*

Figure 6: *Panax variabilis* (Manipur): Whole plant **(a)**; herbarium specimen **(b)**; Inflorescence **(c)**; rhizome **(d)**.

Phenology

The re-emergence of seedling takes place during late March and late May. The vegetative growth occurs between late April and mid July. Flowering occurs during late May and continues up to early July. The fruiting period starts from late June and continues up to mid August. Seed setting starts during late July and continues till mid September. Most of the plants senesce by late September and early October (Table 6).

Table 6: Phenology of *Panax variabilis* from Manipur

	Jan	Feb	Mar	Apr	May	Jun	Jul	Aug	Sept	Oct	Nov	Dec
ANNUAL CYCLE	e ml	e ml	e ml	e ml	e ml	e ml	e ml	e ml	e ml	e ml	e ml	e ml
A												
B												
C												
D												
E												

(A= Emergence of seedlings; B= Establishment of seedlings; C= Flowering; D= Fruiting; E= Seed set) (e = Early; m = Mid; l = Late)

Panax sokpayensis (Courtesy, Sharma & Pandit 2009)

A perennial herb of 80–130 cm tall. Rhizome simple, horizontal, 20–60 cm long with prominent stem scars and rings at nodes; internodes short and thick; tuber single, globose in each rhizome which often decays in old individuals. Stem cylindrical, erect, stout, 50–90 cm long, scales at base, deciduous. Leaves exstipulate, palmately compound, 4–5 per stem, whorled towards the stem apex, 20–38 cm long, petiolate; petioles glabrous, stout, 10–20 cm long; leaflets 5, oblanceolate to narrowly elliptic, 7–16 cm long, 2.5–5 cm broad; petiolules, 1–2.5 cm long; apex caudate, upto 4 cm long; base rounded, attenuate or oblique; margin serrulate, some biserrulate; densely setose on adaxial surface. Inflorescence a terminal umbel; peduncles stout, 15–30 cm long. Flowers 50–65 per umbel, bracteates, andromonoecious, actinomorphic, epigynous, 3–4 mm long; bracts large, leafy or linear up to 0.5–2.0 cm long, pedicellate; pedicels, 1–2.5 cm long, glabrous; sepals 5, lanceolate, glabrous, alternate to petals, green; petals 5, polypetalous, lanceolate, ca. 3.0 mm long and ca. 2.5 mm wide, glabrous, deciduous, white; stamens 5, 3–4 mm long, filament free; anther oblong, ca. 3 mm long and ca. 2 mm wide, dorsifixed, bilobed; carpels 2–3, free; ovary inferior. Fruit a berry, 2–3 lobed in transverse section, subglobose, ca 0.7 cm long and ca. 0.6 cm wide, lower portion red, upper portion black; seeds 2-3, ovoid, ca. 0.4 mm long and 0.3 mm wide (Figure 7).

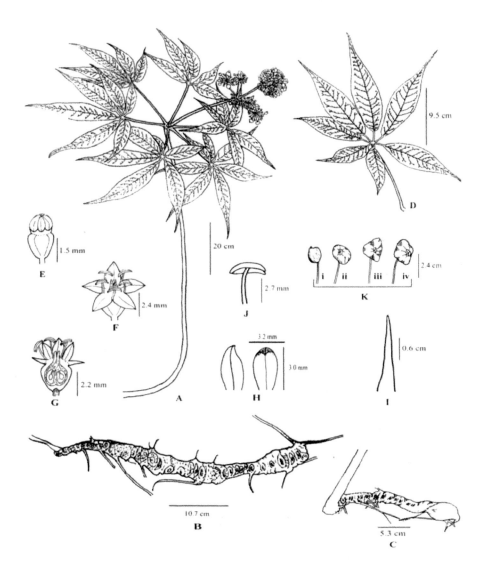

Fig. 7: *Panax sokpayensis* (Sikkim): Different parts of the plant (A-K)
(after Sharma & Pandit 2009).

Ecology

Grows in temperate forests of Sikkim at an elevation range of 1700–2300 m asl. The soil of the habitat is acidic and preferred habitat is shady and moist areas along the stream. The species grows under the canopy of Quercus-Acer trees in association with *Viburnum erubescens*, *Daphne cannabina*, *Urtica parviflora* and *Gerardiana heterophylla*.

Phenology

Seedling emergence starts in the month of March and vegetative growth ends in May. The plant starts flowering in the month of April and extends upto May. The fruiting starts from August and continues till September. Seed setting starts in the month of September and continues till October. The plants senesce during September and October.

DISCUSSION

Araliaceae members are distributed mainly in the tropics and subtropics especially in eastern Asia, southern Asia, south-eastern Asia and Pacific islands, with some genera occurring in the temperate zone which includes north America (Wen et al. 2001). The species growing in north America are distributed within an altitudinal range of 120-1,300 m asl. In eastern Asia, *Panax* species are found in the range of 800-4,000 m asl. In southern Asia which includes north-eastern India, the Himalayan ginseng is distributed at higher altitudinal range i.e. 1,500-3,210 m asl. All the species of *Panax* reported from North-eastern India *viz.*, *P. assamicus*, *P. bipinnatifidus*, *P. bipinnatifidus* var. *angustifolius*, *P. pseudoginseng*, *P. variabilis* and *P. sokpayensis* have restricted distributions. Only *P. pseudoginseng* extended its distribution to Nepal. *P. assamicus* is distributed in two states of the region i.e. Meghalaya and Manipur. These species are quite distinct in their morphological characters but bears high resemblance in many morphological characters. In the present study, taxonomic description for *P. variabilis* reported from Nagaland (Pandey et al. 2009) is given for the first time. Globally, *Panax* exhibits a high level of morphological variation, which was observed in all the Himalayan species. This has created serious problem in the past in establishing the taxonomic identity within the genus in Himalayas.

The phenological events in Himalayan *Panax* species is restricted to a short period of time starting from late March to October. There was overlapping period in all the phenophases but multiple comparison reveals that the difference between the onset and ending of a particular phenophase does not exceed more than 30 days, even if the plants are spatially separated i.e. from different states. The phenological events are also studied in *P. quinquefolius* (Proctor et al. 2003) which revealed similar events in all the phenophases.

Due to their chemical composition which is relatively similar, all *Panax* species have been used in medicine for treatment of various diseases. The pharmacologically active compounds in ginseng are a class of triterpene saponins, commonly known as ginsenosides (Yahara et al. 1979, Kasai et al. 1983, Ma et al. 1999). More than 40 of these saponins have been isolated and identified (Teng et al. 2003), which account for the pharmacological effects of these plants in the modulation of angiogenesis, adaptogenic properties, anti-stress activities, anti-hyperglycemic activities and their effects on the central nervous system. Other activities include, antibiotic, anticancer properties, modulating blood pressure metabolism and immune function (Lee 1992, Liu & Xiao 1992, Fulder 1996, Attele et al. 1999, Chang et al. 2003, Spelman et al. 2006, Xiang et al. 2008).

All *Panax* species in Himalayas have a high value market in nearby countries. Due to this, most of the matured individuals have been harvested from the wild during the past few years. During the present study, over-harvesting, deforestation, forest fragmentation and other anthropogenic activities were identified as factors responsible for making the species threatened. The quantitative data collected from the field indicates that there has been a reduction of 40-50% in the population sizes of four *Panax* species during the past one decade.

The present study highlights that North-eastern India harbours five species of *Panax* which are being used for medicinal purposes by the local people. The populations of these species have been reduced due to over-extraction from the wild which need to be stopped. Local level awareness programmes and development of techniques for propagation of these species (through seeds/rhizomes) are two issues that need to be addressed for effective conservation of *Panax* in India.

ACKNOWLEDGEMENTS

We are thankful to the Joint Directors of Botanical Survey of India, Itanagar and Shillong, Central National Herbarium at Kolkata and Chinese National Herbarium from Beijing for granting permission to consult their herbarium specimens. The financial support received under the research project entitled "Conservation of *Panax* species in North-east India: Characterisation of niche and genetic variability, mapping of potential distributional area, establishment of field gene bank and micropropagation" vide project no. BT/PR-9677/BCE/08/578/2007 dated 25.2.08 from the Department of Biotechnology, is gratefully acknowledged. We thank Dr. Bikarma Singh for his assistance in plant description, Prem Prakash Singh and Raghuvar Tiwary for bringing a few plant specimens. We also thank the village council members of Nagaland, Ukhrul district, Meghalaya and Arunachal Pradesh for their support and help during field.

Competing interest
Authors contributed equally to this work.

LITERATURE CITED

Allen SE, Grimshaw HM, Parkinson JA, Quarmby C. 1974. *Chemical Analysis of Ecological Materials*.Oxford: Blackwell Scientific Publications.

Anderson RC, Fralish JS, Armstrong JE, Benjamin PK. 1993. The Ecology and Biology of *Panax quinquefolium* L. (Araliaceae) in Illinois. *The American Midland Naturalist* 129: 357–372.

Attele AS, Wu JA, Yuan CS. 1999. Ginseng pharmacology Multiple constituents and multiple actions. *Biochemical Pharmacology* 58: 1685-1693.

Banerjee RN. 1968. A taxonomic revision of Indian *Panax* L. (Araliaceae). *Bulletin of the Botanical Survey of India* 40: 20-27.

Burkill IH. 1902. Ginseng in China . *Kew Bulletin* 4: 4-11 .

Chang YS, Seo EK, Gyllenhaal C, Block KI. 2003. *Panax ginseng*: A role in cancer therapy. *Integrated Cancer Therapy* 2: 13-33.

Ernst E. 2002. The risk-benefit profile of commonly used herbal therapies: Ginkgo, St. John's wort, ginseng, Echinacea, saw palmetto, and kava. *Annals of Internal Medicine* 136: 42-53.

Fulder S. 1996. *The Ginseng Book: Natureís Ancient Healer*. New York: Avery Publishing Group.

Hara H. 1970. On the Asiatic species of the genus *Panax*. *Journal of Japanese Botany* 45: 197-212.

Hu SY. 1976. The genus *Panax* (ginseng) in Chinese medicine. *Economic Botany* 30: 11-28.

Kasai R, Besso H, Tanaka O, Saruwatari YI, Fuwa T. 1983. Saponins of red ginseng. *Chemical and Pharmaceutical Bulletin* 31: 2120-2125.

Lee C, Wen J. 2004. Phylogeny of *Panax* using chloroplast trnC-trnD intergenic region and the utility of trnC-trnD in interspecific studies of plants. *Molecular Phylogenetics and Evolution* 31: 894-903.

Lee FC. 1992. *About Ginseng: The Elixir of Life*. New Jersey: Hollyn International Corporation.

Li S. 1956. *Encyclopedia of Herbs*.Vol-2, China.

Linnaeus C. 1735. Systemanaturae. Leyden: Haak.

Linnaeus C. 1753. *Species Plantarum*. Vol. 1-2. Stockholm: Impensis Laurentii Salvii.

Liu CX, Xiao PG. 1992. Recent advances on ginseng research in China. *Journal of Ethnopharmacology* 36: 27-38.

Ma WG, Mizutani M, Malterud KE, Lu SL, Ducrey B, Tahara S. 1999. Saponins from the roots of *Panax notoginseng*. *Phytochemistry* 52: 1133-1139.

Naithani HB. 2013. Occurrence of ginseng (*Panaxpseudoginseng*) in Western Himalaya. *Indian Forester* 139: 473-474.

Pandey AK, AjmalAM, Mao AA. 2007. Genus *Panax* L. (Araliaceae) in India. *Pleione* 1: 46-54.

Pandey AK, Ali MA, Biate DL, Misra AK. 2009. Molecular systematic of *Aralia-Panax* complex (Araliaceae) in India based on ITS sequences of nrDNA. *Proceedings of the National Academy of Sciences, India* 79: 255-261.

Proctor JTA, Bailey WG. 1987. Ginseng: industry, botany, and culture. *Horticulture Review* 9: 187-236.

SeemannB. 1868. Revision of the natural order *Hederaceae*: On the genus *Panax*. *Journal of Botany* 6: 52-58.

Sharma SK, Pandit MK. 2009. A new species of *Panax* L. (Araliaceae) from Sikkim Himalaya, India. *Systematic Botany* 34: 434-438.

Singh NP, Chauhan AS, Mondal MS. 2000. *Flora of Manipur: Vol. 1. Ranunculaceae-Astercaeae*. Calcutta: Botanical Survey of India.

Spelman K, Burns J, Nichols D, Winters N, Ottersberg S, Tenborg M. 2006. Modulation of cytokine expression by traditional medicines: a review of herbal immunomodulators. *Alternative Medicine Review* 11: 128-150.

Teng RW, Ang CS, McManus D, Armstrong D, Mau S, Bacic A. 2003. Regioselective acylation of ginsenosides by Novozyme 435.*Tetrahedron Letters* 44: 5661-5664.

Wallich N. 1829. An account of the Nepal ginseng. Transactions. *Medical and Physical Society of Calcutta* 4: 115-120.

Wen J, Plunkett GM, Mitchell AD, Wagstaff SJ. 2001. The Evolution of Araliaceae: A Phylogenetic Analysis Based on ITS Sequences of Nuclear Ribosomal DNA. *Systematic Botany* 26: 144-167.

Wen J, Zimmer EA. 1996. Phylogeny and biogeography of *Panax* L. (the ginseng genus, Araliaceae): inferences from ITS sequences of nuclear ribosomal RNA. *Molecular Phylogenetics and Evolution* 6: 167-177.

Wen J. 2001.Species diversity, nomenclature, phylogeny, biogeography, and classification of the ginseng genus (*Panax* L., Araliaceae). In: Punja ZK, ed. *Proceedings of the International Ginseng Workshop: Utilization of Biotechnological, Genetic and Cultural Approaches for North American and Asian Ginseng Improvement.* Canada: Simon Fraser University Press, 67-88.

Wu ZY. 1983. On the significance of Pacific intercontinental discontinuity. *Annals of the Missouri Botanical Garden* 70: 577-590.

Xiang YZ, Shang HC, Gao XM, Zhang BL. 2008. A comparison of the ancient use of ginseng in traditional Chinese medicine with modern pharmacological experiments and clinical trials. *Phytotherapy Research* 22: 851-858.

Yahara S, Kaji K, Tanaka O. 1979. Further study on damarane-type saponins of roots, leaves, flower buds, and fruits of *Panax ginseng* C.A. Meyer. *Chemical and Pharmaceutical Bulletin* 27: 99-92.

Yoo KO, Malla KJ, Wen J. 2001. Chloroplast DNA variation of *Panax* in Nepal and its taxonomic implications. *Brittonia* 53: 447-453.

8

Validated LC-MS Method for Quantitation of Phytochemicals of *Butea monosperma*

Vikas Bajpai[1], Bikarma Singh[2] and Brijesh Kumar[1, 3]

[1]*Sophisticated Analytical Instrument Facility, CSIR-Central Drug Research Institute, Lucknow-226001, Uttar Pradesh, INDIA*
[2]*Plant Sciences (Biodiversity and Applied Botany Division), CSIR-Indian Institute of Integrative Medicine, Jammu 180001, J&K, INDIA*
[3]*Academy of Scientific and Innovative Research, New Delhi-110001, INDIA*
Corresponding Email: vikasbajpai23@gmail.com, drbikarma@iim.res.in
brijesh_kumar@cdri.res.in

ABSTRACT

Butea monosperma (Lam.) Taub, known as 'Flame of the Forest' is used in Indian traditional system of medicine, such as Siddha Ayurveda, and Unani due to its medicinal properties. It is used in treatment of diabetes, diarrhoea, sore throat and reported to possess several biological activities such as anti-asthmatic, anti-diabetic, anti-fertility, anti-tumor and osteogenic. Recent studies revealed the phytochemicals responsible for osteogenic activity in the bark of *B. monosperma*. An ultra-performance liquid chromatography tandem mass spectrometry (UPLC-MS/MS) method was developed for identification and simultaneous quantitation of bioactive compounds from *B. monosperma*. Validation of the developed method was performed according to international conference on harmonization guidelines. Quantitative interpretation showed significant variations in the content of bioactive compounds in different parts of the plant and samples collected from three geographical regions. Principal component analysis was also used to study the quantitative variations of markers.

Keywords: Analytical method, Bioactive phytochemicals, *Butea monosperma*, LC-MS, Quantitation, Validation.

INTRODUCTION

Plants are the major source of traditional medicines and drugs. They also play a significant role in modern drug discovery programs (Raskin et al. 2002). *Butea monosperma* (Lam.) Taub, commonly known as "Flame of the Forest", is an important medicinal plant (Burli and Khade 2007) of India. This plant species is reported for the treatment of diabetes, diarrhoea, sore throat and is known to exhibit a wide variety of biological activities such as anti-asthmatic, anti-diabetic, anti-fertility and anti-tumor (Kirtikar and Basu 1935, Mazumder et al. 2011, Harish et al. 2014, Gupta 2015, More et al. 2018). The important bioactive metabolites reported from *B. monosperma* are buteaspermin A, buteaspermin B, buteaspermanol, butin, butrin, biochanin A, calycosin, cladrin, flemmichapparin C, genistein, isobutrin, lupeonone, lupeol and ononin (Sindhia and Bairwa 2010, Fageria and Rao 2015). Bioactive compounds such as daidzein, cajanin, isoformononetin, cladrin, formononetin, methylformononetin, medicarpin, prunetin and buteaspermanol have also been isolated from the bark of *B. monosperma* (Maurya et al. 2009, Bajpai et al. 2018). Medicarpin found in bark has shown significant bone healing property in a dose-dependent manner. Hence it is important to evaluate the quality and the quantity of these compounds in bark and other parts of *B. monosperma* (Tyagi et al. 2012). Several analytical methods, such as thin-layer chromatography (Nile and Park 2014), high-performance liquid chromatography (Gupta et al. 2010) and gas chromatography-mass spectrometry (Patil and Rajput 2012) have been developed earlier for the detection and determination of these phytochemicals in *B. monosperma* (Sharma et al. 2015).

Liquid chromatography technique is an important tool for the separation of organic compounds, while mass spectrometry is used for structure elucidation (Allwood and Goodacre 2010). The hyphenation of both LC and MS serve as the preferred technique for the detection of selected LCMS/MS in plant extracts. Triple quadrupole LC-MS/MS is ideal for the quantitative work. This technique is very sensitive, selective and specific, can reach to a very low limit of detection and quantitation (Schwartz et al. 1990). UPLC-QTRAP MS allows rapid detection of the analytes in a very short run time with low consumption of solvents at very low analyte concentration and has an increasing role in quantitation of plant metabolites directly from crude extract (King and Fernandez-Metzler 2006).

Hence, an UPLC-MS/MS method was developed and validated for the simultaneous quantitation of the main osteogenic compounds in *B. monosperma* leaf, stem and bark samples collected from three different geographical regions.

BOTANICAL DESCRIPTION

Butea monosperma is a small to medium sized deciduous trees (Figure 1) (Ambasta 1994, Orwa et al. 2009). It grows approximately 5-20 m high length. It has curved trunk and ash coloured rough fibrous bark. Leaves are three foliated with 8-15 cm long petioles; stipules are linear-lanceolate, leathery leaflets. Lateral leaves are obliquely ovate, while terminal ones showed rhomboid obovate, which are 12-25 cm long, 10-26 cm broad, with varying 7-8 pairs of lateral veins. Inflorescence is raceme, 5-40 cm long, arise at the top of leafless branchlets. Orange to red coloured flowers with campanulate sepals, tubular, 4 lobes present, having 5-7 cm long petals, wings and keel curved and densely pubescent (Cowen 1984, Mazumder et al. 2011). Fruits appears in the form of an indehiscent pods of 9-24 cm long, 3-6 cm broad, covered with short brown hairs stalked, yellowish to brown, with a single seed

Fig. 1: Butea monosperma, (A) Habitat, (B) Pods, (C) Flowers, (D) Seeds
(*Courtesy*: Google)

near the apex. Seeds are 2.5-3.5 cm long, 1.5-2.0 cm broad, flattened and ellipsoid with reddish brown, wrinkled seed coat (Orwa et al. 2009, Fageria and Rao 2015, More et al. 2018).

COMMON NAMES

According to Kirtikar and Basu (1935) *Butea monosperma* tree is known by many common names in different Indian languages.

- Palash in Sanskrit

- Dhak or chalcha in Hindi

- Bastard Teak or Parrot tree in English

- Polashi in Bengali

- Kakracha in Marathi

- Khakharo in Gujarati

- Parasa in Tamil

SYNONYMOUS NAMES

Butea frondosa Roxb. *B., frondosa* Willd., *B. frondosa* Willd. var. *lutea* (Witt.) Maheshw, *Erythrina monosperma* Lam., *Plaso monosperma* (Lam.) Kuntze, *P. monosperma* (Lam.) Kuntze var. *flava* Kuntze, *P. monosperma* (Lam.) Kuntze var. *rubra* Kuntze, *Rudolphia frondosa* (Willd.) Poir (Burli and Khade 2007).

DIVERSITY AND DISTRIBUTION

The genus *Butea* is mainly represented by four major species, *viz.*, *Butea braamiana* DC, *Butea buteiformis* (Voigt) Mabb; *Butea monosperma* (Lam.) Taub and *Butea superba* Roxb. These species are distributed worldwide. *Butea monosperma* is a native of the tropical region of South Asia and reported from Cambodia, China, India, Indonesia, Japan, Laos, Myanmar, Nepal, Papua New Guinea, Sri Lanka, Thailand and Vietnam (Fig. 2) (Orwa et al. 2009). This species is very common throughout the larger part of India, extending in the North West Himalayas up to 1000 meter above the mean sea level, and higher in the outer Himalaya region, up to 1200 meter above the mean sea level and in South India up to 1300 meter above the mean sea level (Fageria and Rao 2015). In India, this plant is abundantly seen in states like Arunachal Pradesh, Assam, Himachal Pradesh, Jammu & Kashmir, Karnataka, Kerala, Maharashtra, Manipur,

Fig. 2: Distribution of *Butea monosperma* in South Asian region (Orwa et al. 2009)

Meghalaya, Mizoram, Nagaland, Rajasthan, Sikkim, and Tripura (Wealth of India 1988, Gupta 2015, Fageria and Rao 2015).

ECOLOGY AND CLIMATE

The tropical and subtropical climate is suitable for the growth of *Butea monosperma* (Orwa et al. 2009). It is commonly found throughout the drier parts of India; frequently it is gregarious in forests, open grasslands and wastelands (Orwa et al. 2009). It is characterized by forming pure patches in grazing grounds and other exposed places, avoiding extermination owing to its resistance to browsing and its ability to reproduce from seed and root suckers in the community forming. Mean annual temperature ranges between -4 to 49 degrees, while rainfall ranges between 450-4500 mm (Verma et al. 1998). It can grow on a different variety of soils including black cotton soil, clay loams and gravelly sites, shallow, and even saline or waterlogged soils. Seedlings flourish perfectly on a rich loamy soil with pH recorded between 6-7 under high temperature and high relative humidity (Fageria and Rao 2015, More et al. 2018).

PHYTOCHEMISTRY

Butea monosperma is reported to have phytochemicals such as triterpenes, flavonoids, glycosides etc. (Mishra et al. 2000, Sindhia and Bairwa 2010). Compounds such as monospermoside, isomnopermoside, palasonin, were reported from *B. monosperma* seed (Gupta et al. 1970). Bioactive compounds namely daidzein, cajanin, isoformononetin, cladrin, formononetin, methylformononetin, medicarpin, prunetin and buteaspermanol have been isolated as the principal constituents using bioassay-guided fractionation from the ethanolic extract of *B. monosperma* bark. (Maurya et al. 2009). Structures of major bioactive compounds reported from the *B. monosperma* is shown in Fig. 3.

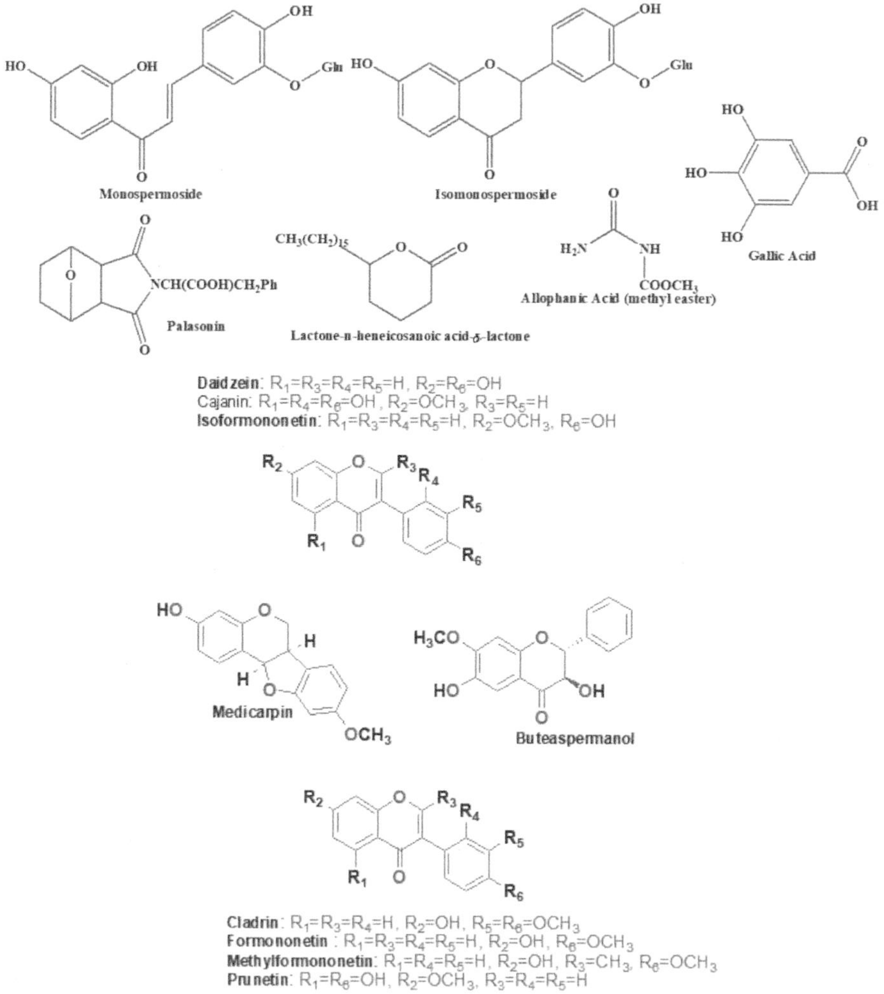
Monospermoside
Isomonospermoside
Gallic Acid
Palasonin
Lactone-n-heneicosanoic acid-δ-lactone
Allophanic Acid (methyl easter)
Daidzein: $R_1=R_3=R_4=R_5=H$, $R_2=R_6=OH$
Cajanin: $R_1=R_4=R_6=OH$, $R_2=OCH_3$, $R_3=R_5=H$
Isoformononetin: $R_1=R_3=R_4=R_5=H$, $R_2=OCH_3$, $R_6=OH$
Medicarpin
Buteaspermanol
Cladrin: $R_1=R_3=R_4=H$, $R_2=OH$, $R_5=R_6=OCH_3$
Formononetin: $R_1=R_3=R_4=R_5=H$, $R_2=OH$, $R_6=OCH_3$
Methylformononetin: $R_1=R_4=R_5=H$, $R_2=OH$, $R_3=CH_3$, $R_6=OCH_3$
Prunetin: $R_1=R_6=OH$, $R_2=OCH_3$, $R_3=R_4=R_5=H$

Fig. 3: Major chemical composition of *Butea monosperma*

Several phytochemicals namely butein, butin , isobutrin, coreopsin, isocoreopsin (butin 7-glucoside), sulphurein, monospermoside (butein3–e-D-glucoside) and isomonospermoside, chalcones, aurones, isobutyine, palasitrin, 3',4',7 -trihydroxy flavones, myricyl alcohol, stearic, palmitic, arachidic and lignoceric acids glucose, fructose, histidine, aspartic acid, alaine and phenylalanine were reported from different parts of *B. monosperma* (Burli and Khade 2007, More et al. 2018) as listed in Table 1.

Table 1: Phytochemicals reported from different parts of *Butea monosperma*

Plant parts	Name of phytochemicals (Sindhia and Bairwa 2010; Mazumder et al. 2011, Fageria and Rao 2015, Jain et al. 2016, More et al. 2018)
Leaves	3- alphahydroxyeuph-25-enylheptacosanoate; 3,9-dimethoxypterocapan; Lignoceric acid; Linoleic acid; Oleic and; Palmitic acid.
Stem	l-3-α-L-arabinopyranoside; 3-methoxy-8,9-methylenedioxypterocarp-6-ene; 21-methylene-22-hydroxy-24-oxooctacosanoic acid me ester; 4-pentacosanylphenol; pentacosanyl- β -D-glucopyranoside; 3-Z-hydroxyeuph-25-ene; 2,14-dihydroxy-11,12-dimethyl-8-oxo-octadec-11-enylcyclohexane; 5-methoxygenistein; 8-C-prenylquercetin 7,4'-di- O-methyl-3-O- α-L-rhamnopyranosyl(1-4)- α-L-rhamnopyranoside; 3-hydroxy-9-methoxypterocarpan [(-)-medicarpin]; Lupenone, Lupeol; Nonacosanoic acid;. Prunetin; Sitosterol; Stigmasterol-e-D-glucopyranoside;
Bark	3, 9-dimethoxypterocarpan; 3 α- hydroxyeuph-25-enyl heptacosanoate; Alanind; Allophanic acid; Buteaspermanol; Butolic acid; Butrin; Cajanin; Cladrin; Cyanidin; Daidzein; Formononetin; Gallic acid; Histidine; Isoformononetin; Kino-tannic acid; Lupenone; Lupeol; (-)-Medicarpin; Methylformononetin; Miroestrol; Palasimide; Palasitrin; Prunetin; Pyrocatechin; Shellolic acid;
Root	Glucose; Glycine; aromatic hydroxyl compounds;
Flower	Alanine; Arachidic acid; Aspartic acid; 3',4',7- trihydroxyflavone; Aurones; Butein; Butin; Chalcones; Coreopsin; Fructose, Glucose, Histidine; Isobutrin; Isobutyine; Isocoreopsin (Butin 7-glucoside); Isomonospermoside; Lignoceric acids; Monospermoside (butein 3-e-D-glucoside); Myricyl alcohol; Palasitrin; Palmitic acid; Phenylalanine; Sulphurein; Stearic acid;
Seed	Allophonic acid, Jalaric ester I; Jalaric ester II; Laccijalaric ester I; Laccijalaric ester II, Monospermosides, Palasonin; Somonospermoside;

UPLC-MS BASED QUANTITATION OF PHYTOCHEMICALS

Sample Preparation for LC-MS/MS Method Development

Good method development approaches require many experimental runs as necessary to achieve the best analysis results. Method development must be as simple as possible. The method development protocols applied here (Fig. 4) is applicable to both known and unknown chemical components.

Raw Material
Part Used
(Root, Leaves, Stem, Seeds, Bark or Whole Plant)
↓
Level of Grinding (Mesh Size)
↓
(1) Plant sample: Solvent (4)
↓
(Repetition)
Three to Six Times (according to sample)
24 to 72 Hrs. (as required)
↓
Concentration
(Rotavapor)
↓
Drying
↓
Quantity of Extract Obtained
LC-MS Method Development for analysis of B. monosperma plant parts
Solvent Selection
↓
UPLC Column Selection
↓
Injection Volume Selection
↓
Method Development
↓
LC-MS Test
Tuning of MS Parameters
Chromatographic Peak Separation
↓
Modifications
↓
Method Reproducibility Test
↓
LC-MS
Quantitation of Osteogenic Phytochemicals of B. monosperma plant parts
Sample preparation (Extract and standards)
↓
MRM parameters optimization
(DP, EP, CE, CXP)
↓
UPLC Column selection
↓
Injection volume selection
↓
LC-MRM Test
Chromatographic peak separation
↓
Modification
↓
Final LC-MRM

Fig. 4: Extraction, LC-MS method development and parameter optimization.

Analytical grade chemicals and materials are required to develop an analytical method. Solvents such as acetonitrile, methanol (LC-MS grade) and formic acid (analytical grade) were purchased from Fluka, Sigma-Aldrich (St. Louis, MO, USA). AR grade ethanol was purchased from Merck Millipore (Darmstadt, Germany). Milli-Q Ultra-pure water was obtained from Millipore water purification system (Millipore, Milford, MA, USA). The reference standards of daidzein, cajanin, isoformononetin, cladrin, formononetin, methylformononetin, medicarpin, prunetin and buteaspermanol were isolated and collected from bark of *B. monosperma* by Maurya et al. (2009). Leaf and bark of *B. monosperma* were collected from Vijaypura and Uttarbehni of Jammu region and from Uttar Pradesh and Mizoram from naturally growing population. Stem and twig of *B. monosperma* were obtained from campus of Mizoram University.

Preparations of Extract

Dry plant parts of *B. monosperma* were homogeneously powdered using a pulverizer. The dried powder of each part (about 15 g) was suspended in 150 mL of 100% ethanol and sonicated for 30 min at 25°C using ultrasonic water bath (53 KHz), kept for 24 hours at room temperature (about 26-28°C) and filtered. This extraction process was repeated three times for each sample. The combined filtrate obtained from all three repeats of each sample was again filtered through Whatman filter paper and dried under reduced pressure using rotavapor (Buchi Rotavapor-R2, Flawil, Switzerland) at 40°C (Bajpai et al. 2018).

Dried extracts (approximately 1 mg) were weighed accurately and dissolved in 1 mL of 100% methanol and simply vortexed (Bandelin SONOREX, Berlin). The solutions were filtered through 0.22 μm syringe filter (Millex-GV, PVDF, Merck Millipore, Darmstadt, Germany). The filtrates were further diluted with methanol to final working concentration. 2 μL aliquot was injected into the UPLC–MS/MS system for analysis. Primary stock solutions of compounds daidzein, cajanin, isoformononetin, cladrin, formononetin, methylformononetin, medicarpin, prunetin and buteaspermanol were prepared by dissolving the compounds in methanol to achieve desired concentration of 1 mg/mL of each compound. A mixed standard stock solution containing all the analytes was prepared in methanol. Now the working standard solutions were prepared by diluting the mixed standard stock solution with methanol to a series of concentrations within the ranges 0.5 to 1500 ng/mL for plotting calibration curves as shown in Table 2. The calibration curves were constructed by plotting the value of peak areas versus the value of concentrations of each compound. All stock solutions were stored at -20 °C until use.

Table 2: Concentration of the analytes used to construct calibration curve

Name	Concentration range (ng/mL)
Daidzein	1, 5, 25, 50 and 100
Cajanin	0.5, 1, 2.5, 5 and 20
Isoformononetin	0.5, 5, 25, 100 and 200
Cladrin	0.5, 1, 2, 10 and 25
Formononetin	0.5, 1, 2.5, 10 and 25
Methylformononetin	0.5, 2, 10, 50 and 100
Medicarpin	0.5, 50, 100, 150 and 250
Prunetin	1, 10, 50, 100 and 125
Buteaspermanol	0.5, 1, 5, 10 and 25

Instrumentation and Analytical Conditions

An acquity ultra-performance liquid chromatography (UPLC) system consisting of an auto sampler and a binary pump (Waters, Milford, MA) equipped with a 10μL loop was used. The compounds were separated on an Acquity UPLC CSH C18 column (1.7 μm, 2.1×100 mm, Waters, Milford, MA) analytical column at 30°C. A gradient elution was achieved using two solvents: 0.1% (v/v) formic acid in water (A) and methanol (B) at a flow rate of 0.35 mL/min. The gradient program consisted of an initial hold to 50% (B) till 2 min then increased from 50% to 53% (B) in 3 min, 53% to 56% (B) in 4.5 min, 56% to 60% (B) in 5.5 min, 60% to 65% (B) in 6 min, 65% to 98% (B) in 6.8 min, 98% to 90% (B) in 7.2 min, 90% to 50% (B) in 7.5 min, followed by initial condition of 50% B till 8.0 min, with a sample injection volume of 2μL (Bajpai et al. 2018).

The UPLC system was interfaced with hybrid linear ion trap triple-quadrupole mass spectrometer (API 4000 QTRAP™ MS/MS system from AB Sciex, Concord, ON, Canada) equipped with electrospray (Turbo V™) ion source. ESI in negative ion mode was used to optimize the parameters which were the ion spray voltage 4200 V, the turbo spray temperature, 550°C; nebulizer gas, 20 psi; heater gas, 20 psi; collision gas, and curtain gas at 20 psi. Optimization of the mass spectrometric conditions were carried out by infusing 100 ng/mL solutions of the compounds dissolved in methanol at 10 μL/min flow rate using a Harvard syringe pump (Harvard Apparatus, South Natick, MA, USA). The full range scan from m/z 100 to 1000 in-ESI-MS analysis was recorded. The precursor ion, product ion, corresponding declustering potential (DP), entrance potential (EP), collision energy (CE) and cell exit potential (CXP) were optimized and the most intense product ion from precursor ion of each compound was chosen for the MRM (Table 3). Analyst 1.5.1 software (AB Sciex) was used for data acquisition and data processing (Bajpai et al. 2018).

Optimization of UPLC-MS Conditions

Two columns, acquity BEH C$_{18}$ column (2.1 mm×50 mm, 1.7μm) and Acquity UPLC CSH C18 column (2.1 mm×100 mm, 1.7 μm), with varying flow rates, mobile phases, and column temperatures were tested ensuing in the best combination, as described in experimental section for resolution and detection of all the analytes.

ESI positive (+) and negative (-) ion modes were evaluated to get better response of analytes. The best response for compounds 1-9 was achieved in the (-) ESI mode. All the compound dependent MS parameters (precursor ion, product ion, declustering potential (DP) and collision energy (CE) were carefully optimized in (-) ESI mode by flow injection analysis (FIA) using individual standard solution. The chemical structures of all nine compounds were characterized based on their retention behaviour and MS information. In the full scan mass spectra, the parent ion of daidzein (*m/z* 252.9), cajanin (*m/z* 299.3), isoformononetin (*m/z* 266.7), cladrin (*m/z* 297.2), formononetin (*m/z* 267.0), methylformononetin (*m/z* 280.9), medicarpin (*m/z* 269.0), prunetin (*m/z* 283.0) and buteaspermanol (*m/z* 284.9) were stable in high abundance. The most intense product ions under the product ion scan mode were for daidzein *at m/z* 223.0, for cajanin at *m/z* 164.9, isoformononetin at *m/z* 251.9, cladrin at *m/z* 267.0, formononetin at *m/z* 251.9, methylformononetin at *m/z* 265.9, medicarpin at *m/z* 254.0, prunetin at *m/z* 267.9 and for buteaspermanol *at m/z* 269.9. The mass spectrometric parameters were optimized to obtain the higher signal of precursor ions and product ions as above mentioned (Bajpai et al. 2018).

Analytical Method Validation (Guideline ICH 2005)

Validation of an analytical method is a set of experiment by which it is recognized by laboratory studies, that the performance characteristics of the analytical method meet the requirements for the indented analytical application. The objective of validation of an analytical procedure is to demonstrate the performance and reliability of a

Table 3: Compound dependent parameters (MRM)

Compound	Retention time (min)	Precursor (Q1) mass (Da)	Product (Q3) mass (Da)	Declustering Potential (eV)	Entrance	Collision Energy (eV)	Cell Exit Potential (eV)
Daidzein	1.5	252.9	223.0	-110.2	-11.0	-45.0	-12.5
Cajanin	3.6	299.3	164.9	-94.5	-8.2	-30.0	-13.1
Isoformononetin	3.4	266.7	251.9	-92.2	-9.0	-29.0	-22.1
Cladrin	2.7	297.2	267.0	-141.0	-8.8	-32.0	-18.1
Formononetin	4.3	267.0	251.9	-108.1	-9.3	-29.0	-23.1
Methylformononetin	4.8	280.9	265.9	-175.0	-9.0	-30.0	-26.0
Medicarpin	4.5	269.0	254.0	-81.0	-9.7	-24.5	-21.3
Prunetin	5.7	283.0	267.9	-131.3	-10.0	-31.0	-17.1
Buteaspermanol	1.9	284.9	269.9	-86.0	-10.5	-26.0	-22.0

method as well as generate confidence on the results. The developed UPLC-MS method was validated according to the international conference on harmonization (ICH, Q2R1) guidelines by specificity, selectivity, linearity, LODs and LOQs, precision, stability and recovery (Guideline 2005).

Specificity and Selectivity

Specificity is the capability to clearly assess the analyte in the presence of other components, which may be expected to be present. Selective refers to a method that provides responses for a number of chemical entities that may or may not be distinguished from each other.

Stock solutions of the extracts and standard compounds were prepared in methanol. The chromatographic interferences were assessed by comparing chromatograms of blank methanol and the methanol samples spiked with all nine compounds. No interfering peaks were observed at the retention time of each compound in blank methanol sample. All compounds were rapidly eluted with retention times of 1.5, 3.6, 3.4, 2.7, 4.3, 4.8, 4.5, 5.7 and 1.9 min for daidzein, cajanin, isoformononetin, cladrin, formononetin, methylformononetin, medicarpin, prunetin and buteaspermanol, respectively (Fig. 5) (Bajpai et al. 2018).

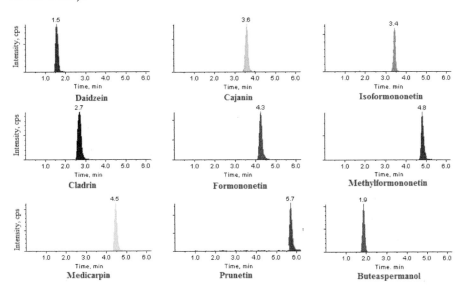

Fig. 5: Extracted ion chromtaogram of analytes

Linearity of Calibration Curves

A linear relationship should be evaluated across the range of the analytical procedure. The linearity of an analytical procedure is its ability to obtain test results which are directly proportional to the concentration of analyte in the sample. The linearity of the method was established by constructing calibration curves over a concentration range from 0.5-1500 ng/mL. Six concentrations of calibration standard solution were used and analyzed in triplicate, to construct the calibration curves by plotting the ratios of the peak areas of each standard to the concentration of each compound. The slope, intercept, and correlation coefficient of each calibration curve were determined by linear regression analysis.

The calibration curves of compounds 1 to 9 showed a linear relationship between peak area and concentration over the range of 1-100 for daidzein, 0.5-20 for cajanin, 0.5-200 for isoformononetin, 0.5-25 for cladrin, 0.5-25 for formononetin, 0.5-100 for methylformononetin, 0.5-250 for medicarpin, 1-125 for prunetin and 0.5-25 for buteaspermanol respectively. The correlation coefficient (R^2) of the calibration curves for all compounds were found more than 0.9995 (Table 4) (Bajpai et al. 2018).

Limits of Detections (LODs) and Limits of Quantitations (LOQs)

The limit of detection of an individual analytical procedure is the lowest amount of analyte in a sample which can be detected but not necessarily quantitated as an exact value. The limit of quantitation of an individual analytical procedure is the lowest amount of analyte in a sample which can be quantitatively determined with required precision and accuracy. The LODs and LOQs were defined as a signal-to-noise ratio (S/N) equal to 3.3 and 10, respectively.

Based on the standard deviation of the response and the slope, the LODs and LOQs for each compound of all nine phytochemicals of *B. monosperma* leaf and bark varied from 0.03-0.30 ng/mL and 0.09 to 0.92 ng/mL, respectively (Table 4) (Bajpai et al. 2018) which are much lower than those reported by earlier HPLC methods (Gupta et al. 2010, Patil and Rajput 2012, Nile and Park 2014).

Precision, Stability and Recovery

Precision of an analytical procedure expresses the closeness between a series of measurements obtained from multiple sampling of the same homogeneous sample under the prescribed conditions. System stability

Table 4: Validation parameters of the developed method

Reference standards	Linearity			LOD (ng)	LOQ (ng)	Precision (%RSD)		Stability % RSD (n=5)	Recovery (n=3)	
	Linear range (ng/mL)	R^2	Regression equation			Intraday (n=6)	Interday (n=6)		Mean	RSD (%)
Daidzein	1.0-100	1.0000	y = 4584*x + 697.9	0.03	0.09	0.12	0.21	1.58	102.3	1.06
Cajanin	0.5-20	0.9999	y = 2622*x − 53.94	0.12	0.36	1.32	0.95	0.78	95.2	1.41
Isoformononetin	0.5-200	0.9995	y = 138*x − 7.64	0.04	0.11	1.74	1.07	2.21	102.7	1.85
Cladrin	0.5-25	1.0000	y = 12057*x − 700.9	0.11	0.32	1.39	1.84	2.14	98.8	0.92
Formononetin	0.5-25	0.9999	y = 297*x + 84.8	0.15	0.45	0.98	1.65	2.06	105.8	1.88
Methylformononetin	0.5-100	1.0000	y = 432.8*x − 0.100	0.30	0.92	0.61	1.20	2.80	101.6	1.95
Medicarpin	0.5-250	0.9998	y = 187*x + 8.15	0.13	0.39	0.51	0.98	1.51	97.1	1.65
Prunetin	1-125	1.0000	y = 300*x + 48.6	0.06	0.18	1.23	0.98	2.40	100.9	1.45
Buteaspermanol	0.5-25	0.9999	y = 10300*x − 1620	0.05	0.14	0.11	1.01	1.56	104.2	1.32

is considered appropriate when the RSD, calculated on the assay results obtained at different time intervals, does not exceed more than 20 percent of the corresponding value of the system precision.

Intra- and inter-day variations were chosen to determine the precision of the developed method. For intra-day precision test, the standards solutions were analyzed for six replicates within one day, while for inter-day precision test, the solutions were examined in duplicate on three consecutive days. In order to again assess the intra- and inter-day precision and accuracy of the assay, samples at low, medium, and high concentrations were prepared as described above. Further, the interday and intraday precisions of the assay were assessed for the analysis of samples in six replicates within a day and three consecutive days, respectively. The accuracy was calculated on the basis of the difference in the mean calculated concentration. In order to investigate the solution stability of the samples, each sample solution was analyzed at 0, 2, 4, 8, 12 and 24 h and variations were expressed in relative standard deviation (RSD). Recovery assays were carried out for compounds 1 to 9 by spiking samples with known amount of compounds at three different concentration levels (120, 100, and 80%), respectively, in triplicate.

Intra-day and inter-day precisions for each compound are given in Table 4. The assay values on both the occasions (intra- and inter-day) were found to be within the accepted variable limits. Intra-day and inter-day precision were less than 1.74 % and 1.84 %, respectively. Stability of sample solutions stored at room temperature (about 26-28 °C) was investigated by replicate injections of the sample solution at 0, 2, 4, 8, 12 and 24 h. The RSDs values of stability of the nine compounds were 2.80 %. A recovery test was applied to evaluate the accuracy of this method. Three different concentration levels (high, middle and low) of the analytical standards were added into the samples in triplicate and mean recoveries were determined. The developed analytical method had shown good accuracy with overall recovery in the range from 95.2-105.8 % (RSD 1.95 %) as shown in Table 4 (Bajpai et al. 2018).

Method Application

The developed and validated UPLC-MS method was used to quantify the selected phytochemicals in ethanolic extract of different parts of B. monosperma and the contents of all compounds are listed in Table 5 (Bajpai et al. 2018).

Table 5: Contents (in mg/kg) of compounds 1-9 in different *Butea monosperma* samples and plant parts (n=3)

Sample	Daidzein	Cajanin	Isoformononetin	Cladrin	Formononetin	Methyl formononetin	Medicarpin	Prunetin	Buteaspermanol
J1 leaf	295	bdl	10600	31.75	bdl	196	1810	10800	820
J2 leaf	1230	2.79	40600	22.95	28.25	39.05	6950	49000	2605
J3 leaf	540	bdl	32050	bdl	bdl	9.3	98	33150	6750
UP leaf	1800	2.45	660	31.95	bdl	9.3	91	1570	795
Mz leaf	131	4.165	3530	147	70	227.5	2495	22300	1895
J1 bark	288	bdl	13900	23.9	bdl	15.8	63.5	70000	235
J2 bark	156.5	bdl	13250	1.11	bdl	18.5	123	57000	98
J3 bark	56	bdl	8000	145	bdl	2.255	67.5	41800	7.55
UP bark	2715	181.5	1480	163	bdl	46.65	233.5	49250	117
Mz bark	411	53.5	24900	1195	bdl	7.65	200.5	145000	48.15

bdl: below detection level; Mz: Mizoram; J1, J2, J3: Jammu; UP: Uttar Pradesh

Notable differences in contents of targeted phytochemicals of *B. monosperma* plant /parts samples collected from different geographical regions were observed during this analysis. Daidzein (2490 mg/kg) detected from stem was found as the most abundant compound among all the parts, while cajanin (53.5 mg/kg), isoformononetin (24900 mg/kg), cladrin (1195 mg/kg) and prunetine (145000 mg/kg) were abundant in the bark of *B. monosperma* collected from Mizoram. The total content of all bioactive compounds was highest in the bark. All these bioactive compounds were first identified and quantitated in the leaf of *B. monosperma* which has the second highest total content. The content of all compounds was insignificant in twig. In all geographical collection of leaf daidzein (1800 mg/kg) was highest in UP region while the isoformononetin (40600 mg/kg) was most abundant in Jammu region. The contents of medicarpin (6950 mg/kg), prunetin (49000 mg/kg) and buteaspermanol (6750 mg/kg) were also high in Jammu region samples (Bajpai et al. 2018).

The total content of selected phytochemicals in leaf collected from Mizoram was higher than UP sample. The total content of these analytes was highest in the leaf samples of Jammu region and lowest in UP region. Formononetin was below detection level in the bark from all geographical regions. Bark was identified as the richest source of prunetin of all samples and also found in highest amount from Mizoram region samples. Total content of all bioactive compounds was highest in bark of Mizoram region.

Principal Component Analysis

Principal component analysis (PCA) is a discrimination technique was carried out on the basis of contents obtained by quantitative analysis of leaf and bark of *B. monosperma* collected from different geographical regions in India. Minitab 17.0 software trial version was used for PCA. It was applied to explore contribution of all nine selected phytochemicals (1-9) for discrimination of *B. monosperma* leaf and bark. The content of all nine compounds in five samples of each leaf and bark were determined simultaneously and three replicate of each sample were subjected to PCA analysis. Hence the data matrix was composed of 15 X 9 for leaf and 15 X 9 for bark samples. The PCA was performed using a correlation matrix. To discriminate the bark samples of all selected geographical region, the content of compounds 1-9 were subjected to PCA analysis in which formononetin (m/z 267.0) was not significant due to zero variance. The PCA score plot from the rest eight variables completely discriminated the bark samples of selected geographical regions (Fig. 6A). In case of leaf out of nine variables (content of compounds) only five variables namely daidzein (*m/z* 252.9), cajanin (*m/z* 299.3), isoformononetin (*m/z* 266.7), cladrin (*m/z* 297.2) and formononetin (*m/z* 267.0) were able to discriminate the cases of geographical regions (Fig. 6B). It indicated that the content of bioactive molecules can discriminate the case of different geographical regions and could be used as bioactive marker compounds for the *B. monosperma* leaf and bark (Bajpai et al. 2018).

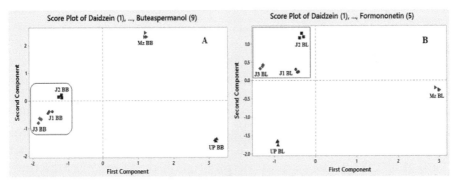

Fig. 6: PCA score plots for discrimination of the bark (A) and leaf (B)

Traditional Uses and Pharmacology Aspects

B. monosperma plant parts have been used since long time as medicine for the treatment of piles, skin diseases, tumors, ulcers and wounds (Mengi and Deshpande 1995, Sindhia and Bairwa 2010). It is extensively used in Ayurveda, Homeopathy and Unani systems of medicine and has become an important source of modern medicine. Commonly in folklores it is used as astringent, aphrodisiac, diuretics and tonic (Sindhia and Bairwa 2010, Rana and Avijit 2012). The plant is a well-known rejuvenator according to the traditional system of medicine.

Due to its multipurpose uses, this plant became a versatile tree with immense medicinal and economical value. Due to the presence of an active constituent (-) medicarpin in the stem bark of *B. monosperma*, it shows antifungal activity. The bark extract express cutaneous wound healing in rats on the topical administration of alcoholic bark extract. The stem bark is also used for the treatment diarrhea, dysentery and dyspepsia (The wealth of India 1988, Sindhia and Bairwa 2010). *B. monosperma* leaves have anticonvulsive and anti-inflammatory activities, and promotes diuresis and menstrual flow (Patil et al. 2006,

Burli and Khade 2007). Extract of *B. monosperma* flower exhibit antidiabetic (Somani et al. 2006), antiesterogenic, antifertility (Sharma and Deshwal 2011) and free radical scavenging activities (Lavhale and Mishra 2007). It is also used as herbal tonic to treat and cure burn sensation, skin disease and gout (Burli and Khade 2007). The flower of this plant is also used in astringent diarrhoea and diuretic (Patil et al. 2006).

The oil of *B. monosperma* seed has significant bactericidal and fungicidal effect (More et al. 2018). The seed has also shown haemagglutinating activity and is used as remedy against intestinal worms (Iqbal et al. 2006, Patil et al. 2006, Burli and Khade 2007). It also used in urinary stones, inflammation condition, eye disease and in treatment of bleeding piles (Burli and Khade 2007). *B. monosperma* roots are useful in curing night blindness and filariasis disease (Mengi and Deshpande 1995, Fageria and Rao 2015). Roots of this plant are bitter, antihelminthic and causes temporary sterility in women (The wealth of India, 1988, Fageria and Rao 2015). Stem of *B. monosperma* is useful as indigenous medicine for the treatment of dyspepsia and sore throat (Patil et al. 2006). The gum of the plant is used in cure of ringworm, septic sore throat and stomatitis. It is also used for the treatment of diarrhea, excessive perspiration and leucorrhoea (Boutelje 1980, Mazumder et al. 2011).

SCIENTIFIC INTERVENTION WITH FUTURE PERSPECTIVES

A precise, fast and validated UPLC-MS analytical method has been developed for quantitation of phytochemicals in *B. monosperma* plant. In this study phytochemicals reported for osteogenic activity from the bark of plants were simultaneously determined in other parts such as leaf, stem, bark, stem and twig by UPLC-MS method. The method is effective and validated according to ICH guidelines. Validation parameters like regression coefficient, linearity, precisions, LOD, LOQ, stability and recovery were completed successfully. This method can be applied for quantification of these phytochemicals in crude extracts from other plants/herbal formulations and dietary supplements. PCA can be effectively used for discriminating among the plants of different geographical regions on the basis of contents of phytochemicals present. These phytochemicals can also be used as bioactive markers. However, much work is still required to identify and quantitate other known and unknown compounds present in this potential medicinal plant. Fingerprint analysis is also the main demand for the standardization and quality control of this plant. The biological activity evaluation according to identified compounds can be further explored.

LITERATURE CITED

Allwood JW, Goodacre R. 2010. An introduction to liquid chromatography–mass spectrometry instrumentation applied in plant metabolomic analyses. *Phytochemical Analysis* 21(1): 33-47.

Ambasta BP. 1994. The useful plants of India. *CSIR, New Delhi* 1 -91.

Bajpai V, Singh A, Singh P, Sharma K, Singh B, Singh BP, Sahai M, Maurya R, Kumar B. 2018. Development of Ultra Performance Liquid Chromatography Tandem Mass Spectrometry Method for Simultaneous Identification and Quantitation of Potential Osteogenic Phytochemicals in *Butea monosperma*. *Journal of Chromatographic Science*. doi.org/10.1093/chromsci/bmy050.

Burli DA, Khade AB. 2007. A Comprehensive review on *Butea monosperma* (Lam.) Kuntze. *Pharmacognosy Reviews* 1(2): 94-99.

Boutelje JB. 1980. Encyclopedia of world timbers: names and technical literature. *Swedish Forest Products Research Laboratory.*

Cowen DV. 1984. Flowering Trees and shrubs in India. Bombay: Thacker and Co.Ltd. *Agroforest Today* 6: 7.

Fageria D, Rao DV. 2015. A review on *Butea monosperma* (Lam.) kuntze: A great therapeutic valuable leguminous plant. *International Journal of Scientific and Research Publication* 5: 1-8.

Gupta SK. 2015. Ethnobotany of *Butea Monosperma* (LAM.) Kuntze. in Jammu and Kashmir, India. *International Journal of Science and Research* 6(6): 333-334.

Gupta SR, Ravindranath B, Seshadri TR. 1970. The glucosides of *Butea monosperma*. *Phytochemistry* 9(10): 2231-2235.

Gupta V, Dwivedi AK, Yadav DK, Kumar M, Maurya R. 2010. Reverse phase-HPLC method for determination of marker compounds in NP-1: an anti-osteoporotic plant product from *Butea monosperma*. *Natural Product Communication* 5: 47-50.

Guidelines, Validation of Analytical Procedures: Text Methodology. 2005. *International Conference on Harmonisation (ICH) Q2 (R1)*. Accessed in February, 2014.

Harish M, Ahmed F, Urooj A. 2014. In vitro hypoglycemic effects of *Butea monosperma* Lam. leaves and bark. *Journal of Food Science and Technology* 51: 308-314.

Iqbal Z, Lateef M, Jabbar A, Ghayur MN, Gilani AH. 2006. *In vivo* anthelmintic activity of Butea monosperma against Trichostrongylid nematodes in sheep. *Fitoterapia* 77(2): 137-140.

Jain D, Jain A, Shrivastava S. 2016. A Review on *Butea Monosperma*: A Pharmacologically Potent Plant. *Asian Journal of Pharmaceutical Education and Research* 5(4): 9-28.

King R, Fernandez-Metzler C. 2006. The use of Qtrap technology in drug metabolism. *Current Drug Metabolism* 7(5): 541-545.

Kirtikar KR. Basu BD. 1935. Indian medicinal plants, *Lalit mohan Basu Allahabad India* 785-788.

Lavhale MS, Mishra SH. 2007. Evaluation of free radical scavenging activity of *Butea monosperma* Lam. Indian. *Journal of Experimental Biology* 45: 376-384.

Maurya R, Yadav DK, Singh G, Bhargavan B, Narayana PS, Sahai M, Singh M. 2009. Osteogenic activity of constituents from *Butea monosperma*. *Bioorganic Medicinal Chemistry Letters* 19: 610-613.

More S, Jadhav VM, Kadam VJ. 2018. A comprehensive review on *Butea monosperma*: A valuable traditional plant. *International Journal of Botany Studies* 3(2): 65-71.

Mazumder PM, Das MK, Das S, Das S. 2011. *Butea monosperma* (LAM.) Kuntze - A Comprehensive Review. *International Journal of Pharmaceutical Science and Nanotechnology* 4(2): 1390-1393.

Mengi SA, Deshpande SG. 1995. Comparative Evaluation of *Butea frondosa* and Flurbiprofen for Ocular Anti inflammatory Activity in Rabbits. *Journal of Pharmacy and Pharmacology* 47: 997-1001.

Mishra M, Shukla YN, Kumar S. 2000. Euphane triterpenoid and lipid constituents from *Butea monosperma*. *Phytochemistry* 54: 835-838.

Nile SH, Park SW. 2014. HPTLC analysis, antioxidant and antigout activity of Indian plants. *Iranian Journal of Pharmaceutical Research* 13: 531-539.

Orwa C, Mutua A, Kindt R, Jamnadass R, Simons A. 2009. Agroforestree database: a tree species reference and selection guide version 4.0. *World Agroforestry Centre ICRAF, Nairobi, KE*, 1-5.

Patil B, Rajput A. 2012. GC-MS analysis of biologically active compounds of chloroform extract of leaves of *Butea monosperma*. *Journal of Pharmacy Research* 5: 1228-1230.

Patil MV, Pawar S, Patil DA. 2006. Ethnobotany of *Butea monosperma* (Lam.) Kuntze in North Maharashtra, India. *Natural Product Radiance* 5(4): 323-325.

Rana F, Avijit M. 2012. Review on *Butea monosperma*. *International Journal of Research in Pharmacy and Chemistry* 2: 1035-9.

Raskin I, Ribnicky DM, Komarnytsky S, Ilic N, Poulev A, Borisjuk N, Brinker A, Moreno DA, Ripoll C, Yakoby N, O'Neal JM. 2002. Plants and human health in the twenty-first century. *Trends in Biotechnology* 20(12): 522-531.

Schwartz JC, Wade AP, Enke CG, Cooks RG. 1990. Systematic delineation of scan modes in multidimensional mass spectrometry. *Analytical Chemistry* 62(17): 1809-1818.

Sharma C, Kumari T, Pant G, Bajpai V, Srivastava M, Mitra K, Kumar B, Arya KR. 2015. Plantlet formation via somatic embryogenesis and LC ESI Q-TOF MS determination of secondary metabolites in *Butea monosperma* (Lam.) Kuntze. *Acta Physiologiae Plantarum* 37(11): 239.

Sharma AK. Deshwal N. 2011. An overview: On phytochemical and pharmacological studies of *Butea monosperma*. *International Journal of Pharmtech Research* 3: 864-871.

Sindhia VR, Bairwa R. 2010. Plant review: *Butea monosperma*. *International Journal of Pharmaceutical and Clinical Research* 2(2): 90-4.

Somani R, Kasture S, Singhai AK. 2006. Antidiabetic potential of *Butea monosperma* in rats. *Fitoterapia* 77: 86-90.

The Wealth of India, A dictionary of India raw material and Industrial products. *Publication and Information Directorate, CSIR, New Delhi* 1988.

Tyagi M, Srivastava K, Kureel J; Kumar A, Raghuvanshi A, Yadav D, Maurya R, Goel A, Singh D. 2012. A premature T cell senescence In Ovx mouse is inhibited by repletion of estrogen and medicarpin: a possible mechanism for alleviating bone loss. *Osteoporosis International* 23: 1151-1161.

Verma M, Shukla YN, Jain SP, Kumar S. 1998. Chemistry and biology of the Indian dhak tree *Butea monosperma*. *Journal of Medicinal and Aromatic Plant Sciences* 20: 85-92.

9

Diversity, Bioprospection and Commercial Importance of Indian Magnolias

Aabid Hussain Mir[1], Licha Jeri[2], Krishna Upadhaya[3*]
Nazir Ahmad Bhat[2], Rajib Borah[3], Hiranjit Choudhury[3]
and Yogendra Kumar[2]

[1]*Department of Environmental Studies, North-Eastern Hill University*
Shilllong - 793 022, Meghalaya, INDIA
[2]*Centre for Advanced Studies in Botany, North-Eastern Hill University*
Shilllong - 793 022, Meghalaya, INDIA
[3]*Department of Basic Sciences and Social Sciences, North-Eastern Hill*
University, Shilllong - 793 022, Meghalaya, INDIA
Email: upkri@yahoo.com

ABSTRACT

The current paper discusses the diversity, distribution and commercial importance of Indian Magnolias. Globally, these species are found in tropical, subtropical and temperate forests of southeastern Asia and tropical America. In India, there are 25 species of Magnolias and most of them are distributed in Northeast India, with highest number in Assam. Of all the species, two (*Magnolia gustavii* and *M. pleiocarpa*) are critically endangered, one (*M. pealiana*) endangered, two (*M. manii* and *M. nilagirica*) vulnerable, nine least concern and ten data deficient at global level. The highly threatened nature of most of the magnolias species calls for their immediate conservation and protection measures. Members of this genus are known to be rich in a wide variety of biologically active compounds including alkaloids, flavonoids, lignans, neolignans and terpenoids. Many of the species have been found to possess potent procognitive activity, anti-oxidative, anti-microbial, anti-inflammatory, anti-angiogenic, diuretic, anti-ulcer, analgesic, anti-helmintholytic, and anti-cancer activities. The species have huge economic potential and are used for a number of purposes including ornamental, medicinal, culinary, timber and joinery works.

Keywords: Conservation, Magnolias, Diversity, Phytochemistry, Pharmacology, India

INTRODUCTION

The family Magnoliaceae is a major group of angiosperms, displaying many characters that are considered as evolutionarily primitive (Qiu et al. 1995). Globally, the members of the family are distributed from tropical to temperate regions of eastern North America, east Asia and south America, with a total of 250 taxa (Azuma et al. 2001). Magnolia is the largest and most important genus in the family, with a total of 247 species (TPL 2018), of which about 25 species are found in India (Kumar 2012). Unlike most angiosperms, whose flower parts are in whorls, the magnolias have their stamens and pistils in spirals on a conical receptacle. This arrangement is found in some fossil plants and is believed to be a basal condition for angiosperms. Another primitive aspect of the species is exhibited in their large, cup shaped flowers that lack distinct petals or sepals and in turn replaced by a large non-specialized flower part, resembling petals called tepals (Kundu 2009). The fruit is an etaerio of follicles which usually become closely appressed as they mature and open along the abaxial surface. Seeds have a fleshy coat and color ranges from red to orange. The members of the genus are pollinated by beetles as the winged insects did not evolve during primitive time, while the dispersal of seeds mainly occurs by birds (Kundu 2009). The members of the family are very important for their use as timber, ornamental, medicinal and for other products (Pandey and Misra 2009). Many species have long been used in folk medicine, which have attracted a great deal of interest in bioprospection studies (Martinez et al. 2006). The bioprospection, which entails the search for economically valuable genetic and biochemical resources (e.g. new medicines) from nature, if well managed, can be advantageous, since it can generate income, and at the same time provides incentives for the conservation of biodiversity (Martinez et al. 2006).

India has been globally recognized as one of the megabiodiverse region, rich in endemic species (Mittermeier et al. 2004). The country has a rich diversity of Magnolia species, of which most are located in Himalayan regions, with many of them having localized distribution

(Kumar 2012). In spite of economic and pharmaceutical importance, Magnolia species of the country are declining at an alarming rate and are facing the danger of extinction and are therefore, the key species for conservation needing immediate attention. There is less information available on diversity, bioprospection and commercial values of Indian magnolias. Therefore, the present study was carried out to investigate the diversity, bioprospection and commercial values of Indian magnolias. The study aimed to provide a detailed database of Magnolia and also to help in introducing these species in the modern clinical preparations. The present review will help in further research, development and conservation of the Indian species.

WORK DONE

Diversity and distribution of the magnolias in India was compiled by using published literature and detailed floristic field surveys carried out in various parts of northeast India. The consulted floristic literature includes Flora of British India (Hooker 1875), Bengal Plants (Prain 1903), Flora of Upper Gangetic Plains (Duthie 1922), Flora of Madras Presidency (Gamble and Fischer 1935), Flora of Assam (Kanjilal et al. 1934), Dicotyledonous Plants of Manipur Territory (Deb 1961), Flora of Jowai (Balakrishnan 1981), Flora of Tripura (Deb 1981), Flora of Himachal Pradesh (Chowdhery and Wadhwa 1984), Flora of Karnataka (Saldanha 1984), Forest Flora of Meghalaya (Haridasan and Rao 1985), Flora of Tamil Nadu (Henry et al. 1989), Materials for the Flora of Arunachal Pradesh (Hajra et al. 1996) and Flora of Andhra Pradesh (Pullaiah and Ali-Mouali 1997). Herbarium specimen housed at Botanical Survey of India (BSI), Eastern Regional Circle, Shillong and BSI Arunachal Pradesh was consulted. Reports from national, regional and international organizations, the Annals of Royal Botanical Gardens (King 1891) were also consulted. Database for bioprospection of the species, including pharmacological studies and phytochemical constituents were compiled using published literatures (Zheng et al. 1999, Sarker and Maruyama 2002, Aruna et al. 2012, Chowdhury et al. 2013, Karthikeyan et al. 2016). The threat status of the species provided is based on IUCN Red Data List (IUCN 2018).

DIVERSITY AND DISTRIBUTION

A total of 25 species of magnolias are found in India, of these one species *M. grandiflora* is a native of south-eastern region of United States. This species has been introduced in India and found in the regions of North-Eastern and Southern India (Kumar 2012). Out of

29 states and 7 Union territories, the species of Magnolia were distributed in 21 States and 1 Union Territory. Majority of the species were reported from the state of Assam (20 species), followed by Arunachal Pradesh (16), Sikkim (14), Meghalaya (13), West Bengal (8), Manipur and Nagaland (6 each), Mizoram (6), Tamil Nadu (5), Tripura (4). Out of the 25 species, *M. champaca* was the widely distributed plant as reported in most of the states (16 states) followed by *M. grandiflora* (13 states). *M. doltsopa* and *M. hodgsonii* were reported from 8 states each, while *M. baillonii*, *M. pealiana* and *M. pleiocarpa* were restricted to one state each. The detailed checklist of Indian Magnolia species along with their distribution and threats is presented in table 1.

The Magnolia species of the country are mostly found in dense forests, with closed canopy. Since last few decades, owing to industrialization, urbanization and population rise, the habitat of the species has been destroyed and severely fragmented. During the current study, it was observed that out of 25 species, two species (*M. gustavii* and *M. pleiocarpa*) are critically endangered, one (*M. pealiana*) as endangered and two (*M.manii* and M. *nilagirica*) are Vulnerable at global level (IUCN 2018). There were nine species that were classified as least Concern and ten as data deficient (IUCN 2018).

Among the 25 species, 8 species (*M. baillonii, M. kingie, M. mannii, M. pealiana, M. pleiocarpa, M. punduana, M. rabaniana* and *M. oblonga*) showed a decreasing population trend, hence these species needs to be evaluated more extensively so as to know its proper threat concerning conservation. Anthropogenic threats like deforestation, forest fragmentation, shifting cultivation, and agricultural expansion are the main factors responsible for their population decline (Rivers et al. 2016). Threatened species with limited geographical ranges are most susceptible to extinction than widely distributed species as they are extremely vulnerable to environmental change and anthropogenic disturbances (Myers 1988). The highly threatened nature of the species calls for their immediate conservation and rescue measures.

PHYTOCHEMICALS IN MAGNOLIAS

Magnolia species are known to contain wide range of biologically active compounds including alkaloids, flavonoids, lignans, neolignans and terpenoids (Pyo et al. 2002). The chemical structures of major compounds found in Indian magnolias are given in figure 1. The flowers are pleasantly scented and essential oils are extracted from many species

Table 1: Diversity, distribution phenology IVCN and status of Indian Magnolias

Species	Global distribution	Distribution in India	IUCN Status	Flowering	Fruiting	Flower color
Magnolia baillonii Pierre	Cambodia, China, India, Myanmar, Thailand and Vietnam	Assam	LC	Mar-May	Aug-Oct	White or cream
Magnolia campbellii Hook. f. & Thomson	Bhutan, China, India, Myanmar and Nepal	Arunachal Pradesh, Assam, Manipur, Tamil Nadu and Sikkim	LC	Feb-Apr	Oct-Dec	White or cream
Magnolia cathcartii (Hook.f. & Thomson) Noot.	Bhutan, China, India, Myanmar, Thailand and Vietnam	Meghalaya, Nagaland, Sikkim and West Bengal	LC	Apr-May	Sept-Oct	White
Magnolia caveana (Hook. f. & Thomson) D.C.S. Raju & M.P. Nayar	China, India and Myanmar	Arunachal Pradesh, Assam, Meghalaya and Nagaland	DD	Mar-Apr	Jun-Jul	White
Magnolia doltsopa (Buch. -Ham. ex DC.) Figlar	Bhutan, China, India, Myanmar and Nepal	Arunachal Pradesh, Assam, Manipur, Meghalaya, Mizoram, Nagaland, Sikkim and West Bengal.	DD	Sept-Oct	Oct-Nov	White
Magnolia globosa Hook. f. & Thomson	Butan, China, India, Myanmar and Nepal	Arunachal Pradesh, Sikkim and West Bengal	LC	May-Jun	Aug-Sept	White or cream
Magnolia griffithii Hook. f. & Thomson	Bangladesh, India and Myanmar	Arunachal Pradesh and Assam	DD	Mar-May	Sept-Nov	White or cream

Contd.

Species	Distribution	Indian distribution	Status	Flowering	Fruiting	Flower colour
Magnolia gustavii King	India, Myanmar and Thailand	Arunachal Pradesh and Assam.	CR	Apr-May	Nov-Jan	White or cream
Magnolia hookeri (Cubitt & W.W.Sm.) D.C.S. Raju & M. P. Nayar	China, India, Myanmar and Thailand	Arunachal Pradesh and Assam	DD	Apr-May	Jun-Jul	White or cream
Magnolia kingii (Dandy) Figlar	India and Bangladesh	Assam, Meghalaya, Sikkim and West Bengal	DD	Aug-Sept	Nov-Dec	White
Magnolia kisopa (Buch.-Ham. ex DC.) Figlar	Bhutan, China, India and Nepal.	Uttar Pradesh, Uttarakhand and Sikkim	DD	Jul-Sept	Oct-Jan	White or cream
Magnolia liliifera (L.) Baill.	Cambodia, China, India, Indonesia, Lao PDR, Malaysia, Papua New Guinea, Philippines, Singapore, Thailand and Vietnam	Andaman Islands, Assam and Sikkim	LC	Apr-May	Nov-Dec	Yellow to pinkish Purple
Magnolia mannii (King) Figlar	India and Bangladesh	Arunachal Pradesh and Assam	VU	Oct-Dec	Apr-May	White or cream
Magnolia nilagirica (Zenker) Figlar	India and Sri Lanka	Tamil Nadu, Karnataka and Kerala	VU	Mar-Apr	Aug-Sept	White
Magnolia pealiana King.	India	Assam	E	Nov-Dec	Apr-May	White or cream
Magnolia pleiocarpa (Dandy) Figlar & Noot.	India	Assam	CR	Apr-May	Aug-Nov	White
Magnolia pterocarpa Roxb.	India and Myanmar	Arunachal Pradesh, Assam, Meghalaya, Tripura and Sikkim.	DD	Apr-May	Oct-Dec	White

Contd.

Magnolia punduana (Hook.f. & Thomson) Figlar	Bangladesh, Bhutan and India	Arunachal Pradesh, Manipur, Meghalaya, Nagaland and Sikkim	DD	Oct-Nov	Aug-Sept	White or cream
Magnolia rabaniana (Hook. f. & Thomson) Raju & Nayar.	India	Arunachal Pradesh, Assam, Meghalaya, Mizoram and Sikkim	DD	Apr-June	Aug-Oct	White
Magnolia champaca (L.) Baill. ex Pierre	Bangladesh, Cambodia, China, India, Indonesia, Lao PDR, Malaysia, Myanmar, Nepal, Thailand and Vietnam	Assam, Andaman and Nicobar Islands - Introduced, Himachal Pradesh, Jammu-Kashmir, Karnataka, Kerala, Manipur, Meghalaya, Mizoram, Nagaland, Orissa, Sikkim, Tamil Nadu, Tripura, Uttar Pradesh and West Bengal	LC	Jun-Sep	Sept-Oct	Shades of Yellow
Magnolia hodgsonii (Hook.f. & Thomson) H.Keng	Bhutan, China, India, Myanmar, Nepal and Thailand	Arunachal Pradesh, Assam, Himachal Pradesh, Manipur, Meghalaya, Mizoram, Sikkim, West Bengal.	LC	Apr-May	Jun-Sept	White to pale pink
Magnolia insignis Wall.	China, India, Myanmar, Nepal, Thailand and Vietnam	Assam, Manipur and Meghalaya and Sikkim	LC	May- Jul	Sept-Jan	White to pinkish
Magnolia lanuginosa (Wall.) Figlar & Noot.	Bhutan, China, India and Nepal	Arunachal Pradesh, Assam, Meghalaya, Nagaland, Sikkim	DD	May-Jun	Sept-Oct	White or cream

Contd.

Botanical name	Distribution	Distribution in India	IUCN status	Flowering	Fruiting	Flower colour
Magnolia oblonga (Wall. ex Hook.f. & Thomson) Figlar	Bangladesh and India	Arunachal Pradesh, Assam, Meghalaya, West Bengal and West Bengal	LC	Feb-May	Jul-Oct	White or cream
Magnolia grandiflora L.	India, Pakistan, Europe, North America: Costa Rica, Honduras, Mexico, United States of America; South America: Columbia, Ecuador	Arunachal Pradesh, Bihar, Jharkhand, Karnataka, Madhya Pradesh, Maharashtra, Meghalaya, Mizoram, Tamil Nadu, Tripura, Uttar Pradesh, Uttarakhand, West Bengal	LC	May-Jul	Oct-Dec	White or cream

such as *M. champaca* and *M. pterocarpa* (Kundu 2009). The chemical constituents of Magnolia bark have been well studied (Zheng et al. 1999) and classified into phenols (magnolol, honokiol, isomagnolol, tetrahydromagnolol, bornylmagnolol, piperitylmagnolol, piperibylholokiol, dipiperitylnolol, magnatriol, magnilognan, randainal, randaiol, syringaresinol, 4-O-methylhonokiol), alkaloids (magnocurarine, magnoflorine, anonaine, michelarbine, liriodenine, salicifoline, tubocurarine) and essential oils (eudesmol, pinene, camphene, limonene, bornyl acetate, caryophyllene epoxide, and cryptomeridiol). Magnolia leaves contain one-fifth of the magnolol and honokiol found in the bark (Zheng et al. 1999). The amount and composition of compounds are governed by the age of the plant or its parts, the geographical source of the plants investigated, and their general habitat (Sarker and Maruyama 2002). Alkaloids of aporphine/ noraporphine type and lignans and neolignans isolated from Magnolia have been used as chemotaxonomic markers for other plant families (Hegnauer 1986).

Different species have been reported to produce different phytochemicals, which vary in amount and concentration (Karthikeyan et al. 2016, Sahakitpichan et al. 2017). *M. lanuginosa* has been shown to produce a number of sesquiterpene lactones of germacrane type, including lanuginolide and dihydroparthenolide (Talapatra et al. 1970). Along with the oxoaporphine alkaloid liriodenine, an oxoaporphine alkaloid lanuginosine and a noraporphine alkaloid michelanugine have also been isolated from the trunk bark of *M. lanuginosa* (Talapatra et al. 1974). A bioassay-directed isolation scheme yielded three germacranolides (dihydroparthenolide, lanuginolide and 11, 13-dehydrolanuginolide) from an ethanol extract of the fruit of the *M. doltsopa* (Cassady et al. 1979). Phytoconstituents reported from *M. champaca* include sterols, tannins, flavonoids (Karthikeyan et al. 2016), sesquiterpenes including, michelia-A, liriodenine, parthenolide and guaianolides (Iida and Kazuoito 1982), and volatile oils containing compounds like benzyl acetate, linalool and isoeugenol (Taprial 2015).

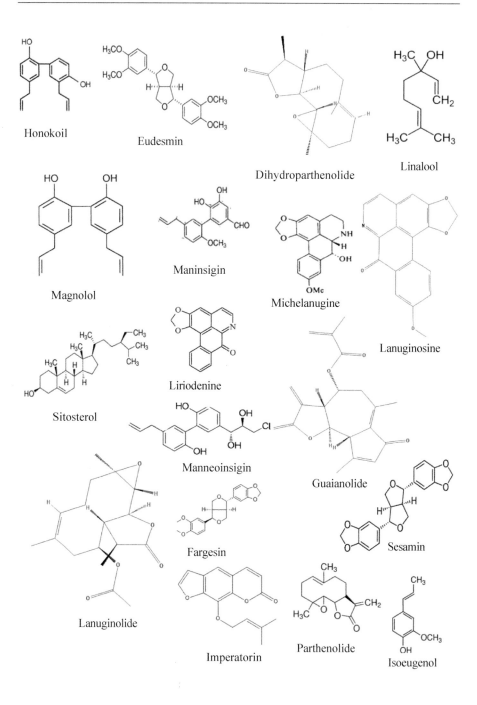

Fig. 1: Chemical structure of major compounds isolated from Indian magnolias

Magnolia pterocarpa bark and leaves contain number of coumarins and lignans such as sesamin, eudesmin, fargesin, imperatorin, dimethyl teraphthalate and β-sitosterol (Talpatra et al. 1983). *M. baillonii* bark contains germacranolide epoxides, dihydroparthenolide, parthenolide, liriodenine (Ruangrungsi et al. 1987). *M. hookeri* twigs consist of various sesquiterpene-neolignans namely 5-allyl-2-(4-allyl-phenoxy)-3-[7-(1-hydroxy-1-methyl-ethyl)-1, 4a-dimethyl-decahydro-naphthalen-1-yloxy]-phenol, eudesobovatol A and eudesobovatol B, three lignans-obovatol, honokiol and magnolol, along with trans-eudesmane-4, 11-diol, β-eudesmol, (-)-10-epi-5 β-hydroxy-β-eudesmol, epi-carrisone and gynurenol (Hu et al. 2015, Qi et al. 2015). Different types of maninsigins, neolignans, sesquiterpenes, manneoinsigins and lignans (Shang et al. 2013) have been isolated from the leaves and stems of *M. insignis.* Three phenylethanoid glycosides, 2-(3-hydroxy-4-methoxyphenyl) ethyl 1-*O*-β-D-allopyranoside (hodgsonialloside A), 2-(3-hydroxy-4-methoxyphenyl) ethyl 1-*O*-β-D-glucopyranosyl-(1 → 4)-β-D-allopyranoside (hodgsonialloside B) and 2-(3-methoxy-4-hydroxyphenyl) ethyl 1-*O*-β-D-allopyranoside (hodgsonialloside C) have been isolated from the leaves of *M. hodgsonii,* whereas from both bark and leaves other compounds like tyrosol 4-*O*-β-D-xylopyranosyl-(1→6)-β-D-glucopyranoside, kaempferol 3-*O*-neohesperidoside, kaempferol 3-*O*-rutinoside, kaempferol 3-*O*-α-L-rhamnopyranosyl-(1 → 2)-[α-L-rhamnopyranosyl-(1 → 6)]-β-D-glucopyranoside, (+)-syringaresinol *O*-β-D- and glucopyranoside oblongionoside C have been isolated (Sahakitpichan et al. 2017).

PHARMACOLOGICAL ASPECTS

Many Magnolia species have traditionally been used since centuries, largely to treat various illnesses ranging from simple headaches to complicated cancers. For example, *M. champaca* leaves are used for the treatment of fever, swelling, vaginal infections, colic, leprosy and in eye disorders (Kirtikar and Basu 1984; Aruna et al. 2012). The flowers find use in dyspepsia, nausea, and fever, also useful as a diuretic in renal diseases (Karthikeyan et al. 2016). Magnolia bark is believed to be a drug that replenishes energy, the basic function for existence. The common belief is that the plant can strengthen a patient's life energy by activating circulation of vital energy and eliminating dampness (Sarker and Maruyama 2002). The bark is used for pain in the abdomen due to entrapped gas and a feeling of congestion in the chest. It is also used in psychological disorders, relieves accumulation of damp in the spleen and the stomach marked by epigastric stuffiness, vomiting and diarrhea.

Table 2: Major phytochemicals in Magnolia species of India

Species	Major compounds	Formulae	Properties	Sources
M. campbellii	Lanuginosine	$C_{18}H_{11}NO_4$	Cytotoxic, antiviral and antiplatelets activities	Talapatra et al. (1970); Ulubelen et al. (1990)
	Liriodenine	$C_{17}H_9NO_3$	Trypanocidal activity, antitumor, antimicrobial, Inhibits dopamine biosynthesis	
	Magnolol	$C_{18}H_{18}O_2$	Anti-arrhythmania, platelet aggregation inhibition, anti-inflammatory, antiviral	
	Sesamin	$C_{20}H_{18}O_6$	Anti-cholesteremic, antihypertensive, antioxidant, anti-bacterial activities	
M. lanuginosa	Lanuginolide	$C_{17}H_{24}O_5$	Anti-inflammatory and anti-hyperalgesic	Talapatra et al. (1970); (1974); Feltenstein et al. (2004); Rummel et al. (2011); Ulubelen et al. (1990); Talapatra et al. (1970)
	Dihydroparthenolide	$C_{15}H_{22}O_3$	Anti-inflammatory, anti-hyperalgesic, fever-reducing action	
	Liriodenine	$C_{17}H_9NO_3$	-	
	Lanuginosine	$C_{18}H_{11}NO_4$	-	
	Michelanugine	$C_{18}H_{17}NO_4$	Antitumor	
M. pterocarpa	Imperatorin	$\acute{y}C_{16}H_{14}O_4$	Antibacterial, antiviral, anticancer, bone loss inhibitor, hepatoprotective action, anti-inflammatory	Talapatra et al. (1983); Cho et al. (1999); Hu et al. (2015)
	Eudesmin	$C_{22}H_{26}O_6$	Anticancer, TNF-á production or T lymphocyte proliferation, induces vascular relaxation and induce neurite outgrowth from PC12 cells	

Contd.

Species	Compound	Formula	Activity	References
	Fargesin	$C_{21}H_{22}O_6$	Anti-trypanosomal, anti-inflammatory, used in the treatment of nasal congestion and sinusitis, stimulates glucose uptake in L6 myotubes	Cassady et al. (1979)
	Magnolol	$C_{18}H_{18}O_2$	-	
	Sitosterol	$C_{29}H_{50}O$	Anti-pyretic, anti-cancer, angiogenic, antioxidant, immunomodulatory, anti-hyperlipidemic, anti-atherosclerotic, and plays important role in lowering of cholesterol levels	
M. doltsopa	Sesamin	$C_{20}H_{18}O_6$	-	
	Dihydroparthenolide	$C_{15}H_{22}O_3$	-	
	Lanuginolide	$C_{17}H_{24}O_5$	-	Karthikeyan et al. (2016); Iida and Kazuoito (1982); Taprial (2015)
M. champaca	Liriodenine	$C_{17}H_9NO_3$	-	
	Parthenolide	$C_{15}H_{20}O_3$	Anti-inflammatory, anti-tumor, treats fever, migraine headache, rheumatoid arthritis, anti-cancer, analgesia	
	Magnolol	$C_{18}H_{18}O_2$	-	
	Honokiol	$C_{18}H_{18}O_2$	Anti-tumerogenic, neurotrophic activity, anti-thrombotic, anti-inflammatory, anti-oxidant, cytotoxicity inhibition, GABAA modulation, CA2+ inhibition, anti-viral, metabolic activity	
	Guaianolide	$C_{19}H_{20}O_5$	Anti-inflammatory, anti-proliferative activity on human tumor cell lines, inhibition of nuclear factor kappa B DNA binding	
	Linalool	$C_{10}H_{18}O$	Anti-oxidant, supresses voltage-gated currents in sensory neurons and cerebellar purkinje cells, insecticidal activity,	

Contd.

Species	Compound	Formula	Activity / Use	References
M. baillonii	Isoeugenol	$C_{10}H_{12}O_2$	an important fragrance chemical	Ruangrungsi et al. (1987); Feltenstein et al. (2004); Rummel et al. (2011)
	Dihydroparthenolide	$C_{15}H_{22}O_3$	Approved flavoring agent for foods, anti-microbial, anti-oxidant, effective constituent against Black plague	
	Parthenolide	$C_{15}H_{20}O_3$	-	
	Liriodenine	$C_{17}H_9NO_3$	-	
M. hookeri	Eudesmin	$C_{22}H_{26}O_6$	-	Qi et al. (2015); Hu et al. (2015)
	Carrisone	$C_{15}H_{24}O_2$	Anti-neoplastic, reduce the damage to DNA induced by UV	
M. insignis	Maninsigin	$C_{17}H_{16}O_4$	Cytotoxic and neurite outgrowth-promoting activities	Shang et al. (2012); (2013)
	Manneoinsigins	$C_{18}H_{19}O_4Cl$	Cytotoxic activity against the HL-60 human tumor cell line	
M. hodgsonii	Kaempferol	$C_{15}H_{10}O_6$	Anti-oxidant, anti-inflammatory, anti-microbial, anti-diabetic, and antitumor	Sahakitpichan et al. (2017); Shields, (2017); Fu et al. (2008)
	Oblongionoside	$C_{24}H_{44}O_{11}$	Anti-bacterial, anti-tumor, anti-viral, anti-inflammatory, neuro-protective, anti-oxidant, hepatoprotective, immunomodulatory, and tyrosinase inhibitory actions.	

Owing to their high significance in traditional herbalism, a number of Magnolia species have been exposed to modern drug discovery screening, and as a result, several drugs have been developed from the compounds isolated from these species. The pharmacological properties of Magnolia species is given in Figure 2. Many of the species have been found to possess potent procognitive, anti-oxidative, anti-microbial, anti-inflammatory, anti-angiogenic, diuretic, anti-ulcer, analgesic, anti-helmintholytic and anti-cancer activities (Taprial 2015). *Michelia champaca* leaves proven to have many properties including anti-inflammatory (Gupta et al. 2011a), anti-fertility (Taprial et al. 2013), anti-ulcer (Kumar et al. 2011, Mullaicharam et al. 2011), anthelmintic (Dama et al. 2011), anti-microbial (Khan et al. 2002), analgesic (Mohamed et al. 2009), diuretic (Ahmad et al. 2011), cytotoxic (Balurgi et al. 1997) and anti-diabetic (Gupta et al. 2011b). Germacranolide compound, dehydrolanuginolide present in fruit of *M. doltsopa* has been found to have cytotoxic activity, hence a possible cure for cancer (Cassady et al. 1979). The leaf methanolic extracts of *M. pterocarpa* were found to have thrombolytic and membrane stabilizing properties (Chowdhury et al. 2013). Parthenolide found in *M. baillonii* possess anti-tumor activity (Ruangrungsi et al. 1987). *M. insignis* bark is reported to show significant cytotoxic and neurite outgrowth-promoting activities, as well as their antagonistic activity toward Farnesoid X receptor (FXR) ligand (Shang et al. 2013).

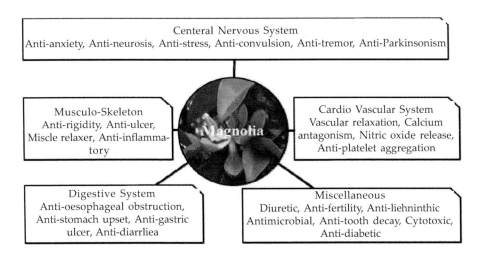

Fig. 2: Pharmacological properties of Magnolia species (Sarker and Maruyama 2002).

COMMERCIAL VALUES

Most of the Magnolia species have economic importance (Wiersema and Leon 1999). Many species have large, fragrant and beautiful flowers, which are widely cultivated for ornamental purposes. *M. campbellii, M. grandiflora, M. pterocarpa* are beautiful varieties and were favorites with British gardeners in India (Randhawa 1961). However, fewer parks and gardens have them on their list of favorites these days and a revival of interest is recommended, especially for wild varieties (Sarker and Maruyama 2002). *M. campbellii* becomes leafless in winters, but is much pretty with rose pink, cup-shaped, scented flowers, which adorn the tree in spring. In Manipur, magnolias are commonly grown in gardens, with a local name *ootahmbal* means "tree lotus" and flowers are used in rites and rituals (offering) during religious functions (Sivkishen 2017). In many states, including Meghalaya, Assam, Sikkim, Himachal Pradesh, *M. grandiflora* is widely cultivated along roadsides, in parks and home gardens as an ornamental tree. Recently, hybridization technique have been successfully used in combining the best aspects of different Magnolia species to give plants which flower at an earlier age than the parent species, as well as having more impressive flowers. Examples include *M. soulangeana*, produced by hybridridization of *M. liliiflora* and *M. denudata* (Biedermann 1987, Chen and Zeng 1998, Seth 2004). Other Magnolia species of the country has also the potential for use in floriculture market.

Apart from their aesthetic value, the species are known for their fragrant oil, widely used in perfume and cosmetics industry (Sarker and Maruyama 2002). Flowers of *M. champaca* are source of champa oil, used in perfumery and for preparation of attars and perfumed hair oils (Karthikeyan et al. 2016). In Southeast Asia, flowers are floated in bowls of water to scent the room, as a fragrant decoration for bridal beds, and for garlands. Commercially, the perfume is sold as "joy perfume". They also yield a yellow dye, which can be exploited on commercial scale.

Magnolia flowers have been used for culinary purposes (Seth 2004). For example, in some Asian cuisines, petals of *M. grandiflora* are pickled and used as a spicy condiment and buds are pickled and used to flavor rice and scent tea (Cornucopia 1990). In Japan, the young leaves and flower buds of *M. hypoleuca* are boiled and eaten as a vegetable, and the old leaves are made into a powder and used as flavor (McMinn and Maino 1981). In Arunachal Pradesh, *M. campbellii* and *M. dolstopa* flower buds are grinded, mixed with locally prepared food and eaten. In India, Magnolia leaves (e.g. *M. pterocarpa, M. champaca*) are used as food plants by the larvae of some Lepidoptera species (e.g. *Antheraea assama, A. asamensis, A. mejankari*) for production of muga silk, thus have a huge potential to be used in the industry commercially (Seth 2004, Nath et al. 2008). Moreover, the market for medicinal plants as a whole is ever-growing, and Magnolia certainly is one of the commercially important genera. Owing to its versatile medicinal properties as discussed in detail in previous section, several species have been incorporated and used in different commercially successful medicine preparations. Many Magnolia medicinal products are available in the market (e.g. *Adrenal Health* and *Sleep Thru*).

Nearly two-thirds of commercial Magnolia species can be used as timber and in making furniture products (Priester 1990). The sapwood of Magnolia is creamy white, while the heartwood is light to dark brown, often with greenish to purple-black streaks or patches (McLaughlin 1933). The high-quality wood of Magnolia is even-textured and moderately heavy, fairly hard and straight grained. It is resistant to heavy shrinkage, is highly shock absorbent, and has a relatively low bending and compression strength. It takes glue well, has a good nailing quality, and stains and varnishes easily (Godfrey 1988). Magnolia wood is used by the food industry for making cherry boxes, flats, and baskets, and is used for tongue depressers, broom handles, veneers, and venetian blinds (Duncan and Duncan 1988, Godfrey 1988).

Among Indian species, wood of *M. champaca* is used for posts, boards, veneers, furniture, decorative fittings, joinery work, carriage and ship building, and carving (http://greencleanguide.com). The wood of *M. rabaniana* is of great economic value, and is used for making boxes, musical instruments, and yields an excellent commercial timber called "white wood" (Mir et al. 2017). *M. cathcartii* wood is used for planking and for joinery works. *M. lanuginosa* and *M. punduana* is a valued timber plant and is used in house construction and making furniture (Mir et al. 2016, Iralu and Upadhaya 2015). *M. oblonga* wood is employed for planking, rough furniture, cabinet work, and canoes and for tea chests (Sarker and Maruyama 2002). In addition, *M. insignis* and *M. caveana* is also used as fire and fuel wood in many parts of Meghalaya.

There are many agencies, which are selling Magnolia products commercially. The notable ones are, *Plantago*, (https://plantago.nl/home.html), Gaia herbs, selling medicinal products (https://www.gaiaherbs.com/) and Magnolia home, selling magnolia wood products (https://www.magnoliahomefurniture.com/home).

CONCLUSION

It may be concluded that the Magnolia species are under severe threat due to a number of human activities and warrants urgent conservation initiatives in order to prevent them from extinction. The habitat of the species needs to be strictly protected and the population need to be monitored from time to time. Forest fires, illegal timber extraction, and agricultural expansion, which are rapidly contributing to the forest degradation and fragmentation in the country, need to be checked. The species need to be brought under *in-situ* and *ex-situ* conservation programs. The Magnolia species have a huge potential for commerce and improve the economic security of many people. But the horticultural as well as economical potential of the species needs to be harnessed properly. In order to reduce the pressure on the species in the wild, emphasis should be given for mass cultivation followed by introduction in home gardens and agroforestry. Although magnolia species are globally known for their phytochemicals and the role in modern medicine, very less information is available on the phytochemical and pharmacological aspects of Indian magnolias. Thus these species need to be scientifically studied in order to harness their full potential as medicinal plants of therapeutic value. Government should provide necessary institutional and financial support to scientific research in order to promote the potential role of magnolias in herbal medicine and pave the way to modern clinical applications.

The pharmacological properties of Magnolia species are because of the wide variety of compounds present in the species. It has been found to be useful in treatment of sepsis as experimented in rats (Kong et al. 2000). Watanabe et al. (1983) found a distinct sedative and centrally acting muscle relaxant action of the ether extract of Magnolia bark, as well as an anti-ulcer effect in ulcers induced by stressful treatment in experimental animals. Wang and Chen (1998) explored the role of inositol trisphosphate in signaling pathway that leads to the elevation of cytosolic-free Ca^{2+} in rat neutrophils stimulated with magnolol. They suggested that a pertussis toxin-insensitive inositol trisphosphate signaling pathway is involved in magnolol-induced (Ca^{2+})

elevation in rat neutrophils. Both magnolol and honokiol also exhibited free radical scavenging activities in the diphenyl-p-picrylhydrazyl assay (Lo et al. 1994). Honokiol, is reported to have central depressant action and at much lower doses, anxiolytic activity (Kuribara et al. 1998).

The magnolol exhibited remarkable inhibitory effects on mouse skin tumor promotion in an *in-vivo* carcinogenesis test, suggesting that magnolol might be valuable anti-tumor promoters (Konoshima et al. 1991). Magnolol and honokiol have significant activity against gram-positive and acid-fast bacteria and fungi (Clark et al. 1981). Magnolol and honokiol also have anti-platelet activity as it relaxes muscle by releasing endothelium-derived relaxing factor and by inhibiting calcium influx through voltage-gated calcium channels (Teng et al. 1990).

ACKNOWLEGEMENTS

Authors are thankful to village Heads (Dolloi, Syiem, Sordar and Headman) of various localities for allowing to conduct this study. Help and support received from the Botanical Survey of India, Eastern Regional Circle and the State Forest Departments of various States is also acknowledged.

REFERENCES CITED

Ahmad H, Saxena V, Mishra A, Gupta R. 2011. Diuretic activity of aq. extract of *M. champaca* Linn. Leaves and stembark in rats. *Pharmacology Online* 2: 568-574.

Aruna G, Praveenkumar A, Munisekhar P. 2012. Review on *Michelia champaca* Linn. *International Journal of Phytopharmacy Research* 3(1): 32-4.

Azuma H, García-Franco JG, Rico-Gray V, Thien LB. 2001. Molecular phylogeny of the Magnoliaceae: the biogeography of tropical and temperate disjunctions. *American Journal of Botany* 88: 2275-2285.

Balakrishnan NP. 1981. *Flora of Jowai*. Vol. I. Botanical Survey of India: Howrah.

Balurgi VC, Rojatkar SR, Pujar PP, Patwardhan BK, Nagrasampagi BA. 1997. Isolation of parthinolide from the leavesof *Michelia champaca* L. *Journal of Indian Drugs* 34(6): 409-15.

Biedermann IE. 1987. Factors affecting establishment and development of Magnolia hybrids in vitro. *Acta Horticulturae* 212: 625-629.

Cassady J, Nobutoshi M, Ching-Jer O, Chang J, McLaughlin L. 1979. Dehydrolanuginolide, a cytotoxic constituent from the fruits of *Michelia doltsopa*. *Phytochemistry* 18(9): 1569-1570.

Chen W, Zeng Q. 1998. The grafting propagation of Magnoliaceae. *Journal of Tropical and Subtropical Botany* 6: 70-78.

Cho JY, Voo ES, Baik KU, Park MH. 1999. Eudesmin Inhibits Tumor Necrosis Factor-Production and T cell Proliferation. *Archives of Pharmacal Research* 22(4): 348-353.

Chowdhery HJ, Wadhwa BM. 1984. *Flora of Himachal Pradesh*. Botanical Survey of India: Calcutta.

Chowdhury SR, Sharmin T, Hoque M, Md. Sumsujjaman, Das M, Nahar F. 2013. Evaluation of thrombolytic and membrane stabilizing activities of four medicinal plants of Bangladesh. *International Journal of Pharmaceutical Sciences and Research* 4(11): 4223-27.

Clark AM, El-Feraly FS, Li WS. 1981. Antimicrobial activity of phenolic constituents of *Magnolia grandiflora* L. *Journal of Pharmaceutical Sciences* 70: 951-952.

Cornucopia FS. 1990. *A Source Book of Edible Plants*. Kampong Publications.

Dama G, Bidkar J, Deore S, Jori M, Joshi P. 2011. Helmintholytic activity of the methanolic and aqueous extracts of leaves of *Michelia champaca*. *Research Journal of Pharmacology and Pharmacodynamics* 3(1): 25-6.

Deb DB. 1961. Dicotyledonous plants of Manipur Territory. *Bulletin of Botanical Survey of India* 3(3&4): 253-350.

Deb DB. 1981. *The Flora of Tripura State*. New Delhi: Today and Tomorrow Printer and Publisher.

Duncan WH, Duncan MB. 1988. *Trees of the Southeastern United States*. Athens, GA: The University of Georgia Press.

Duthie JF. 1922. *Flora of Upper Gangetic plain and the adjacent Siwalik and sub-Himalaya tracts*. Vol. I-III, Botanical survey of India: Calcutta.

Feltenstein MW, Schuhly W, Warnick JE, Fischer NH, Sufka KJ. 2004. Anti-inflammatory and anti-hyperalgesic effects of sesquiterpene lactones from magnolia and bear's foot. *Pharmacology Biochemistry Behavior* 79:299-302.

Fu G, Pang H, Wong YH. 2008. Naturally occurring phenylethanoid glycosides: potential leads for new therapeutics. *Current Medicinal Chemistry* 15(25): 2592-613.

Gamble GS, Fischer CEC. 1935. *Flora of Presidency of Madras*. Vol. I-III. Botanical survey of India: Calcutta.

Godfrey RK. 1988. *Trees, shrubs, and woody vines of Northern Florida and adjacent Georgia and Alabama*. Athens, GA: The University of Georgia Press.

Gupta S. Mehta K, Chauhan D, Nair A. 2011a. Anti-inflammatory activity of leaves of *Michelia champaca* investigated on acute inflammation induced rats. *Latin American Journal of Pharmacy* 30(4): 819-22.

Gupta S, Mehta K, Chauhan D, Kumar S, Nair A. 2011b. Morphological changes and antihyperglycemic effect of *M.champaca* leaves extract on β- cell in Alloxan induced diabetic rats. *Recent Research in Science and Technology* 3(1): 81-7.

Hajra PK, Verma DM, Giri GS. 1996. *Materials for the Flora of Arunachal Pradesh*. Vol. I. Botanical Survey of India: Dehra Dun, 534-544.

Haridasan K, Rao RR. 1985. *Forest Flora of Meghalaya*. Vol. I. India, Dehradun: Bishen Singh Mahendra Pal Singh.

Hegnauer R. 1986. Phytochemistry and plant taxonomy- an essay on the chemotaxonomy of higher plants. *Phytochemistry* 25(7): 1519-1535.

Henry AN, Chitra V, Balakrishnan NP. 1989. *Flora of Tamil Nadu*. Vol. I-III. Botanical survey of India: Coimbatore.

Hooker JD. 1875. *The Flora of British India*. Vol. I. A shrford, L. Reeve & Co.Ltd, London.

Hu X, Sui X, Wang Y, Wang W, Wu H, Zhang F, Tan Y, Zhang F. 2015. Sesquiterpene-neolignans from *Manglietia hookeri*. *Natural Product Research* 30(13): 1477-1483.

Iida T, Kazuoito. 1982. Sesquiterpene lactone from *Michelia champaca*. *Phytochemistry* 21(3): 701-703.

Iralu V, Upadhaya K. 2015. Notes on *Magnolia punduana* Hook.f.sthomson (Magnoliopsida: Magnoliales: Magnoliaceae): an endemic and threatened tree species of northeastern India. *Journal of Threatened Taxa* 7(9): 7573-7576.

IUCN. 2018. The IUCN Red List of Threatened Species. http:// www.iucnredlist.org. accessed on 05 April 2018.

Kanjilal UN, Kanjilal PC, Das A. 1934. *Flora of Assam*. Vol. I. Government Press, Shillong.

Karthikeyan V, Balakrishnan BR, Senniappan P, Janarthanan L, Anandharaj G, Jaykar B. 2016. Pharmacognostical, Phyto-Physicochemical Profile of the Leaves of *Michelia champaca* Linn. *International Journal of Pharmacy and Pharmaceutical Research* 7 (1): 331-344.

Khan MR, Kihara M, Omoloso AD. 2002. Anti-microbial activity of *Michelia champaca*. *Fitoterapia* 73(7-8): 744-748.

King G. 1891. *The Magnoliaceae of British India*. Annual Royal Botanical Garden: Calcutta.

Kirtikar KR, Basu BD. 1984. *Indian Medicinal Plants*. Vol. 1, 2nd ed. Dehradun, India: Bishan Singh Mahender Pal Singh.

Kong CW, Tsai K, Chin JH, Chan WL, Hong CY. 2000. Magnolol attenuates peroxidative damage and improves survival of rats with sepsis. *Shock* 13(1): 24-28.

Konoshima T, Kozuka M, Tokuda H, Nishino H, Iwashima A, Haruna M, Ito K, Tanabe M. 1991. Studies on inhibitors of skin tumor promotion, IX. Neolignans from magnolia officinalis. *Journal of Natural Products* 54(3): 816-822.

Kumar VS. 2012. Magnoliaceae of Indian region-an appraisal. In: Panda S and Ghosh C (Eds.). Diversity and conservation of plants and traditional knowledge. Bishen Singh and Mahendra Pal Singh, Dehradun.

Kumar S M, Aparna P, Poojitha K, Karishma SK, Astalakshmi N. 2011. A comparative study of *Michelia champaca* L. flower and leaves for anti-ulcer activity. *International Journal of Pharmaceuticals Sciences and Research* 2(6): 1554-8.

Kundu SR. 2009. A synopsis on distribution and endemism of Magnoliaceaes in Indian Sub-continent. *Thaiszia Journal of Botany* 19: 47-60.

Kuribara H, Stavinoha WB, Maruyama Y. 1998. Behavioral pharmacological characteristics of honokiol, an anxiolytic agent present in extracts of Magnolia bark, evaluated by an elevated plus-maze test in mice. *Journal of Pharmacy and Pharmacology* 50(7): 819-826.

Lo YC, Teng CM, Chen CF, Chen CC, Hong CY. 1994. Magnolol and honokiol isolated from Magnolia officinalis protect rat heart mitochondria against lipid peroxidation. *Biochemical Pharmacology* 47(3): 549-553.

Martinez GJ, Planchuelo AM, Fuentes E, Ojeda M. 2006. A numeric system to establish conservation priorities for medicinal plants in the Paravachasca Valley, Cordoba, Argentina. *Biodiversity and Conservation* 15: 2457-2475.

McLaughlin RP. 1933. Systematic anatomy of the woods of the Magnoliaceae. *Tropical Woods* 34: 3-38.

McMinn H, Maino E. 1981. Pacific Coast Trees. University of California Press, 432.

Mir AH, Iralu V, Pao NT, Chaudhury G, Khonglah CG, Chaudhary KL, Tiwari, BK, Upadhaya K .2016. *Magnolia lanuginosa* (Wall.) Figlar & Noot. in West Khasi Hills of Meghalaya, Northeastern India: Re-collection and implications for conservation. *Journal of Threatened Taxa* 8: 8398-8402.

Mir AH, Upadhaya K, Odyuo N, Tiwari BK. 2017. Rediscovery of *Magnolia rabaniana* (Magnoliaceae): A threatened tree species of Meghalaya, northeast India. *Journal of Asia Pacific Biodiversity* 10: 127-131.

Mittermeier RA, Gils PR, Hoffman M, Pilgrim J, Brooks T, Mittermeier CG, Lamoreaux J, Da Fonseca GAB. 2004. (eds.). Hotspots Revisited. Earth's Biologically Richest and Most Endnagered Terrestrial Ecoregions. USA: CEMEX.

Mohamed HM, Jahangir R, Hasan SMR, Akter R, Ahmed T, Md. Islam I, Faruque A. 2009. Anti-oxidant, analgesicand cytotoxic activity of *M. champaca* Linn. Leaf. *Stamford Journal of Pharmaceutial Sciences* 2(2): 1-7.

Mullaicharam AR, Surendra Kumar M. 2011. Effect of *Michelia champaca* Linn. on pylorus ligated rats. *Journal of Applied Pharmaceutical Science* 1(2): 60-4.

Myers N. 1988. Threatened biotas: "Hot spots" in tropical forests. *The Environmentalist* 8: 1-20.

Nath R, Nath SK, and Devi D. 2008. Study and conservation of host food plants of muga silkworm, *Antheraea assamensis* (helfer), of Assam. *Nature Environment and Pollution Technology* 7 (1): 83-92.

Pandey SN, Misra SP. 2009. *Taxonomy of Angiosperms*. New Delhi: Ane Books Pvt. Ltd., 341-342.

Prain D. 1903. *Bengal Plants*. Vol. I-II. (Eds.) DehraDun: Bishen Singh Mahendra Pal Singh.

Priester DS. 1990. *Magnolia virginiana* L. sweetbay. In *Silvics of North America*, vol. 2, *Hardwoods*, edited by R.M. Burns and B.H. Honkala Washington DC: US Department of Agriculture, Forest Service, 445-448.

Pullaiah T, Ali-Mouali D. 1997. *Flora of Andhra Pradesh*. Vol. II. Jodhpur: Scientific Publisher.

Pyo MK, Koo YK, Yun-Choi HS. 2002. Anti-platelet effect of phenolic constituents isolated from the leaves of *Magnolia obovata*. *Natural Products Sciences* 8(4): 147-151.

Qi MG, Zhang F, Wang WS, Wu HB, Yuan HC, Jiao YG, Dong XJ. 2015. Eudesmane sesquiterpenes from twigs of *Manglietia hookeri*. *China journal of Chinese material media* 40(16): 3229-3232.

Qiu Y-L, Parks CR, Chase MW. 1995. Molecular divergence in the eastern Asia-eastern North America disjunct section Rytidospermum of Magnolia (Magnoliaceae). *American Journal of Botany* 82: 1589-1598.

Randhawa MS. 1961. *Beautiful trees and gardens*. New Delhi: Indian Council of Agricultural Research.

Rivers M, Beech E, Murphy L, Oldfield S. 2016. *The Red List of Magnoliaceae* – revised and extended. Richmond, UK: BGCI.

Ruangrungsi N, Rivepiboon A, Lange GL, Lee M, Decicco CP, Picha P, Preechanukool K. 1987. Constituents of Paramicheliabaillonii: a new antitumor germacranolide alkaloid. *Journal of Natural Product* 50(5): 891-896.

Rummel C, Gerstberger R, Roth J, Hubschle T. 2011. Parthenolide attenuates LPS-induced fever, circulating cytokines and markers of brain inflammation in rats. *Cytokine* 56: 739-748.

Sahakitpichan P, Chimnoi N, Wongbundit S, Vorasingha A, Ruchirawat S, Kanchanapoom T. 2017. Phenylethanoid glycosides from the leaves of *Magnolia hodgsonii*. *Phytochemistry Letters* 21: 269-272.

Saldanha CJ. 1984. *Flora of Karnataka*. New Delhi: Oxford and Publishing Co.

Sarker SD, Maruyama Y. 2002. Magnolia: The Genus Magnolia, London: Taylor and Francis.

Seth MK. 2004. Trees and their economic importance. *The Botanical Review* 69(4): 321-376.

Shang S, Kong L, Yang L, Jiang J, Huang J, Zhang H, Shi M, Zhao W, Li H, Luo H, Li Y, Xiao W, Sun H. 2013. Bioactive phenolics and terpenoids from *Manglietia insignis Fitoterapia* 84: 58-63.

Shang SZ, Yan JM, Zhang HB, Shi YM, Gao ZH, Du X, Li Y, Xiao WL, Sun HD. 2012. Two new neolignans from *Manglietia insignis*. *Natural Products and Bioprospecting* 2(5): 227-230

Shields M. 2017. Chapter 14 – Chemotherapeutics in Pharmacognosy Fundamentals, Applications and Strategies. Pages 295-313 https://doi.org/10.1016/B978-0-12-802104-0.00014-7

Sivkishen J. 2017. One thought on "Magnolia". wordpress.com (accessed on 25-04-2018).

Talapatra SK, Patra A, Talapatra B. 1970. Lanuginolide and Dihydroparthenolide, Two New Sesquiterpenoid Lactones from *Michelia lanuginosa*. The Structure, Absolute Configuration, and a Novel Rearrangement of Lanuginolide? Chemical Communications, 1534.

Talapatra SK, Patra A, Talapatra B. 1974. Alkaloids of *Michelia lanuginosa* Wall. *Tetrahedron* 31: 1105-1107.

Talpatra B, Ray G, Talpatra SK. 1983. Polyphenolic constituents of *Magnolia pterocarpa* Roxb. *Journal of the Indian Chemical Society* 60: 96-98.

Taprial S, Kashyap D, Mehta V, Kumar S, Kumar D. 2013. Antifertility effect of hydroalcoholic leaves extract of *Michelia champaca* L.: an ethnomedicine used by Bhatra women in Chhattisgarh state of India. *Journal of Ethnopharmacology* 147(3): 671-675.

Taprial S. 2015. A review on phytochemical and pharmacological properties of *Michelia champaca* Linn. Family: Magnoliaceae. *International Journal of Pharmacognosy* 2(9): 430-6.

Teng CM, Yu SM, Chen CC, Huang YL, Huang TF. 1990. Inhibition of thrombin- and collagen-induced phosphoinositides breakdown in rabbit platelets by a PAF antagonist—denudatin B, an isomer of kadsurenone. *Thrombosis Research* 59(1): 121-130.

TPL. 2018. The Plant List: a working list of all known plant species. version 1.1, released in september 2013. www.theplantlist.org.

Ulubelen A, Oksuz S, Schuster A. 1990. A sesquiterpene lactone from *Achillea millefolium* subsp. *millefolium*. *Phytochemistry* 29: 3948-3949.

Wang JP, Chen CC. 1998. Magnolol induces cytosolic-free Ca2+ elevation in rat neutrophils primarily via inositol trisphosphate signalling pathway. *European Journal of Pharmacology* 352: 329-334.

Watanabe K, Watanabe H, Goto Y, Yamaguchi M, Yamamoto N, Hagino K. 1983. Pharmacological properties of magnolol and honokiol, neolignane derivatives, extracted from *Magnolia officinalis*: central depressant effects. *Planta Medica* 49: 103-108.

Wiersema JH, Leon B. 1999. *World Economic Plants*: A Standard Reference. Boca Raton, FL: CRC Press.

Zheng HZ, Dong CH, She J. 1999. Modern Study of Traditional Chinese Medicine. Beijing: Xue Yuan Publisher, 3280-3304.

10

Traditional Medicinal Plants Used to Treat Dermatological Disorders in District Udhampur, Jammu and Kashmir, India

Harpreet Bhatia[1†], Yash Pal Sharma[1] and Rajesh Kumar Manhas[2†]

[1]*Department of Botany, University of Jammu, Jammu– 180001
Jammu & Kashmir, INDIA*
[2]*Department of Botany, Govt. Degree College, Kathua– 184104
Jammu & Kashmir, INDIA*
Email: harpreetbhatia2@gmail.com; yashdbm3@yahoo.co.in

ABSTRACT

Dermatology is the branch of medicine concerned with the diagnosis and treatment of skin disorders. Owing to the lack of proper hygiene and sanitation, incidence of skin diseases is a major health burden particularly in developing and under-developed countries. The study aimed at documenting the plants traditionally used to cure different types of dermatological disorders by the tribals and natives of district Udhampur. Ethnomedicinal data was collected by interviewing 91 infomants between the age group 26-89 years. Quantitative approaches were used to determine Use value (UV). The study revealed that a total of 64 plant species belonging to 59 genera and 43 families were used by the tribals and natives of district Udhampur for the treatment of 25 ailments of skin disorders. The most prevalent families were Asteraceae, Convolvulaceae, Rosaceae, Solanaceae and Euphorbiaceae. Majority of the taxa were growing in wild. Among the different plant parts used, leaf was the most frequently used. About 90% of medicinal preparations were applied topically. The most important medicinal species as per the use-value (UV) included *Phyllanthus emblica, Aloe vera, Ficus carica,* and *Cannabis sativa.* The present investigation may serve as a baseline data to initiate further research for the discovery of new bioactive compounds, thereby, providing important leads in the development of novel drugs for the treatment of such disorders.

Keywords: Ethnomedicine, Dermatological disorders, Use value, Western Himalaya, India

INTRODUCTION

Dermatology is the branch of medicine concerned with the diagnosis and treatment of skin disorders. Skin is the largest organ of the human body, both in terms of surface area and weight. It is made up of three layers: epidermis, dermis and hypodermis and acts as the primary line of defense protecting the body from pathogens while sustaining micro-organisms that influence human health (Grice et al. 2009). Skin diseases affect people of all ages from neonates to the elderly accounting for 34% of all occupational diseases encountered worldwide (Spiewak, 2000). The common dermatological disorders include abscess, ringworm, dandruff, hair fall, burns, boils, wounds, eczema, psoriasis, leprosy, acne, blemishes, warts, rashes, sores, alopecia, scabies, leucoderma and acne. These disorders may be related to bacterial, viral or fungal infections, pigmentation complications, trauma, etc. (Abbasi et al. 2010).

Owing to the lack of proper hygiene and sanitation, incidence of skin disease is a major health burden particularly in developing and under-developed countries. Since antiquity, traditional medicinal plants have been employed in the treatment of dermatological ailments. Being safe, cost effective and easily available, there is an attempt to incorporate them in the primary health care in many countries (Saikia et al. 2006).

India, with its varied climatic conditions, harbours a rich plethora of medicinal plants. It has a strong base of many systems of medicines including Ayurveda, Unani, Sidha and other local health practices. Charak Samhita (700 BC) and Sushruta Samhita (200 BC) are regarded as the traditional medicinal records (Sinha 2002). Few studies have been carried out to investigate and document the dermatological disorders throughout India (Sivarajani and Ramakrishanan 2012, Sharma et al. 2014, Agarwal and Chouhan 2014, Tiwari 2015, Anamika and Kumar 2016, Panda et al. 2016, Nuzhat and Vidyasagar 2017, Gupta and Gupta 2018).

District Udhampur in Jammu and Kashmir State of India, is a rich reservoir of biodiversity including medicinal flora. Due to poor financial status and lack of modern health care provisions, the local population inhabiting the rural belts of the district relies on knowledge-base which is inherited and passed on from one generation to another as a verbal tradition (Rao and Shanpru 1981, Chhetri 1994). This knowledge is usually confined to local healers and elderly community members and is in danger of being lost forever. Realizing the importance of medicinal plants in healing ailments and rapidly depleting traditional knowledge, the present study was the first effort to investigate and document the ethnomedicinal usage of plants of the area for the treatment of dermatological disorders.

Study Location

District Udhampur of Jammu division of J & K state lies between 32°34' & 39°30' North latitude and 74°16' & 75°38' East longitude. It embodies a mountainous terrain with an altitudinal variation from 600 to 2900 m above mean sea level and has a total area of 2380 sq. km. The present study was conducted in different villages of all the 4 tehsils *viz.*, Udhampur, Chenani, Ramnagar and Majalta. The district is divided into three climatic zones (a) Temperate zone (b) Sub-tropical zone and (c) Intermediate zone. Besides this, alpine zone occurs in the higher regions of the district (Swami and Gupta 1998, Khan 2010). Three nomadic tribes *viz.*, Gujjars, Bakkarwals and Gaddis inhabit district Udhampur and are either permanently settled in the study area or practice trans-humance. Agriculture and livestock rearing are the main occupations of the rural folk.

Data Collection

Frequent field trips were undertaken in different seasons in 32 villages of the four tehsils of district Udhampur to gather information on ethnomedicinal plants effective in the treatment of dermatological disorders. A total of 91 infomants (66 males and 25 females) between the age group 26-89 years were selected. Verbal consent was taken from the participants and objectives of the study were made clear to them. Information was obtained by conducting interviews (through questionnaire) and group discussions with the informants (local healers, nomads, community members) in their local language. Data of medicinal plants collected included common dermatological ailments prevalent in the study area, local name of plant species, parts used, methods of preparation, route of administration, dose. The medicinal property of each plant species was confirmed, if atleast five separate informants had a similar opinion.

The collected plant specimens were dried and mounted on the herbarium sheets using standard protocols. For plant identification, Flora of Jammu and Flora of Udhampur were consulted. For confirmation of identification, herbaria of IIIM, Jammu and Department of Botany, University of Jammu were consulted. Valid botanical names with author citations were verified from The Plant List, Version 1.1 (TPL 2013). The gathered field information was analyzed and compiled to draw a clear picture of the ethnomedicinal plants effective in the treatment of dermatological disorders in district Udhampur.

Data Analysis

Use Value (UV)

The relative importance was calculated employing the use-value (Phillips et al. 1994), a quantitative measure for the relative importance of species known locally:

$$UV = \Sigma U/n$$

where, U is the number of use-reports cited by each informant for a given species and n refers to the total number of informants. Use values are high when there are many use-reports for a plant, implying that the plant is important, and approach zero (0) when there are few reports related to its use. The use value, however, does not distinguish whether a plant is used for single or multiple purposes.

DIVERSITY OF MEDICINAL WEALTH

A total of 64 plant species belonging to 59 genera and 43 families were used by the local populace of district Udhampur for the treatment of 25 categories of dermatological disorders (Table1). Frequently used plant families in skin care include Asteraceae (5 species), Convolvulaceae (4 species), Rosaceae, Solanaceae, Euphorbiaceae, Apocynaceae and Moraceae (3 species each), Fabaceae, Asclepiadaceae, Ranunculaceae and Meliaceae (2 species each). Rest of the families were represented by one species only. The dominance of medicinal plant species from Asteraceae may be attributed to their good seed dispersion capacity, longevity of seeds for long distance dispersal, perennation, complex phytochemistry, high fecundity and agility (Tremetsberger et al. 2005).

Table 1: Traditional medicinal plant species used for curing dermatological disorders along with their use-value

Botanical name / Family	Local name	Habit/ source	Plant part used	Mode of use	Drug Administration route	Diseases cured (citations)	Use-value (UV)
Abrus precatorius L./ Fabaceae	Ratti	Cl, W	Seeds	Infusion, paste	Topical	Abscess (9) and ringworm (14)	0.25
Aesculus indica (Wall. ex Cambess.) Hook./ Hippocastanaceae	Goon	T, W	Seeds	Paste / Decoction	Topical / Topical	Dandruff (15) / Frostbite (24) and abscess (6)	0.49
Allium cepa L./ Alliaceae	Pyaaz, Ganda	H, C	Bulb	Paste	Topical	Abscess (29)	0.32
Aloe vera (L.) Burm. f./ Aloaceae	Kuaargandal	H, C	Latex	As such	Oral, topical	Abscess (20), burns (25), boils (15), wounds (6), eczema (11) and psoriasis (14)	1
Anagallis arvensis L./ Primulaceae	Jonkmari	H, W	Whole plant	Extract	Topical	Dermatitis (6) and leprosy (13)	0.21
Argemone mexicana L./ Papaveraceae	Peeli Kandiari, Satyanashi	H, W	Leaf / Seed	Extract / Paste	Topical / Topical	Ringworm (22) / Eczema (16)	0.42
Arnebia euchroma (Royle) I.M. Johnston / Boraginaceae	Ratanjot, Lalmundi	H, W	Latex	As such	Topical	Alopecia (10)	0.11
Artemisia maritima L./Asteraceae	Sesaki	H, W	Leaf	Extract	Topical	Abscess (15)	0.16
Azadirachta indica A. Juss./ Meliaceae	Nimm	T, W/C	Leaf / Bark / Seed	Juice / Powder / Powder	Topical / Oral / Topical	Acne (15) / Eczema (12) / and psoriasis (24)	0.80

Contd.

Botanical name/Family	Local name	Habit	Part used	Form	Route	Ailments (No.)	Value
Berberis lycium Royle/ Berberidaceae	Kamble, Kamblu	S, W	Leaf	Paste	Topical	Leprosy (22), Blemishes (32)	0.35
Calotropis procera (Aiton) W.T.Aiton/ Asclepiadaceae	Nikka aak, Desi aak	S, W	Latex	As such	Topical	Ringworm (30), psoriasis (23) and abscess (19).	0.79
Cannabis sativa L./ Cannabaceae	Bhang	H, W	Leaf	Paste	Topical	Alopecia (32), dermatitis (26), eczema (6) and psoriasis (12)	0.84
Carissa spinarum L./ Apocynaceae	Garna	S, W	Leaf	Paste	Topical	Herpes (17)	0.09
Catharanthus roseus (L.) G.Don / Apocynaceae	Sadabahar	H, C	Latex	As such	Topical	Abscess (13)	0.01
			Leaf	As such	Topical	Abscess (10)	
Cedrus deodara (Roxb.ex D.Don) G.Don/Pinaceae	Deodar	T, W	Bark	Powder	Topical	Dermatitis (9)	0.33
Citrus limon (L.) Osbeck / Rutaceae	Nimbu	T, C	Leaf	Paste	Topical	Herpes (9)	0.55
			Fruit	Juice	Topical	Hair-fall (25) and dandruff (25)	
Clematis grata Wall./ Ranunculaceae	Nikki Totar	Cl, W	Leaf	Paste	Topical	Ringworm (16)	0.18
Colebrookea oppositifolia Smith/ Lamiaceae	Chitti Suaali	S, W	Leaf	As such	Topical	Abscess (12)	0.13
Convolvulus arvensis L./ Convolvulaceae	Hiran Padi, Hiran Khuri	Cl, W	Leaf	Paste	Topical	Acne (6)	0.07
Cryptolepis dubia (Burm.f.) M.R. Almeida/ Asclepiadaceae	Kali Tairni	Cl, W	Latex	As such	Topical	Ringworm (24)	0.26
Cuscuta reflexa Roxb./ Convolvulaceae	Aandal-Kaandal	Cl, W	Whole plant	Paste	Topical	Ringworm (21) and warts (15)	0.40
Cynodon dactylon (L.)Pers./	Khabbal	H, W	Leaf	Paste	Topical	Warts (11),	0.38

Contd.

Botanical name/ Family	Local name	Habit	Part used	Preparation	Mode	Uses	Value
Poaceae			Leaf	Extract	Oral	wounds (12) and eczema (6) Dermatitis (6)	0.34
Cyperus rotundus L./ Cyperaceae	Deela	H,W	Rhizome	Paste	Topical	Abscess (14), inflammation (11) and itching (6)	0.13
Daphne oleoides Schreb/ Thymelaeaceae	Kaag sadi	H, W	Leaf	Paste	Topical	Abscess (6) and rashes (6)	0.11
Diospyros montana Roxb./ Ebenaceae	Rajaan	T, W	Leaf	Paste	Topical	Abscess (10)	0.41
Eclipta prostrata (L.) L./ Asteraceae	Bhringraaj	H, W	Leaf	Decoction	Oral	Abscess (15), boils (6) and leprosy (16)	0.57
Euphorbia helioscopia L./ Euphorbiaceae	Dudhal-patal	H, W	Latex	As such	Topical	Ringworm (21), abscess (16) and warts (15)	0.40
E. royleana Boiss./ Euphorbiaceae	Sula, Thor	S, W	Whole plant / Latex	Paste / As such	Topical / Topical	Leucoderma (6) / Abscess (19) and boils (11)	0.91
Ficus carica L./Moraceae	Anjeer	T, W	Fruits / Latex / Leaf	Infusion / As such / Juice	Oral / Topical / Topical	Leucoderma (15), leprosy (6) / Warts (24), acne (6), scabies (25) and dermatitis (7)	0.49
F. palmata Forssk./ Moraceae	Sula, Thor	S/ T, W	Leaf / Latex	Paste / As such	Topical / Topical	Abscess (10) / To pullout thorn from skin (35)	

Contd.

Botanical name / Family	Local name	Code	Part	Preparation	Route	Ailment (number)	Value
F. religiosa L./ Moraceae	Bar, Peepal	T, W	Leaf Bark	Decoction Powder	Topical Topical	Hair-fall (7), dandruff (7) and scabies (6)	0.30
Fragaria vesca L./ Rosaceae	Phulu	H, W	Fruits	Juice	Topical	Sunburns (12) and blemishes (6)	0.20
Geranium wallichianum D.Don ex Sweet/Geraniaceae	Rattan Jot, Lal Jari	H, W	Bark	Powder	Topical	Hair-fall (16), dandruff (16), dry and scaly skin (12)	0.48
Hedera nepalensis K.Koch/ Araliaceae	Kathimbri bel	Cl, W	Leaf	Paste	Topical	Abscess (7)	0.08
Hibiscus rosa-sinensis L./ Malvaceae	Gudaal	S, C	Flowers	Paste	Topical	Alopecia (13)	0.14
Impatiens balsamina L./ Balsaminaceae	Teera	H, W/C	Leaf	Paste	Topical	Dermatitis (28)	0.31
Ipomoea carnea Jacq./ Convolvulaceae	Bilaitti Aak	S, W	Leaf	Paste	Topical	Abscess (9), ringworm (16) and scabies (11)	0.40
I. pes-tigridis L./ Convolvulaceae	Panja Bel	Cl, W	Leaf	Paste	Topical	Acne (6), wounds (6) and cracked heels (11)	0.25
Juglans regia L./ Juglandaceae	Akhroat	T, C	Unripe fruit	Paste	Topical	Alopecia(23)	0.25
Justicia adhatoda L./ Acanthaceae	Brenkad	S, C	Leaf	Extract	Oral	Abscess (28) and scabies (19)	0.52
Lawsonia inermis L./ Lythraceae	Mahendi	S, W/C	Leaf Leaf Seeds	Paste As such Paste	Topical Oral Topical	Dermatitis (25) Leucoderma (13). Leprosy (15)	0.58
Linum usitatissimum L./ Linaceae	Alsi	H, W	Seeds	Paste	Topical	Abscess (9)	0.10
Lycopersicon esculentum Mill./	Tamatar	H, C	Leaf	Juice	Topical	Ring worm (22)	0.24

Contd.

Family / Species	Local name	Code	Part used	Preparation	Route	Uses	FL
Solanaceae *Melia azedarach* L./ Meliaceae	Daraink	T, W	Leaf	Decoction	Topical	Hair-fall (29) and dandruff (29)	0.64
Mirabilis jalapa L./ Nyctaginaceae	Galwasi	H, W	Leaf	Paste	Topical	Abscess (26) and leprosy (12)	0.42
Narcissus tazetta L./ Amaryllidaceae	Heermanjar	H, W/C	Flowers, bulb	Paste	Topical	Abscess (34), boils (21) and wounds (16)	0.78
Nerium oleander L./ Apocynaceae	Lal gandila	S, W	Latex	As such	Topical	Abscess (6) and alopecia (19)	0.27
Oxalis corniculata L./ Oxalidaceae	Khattibooti, Nikki Ammi	H, W	Leaf	Juice	Topical	Warts (13)	0.14
Phyllanthus emblica L./ Euphorbiaceae	Amla	T, W/C	Fruits / Fruits	As such / Infusion	Oral / Topical	Acne (14) Hair-fall (32), dandruff (32) and graying of hair (15)	1.02
Plantago lanceolata L./ Plantaginaceae	Challa	H, W	Leaf	Paste	Topical	Abscess (6), boils (9) and wounds (6)	0.23
Prunus armeniaca L. /Rosaceae	Saadi, Haadi	T, W/C	Ash of seeds	Paste	Topical	Ringworm (15)	0.16
Ranunculus laetus Wall. ex Hook.f. & J.W. Thomson/ Ranunculaceae	Darilli	H, W	Latex	As such	Topical	Alopecia (24)	0.26
Robinia pseudoacacia L. / Fabaceae	Kikar	T, W	Ash of leaves	Paste	Topical	Dermatitis (11)	0.12
Rosa moschata Herrm./Rosaceae	Karir	Cl, W	Petals	Paste	Topical	Acne (13) and blemishes (12)	0.27
Rumex hastatus D.Don/	Ammi	H, W	Flowers,	Juice	Topical	Acne (10)	0.11

Contd.

Polygonaceae

Species / Family	Local name	Code	Part used	Preparation	Route	Ailment (no.)	Value
Senecio vulgaris L./ Asteraceae	Jari	H, W	stem, Leaf	Paste	Topical	Dermatitis (14)	0.15
Senna tora (L.) Roxb./ Caesalpiniaceae	Loki Haedma, Aayroun	H, W	Seeds, Leaf	Paste, Juice	Topical	Leucoderma (16) and leprosy (19), Leprosy (19)	0.59
Solanum americanum Mill./ Solanaceae	Kaayankothi, Peelkaan, Makoy	H, W	Leaf, Leaf	Paste, Juice	Topical	Acne (12), Dermatitis (6)	0.20
S. tuberosum L./ Solanaceae	Aalu	H, C	Tuber	Paste	Topical	Blemishes (11)	0.12
Sonchus wightianus DC./ Asteraceae	Dudhli	H, W	Leaf	Paste	Topical	Abscess (6) and boils (6)	0.13
Urtica dioica L./ Urticaceae	Saddar	H, W	Leaf	Paste	Topical	Dandruff (13)	0.14
Vitex negundo L./ Verbenaceae	Bana	S, W	Leaf	Paste	Topical	Abscess (25), ringworm (28) and scabies (21)	0.81
Xanthium strumarium L./ Asteraceae	Jojra	H, W	Leaf	Paste	Topical	Alopecia (38), eczema (25)	0.69
Ziziphus jujuba Mill./ Rhamnaceae	Bair	T, W/C	Fruits	Decoction	Topical	Hair-fall (11) and dandruff (13)	0.26

With respect to various sources of ethnomedicinal plants effective against dermatological disorders in the study area, majority of the taxa (75%) were growing in wild, whereas 14.06% species were purely cultivated and 10.94% species existed in both wild and cultivated forms. Yinegar et al. (2007) reported that in Ethiopia, traditional healers usually collected medicinal plants from the wild indicating that the respondents have yet not started the cultivation of plant species they are using as medicine.

Out of the various life forms, herbs were the most dominant (48.44%), followed by trees (21.88%), shrubs (18.75%) and climbers (10.93%). Simbo (2010) attributed the dominance of herbs in medicinal flora to their easy accessibility in the nearby areas as compared to trees and shrubs that are often harvested from forest patches that are distant from residential areas.

Among the different plant parts used, leaf (48.19%) was the most frequently used plant part, followed by latex (12.05%), seed (10.84%), fruit (9.64%), bark and flower (4.83% each), whole plant (3.61%), bulb (2.41%), rhizome, tuber and stem (1.20% each). The preference of leaves in the preparation of remedies by the rural populace was ascribed to ease of preparation (Gazzaneo et al. 2005) and the presence of more bioactive ingredients developed in them in response to phytophagous organisms since they are the most vulnerable parts of a plant (Bhattarai et al. 2006).

Herbal medicines were prepared in the form of paste (48.74%), taken as such (17.07%), juice (10.98%), decoction (7.31%), powder and extract (6.10% each), infusion (3.66%). Out of these, 90% medicinal preparations were applied topically, whereas only 10% were administered orally.

A total of 25 dermatological ailments viz., abscess, ringworm, dandruff, frost bite, burns, boils, wounds, eczema, psoriasis, dermatitis, leprosy, blemishes, acne, herpes, hairfall, warts, inflammation, rashes, scabies, dry scaly skin, alopecia, sores, cracked heels, leucoderma and graying of hair were reported in the present study.

In order to find out the relative importance of various ethnomedicinal plant species effective in the treatment of dermatological disorders in the study area, a quantitative index i.e. use-value (UV) was employed. The use-value ranged from 0.01 -1.02. Plants with the highest use values included *Phyllanthus emblica* (UV=1.02), *Aloe vera* (UV=1), *Ficus carica* (UV=0.91), *Cannabis sativa* (UV=0.84), *Vitex negundo* (UV=0.81),

Azadirachta indica (UV=0.80), *Calotropis procera* (UV=0.79), *Narcissus tazetta* (UV=0.78), *Melia azedarach* (UV=0.64), *Euphorbia helioscopia* (UV=0.57), *Citrus limon* (UV=0.55).

The locals used fruits of *Phyllanthus emblica* (UV=1.02) to cure dermatlological disorders like acne, hair-fall, dandruff and graying of hair. Panda et al. (2016) have reported its bark decoction in coconut oil to cure scabies whereas others reported it as a cure against dandruff and graying of hair in Assam and Manipur (Hazarika 2012, Nuzhat and Vidyasagar 2017).

Aloe vera (UV=1), was another important ethnomedicinal plant, used either orally or topically by the locals to cure abscess, burns, boils, wounds, eczema and psoriasis. *A. vera* finds mention as a multipurpose skin treatment plant in Ebers Papyrus, De Materia Medica and traditional Ayurvedic medicine (Quattrocchi 2012) Many active ingredients including aloesin, aloeemodin, acemannan, aloeride, methylchromones, flavonoids, saponin, amino acids, vitamins, and minerals are responsible for the medicinal property of this plant. A cream containing 0.5% aloe has been reported to lessen skin plaques linked with psoriasis (Syed et al. 1996). Gel application is known to improve partial thickness burns (Kaufman et al. 1988) and also aids the skin to survive frostbite injury (Miller and Koltai 1995). The study revealed that most of the patients were fully satisfied with the traditional treatment given to them.

CONCLUSION

Dermatological disorders are highly prevalent particularly among the rural populace of district Udhampur who have inadequate access to hygienic levels of sanitation thereby causing great health loss. The present study reported 64 species of medicinal plants commonly used against dermatological disorders. Locals of the study area are heavily dependent on the medicinal flora. But, in the present scenario, ethnomedicinal plant usage is dwindling fast as the younger generation is less interested in inheriting this legacy due to fascination towards western lifestyle, allopathic drugs, modernization and acculturation. The reckless usage of these plants without any heed towards their sustainable utilization is also posing serious threats to their abundance. Therefore, there is a dire need to protect this medicinal wealth before it is lost forever. Formulation and implementation of effective management plans is urgently needed to ensure the lasting presence of these plant species.

Conflict of Interests

The authors declare that they have no competing conflict of interest.

ACKNOWLEDGEMENTS

The authors would like to extend their sincere thanks and appreciation to UGC, New Delhi, India for funding this work through UGC-BSR fellowship and also to DST, PURSE, SAP funded Department of Botany, University of Jammu for providing necessary laboratory facilities.

LITERATURE CITED

Abbasi AM, Khan MA, Ahmad M, Zafar M, Jahan S, Sultana S. 2010. Ethnopharmacological application of medicinal plants to cure skin diseases and in folk cosmetics among the tribal communities of North-West Frontier Province, Pakistan. *Journal of Ethnopharmacology* 128: 322–335.

Agarwal R, Chouhan D. 2014. Indigenous medicinal herbs used by tribal of c.g. for skin diseases treatment. *Indian Journal of Scientific Research* 4: 108-11.

Gupta DK, Gupta G. 2018. Endemic Use of Medicinal Plants for the Treatment of Skin Diseases in the Balod district *IOSR Journal of Pharmacy* 8: 18-24.

Anamika, Kumar K. 2016. Ethno medicinal plants used in the treatment of skin diseases by the tribals of topchanchi wild life sanctuary area, Dhanbad, Jharkhand, India. *International Journal of Bioassays* 5: 4902-4904.

Bhattarai S, Chaudhary RP, Taylor RSL. 2006. Ethnomedicinal plants used by the people of Manang district, central Nepal. *Journal of Ethnobiology and Ethnomedicne* 2: 41.

Gazzaneo LRS, Lucena RFP, Albuquerque UP. 2005. Knowledge and use of medicinal plants by local specialists in a region of Atlantic forest in the state of Pernambuco (North eastern Brazil). *Journal of Ethnobiology and Ethnomedicine* 1: 1-9.

Grice EA, Kong HH, Conlan S, Deming CB, Davis J, Young AC, Bouffard GG, Blakesley RW, Murray PR. 2009. Topographical and temporal diversity of the human skin microbiome. *Science* 324: 1190-1192.

Hazarika R, Santoshkumar SA, Bijoy N. 2012. Ethno Medicinal Studies of Common Plants of Assam and Manipur. *International Journal of Pharmaceutical and Biological Archives* 3: 809-815.

Kaufman T, Kalderon N, Ullmann Y, Berger J. 1988. *Aloe vera* gel hindered wound healing of experimental second-degree burns: A quantitative controlled study. *Journal of Burn Care and Rehabilitation* 9: 156-159.

Khan BA. 2010. *District disaster management plan, Udhampur*. District Disaster Management Committee, Udhampur.

Miller MB, Koltai PJ. 1995. Treatment of experimental frostbite with pentoxifylline and *Aloe vera* cream. *Archives of Otolaryngology- Head and Neck Surgery* 121: 678-680.

Nuzhat T, Vidyasagar GM. 2017. Ethnomedicinal Oil Plants used in Treating Skin Diseases in Hyderabad Karnataka Region, Karnataka, India. *International Journal of ChemTech Research* 10: 770-784.

Panda T, Mishra N, Pradhan BK. 2016. Folk Knowledge on Medicinal Plants Used for the Treatment of Skin Diseases in Bhadrak District of Odisha, India. *Medicinal and Aromatic Plants* 5: 2-7.

Phillips O, Gentry AH, Reynel C, Wilki P, Gavez-Durand CB. 1994. Quantitative ethnobotany and Amazonian conservation. *Conservation Biology* 8: 225-248.

Quattrocchi U. 2012. CRC World Dictionary of Medicinal and Poisonous Plants: Common Names, Scientific Names, Eponyms, Synonyms, and Etymology. *Chemical Rubber Company Press.*

Saikia AP, Ryakala VK, Sharma P, Goswami P, Bora U. 2006. Ethnobotany of medicinal plants used by Assamese people for various skin ailments and cosmetics. *Journal of Ethnopharmacology* 106: 149–157.

Sharma J, Gairola S, SharmaYP, Gaur RD. 2014. Ethnomedicinal plants used to treat skin diseases by Tharu community of district Udham Singh Nagar, Uttarakhand, India. *Journal of Ethnopharmacology* 158: 140–206.

Simbo DJ. 2010. An ethnobotanical survey of medicinal plants in Babungo, north-west region, Cameroon. *Journal of Ethnobiology and Ethnomedicine* 6: 8.

Sinha P. 2002. Overview on diseases of medicinal plants and their management. In: Govil JN, Singh VK, eds. *Recent Progress in Medicinal Plants-diseases and Their Management.* Sci tech Publishing, USA, pp. 212.

Sivarajani R, Ramakrishanan K. 2012. Traditional uses of medicinal plants in treating skin diseases in Nagapattanam district of Tamil Nadu, India. *International Research Journal of Pharmacy* 3: 5-12.

Spiewak R. 2000. Occupational skin diseases among farmers. In: Zagorski, J. (Ed.), Occupational and Para-Occupational Diseases in Agriculture. *Institute of Agricultural Medicine, Lublin,* pp. 42–152.

Swami A, Gupta BK. 1998. *Flora of Udhampur.* Bishen Singh Mahendra Pal Singh, Dehradun, India, pp. 455.

Syed TA, Ahmad SA, Holt AH. 1996. Management of psoriasis with *Aloe vera* extract in a hydrophilic cream: a placebo-controlled, double-blind study. *Tropical Medicine and International Health* 1: 505-509.

Tiwari AK. 2015. Indigenous knowledge for treating skin disease in some selected districts of Chhattisgarh (India) *International Journal of Recent Scientific Research* 6: 2654-2657.

Tremetsberger KH, Weiss-Schneeweiss, Stuessy TF, Samuel R, Kadlee, Ortiz GMA, Talavera S. 2005. Nuclear ribosomal DNA and karyotypes indicate a NW-African origin of South American *Hypochaeris* (Asteraceae, Cichorieae). *Molecular Phylogenetics and Evolution* 35: 102-116.

Yineger H, Kelbessa E, Bekele T, Lulekal E. 2007. Ethnoveterinary medicinal plants at Bale Mountains national park, Ethiopia. *Journal of Ethnopharmacology* 112: 55-70.

11

Picrorhiza kurroa Royle ex Benth, a Therapeutically Important Himalayan Plant for Drug Discovery

Simmi Sharma

CSIR- Institute of Himalayan Bioresource Technology, Palampur - 176 061
Himachal Pradesh, INDIA
Email: Simmi17jan@gmail.com

ABSTRACT

Picrorhiza kurroa Royle ex Benth. is a perennial herbaceous plant species of the family Scrophulariaceae growing at high altitude regions of Himalayas, and recorded to have a wide application in Ayurvedic system of medicine known by the name of kutki. In traditional system, this plant species is used to cure disorders of liver, respiratory tract fever, dyspepsia, chronic diarrhea and scorpion sting. *P. kurroa* is a growing source of interest because of having number of interesting activities such as immnunomudulatory, hepatoprotective and anti-cancer, highly used by pharmaceutical companies for development of new molecules and herbal botanical formulations. It has been scientifically validated that this species is a repository of bioactive chemicals such as picrosides, kutkoside, and cucurbitacins isolated and characterized mainly from the roots of this plant. This communication deals with the chemistry and biological activities associated with its main chemical constituents, and can be helpful in development of new drug as medicine. This communication will provide a baseline information and useful tool for researchers who work on the area of natural products, and look for new direction in drug discovery.

Keywords: Medicinal plant, *Picrorhiza kurroa*, Scrophulariaceae, Irioid, Picrosides, Kutkin, Picroliv, Drug discovery.

INTRODUCTION

Genus *Picrorhiza* belongs to the family Scrophulariaceae whose members grow at high altitude resions. There are more than 30 species of this genus which are widely distributed in the hilly areas of Bhutan, China, Nepal and Pakistan. In India, *Picrorhiza kurroa* Royle ex Benth and *Picrorhiza scrophulariflora* Pennell are two species reported under this genus (Ansari et al. 1988) growing between altitude of 2000-4500m above mean sea levels (Sultan et al. 2017). *P. scrophulariflora* is predominant in Himalayan regions of Nepal, Sikkim and Tibet while *P. kurroa* is mainly present in the North-Western Himalayas regions of India distributed from Kashmir to Kumaon, Garhwal and Sikkim regions (Masood et al. 2015). From this species, 132 constituents were isolated and characterized from different parts of plants such as seeds, rhizomes, roots, stem and leaves (Jitand Varshney 2013).

Picrorhiza kurroa is a small perennial herb (Sood and Chauhan 2010). The name of this plant arises because of its bitter taste (in Greek *'picros'* means 'bitter') and it falls in the category of bitter drugs. Other common name of this plant is Kutki (Dorsch et al. 1991). *Picrorhiza kurroa* became a critically endangered plant species because of its growing demand, reckless assortment form wild and inadequate cultivation (Rai et al. 2000, Mehra et al. 2011).

Fig. 1: Wild habitat of *Picrorhiza kurroa*

Preparation of Extract

The plant rhizomes were dried and powdered with the help of grinder as sample material for investigation of compounds. The extraction of these powdered rhizomes were done in Soxhlet apparatus using methanol as solvent. After the samples were concentrate this methanolic extract under reduced pressure. A dark brown viscous mass was obtained. A small amount of the extract was examined chemically to determine the existence of various chemical constituents (Ali et al. 2017).

Isolation of Phytoconstituents

For the isolation of different compounds slurry of viscous dark brown mass was prepared by dissolving of different compounds in minimum quantity of methanol, and then adsorbing on silica gel. Air dried slurry was subjected to silica gel column chromatography in petroleum ether. The column was eluted successively with petroleum ether, mixture of petroleum ether - chloroform in ratios (9:1, 3:1, 1:1, 1:3), chloroform and the mixture of chloroform - methanol (99:1, 97:3, 95:5, 92:8, 9:1, 3:1, 1:1, 1:3). The fractions were collected and the homogeneity of spots were checked through TLC. The fractions having same R_f values were pooled and the crystallized. Then, recrystallization was performed to obtain the following compounds (Ali et al. 2017).

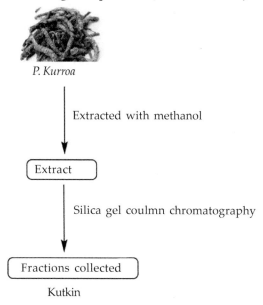

Fig. 2: Protocol for isolation of phytochemicals from *Picrorhiza kurroa*.

PHYTOCHEMISTRY

Picrorhiza kurroa has a complex chemistry. The chemical investigations on the leaves roots/rhizomes and seeds of the plant shows the existence of iridoids (Gupta 2001), acetophenones (Basu et al. 1971), cucurbitacins (Stuppner and Wagner 1989, Stuppner and Wagner 1989a, Stuppner et al. 1990) and triterpenoids.

Kutkoside, picroside I and picroside II are major bioactive components isolated from *P. kurroa* plant (Singh and Rastogi 1972). Moreover, picrohizin, apocyanin, androsin, picroside III, veronicoside, picein, pikuroside, minecoside, rosin and cucurbitacin glycosides are other constituents reported from *P. kurroa* (Ali et al. 2017). The main bioactive constituent form this plant is Kutkin, which is responsible for bitter taste of this plant. Kutkin is a mixture of two major compounds i.e., picroside I and kutakoside in the ratio of 1:2 (Singh and Rastogi 1972, Ansari et al. 1988). Picroside I is cinnamoyl ester of cataplol and kutkoside is vanilloyl esters of catalpol. Three picrosides named picroside I (P-I), picroside II (P-II) and pichroside III (P-III) are reported (Sharma et al. 2012). All these are irioid glycosides. Structures of these glycosides and other compounds isolated is shown in Fig. 3. These are one of the most active components of *Picrorhiza kurroa*.

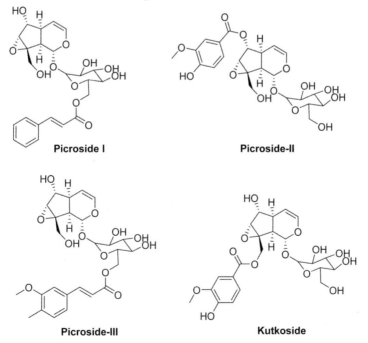

Fig. 3: Structure of Picrosides isolated from *Picrorhiza kurroa*.

A herbal drug formulation named picroliv is made by mixing picroside-1 and kutkoside in the ratio 1.0: 1.5 (w/w). The concentration of kutkoside is somewhat higher in picrolive than kutkin. There are many other drug formulations that are available commercially contain P-I and P-II in different ratios (Verma et al. 2009).

Other active constituent isolated from *P. kurroa* is apocynin which is a phenolic glycoside. Apocynin is named as acetovanillone. It is basically a α-methoxy substituted catechol and structurally related to vanillin. Apocynin is potential anti-inflammatory agent. Other phonelic glycosides isolated from this plant are picein and androsin. Androsin is an acetophenone glucoside and known to possess anti-asthmatic activity (Sood et al. 2010).

Cucurbitacin B, D and Q were also isolated from *P. Kurroa* plant (Stuppner and Wagner 1989a). These are triterpenes with basic 19-(10→9β)–abeo−10α–lanost–5−ene ring skeleton. The presence of 5, (6)−double bond is the common feature among their skeleton (Dinan et al. 2001).Various kind of activities associated with cucurbitacins. It was reported that cucurbitacin B, inhibits cyclooxygenase (COX)-2 enzymes without effecting COX-1 enzymes. It was also known to possess anti-artherosclerotic and anti-inflammatory activity (Kushik

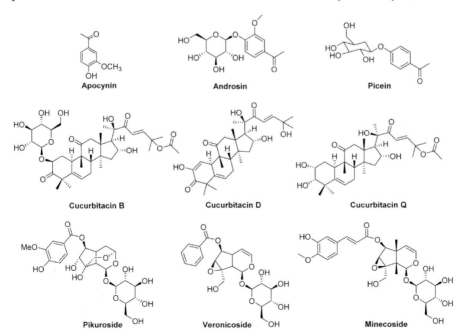

Fig. 4: Chemical constituents isolated from *Picrorhiza kurroa.*

et al. 2015). Furthermore, Cucurbitacin D inhibits ovulation in mice. In addition, Cucurbitacins Q exhibits antitumour activity (Ayaprakasam et al. 2003). Veronicoside is one more bioactive compound isolated from this plant. This compound is having antioxidant potential (Yin et al. 2013) (Fig. 4).

ETHNOBOTANICAL INFORMATION

According to Masood et al. (2015), *Picrorhiza kurroa* is extensively used by local Himalayan people for curing diseases such as stomach ache and high fever. To cure stomachache, 10g root is boiled and taken with honey. Root powder (10g) mixed with honey and 1 g black pepper is given to the adult patient for the treatment of fever. It is advised to take kutki powder (0.25g) with mother's milk in case of stomach ache in newborns (Arya et al. 2013). It is also for used for veterinary purposes in Kashmir. In India, Kutki is used in traditional Ayurvedic and Unani system of medicine as its rhizomes have antibiotic activity. The ayurvedic preparation named arogyavardhini, used to cure liver disorders and kutki is used as the major component for this preparation. Kutki is also used as a substitute for *Gentiana kurroo* (Khan and Ziadi 1989, Zarman and Khan 1970). Also many therapeutic uses of kutki are being explored in India. In Pakistan, kutki is used in the ayurvedic and Greek-Arab systems of medicine. Generally it is used as as an aromatic, stimulant and carminative agent, as a therapy for bronchial asthma, cough and disease of liver, blood, kidney and skin. Under the name of Qusttalakh, *P. kurroa* is used in two herbal preparations (Roghane-Qust-Talakh, an essential oil and *Majoon-E-Murravehulazwah*), which are used to treat Tumors, hypothermia, debility, tetanus and gout (Hamdard 1968).

Rhizome of *P. kurroa* is widely used to cure skin disease, cough, fever, liver disease, indigestion, hepatitis, jaundice and metabolic disorders in Nepal. Plant rhizome is used to treat high blood pressure, intestinal pain, gastritis, bile disease, eye disease, sore throat, lung and blood fever (Lama et al. 2001). It is also described by manufacturers of Kathmandu that rhizome of *P. kurroa* used as purgative and for the treatment of scorpion bites (Amatya 2005). It is considered as a bitter tonic and used as a cholagogue (promoting the flow of bile from the gall bladder), stomachic (stimulating gastric activity) and cathartic (purgative) (IUCN Nepal 2004). In Nepal, kutki is used both by amchi (specialists trained in the Tibetan medical system or Sowa Rigpa) and by non-specialists. In the later case, largely used for treating cough and cold (Ghimire et al. 2005).

In Tibet, Himalayan people used this plant mainly in local herbal praparation as traditional medicine. In China, it is used as remedy for jaundice, fever, malnutrition due to digestive disorders, dysentery and diarrhoea (Zhang et al. 1994). It is used as a drug for cold, cough, and fever in Bhutan. The National Institute of Traditional Medicines and other local hospitals in Bhutan uses the rhizome as an ingredient in developing medicines (Mulliken 2000). Due to choleretic effect of *P. kurroa*, rhizome extract can considerably inhibit the hepatic injury. The plant is bitter in taste and has cold properties. The plant species is used in combination with Chinese herbs including *Corium erinacei* and navel gland secretions of musk deer to treat damp-fire accumulation in the large intestine, hemorrhoid and anal fistula. The dosage of plant used must be in the range of 6-12 g (http://www.chineseherbshealing.com).

PHARMACOLOGY ASPECTS

This is a medicinal herb, usually used to treat liver disorders (Thyagarajan et al. 2002) The rhizome used to treat anaemia, appetite, asthma, alcoholic toxicity, diarrhea, cold, diabetes, high fever cough, dysentery, jaundice, digestive disorders, skin diseases and metabolic disorders (Bhandari et al. 2008, Smit et al. 2000). The rhizomes extract of this plant exhibit anti-malarial activity (Masood et al. 2015). Further, a broad array of biological activities associated with iridoids, such as choleretic, antihepatotoxic, hypolipidemic, antispasmodic, anti-inflammatory, antitumor, antiviral, immunomodulatory, antioxidant, purgative, anti-phosphodiesterase, antiasthmatic, antidiabetic, cardioprotective, neuritogenic, leishmanicidal and molluscicidal activities (Ghisalberti et al. 1998, Indian Herbal Pharmacopoeia, 2002, Joy et al. 2000, Chauhan et al. 2008). Picroliv have shown similar or more potent activity as silymarin, which is generally used to treat liver disorders (Dhawan 1995). In addition, they are also used in drugs prescribed for treatment of respiratory and allergic diseases (Jitand Varshney 2013).

CONCLUSION

Picrorhiza kurroa is an important Himalayan plant species and several variety of biological activities associated with this plant. Present results spotlights information of various chemical constituents biological activity and ethnobotanical uses of this plant. It is extensively used as traditional medicine in India, China, Pakistan, Nepal and Tibet for centuries. Plenty of ethnobotanical and socio-economic uses associated with this plant make it more noteworthy. The chemical constituents isolated from this plant proven to have significant biological activities. So this prominence builds a pressure on the plant regarding its use. According to IUCN it is under threat, and therefore we need to conserve this plant species either by *ex-situ* or *in-situ* approaches.

Abbreviations
IUCN: International Union for Conservation of Nature

LITERATURES CITED

Ali M, Sultana S, Rasool SM. 2017. Chemical Constituents from the Roots of *Picrorhiza kurroa* Royle ex Benth. *Ijppr. Human* 9: 25-35.

Amatya G. 2005. Project Manager, Medicinal and Aromatic Plants and NTFPs Project, IUCN Nepal in litt. to TRAFFIC International.

Ansari RA, Aswal BS, Chander R, Dhawan BN, Garg NK, Kapoor NK, Sharma SK. 1988. Hepatoprotective activity of kutkin the iridoid glycoside mixture of *Picrorrhiza kurroa*. *Ind. J. Med. Res.* 6: 401-404.

Arya D, Bhatt D, Kumar R, Tewari LM, Kishor K, Joshi GC. 2013. Studies on natural resources, trade and conservation of Kutki (*Picrorhiza kurroa* Royle ex Benth., Scrophulariaceae) from Kumaun Himalaya. *Scientific Res. Essays* 8: 575-580.

Basu K, Dasgupta B, Bhattacharya SK, Debnath PK. 1971. Chemistry and pharmacology of apocynin, isolated from *Picrorhiza kurrooa* Royle ex Benth. *Curr. Sci.* 40: 603-604.

Bhandari P, Kumar N, Singh B, Kaul VK. 2008. Simultaneous determination of sugars and picrosides in *Picrorhizakurrooa* species using ultrasonic extraction and high- performance liquid chromatography with evaporative light scattering detection. *J. Chromatogr. A*, 1194: 257-261

Chauhan S, Nath N, Tule V. 2008. Antidiabetic and antioxidant effects of *Picrorhizakurrooa* rhizome extracts in diabetic rats. *Indian J. Clin. Biochem.*23: 238-242.

Dhawan BN. 1995. Picroliv a New Hepatoprotective Agent from an Indian Medicinal Plant, *Picrorhiza kurrooa*. *Med. Chem. Res.*2: 595-605.

Dinan L, Harmatha J, Lafont R. 2001. Chromatographic procedures for the isolation of plant steroids. *J. Chromatogr A*. 935: 105–123.

Dorsch W, Stuppner H, Wagner H, Gropp M, Demoulin S, Ring J. 1991. New antiasthmatic drugs from traditional medicine? *Int. Arch. Allergy Appl. Immunol.* 94: 262-265.

Ghimire SK, Mckey D, Aumeeruddy-Thomas Y. 2005. Conservation of Himalayan medicinal plants: harvesting patterns and ecology of two threatened species, *Nardostachys grandiflora* DC. and *Neopicrorhiza scrophulariiflora* (Pennell) Hong.*Pure Appl. Biol.* 4: 407-417.

Ghisalberti EL. 1998. Biological and pharmacological activity of naturally occurring iridoids and secoiridoids. *Phytomedicine* 5: 147-163.

Gupta P P. 2001. Picroliv: hepatoprotective, immunomodulator. *Drugs Future*, 26: 25-31

Hamdard. 1968. *Qarabadain-eHamdard*. Pharmaceutical Advisory Council, Hamdard Academy, Karachi, Pakistan.

http://www.chineseherbshealing.com/picrorhiza-kurroa.

Indian Herbal Pharmacopoeia, a Revised New Edition, 2002. Indian Drugs Manufacturers Association: Mumbai, 289-297.

IUCN Nepal 2004. *National Register of Medicinal and Aromatic Plants* (Revised and updated). xiii +202 pp. IUCN, Nepal.

Jayaprakasam B, Seeram NP, Nair MG. 2003. Anticancer and antiinflammatory activities of cucurbitacins from Cucurbitaandreana.*Cancer Lett.* 189:11–16.

Joy KL, Rajeshkumar NV, Kuttan G, Kuttan R. 2000. Effect of *Picrorhizakurrooa* extract on transplanted tumours and chemical carcinogenesis in mice. *J. Ethnopharmacol.* 71: 261-266.

Kaushik U, Aeri V, Showkat R. Mir. 2015. Cucurbitacins – An insight into medicinal leads from nature. *Pharmacogn. Rev.*9: 12–18.

Khan AH, Zaidi SH. 1989. *Propagation and Regeneration Technology of Pharmacopeial Medicinal Plants of the Temperate Regions of Pakistan.* Bulletin No. 8. Biological Sciences Research Division, Pakistan Forest Institute, Peshawar, Pakistan.

Lama YC, Ghimire SK, Thomas YA. 2001. *Medicinal Plants of Dolpa: Amchis Knowledge and Conservation.* 150 pp., WWF Nepal Program and People & Plant Initiative, Kathmandu, Nepal.

Masood M, Arshad M, Qureshi R, Sabir S, Amjad MS, HumaQureshi H, Tahir Z. 2015. *Picrorhizakurroa*: An ethnopharmacologically important plant species of Himalayan region. *Pure Appl. Biol.* 4: 407-417.

Mehra TS, Chand R, Sharma YP. 2011. Reproductive biology of Picrorhizakurroa – a critically endangered high value temperate medicinal plant. *J. Med. Arom. Plants* 1: 40–43.

Mulliken TA. 2000. Implementing CITES for Himalayan Medicinal Plants *Nardostachysgrandiflora* and *Picrorhizakurroa.* In: TRAFFIC Bulletin 18: 63-72.

Rai LK, Prasad P, Sharma E. 2000. Conservation threats to some important medicinal plants of the Sikkim Himalaya. *Biol. Conserv* 93: 27–33.

Sah JN, Varshney VK. 2013. Chemical constituents of Picrorhiza genus: a review. *AJEONP.* 1:22-37.

Sharma N, Pathania V, Singh B, Gupta RC. 2012. Intraspecific variability of main phytochemical compounds in *Picrorhiza kurroa* Royle ex Benth. from North Indian higher altitude Himalayas using reversed phase high-performance liquid chromatography. *J. Med. Plants Res* 6: 3181-3187.

Singh B, Rastogi RP. 1972. Chemical examination of *P. kurroa* part II: Re-investigation of kutkin. *Indian J. Chem* 10:29–31.

Smit HF, Berg AJJvan den, Kroes BH, Beukelman CJ, Ufford HCQ van, Dijk H Van, Labadie RP. 2000. Inhibition of Tlymphoncyte proliferation by cucurbitacines from *P. scrofhulariiflora. J. Nat. Prod* 63: 1300-1302

Sood H, Chauhan R. 2010. Biosynthesis and accumulation of a medicinal compound, picroside-I, in cultures of *Picrorhiza kurroa* Royle ex Benth. *Plant Cell, Tissue Organ Cult.*100: 113–117.

Stuppner H, Kahling HP, Seligmann O, Wagner H. 1990. Minor cucurbitacin glycosides from *Picrorhiza kurroa. Phytochemistry* 29: 1633-1637.

Stuppner H, Wagner H. 1989a. New cucurbitacin glycosides from Picrorhiza kurroa. *Planta Med.* 55: 559-563

Stuppner H, Wagner H. 1989b. Minor iridoid and phenol glycosides of *Picrorhiza kurrooa. Planta Med.* 55: 467-469

Sultan P, Rasool S, Hassan OP, 2017. *Picrotrhiza Kurroa* Royle on Bentho. a plant of diverse pharmacolocial potential. *Ann. Phytomed.* 6(1): 63-67.

Thyagarajan P, Jayaram S, Gopalakrishnan V, Hari V, Jeyakumar P, Sripathi M. 2000. Herbal medicines for liver diseases in India. *J. Gastroenterol Hepatol* 17: 370–376.

Verma PC, Basu V, Gupta V, Saxena G, Rahman LU. 2009. Pharmacology and Chemistry of a Potent Hepatoprotective Compound Picroliv Isolated from the Roots and Rhizomes of Picrorhiza kurroa Royle ex Benth. (Kutki). *Curr. Pharm. Biotechnol* 10: 641-649.

Yin L, Lu Q, Tan S, Ding L, Guo Y, Chen F, Tang L. 2016.Bioactivity-guided isolation of antioxidant and anti-hepatocarcinoma constituents from *Veronica ciliate.Chem. Cent. J.* 10: 27.

Zaman MB, Khan MS. 1970. *Hundred Drug Plants of West Pakistan*. Medicinal Plant Branch of the Pakistan Forest Institute, Peshawar, Pakistan.

Zhang YJ, Li W, Zheng M. 1994. *Chinese-English Chinese Traditional Medicine World Dictionary*. Shanxi People's Press, Shanxi, China.

12

Plant Resource Used by Riparian Communities of Dhir and Diplai Wetlands in Assam, Northeastern India

Miniswrang Basumatary and Tapati Das*

Department of Ecology and Environmental Science, Assam University
Silchar - 788011, Assam, INDIA
*Email: das.tapati@gmail.com

ABSTRACT

Wetlands play an important role in provisioning of numerous goods and services for mankind. However, the type and the quantum of ecological services differs amongst wetlands. In the present study, we documented the various plant resources used by the riparian communities of two well-recognized wetlands, Dhir beel and Diplai *beel*, located in Brahmaputra Valley of Assam. The findings of study indicate that the riparian communities are dependent on wetlands for various purposes. The riparian communities uses 48 plant species as vegetables, fruits, fodder, fuel wood, medicine, roofing materials, craft making materials, bio-fertilizer and fish poisoning agent. Some of the plant species such as *Enhydra fluctuans, Lasia spinosa, Colocasia esculenta, Diplazium esculentum, Ipomoea aquatica* and *Monocharia hastata* are sold in the village market as local vegetables. The wetlands generate income for the riparian communities and save cash of many households settled plant near the wetland. Nevertheless, this study also revealed that some near the wetland resources are declining due to population pressure, encroachment near the wetlands, deforestation and over-exploitation of the wetland habitats for creation of many fishery ponds within the wetland and cultivation practices on wetland edges. The socio-economic condition, unemployment, easy accessibility and lack of alternative sources are plausibly the major factors, which are leading to the decline and depletion of the wetland resources. Therefore, appropriate management strategies are required for the conservation and the sustainable utilization of these plant resources.

Keywords: Plant diversity, Utilization pattern, Wetlands, Northeastern India

INTRODUCTION

Wetlands provide ecological goods and services, which support rural socio-economy and human well-being (RCS 2007). They include provisioning of resources such as fish, fiber, drinking water, and other ecological services like water purification, climate regulation, flood regulation, coastal protection, recreational opportunities, and tourism (MEA 2005). Historically, wetlands have contributed greatly to the livelihood of riparian communities through its various ecosystem services thus contributing to poverty alleviation (Wang et al. 2008). However, the knowledge on the use of wetland resources is gradually eroding because of social transformations and disappearance of the wetlands due to their conversion to other land use types. It is of the record that India has lost an estimated 38% of its wetlands because of similar reasons (Vijayan et al. 2004). This necessitates documentation of the use pattern of natural resources occurring in wetlands. Such documentation should help in sensitizing the people as well as scientific community for conservation and protection of wetlands.

The northeastern region of India has a large number of floodplain as well as other seasonal wetlands. With the exception of fish and fishery, these wetlands have been comparatively less assessed from the angle of other resources like plants. Wild plant resources play a very important role in the livelihoods of local communities. These plants serve as alternatives to staple food crops, and are a good source of income for numerous poor families in rural areas (Narzary et al. 2013). Studying the utilization patterns of plant resources may thus help in designing equitable and effective sustainable natural resource use systems (Thondhlana et al. 2012). A better understanding of the benefits of utilizing wetland resources also provide important information for understanding and addressing the causes of wetland degradation and loss (Kakuru et al. 2013).

Therefore, the present study deals with documentation of the plant resources used by the riparian communities in two typical well recognized wetlands located in the Brahmaputra Valley of Assam in northeast India.

Study Area

The present study area comprises of two wetlands, Dhir *beel* and Diplai *beel*, located in Dhubri district coming under Brahmaputra Valley of Assam. These two wetlands lie in the foothills of Chakrashila Wildlife Sanctuary, falling partly in Kokrajhar (BTAD) district and partly in

Dhubri district of Assam. Due to the hilly terrain, most of the rainwater is discharged through numerous perennial springs that feed Dhir *beel* and Diplai *beel* that lie on the southern part of the sanctuary. These two wetlands are approximately 5 km apart from each other. Both the wetlands serve as an important catchment area for some of the streams and other perennial springs such as Howhawi Jhora, Mauriya Jhora and Sikri-Sikla Dwisa that enter into the mighty river, Brahmaputra (www.birdlife.org). These wetlands and streams also serve as major sources of irrigation for the agricultural fields in the nearby villages.

The Dhir *beel* located between 26°16'55"N and 90°21'10"E is 30 km away from Kokrajhar town and from the base of Chakrashila Wildlife Sanctuary. Total area under Dhir *beel* is 1003 ha (NWA 2010). During the rainy season, this wetland is connected with the Brahmaputra river through a small river called Dhir River intersecting the National Highway-31. The wetland is rich in aquatic fauna and serves as a breeding ground for variety of fish. Diplai *beel* is located between 26° 17'16.3"N and 90°18'50.1"E with an approximate size of 450 ha. The wetland is highly infested with water hyacinth (*Eichhornia crassipes*) and hairy water lily (*Nymphea pubescens*). Both the wetlands attract migratory birds in winter including two globally threatened species, Ferruginous Duck *Aythy anyroca*, and Baer's Pochard *A. baeri* (Sinha et al. 2015).

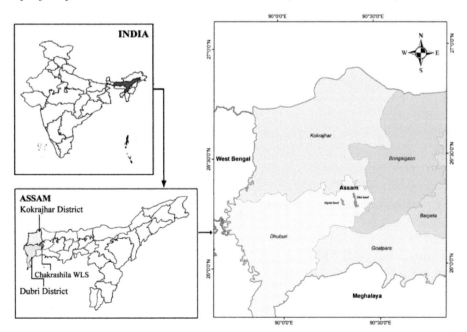

Fig. 1: Geographical location of study area

Survey and Documentation of Plant Resources

Field surveys were carried out from October 2016 to September 2017 in riparian villages of Dhir *beel* and Diplai *beel* to document the wetland resources particularly the plant species that are being used by the riparian communities for various purposes. For the present study, nine villages from Dhir *beel* and eight from Diplai *beel* were selected. For the survey, 229 households from riparian villages of Dhir *beel* and 161 households from riparian villages of Diplai *beel* were selected through systematic random sampling. Information on use pattern of the wetland vegetation was recorded through key informant interview with traditional healers and village elders. Household surveys using semi-structured questionnaires were done to record information on the availability of various wetland resources. Market surveys were undertaken to document various wetland resources particularly the plant species which are sold locally by the riparian communities. Field visits in and around the wetland was carried out to see the type of plant resources present, and the various process of resource extraction and to take their photographs. The plant species were identified using standard literatures (Kanjilal et al. 1934-40, Barooah and Ahmed 2014) and their scientific names were confirmed from the website (www.theplantlist.org.)

Socio-Economic Condition of the Riparian Communities

Boro, Garo, Rabha, Rajbongshi and *Bengali speaking Muslims* are the resident communities of the *beels*. However, the dominant communities vary from one wetland to the other. In Dhir *beel*, the dominant communities were the *Bengali speaking Muslims* followed by the *Boro*. On the other hand, in Diplai *beel* the dominant communities were the *Rajbongshi* followed by the *Garo*. Large numbers of families are living below the poverty line and most of the inhabitants are working as fishermen, occasional labourers, and farmers (Department of Forest, Chakrashila Wildlife Division 2015). The riparian communities of the selected wetlands are dependent on the wetland resources to a varying extent for their livelihood sustenance.

Resources Available and their Utilization Patterns

Both the wetlands provide a number of wetland goods viz., fish, shrimp, snail, wild vegetable, fodder, fire wood, soil and gravels, potable water, timber and non-timber forest products (NTFPs) such as thatching material, medicinal plants; and services like area for cattle grazing, fish farming ponds, and wildlife for hunting. Surface water of the wetlands are used for multiple purposes like washing, bathing, fishery, livestock rearing, jute retting, trans-boundary transportation, irrigation in agricultural land, and recreation like boating. Such wetland products and services encourage a number of human activities, which contribute to the national product and to the welfare of the local people (Devis 1993). No variations in resource types collected from both the wetlands have been observed.

Useful Plant Diversity *vis-à-vis* Habitat Diversity

A total of 48 diffent plant species belonging to 44 genera and 27 families were documented as useful from different habitat types in and around the wetlands (Photoplate 1, Table 1). High plant diversity in the wetlands may be attributed to the heterogeneous habitat conditions. Nevertheless, the Diplai *beel*, which was smaller, was found to have more plant resources as compared to the Dhir *beel*, which was much larger in size. This may be attributed to the irregular topography with numerous hillocks within the Diplai *beel*, which provided diverse niche to supports diverse plant species within the wetland. Apart from these, the quantity of resource extraction and uses also vary among different communities. It was observed that amongst all the riparian communities - *Boro* and *Garo*, the dominant riparian communities of Diplai *beel* - had more knowledge about the usage of various plant species in comparison to other tribal peoples. These communities collect maximum plant resources from different habitat conditions of the wetland and its nearby areas.

Lagerstroemia speciosa (L.) Pers   *Imperata cylindrica* (L.) Raeusch   *Lippia geminata* Kunth

Photoplate 1: Pictorials of valuable plant species found in the two wetlands.

Chrysopogon zizanoides (L.) Roberty.

Centella asiatica (L.) Urban

Hygroryza aristata (Retz.) Nees.

Eleocharis palustris (L.) Roem.

Enhydra fluctuans Lour.

Ipomoea aquatica Forssk.

Lasia spinosa (L.) Thwaites

Monochoria hastata (L.) Solms.

Nymphoides cristata (Roxb.) Kuntze

Euryale ferox Salisb.

Hydrocotyle sibthoxrpoides Lam.

Oxalis corniculata L.

Photoplate 1: Pictorials of valuable plant species found in the two wetlands.

Table 1: Family, latin name, common/vernacular name and the habitat types of the commonly used wetland plant species

Family	Scientific name	Common/ local name	Habitat
Asteraceae	*Enhydra fluctuans* Lour.	Water cress; Helencha; Helonchi; Elanchi	Marsh/ shallow aquatic bodies
Araceae	*Lasia spinosa* (L.) Thwaites	Lasia; Chengmora kochu, Biskochu, Sibru	Moist places
	Amorphophallus sylviticus (Roxb.) Kunth	Olodor	Hillocks within wetland
	Colocasia esculenta (L.) Schott	Kola kochu; Taso gwswm	Moist places
Apiaceae	*Centella asiatica* (L.) Urban	Indian pennywort; Boro manimuni; Manimuni gidir	Moist places, & marsh
Arecaceae	*Calamus tenuis* Roxb.	Cane; Jali bet; Raidwng	Hillocks within wetland
Athyriaceae	*Diplazium esculentum* (Retz.) Sw.	Vegetable fern; Dhekia; Dinkina	Upland areas along the wetland
Bignoniaceae	*Oroxylum indicum* (L.) Kurz.	Indian trumpet flower; Dingdiga; Karokandai,	Hillocks within wetland
	Stereospermum chelonoides (L.f) DC.	Fragrant Padri Tree; Paroli; Serfang; Bosil	Hillocks within wetland
Bixaceae	*Bixa orellana* L.	Lipstick tree; Khendur-gosh; Sindur donfang,	Hillocks within wetland
Convolvulaceae	*Ipomoea aquatica* Forssk.	Water spinach; Kolmou; Mondey	Aquatic bodies within the wetland
	Ipomoea carnea Jacq.	Bush morning glory; Behayaghash; Lota; Deo Komli; Mondey gidir	Aquatic bodies and marshes within the wetland, and wetland bank
Cyperaceae	*Eleocharis palustris* (L.) Roem. &Schult.	Spike grass, Seseri	Marsh areas/ agricultural fields within the wetland
Combretaceae	*Terminalia arjuna* (Roxb. ex DC.) Wight &Arn.	Arjun tree; Arjun gosh	Terrestrial
	Terminalia chebula Retz.	Chebulic Myrobalan; Hilika; Seleka; Haritaki	Terrestrial

Contd.

Family	Scientific name	Common names	Habitat
	Terminalia bellirica (Gaertn.) Roxb.	Belliric myrobalan; Bohera; Bhaora; Surei	Terrestrial
Dipterocarpaceae	*Shorea robusta* Gaertn.	Sal tree; Sal ghos; Sal; Borsa	Terrestrial
Lauraceae	*Litsea glutinosa* (Lour.) C.B. Rob.	Soft Bollygum; Baghnaola; Heluk	Terrestrial
Lythraceae	*Lagerstroemia speciosa* (L.) Pers.	Pride of India; Jarul, Aojhar, Ajagari	Wetland bank, seasonally flooded areas
Moraceae	*Duabanga grandiflora* (DC.) Walp.	Ramdala; Hokolo gash; Khokan	Terrestrial, moist areas
	Streblus asper Lour.	Toothbrush tree; Demon tree; Seora; Shoombara	Moist places, wetland bank
	Ficus benjamina L. var. *benjamina*	Ficus tree; Jorigash; Hengul; Hendla; Kari	Terrestrial
Menyanthaceae	*Nymphoides cristata* (Roxb.) Kuntze	Crested floating heart; Rupa puli	Aquatic bodies within the wetland
Nymphaeaceae	*Euryale ferox* Salisb.	Foxnut; Nikharu; Makhana.	Aquatic bodies within the wetland
	Nymphaea pubescens Willd.	Red Indian water lily, Ronga bhet, Silung	Aquatic bodies within the wetland
	Nymphaea nouchali Burmf.f.	White water lily; Bogabhet; Sadabhet	Aquatic bodies within the wetland
Oxalidaceae	*Oxalis corniculata* L.	Creeping woodsorrel; Panitengisi; Singri	Moist places
Poaceae	*Saccharum ravannae* (L.) L.	Ravenna grass; Ecer; Birnah gidir	Wetland banks, moist places
	Hymenachne amplexicaulis (Rudge) Nees	Bamboo grass; Dolghanh; Dholhagra	Marshy places of the wetlands
	Leersia hexendra Swartz.	Cutgrass; Aralighanh, Arali	Swampy areas of the wetland
	Paspalidium flavidum (Retz.) A.Camus.	Yellow watercrown grass; Chaprighanh	Marshy places
	Hygroryza aristata (Retz.) Nees ex Wight &Arn.	Water grass; Petuli-dhol; Tepadhwl	Aquatic bodies within the wetland
	Imperata cylindrica (L.) Raeusch.	Cogon grass; Kashihagra; Hamfang	Hillocks within wetland

Contd.

Family	Scientific name	Common name(s)	Habitat
	Chrysopogon zizanoides (L.) Roberty.	Vetiver; Mutaghanh; Birnah	Wetland banks, upland areas
	Bambusa assamica Barooah & Borthkur	Bamboo; Owya Gubwi	Hillocks within wetland
Pontederiaceae	Eichornia crassipes (Mart) solms.	Water hyacinth; Meteka; Parna	Wetland water body, marsh areas
	Monochoria hastata (L.) Solms.	Leaf Pondweed; Khar/ Nara meteka; Ajwnai; Sambusok	Marshy places
Polygonaceae	Polygonum hydropiper L.	Marsh pepper knot weed; Bihlongoni; Bisongali	Moist places along wetland banks
Phyllanthaceae	Phyllanthus emblica L.	Indian gooseberry; Amlokhi; Amlai	Terrestrial
Rubiaceae	Paederia foetida L.	Skunkvine;Bhedai-lata; Kipibendwng	Hillocks within wetland/adjacent forested land
	Neolamarckia cadamba (Roxb.) Bosser	Burflower-tree; Kadam	Hillocks within wetland
	Cephalanthus occidentalis L.	Button bush; Hagrani kwdwm; panikwdwm; patamari	Marshes
Rutaceae	Aegle marmelos (L.) Correa	Stone Apple, Behel; Belli	Terrestrial
Sapindaceae	Blighia sapida K.D. Koenig	Ackee; Akhi donfang; Tidumpa	Terrestrial
Trapaceae	Trapa natans L. var. bispinosa (Roxb.) Makino	Water chestnut; Panihingori; Sugreng	Aquatic bodies within the wetland
Umbelliferae	Hydrocotyle sibthorpoides Lam.	Lawn marsh pennywort; Horu-manimuni, manimunipisa	Moist places along the wetland banks
Verbenaceae	Gmelina arborea Roxb.	Gmelina; Gamari; Gambari	Hillocks within wetland
	Lippia geminata Kunth	Bushy matgrass; Motmotiapul; Ontaibajab	Wetland banks, marshes

The 48 identified plant species comprised of vegetables (11 species), fruits (7 species), fodder (5 species), fuel wood (10 species), medicine (7 species), roofing material (2 species), craft making material (4 species), bio-fertilizer (1 species), and fish poisoning agent (1 species) (Table 2, Figure 2). Many wetland plants e.g., *Ipomoea aquatica*, *Colocasia esculenta*, *Lasia spinosa*, *Enhydra fluctuans*, *Diplazium esculentum* and several others were collected for consumption as vegetable and for selling the surplus to the local market. The local farmers also collects grass species like *Leersia hexendra*, *Hymenachne amplexicaulis*, *Hygroryza aristata*, *Eleocharis palustris* and several other species and supplement as fodder to their livestock to reduce the cost for bearing the price of fodder if they had to purchase from the nearby market. They collect fodder from the wetland for 3-4 months (December to March) when grasses on the nearby field disappear due to the effect of dry season. The riparian communities in Dhir *beel* mostly collect *Eleocharis palustris* as fodder, while the riparian communities in Diplai *beel* mostly collect *Hymenachne amplexicaulis* as fodder.

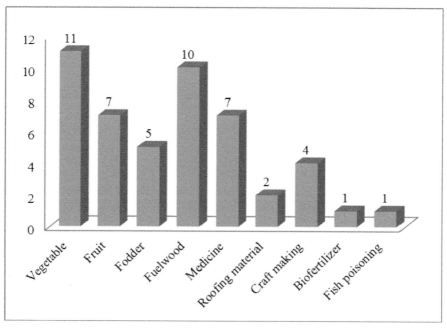

Fig. 2: Use pattern of wetland vegetation by the riparian communities of Dhir *beel* and Diplai *beel* for various purposes

Table 2: Uses of wetland vegetation by the riparian communities of Dhir *beel* and Diplai *beel*

Sl.No.	Scientific Name	Usages	Availability in Dhir *beel*	Availability in Diplai *beel*
1	*Enhydra fluctuans* Lour.	Leave and stems are consumed as vegetable. This plant is considered to be good for diabetic patient. Also consumed for treating worm, stomach trouble, and skin diseases	+	+
2	*Lasia spinosa* (L.) Thwaites	Young stem, leaves and flower spadix are consumed as vegetable. The plant is considered to be good to eat with pork and potato while suffering from chicken pox. Underground stem is eaten to heal body pain	+	+
3	*Amorphophallus sylviticus* (Roxb.) Kunth	Used as vegetable. The plant paste is used with other plant to cure centipede bite	+	+
4	*Calamus tenuis* Roxb.	Used for making craft. Young shoots are eaten as vegetables during *Bohag bihu*. Plant is also used in the treatment of Jaundice with other plant	-	+
5	*Colocasia esculenta* (L.) Schott	Tender leaves, stem, stolon and flowers are eaten as vegetable. Stem are also used for feeding pigs. Used as medicinal for treating swollen body	+	+
6	*Centella asiatica* (L.) Urban	Considered as medicinal for treatment of stomach and liver problem. Leaf juice is extracted by boiling with water and taken orally for the treatment of throat problem	+	+
7	*Diplazium esculentum* (Retz.) Sw.	Consumed as vegetables. The plant is considered to be good for treating iron deficiency	+	+
8	*Oroxylum indicum*(L.) Kurz.	Flowers are eaten fried	-	+
9	*Stereospermum chelonoides* (L.f.) DC.	Used as fuel wood	-	+
10	*Bixa orellana* L.	Used as fuel wood	+	+
11	*Ipomoea aquatica* Forssk.	Leaves and stem are consumed as vegetable. Also considered to be good for patients suffering from Jaundice. The plant is	+	+

Contd.

No.	Botanical name	Uses		
12	Eleocharis palustris (L.) Roem. & Schult.	considered to be good for curing calcium deficiency in patients. Leave juice is drunk for blood purification. Used as fodder for livestock	+	-
13	Terminalia arjuna (Roxb. ex DC.) Wight & Arn.	Plant bark is considered highly medicinal for patients with digestion problem. Tree bark is grinded and soaked in water and the juice extract is given for the treatment of low pressure, gastric problem and liver problem	-	+
14	Terminalia chebula Retz.	Fruits are eaten fresh or dried. It is also used for preparing medicine for treating typhoid patients	-	+
15	Terminalia bellirica (Gaertn.) Roxb.	Fruits are eaten fresh. The plant is used as fuel wood. Its fruits are considered medicinal for treating piles and curing intestinal worm	-	+
16	Ipomoea carnea Jacq.	Used as fuel wood. It is also planted near the roadside to control erosion. Leaf paste is applied in the injured spot to relieve pain. Young shoots are also eaten as vegetable. Flower stalk are used in making medicine for curing throat problem	+	+
17	Shorea robusta Gaertn.	Dried branches are used as fuel wood. Tree trunk are used as pillar in mud house construction	+	+
18	Saccharum ravannae (L.) L.	Used for household fencing and 'broom making and making of various handicraft items	+	+
19	Litsea glutinosa (Lour.) C.B.Rob.	Leaf juice is consumed to reduce anxiety and weakness. Leaf paste is applied on the surface of boil to make a pore and subsequent bursting out of pus and dead cells from the infected portion of the skin	+	+
20	Lagerstroemia speciosa (L.) Pers.	Used as fuel wood. It is also used for making wooden house	+	+
21	Duabanga grandiflora (DC.) Walp.	Plant bark is grinded and juice is extracted. The extracted juice after mixing with water is drunk for the treatment of jaundice	-	+
22	Streblus asper Lour.	Small branches are used as natural tooth brush	+	+
23	Ficusbenjamina L. var. benjamina	Used as fuel wood	+	+

Contd.

No.	Species	Uses		
24	*Nymphoides cristata* (Roxb.) Kuntze	Leaf paste is applied to head to cool head	-	+
25	*Euryale ferox* Salisb.	It is used in treating typhoid. Seeds are consumed fresh or dried. Roots are used for the treatment of kidney stone and for clear urination	-	+
26	*Nymphaea pubescens* Willd.	Stalks are eaten as vegetable to increase blood. The stalk is also used in making herbal formulation known as "borom" with other plant that is applied on head to make the head cool. Also eaten for the treatment of bloody stool and dysentery. Mixture of this plant with the fruit of water chestnut and water cress is used to treat gastric ulcer. Flowers are used to make medicine for piles and liver problem	+	+
27	*Nymphaea nouchali* Burm.f.	Flower stalk are used by the local practitioners for curing swollen body. Sometimes the seeds are also consumed raw or boiled as vegetables.	-	+
28	*Oxalis corniculata* L.	Young shoots and leaves are eaten as vegetables	+	+
29	*Hymenachne amplexicaulis* (Rudge) Nees	Used as fodder for cattle and buffalo	+	+
30	*Leersia hexendra* Swartz.	Used as fodder for livestock. Leaf is used for the treatment of allergic disorder	+	+
31	*Paspalidium flavidum* (Retz.) A.Camus.	Used as fodder for livestock	+	+
32	*Hygroryza aristata* (Retz.) Nees ex Wight & Arn.	Used as fodder. Its roots and young leaves are used for treatment of Tuberculosis.	+	+
33	*Imperata cylindrica* (L.) Raeusch.	Used as roofing material, and in 'broom making. Also used in worshipping during "Kerai" festival of *Bodo* tribe	-	+
34	*Chrysopogon zizanoides* (L.) Roberty.	Used as roofing materials. Roots are used in making herbal mixture known as "borom" that is applied on head to make the head cool	+	+
35	*Bambusa assamica* Barooah & Borthkur	For housing/ household area & backyard fencing and weaving craft etc.	-	+

Contd.

No.	Species	Availability 1	Availability 2	Uses
36	Eichornia crassipes (Mart) solms.	+	+	Dried plants are used as green manure; young leaves are used for healing swollen bodies. Roots are also used in the treatment of tooth decay
37	*Monocharia hastata (L.) Solms.	–	+	Flowers and buds are eaten as vegetable. Leaf paste is used for making medicine for healing swollen bodies. Leaf juice is used in curing boils
38	Polygonum hydropiper L.	+	+	Plant paste is used for fish poisoning. Plant paste of young shoot is used in making herbal mixture known as "borom" that is applied on head to make head cool
39	Phyllanthus emblica L.	–	+	Fruits are eaten fresh. Fruits are also used in preparing medicine for curing typhoid. Dried branches are also used as fuel wood
40	Paederia foetida L.	–	+	Leaves and tender twigs are used as vegetable. Leaves and tender twigs are also used for curing stomach ache and gastric problems
41	Neolamarckia cadamba (Roxb.) Bosser	+	+	Used as fuel wood
42	Cephalanthus occidentalis L.	–	+	Used as fuel wood
43	Aegle marmelos (L.) Correa	–	+	Fruits are eaten when ripe
44	Blighia sapida K.D. Koenig	–	+	Used as fuel wood
45	Trapa natans L. var. bispinosa (Roxb.) Makino	+	+	Fruits are eaten raw. Leaf juice is extracted by boiling with water and taken orally for the treatment of throat problem
46	Hydrocotyle sibthorpoides Lam.	+	+	Leaves and young shoots are consumed as vegetable and also for treating throat problem. Leaves are also used for healing wounds. The plant is also eaten for the treatment of gastric problem
47	Gmelina arborea Roxb.	–		Used as fuel wood
48	Lippia geminate Kunth	+	+	Leaf paste is used for treating head spinning sensation. Roots are used in making herbal mixture known as "borom" with other plant for the treatment of jaundice. It is also used as flavouring agent in some curry
	Total available plant resource:	28	46	

'+' indicates availability and '-' indicates non-availability of a particular wetland resource; '*' indicates that the resources or the manufactured goods using those resources are also sold in the local market by the riparian communities of the wetlands besides their household use or consumption.

The local communities of both the *beels* collect fire wood from the small hillocks located within the wetlands and from the edges of the wetlands to the meet the requirement of fuel wood for cooking purposes. The local communities also extract fibre grass like *Imperata cylindrica, Saccharum ravannae* for making broom and roofing houses. Apart from these the traditional healers of the riparian villages also prepare a diverse form of medicine from plant species like *Euryle ferox, Centella asiatica, Hydrocotyl sibthorphoides, Nymphea pubescens, Nymphoides cristata, Lippia geminata, Enhydra fluctuans* to cure various ailments of the riparian communities. Besides, there are some other plant species of the wetland like *Ipomoea aquatica, Lasia spinosa, Amorphophallus sylviticus, Monocharia hastata, Centella asiatica, Hydrocotyl sibthorphoides, Nymphea pubescens, Lippia geminate*, and *Enhydra fluctuans* which are consumed as vegetable and herbal medicine. The use of wetland plants in medicine preparation varies from one practitioner to the other belonging to different communities.

Issues Related to Plant Resource Extraction from Wetlands

Both the wetlands are considered as common property resource of the riparian communities, and are accessible to all the resident communities of the wetlands. This has led to overexploitation of the wetland resources. Interaction with the riparian people revealed that thatch grass (*Imperata cylindrica*), which once was reported to be abundant in the area, has now dwindled due to habitat modification through burning, cutting and clearing of land for agriculture and human settlement. Another major ecological issue observed in the Diplai *beel* is that, the wetland is heavily infested by *Eichornia crassipes*, an invasive species. This is affecting fishing and other economic activities of the riparian communities. However, some villagers reported that they utilize *Eichornia crassipes* as fertilizer, particularly for potato cultivation. During the field observation, it was noticed that both the *beels* are under great pressure for a number reasons such as increasing population, overexploitation of wetland resources, chemical contamination of the wetland through the use of agro-chemicals for agricultural production, market pressure for fish, encroachment of invasive species like *Eichornia crassipes*, extensive land mining of the hillocks located within the wetlands and its conversion into agricultural land, overgrazing, burning of grasslands for agriculture, siltation during rainy seasons, fuelwood demand for increasing population, illegal encroachment and alteration of natural land cover of the wetland for creation of culture fishery ponds within the wetlands. All these anthropogenic interferences are responsible for the decline of some native vegetation in the wetlands such as *Lagerstroemia speciosa, Stereospermum chelonoides, Ficus benjamina, Euryale ferox, Imperata cylindrica, Blighia sapida, Calamus tenuis* and *Hymenachne amplexicaulis.*

CONCLUSION

In spite of the major ecological and environmental challenges faced by both the wetlands, they contribute to the livelihood of many household largely through provision of numerous natural goods and ecosystem services. The riparian communities are extracting the wetland resources free of cost. Therefore, they are not aware of the value attached with the wetland in terms of their livelihood sustenance. The plant resources present in the larger wetland is less than the smaller wetland. Therefore, the present study highlights that wetlands irrespective of their size have equal importance in the livelihood sustenance of the riparian communities. As the livelihood strategies of the wetland-dependent communities in developing countries are centred on wetland resources (Wang et al. 2008), it is necessary to provide scientific guidance to the stakeholders of these wetlands. The sustainable use of wetlands is the best way for developing countries like India to protect it. In this regard, the best options may be implementation of resource extraction planning and conservation measures to protect these wetlands from further degradation.

Abbreviations

RCS= Ramsar Convention Secretariat

Conflict of Interest

Authors have no conflict of interest.

ACKNOWLEDGEMENTS

We are thankful to the villagers of Dhir *beel* and Diplai *beel* for their kind cooperation and sharing of knowledge/information during the study. The first author is also thankful to the Institutional Level Biotech Hub, Science College, Kokrajhar for providing necessary facility and time for the survey work.

REFERENCES CITED

Barooah C, Ahmed I. 2014. *Plant diversity of Assam: A checklist of Angiosperm and Gymnosperms*. Assam Science Technology and Environment Council, Assam, p. 599.

Davis, TJ. 1993. Towards the wise use of wetlands. Wise use project, Ramsar Convention Bureau, Gland, Switzerland.

Department of Forest, Chakrashila Wildlife Division. 2015. A report proposal for Ramsar site. Department of Forest, Chakrashila Wildlife Division p. 23.

Kanjilal UN, Kanjilal PC, Das A. 1934. Flora of Assam, Vol-1. Government of Assam, Assam.

Kanjilal UN, Kanjilal PC, Das A. 1936. Flora of Assam, Vol-2, Government of Assam, Shillong.

Kanjilal UN, Kanjilal PC, Das A, De R N. 1938. Flora of Assam, Vol-2. The Authority of the Government of Assam.

Kanjilal UN, Kanjilal PC, De RN, Das A. 1940. Flora of India, Vol 4. Government of Assam, Shillong.

Kakuru W, Turyahabwe N, Mugisha J. 2013. Total Economic Value of Wetlands Products and Services in Uganda. *The Scientific World Journal* p. 13

Millennium Ecosystem Assessment, 2005. *Ecosystems and Human Well-being: Synthesis*. Island Press, Washington, DC. Available at: http://www. millenniu massessment.org/ documents/document.356.aspx.pdf, Accessed on: 16/08/ 2016

National Wetland Atlas: Assam, SAC/RESA/AFEG/NWIA/ATLAS/18/2010, Space Applications Centre (ISRO), Ahmedabad, India, p. 174. Available at: http:// envfor.nic.in/downloads/public-information/NWIA_Assam_Atlas.pdf. Accessed on: 16/08/2016.

Narzary H, Brahma S, Basumatary S. 2013. Wild Edible Vegetables Consumed by *Bodo* Tribe of Kokrajhar District (Assam), North-East India. *Archives of Applied Science Research* 5(5): 182-190.

Ramsar Convention Secretariat (RCS). 2007. Wise use of wetlands: A conceptual framework for the wise use of wetlands. Ramsar Handbook for wise use of wetlands, 3rd edition, Vol. 1. Ramsar Convention Secretariat, Gland, Switzerland. p. 24.

Sinha A, Talukdar S, Das GC, Sarma PK, Singha H. 2015. Diversity of winter avifauna in Dheer beel, Assam, India. *Indian BIRDS* 10(3&4): 99–103.

Thondhlana G, Vedeld P, Shackleton S. 2012. Natural resource use, income and dependence among San and Mier communities bordering Kgalagadi Transfrontier Park, southern Kalahari, South Africa. *International Journal of Sustainable Development & World Ecology* 19(5): 460–470.

Vijayan VS, Prasad SN, Vijayan L, Muralidharan S. 2004. *Inland Wetlands of India: Conservation Priorities*, Salim Ali Centre for Ornithology and Natural History, Coimbatore. p. 532.

Wang Y, Yao Y, Ju M. 2008. Wise use of wetlands: current state of protection and utilization of Chinese wetlands and recommendations for improvement. *Environmental Management* 41: 793–808.

13

Boswellia serrata Roxb. ex Colebr., a Multiferous Medicinal Plant Used as Anti-Cancerous and Anti-Inflammatory in Drug Discovery

Rakesh Kumar Nagar[1], Govind Yadav[1,3]* and Bikarma Singh[2,3]*

[1]*Mutagenesis laboratory (Animal House Division),* [2]*Plant Sciences (Biodiversity and Applied Botany), CSIR-Indian Institute of Integrative Medicine Jammu-180001, Jammu & Kashmir, INDIA*
[3]*Academy of Scientific and Innovative Research, Anusandhan Bhawan New Delhi - 110025, INDIA*
*Email: gyadav@iiim.ac.in, r.nagar8788@gmail.com, drbikarma@iiim.ac.in

ABSTRACT

Boswellia serrata Roxb. ex Colebr. known as 'Indian Frankincense' is a multiferous medicinal plants known for boswellic acid and it has application in traditional ayurvedic formulation. The plant possess anti-rheumatic, anti-pyretic, anti-cancerous, anti-inflammatory, anti-hyperlipidemic, anti-coronary, analgesic and hepatoprotective properties due to the presence of several loaded bioactive compounds such as α-amyrins, β-boswellic acid, acetyl-β-boswellic acid, 11-keto-β-boswellic acid, acetyl-11-keto-β-boswellic acid, tetracyclic triterpenoic acids and several others in minor quantities in different parts of plants. Mostly oleo-gum resin is usually used in Indian system of medicine, and acetyl-11-keto-β-boswellic acid (AKBA) is the most potent 5-lipoxygenase enzyme inhibitor responsible for inflammation, anti-cancer (G1 phase arrest) and other group of enzymes liable for allied activities. Besides, it yields essential oils and its major constituents are monoterpenoids such as pinene (2.05-64.7%), cis-verbenol (1.97%), trans-pinocarveol (1.80%), borneol (1.78%), myrcene (1.71%), verbenone (1.71%), limonene (1.42%), thuja-2,4(10)-diene (1.18%) and p-cymene (1.0%). The copaene is the only sesquiterpene reported in minor quantities from essential oils of this species. Its essential show different actvities in relation to anti-cancer and anti-inflammatory. From future perspectives, there is a need to carry out more research on its constituents for the development of value added products and drugs. Investigation on ecology and application of biotechnological intervention would helps in varities improvememnt, cultivation and conservation of this species.

Keywords: Indian Medicinal Plant, *Boswellia serrata*, Anti-cancer, Anti-inflammatory, Value Addition and Drug Discovery.

INTRODUCTION

Family Burseraceae represented by 18 genera and 649 species (TPL 2018) distributed world-wide in tropical to temperate climate. Most of the species under the family are either trees or shrubs, characterized by spines, latex, gum-resins and essential oils , which can be strongly aromatic. *Boswellia* one of the genus under this family comprised of 32 species (TPL 2018), first described by William Roxburgh and Henry Thomas Colebrooke extended its taxonomical description and published it in 'Asiatic Researches' in the year 1807. The centre of origin of this genus is North-East Africa, where 75% of the species are endemic, followed by Madagascar (20 species) and distributed to tropical climate. Boswellia trees when cut, exudes a gum resin, which is gathered by making scraps in the bark of the tree. In Oman, boswellia trees are cultivated for pale whitish resin (Moussaieff et al. 2009). Published literature revealed that 200 compounds have been identified from resins of *Boswellia* spp. (Hamm et al. 2005, Ammon 2005).

The plant possess anti-rheumatic, anti-pyretic, anti-cancerous, anti-inflammatory, anti-hyperlipidemic, anti-coronary, analgesic and hepatoprotective properties, due to the presence several loaded bioactive chemical compounds such as α-amyrins, β-boswellic acid, acetyl-β-boswellic acid, 11-keto-β-boswellic acid, acetyl-11-keto-β-boswellic acid, tetracyclic triterpenoic acids and several others in minor quantities in different parts of plants and resins (Shah et al. 2008). Mostly oleo-gum resin is usually used in Indian system of medicine, and acetyl-11-keto-β-boswellic acid (AKBA) is the most potent 5-lipoxygenase enzyme inhibitor responsible for inflammation (Iram et al. 2017), anticancer (McCarty 2004, Pathania et al. 2015, Park et al. 2011, Yuan et al. 2013) and other group of enzymes liable for allied activities. Besides, it also yield essential oils and major constituents are monoterpenoids such as pinene (2.05%), cis-verbenol (1.97%), trans-pinocarveol (1.80%), borneol (1.78%), myrcene (1.71%), verbenone (1.71%), limonene (1.42%), thuja-2,4(10)-diene (1.18%) and p-cymene (1.0%) (Hamm et al. 2005, Kasali et al. 2002). Copaene, the only sesquiterpene, is reported in minor quantities from essential oils of *B. serrata* (Kasali et al. 2002). These constituents of oil also show different activities in relation to anti-cancer and anti-inflammatory (Takada 2018).

Considering the importance of *Boswellia serrata* in value addition and drug discovery, a project has been sanctioned by DBT 'Development of Phytopharmaceutical Product for Bovine Mastitis'. The present communication deals with biology, chemistry, pharmacology and conservation aspects of *Boswellia serrata* as potent multiferous medicinal plants of India.

BIOLOGICAL ASPECTS

Classification

- Kingdom: Plantae
- Subkingdom: Tracheobionta
- Division: Magnoliophyta
- Class: Magnoliopsida
- Order: Sapindales
- Family: Burseraceae
- Genus: *Boswellia* Roxb. ex Colebr.
- Species: *Boswellia serrata* Roxb. ex Colebr.

Common Names

The generic name *Boswellia* is given after Dr. James Boswell of Edinbergh Botanical Garden and friend of William Roxburgh, Director of Indian Botanical Garden, Kolkata (former Calcutta). The specific name '*serrata*' comes from serra (a saw) referring to the toothed leaf margins.

- English: Indian frankincense
- Hindi: Kundur, Salal
- Bengali: Kundur, Salal

- Gujrati: Dhuph, Gugali

- Kannada: Chitta, Guguladhuph

- Malyalam: Parangi, Saambraani

- Tamil: Parangi, Saambraani

- Telugu: Phirangi, saambraani

- Sanskrit: Ashvamutri, Kundara, Shallaki

- Ayurvedic name: Shallaki

- Unani name: Kundur

Taxonomical Enumerations

Boswellia serrata is a medium-sized deciduous tree, 8-15 m tall, having spreading and drooping branches. Barks are thin, grayish-green, slightly pinkish, easily peels off; bark yield resin or gum when cut. Leaves are alternate, imparipinnate, 20-45 cm in length, crowded towards ends of branches; leaflets opposite, 2.5-7 cm long, 0.5-1.2 cm broad, sessile, ovate-lanceolate, crenate. Flowers are white or slightly yellowish-white, racemes 8-15 cm long, crowded towards the end of branches; sepals persistent, deltoid, pubescent outside, 5 to 7-toothed; petals 5-7, erect, free, 0.5 cm long. Fruits 1-1.3 cm long, 0.3-0.4 cm broad, usually trigonous, with three valves, 1-seeded; pyrenes winged, along the margins.

Diversity and Distribution

Wide distribution of genus *Boswellia* reported from dry regions of tropical Africa, Arabia and India. In Africa, it is distributed in Somalia, Ethiopia, Eritrea, Kenya, Sudan, Tanzania, Madagascar and some other countries (Iram et al. 2017). In Arabia, it is mainly restricted to Yemen, Oman and Socotra (Attorre et al. 2011). In India, it is distributed in some regions of Rajasthan, South East Punjab, Danwara, and Madras (Siddiqui 2011). According to TPL (2018), thirty two species of *Boswellia* recorded to be distributed across the globe (Table 1).

Table 1: World-wide distribution of the genus Boswellia

Sr. No.	Botanical name	Habit	Distribution
1.	*Boswellia serrata* Roxb. ex Colebr.	Tree	India, Africa, China, Pakistan, Myanmar
2.	*Boswellia boranensis* Engl	Tree	Somalia
3.	*Boswellia ameero* Balf.f.	Tree	Island of Socotra (Yemen)
4.	*Boswellia bricchettii* (Chiov.) Chiov.		Australia
5.	*Boswellia bullata* Thulin	Tree	Island of Socotra (Yemen)
6.	*Boswellia chariensis* Guillaumin	Tree	France
7.	*Boswellia dalzielii* Hutch	Tree	Africa
8.	*Boswellia dioscoridis* Thulin	Tree	Island of Socotra (Yemen)
9.	*Boswellia elegans* Engl.	Tree	Africa
10.	*Boswellia elongata* Balf.f.	Tree	Yemen
12.	*Boswellia frereana* Birdw.	Tree	Somalia, Arab
13.	*Boswellia globosa* Thulin	Tree	Native of Somalia
14.	*Boswellia hildebrandtii* Engl.	Tree	Africa
15.	*Boswellia holstii* Engl.	Tree	Australia
16.	*Boswellia madagascariensis* Capuron	Tree	Australia
17.	*Boswellia microphylla* Chiov.	Tree or Shrub	Somalia
18.	*Boswellia multifoliolata* Engl.	Tree or Shrub	Somalia
19.	*Boswellia nana* Hepper	Tree or Shrub	Island of Socotra (Yemen)
20.	*Boswellia neglecta* S.Moore	Tree or Shrub	Somalia
21.	*Boswellia occidentalis* Engl.	Tree	Ethiopia, Africa
22.	*Boswellia odorata* Hutch.	Tree	Africa
23.	*Boswellia ogadensis* Vollesen	Tree	Ethiopia
24.	*Boswellia ovalifoliolata* N.P.L Balakr. & A.N. Henry	Tree	India
25.	*Boswellia papayrifera* (Del.) Hochst.	Tree	Ethiopia, Eritrea, Sudan
26.	*Boswellia pirottae* Chiov.	Tree	Ethiopia
27.	*Boswellia popoviana* Hepper	Tree	Yemen
28.	*Boswellia rivae* Engl.	Tree	Somalia, Ethiopia
29.	*Boswellia ruspoliana* Engl.	Tree	Africa
30.	*Boswellia sacra* Flueck	Tree	Somalia, Arab, Oman
32.	*Boswellia socotrana* Balf.f.	Tree	Socotra in Yemen

In India, two species of *Boswellia* are reported. These include *B. serrata* and *B. ovalifoliolata* have been reported to be distributed in India. Sunnichan et al. (2005) mentioned that *B. serrata* is the only species found in India. But other researchers reported that *B. ovalifoliolata* also occurs on the foothills of Seshachalam hill ranges of Eastern Ghats in Chittoor, Kadapa and Kurnool Districts of Andhra Pradesh up to an altitude of about 600-900 m above mean sea level.

Phenology

On recurring seasonal vegetative and reproductive events, the periodical annotation was made. According to this, the total number of inflorescences were counted randomly from selected branch (N=30) and the number of flowers buds or flowers per inflorescence (N=15) were recorded for each chosen tree.

Reproductive Biology

Pollen ratio was determined according to Cruden (1977). With the help of calibrated ocular micrometer the diameter of fresh pollen grains were measured and mounted in a drop of 50% aqueous glycerine. 1% acetocarmine was used for assessment of pollen fertility. The fluorescein diacetate test introduced by Heslop-Harrison (1970) was used for estimation of pollen viability. Iodine-potassium iodide solution and Sudan III were used respectively for identification of presence of starch and lipids in pollen grains. The stain Hoechst 33258 was used for staining of pollen nuclei (Hough et al. 1985).

MEDICINAL AND TRADITIONAL USAGES

Traditionally, gum resin, ethanol extract, aqueous extract, herbal formulation, essential oils of *Boswellia serrata* that possess medicinal properties have been used as a remedy for treatment of various diseases (Figure 1). The oleo resin is produced by tree trunk and is used for medical application after purification of it.

Fig. 1: Gum resine of *Boswellia serrata* (www.Google.com)

The gum resin obtained from *B. serrata* is being used traditionally in Ayurveda as remedy for many diseases e.g. arthritis, diarrhea, Crohn's disease, dysentery, lung disease, and worms etc. the boswellic acid isolated from the *Boswellia* is a potent anti-inflammatory agent and also has the analgesic property, that make it useful in the treatment of rheumatoid arthritis, and has hypolipidemic property. In addition, this plant has been known for many pharmacological actions such as Anti-asthmatic property by relieving bronchial passageways, immuno-modulation action. *B. serrata* is also used for the treatment of Crohn's disease and ulcerative colitis.

PHYTOCHEMISTRY

Major Chemical Constituents

A number of triterpenoids have been isolated from gum-resin of *Boswellia serrata*. The boswellic acid is a triterpenoids that is isolated from *Boswellia* has shown many pharmacological actions. Due to its wide pharmacological action, many chemists worked on its structure elucidation. There are several other triterpenoids isolated from the gum resin e.g. α-amyrins, 11-keto-α-boswellic acid, 3'hydroxyl urs-9,11-dien-24-oic acid. 3'acetoxy urs-9,11-dien-24-oic-acid. Presence of some tetracyclic triterpenoids such as 3'-hydroxy-tirucall-8, 24-dien-21-oic acid, 3'hydroxy-tirucall-8, 24-dien-21-oic acid, 3-keto-tirucall-8, 24-dien-21-oic acid, and 3'acetoxy tirucall-8,24-dien-21-oic acid were also reported from this species.

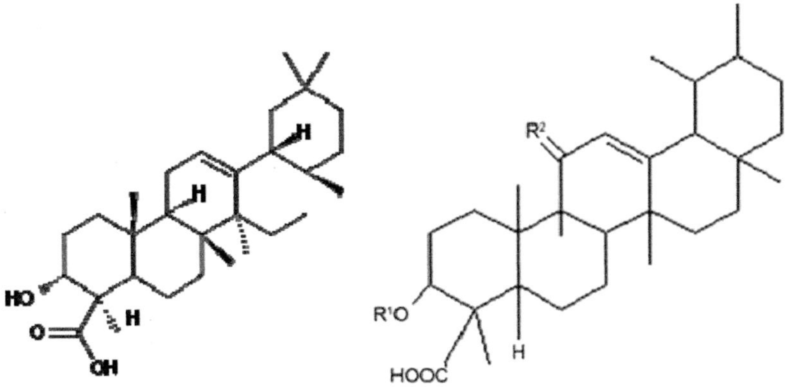

Fig. 2: Chemistry of *Boswellia serrata*, (a) *Boswellic acid* (left), (b) four pentacyclic triterpenoic acid (right:) beta Boswellic acid, R1=H, R2=H$_2$; acetyl-beta-boswellic acid, R1=Ac, R2=H$_2$; 11-keto-beta-boswellic acid, R1=H, R2=O; acetyl-11-keto-beta-boswellic acid, R1=Ac, R2=O).

Table 2: Comparison of chemistry of four pentacyclic triterpenoic acid (Boswellic acid)

Properties	β-Boswellic acid	Acetyl-β-Boswellic acid	11-keto-β-Boswellic acid	Acetyl-11-keto-β-Boswellic acid
Molecular formula	$C_{30}H_{48}O_3$	$C_{32}H_{50}O_4$	$C_{30}H_{48}O_4$	$C_{32}H_{48}O_5$
Molecular weight	456.7	498.74	470.69	512.73
Chemical name	3α-Hydroxy-urs-12-en-23-oic-acid	3α-Acetoxy-urs-12-en-23-oic acid	3α-Hydroxy-urs-12-en-11-keto-23-oic acid	3α-Acetoxy-urs-12-en-11-keto-23-oic acid
Melting point	226-228°	252-255°	195-197°	271-274°
Specific rotation	+106.8°	+138°	+78.5°	+88.5°
UV-MeOH	Maxima at 208nm	Maxima at 208nm	Maxima at 250nm	Maxima at 250nm
NMR (in $CDCl_3$, δ PPM)	5.15, CH=C;4.08, CH-OH;2.3-1.1, Methylenes and methines, 23 protons; 1.1-0.7 Methyls, 21 protons	5.31, CH=C; 5.2 CH-O Ac 2.1, $COCH_3$,1.9-1.25, Methylenes and methines23 proton; 1.2-0.7, Methyls 21 proton	5.55, CH=C; 4.08, CH-OH; 2.6-1.4, Methylenes and methines, 21 proton; 1.25-0.75, Methyls 21 protons	5.55, CH=C; 5.2, CH-OAc; 2.6-1.4, Methylenes and Methines, 21 protons; 1.25-0.75, Methyls 21 protons
FTIR (in KBr. Cm^{-1})	3500 (OH), 1699.5 (COOH)	1732 (OAc), 1701 (COOH)	3460 (OH)1693 (COOH)	1740 (Ac), 1701 (COOH), 647 (α,β-unsaturated carbonyl).
GC-MS	394 (M-68[44 due to –CO_2 and 18 due to –H_2O]) other fragments: 203,189,175, 161	394 (M-104[44 due to –CO_2 and60 due to –HOAc]); 218 (base peak)	408 (M-68[44 due to –H_2O]); 232 base peak Other fragments; 217,175,161,135	408 (M-63[44 due to –CO_2 and 18 due to–CO_2 and 18 due to –HOAc]): 232 base peak; Other fragment; 217, 175,161,135

Essential Oil Constituents

Oleo-gum-resin of *Boswellia serrata* contains many essential oils that includes monoterpenoids and sesquiterpene. The pinene is the major monoterpenoids constituent of essential oils. Other monoterpenoids are listed in Table 3

Table 3: Chemical constituents of essential oils in *Boswellia serrata*

Sr.no.	Name	Percentage composition	Category
1.	Pinene	2.05-64.7	Monoterpenoids
2.	cis-Verbenol	1.97	Monoterpenoids
3.	trans-Pinocarveol	1.8	Monoterpenoids
4.	Borneol	1.78	Monoterpenoids
5.	Myrcene	1.71	Monoterpenoids
6.	Verbenone	1.71	Monoterpenoids
7.	Limonene	1.42	Monoterpenoids
8.	Thuja-2,4(10)-diene	1.18	Monoterpenoids
9.	p-Cymene	1.0	Monoterpenoids
10.	copane	0.3	Sesquiterpene

PHARMACOLOGY ASPECT

1. Anti-inflammatory Activity

Leukotriene Inhibition: Leukotriene such as LTB4 is a key mediator of inflammation that is synthesized by neutrophills, eosinophills, macrophages and mast cells. Ammon et al. (1991) reported that inhibition of LTB4 by ethanolic extract of gum-resin of *Boswellia serrata*. Ammon et al. reported that extract of salai guggul in a dose of 100 mg/kg orally exert a marked inhibition of pleural exudates volume (47.93%).

Anti-arthritic activity: Rheumatoid arthritis is an auto-immune disease which can be identified by the inflamed joints and destruction of cartilage. The presence of cytokines such as Interleukins and TNF-α in the synovial fluid of inflamed joints also causes destruction of cartilage. The extract of oleo-gum-resin of boswellia serrata and its active constituent such as acetyl-11-keto-β-boswellic acid exert a potent pharmacological action on arthritis and inhibition of human leulocyte elastase (Safayhi et al. 1997).

2. Immunomodulatory activity

The oleo-gum-resin extract of boswellia serrata and its constituent such as acetyl-11-keto-β-boswellic acid (AKBA) along with other constituents exerts immuno-modulation activity at higher dose the extract diminish the primary antibody titre in humoral system, while at lower doses it shows increased secondary antibody titre. In cell mediated immunity, boswellic acid enhances the lymphocyte proliferation, while at higher concentration the action is vice versa. The extract also inhibits the paw edema in dose dependent manner (Gupta et al. 2011, Pungle et al. 2003).

Anti-cancer Activity

Boswellia serrata extract that contains around 60% of AKBA possess a potent anticancer property. Boswellic acids show its anticancer property by inhibiting the cell proliferation and inducing apoptosis (caspase-8 pathway) in different cancer cell lines such as human leukemia, colon and hepatoma (Xia et al. 2005).

According to Pang et al. (2009) AKBA act by suppression of angiogenesis via VEGFR2 and mTOR pathway and inhibit the prostate cancer tumor progression.

Liu et al. (2006) reported the anti-cancer property of AKBA against colon cancer cell lines. According to Liu et al the anti-cancer property of AKBA in colon cancer lines is due to G1 phase arrest in AKBA treated cell and decrease in cyclin D1 and E, CDK2 and 4, While the p-21 expression was enhanced after treatment with AKBA. This concludes that suppression of colon cancer may be due to p21by inducing apoptosis.

A clinical study was also performed by Streffer et al. (2001). In this study, the patients receiving higher dose of *Boswellia* serrata extract for seven days, before surgical procedure, the size of perifocal oedma was reduced.

Hypolipidemic and Hepatoprotective Activity

Many scientific reports indicate that hypolipidemic property of *B. serrata*. Zutshi et al, (1980) reported the hypolipidemic property of this plant's extract of oleo-gum-resin. Water soluble extract of *B. serrata* also possess hypolipidemic property and reduce the total cholesterol approximate 38-48% in rats. The *B. serrata* also possess hepatoprotective activity. Zaitone et al. (2015) reported that the hepatoprotective property of boswellic acid. He observed that rat treated with boswellic acid at the dose of 125 or 250 mg/kg or pioglitazone enhance the insulin sensitivity as well as reduction in liver index and IL-6 and TNF-α expression.

Anti-diarrhoeal Activity

Boswellia serrata is extensively used as a remedy for the treatment of diarrhoea and inflammatory bowel syndrome. Borelli (2006) reported its anti-diarrhoeal property without causing constipation. In this study, the plant extract exerts its anti-diarrheal property by decreasing the intestinal mortility.

Anti-microbial Activity

Gum resin of *Boswellia serrata* possesses significant anti-bacterial activity against different gram positive and gram negative pathogens. Ismail (2014), reported the anti-bacterial activity of this species gum-resin at the dose of 25, 50, 75 and 100 mg/ml against gram positive *(Bacillus subtilis, staphylococcus aureus* and *streptococcus pneumonia)* and gram negative *(E. coli, Klebisiella pneumonia, Enterobacetr aerogenes, pseudomonas aeruginosa and Proteus vulgaris).*

The boswellic acid has potent anti-bacterial activity against gram positive and gram negative bacteria. As per Raja et al. (2011), MIC of boswellic acid (AKBA) reported was to be between 2-4 µg/ml.

Synergistic Anti-microbial Activity

Sadhasivam S. et al. (2016) showed synergistic anti-microbial activity of Boswellia serrata essential oils with various azoles against pathogen associated with skin, scalp and nail infections specialy synergisticaly antifungal activity in combination azoles against rsistant candida albicans strain.

Anti-asthmatic Activity

Boswellia serrata has promising role on respiratory diseases. Boswellia is being used since ancient in steam inhalation, to treat cough, bronchitis, and asthma. The alcoholic extract of salai guggal inhibits the leukotriene production in patient suffering from asthma. Ali et al. (2011) reported that boswellic acid exert its effect by suppressing 5-lipo-oxygenase results in reduction of infiltration of cells. Liu etal reported the anti-asthmatic property of boswellic acid-in mariner model and found that after treatment with boswellic acid, the animals suppress allergic airway inflammation.

Safety and Toxicology

The safety study of *Boswellia serrata* was carried out by Singh et al (2012). According to this study, this species was found to be safe upto the concentration of 500 mg/kg body weight. During treatment period, the animal gain increase in weight but less as compared to control group.

CONCLUSION

The *Boswellia* has been used as a remedy for the treatment of various diseases in traditional system. This review covered the potential of *Boswellia serrata* in various ailments of the human *viz.* anti-infective, anti-inflammatory, anti-arthritic anti-cancer, hypolipidemic, hepatoprotective immuno-modulation, anti-diarrheal, anti-asthmatic activity. Its application is not only in traditional ayurvedic formulation, pure and synthetic, semisynthetic derivative has proven potential. For future perspectives, there is need to carry out more extensive research on its constituents for the development of value added products and drugs. It is an alternative to various anti-inflammatory drugs such as NSAIDs. Investigation on ecology and application of biotechnological intervention would be helpful in cultivation and conservation of this species.

Conflict of Interest

Authors declare no conflict of interest

ACKNOWLEDGEMENT

Authors would like to thank Director IIIM, Dr. Ram A. Vishwakarma for necessary facilities and encouragement.

REFERENCES CITED

Ali EN, Mansour SZ. 2011. Boswellic acid extract attenuates pulmonary fibrosis induced by bleomycin and oxidative stress from gamma irradiation in rats. *Chinese Medicne* 6: 36. DOI: 10.1186/1749-8546-6-36.

Ammon HP, Mack T, Singh GB, Safayhi H. 1991. Inhibition of leukotriene B4 formation in rat peritoneal neutrophills by an ethanolic extract of the gum-resin exudate of *Boswellia serrata*. *Planta Medica* 57 (3): 203-207.

Ammon HP. 2006. Boswellic acid in chronic inflammatory diseases. *Planta Medica* 72(12): 1100-1116.

Attorre F, Taleb N, Sanctis MD, Farcomeni A, Guillet A, Vitale M. 2011. Developing conservation strategies for endemic tree species when faced with time and data constrains: *Boswellia* spp. On Socotra (Yemen). *Biodiversity and Conservation* 20 (7):1483-1499.

Borrelii F, Capasso F, Capasso R, Ascione V, Aviello G, Longo R, Izzo AA. 2006. Effect of *Boswellia serrata* on intestinal motility in rodents: Inhibition of diarrhoea without constipation. *British Journal of Pharmacology* 148: 553-560.

Cruden RW. 1977. Pollen Ovule Ratio: A Conservative indicator of breeding system in flowering plants. *International Journal of Organic Evolution* 31(1): 32-46.

Gupta A, Khajuria A, Singh J, Singh S, Suri KA, Qazi GN. 2011. Immunological adjuvant effect of *Boswellia serrata* (BOS-2000) on specific antibody and cellular response to ovalbumin in mice. *International Immunopharmacology* 11(8): 968-975.

Hamm S, Bleton J, Connan J, Tchapla A. 2005. A chemocal investigation by headspace SPME and GC-MS of volatile and semi-volatile terpenes in various olibanum samples. *Phytochemistry* 66(12): 1499-1514.

Heslop-Harrison J, Heslop-Harrison Y. 1970. Evaluation of pollen viability by enzymatically induced fluorescence; intracellular hydrolysis of fluorescein diacetate. *Biotechnic and Histochemistry* 45(3): 115-120.

Hough T, Bernhardt P, Knox RB, Williams EG. 1985. Application of fluorochrome to pollen biology. II. The DNA probes ethidium bromide and Hoechst 33258 in conjunction with the callose-specific aniline blue fluorochrome. *Biotechnic and Histochemistry* 60(3): 155-162.

Iram F, Khan SA, Hussain A. 2017. Phytochemistry and potential therapeutic actions of boswellic acids: A mini-review. *Asian Pacific Journal of Tropical Biomedicine* 7(6): 513-523.

Ismail SM, Aluru S, Sambasivarao KRS, Matcha B. 2014. Antimicrobial activity of frankincense of *Boswellia serrata*. *International Journal of Current Microbiology and Applied Sciences* 3(10): 1095-1101.

Kasali AA, Adio AM, Oyedeji AO, Eshilokun AO, Adefenwa M. 2002. Volatile constituents of *Boswellia serrata* Roxb. (Bursereceae) Bark. *Flavour and Fragrance Journal* 17: 462-464.

Liu J, Huang B, Hoo SC. 2006. Acetyl-keto-β-boswellic acid inhibits cellular proliferation through a p21-dependent pathway in coloncancer cells. *British Journal of Pharmacology* 148(8): 1099-1107.

Liy Z, Liu X, Sang L, Liu H, Xu Q, Liu Z. 2015. Boswellic acid attenuates asthma phenotypes by downregulation of GATA3 via pSTAT6 inhibition in a murine model of asthma. *International Journal of Clinical and Experimental Pathology* 8(1): 236-243.

McCarty MF. 2004. Targeting multiple signaling pathways as a strategy for managing prostate cancer: Multifocal signal modulation therapy. *Integrative Cancer Therapies* 3(4): 349-380.

Moussaieff A, Mechoulam R. 2009. Boswellia resin: from religious ceremonies to medical uses; a review of in-vitro, in-vivo and clinical trials. *Journal of Pharmacy and Pharmacology* 61(10): 1281-1293.

Pang X, Yi Z, Zhang X, Sung B, Qu W, Lian X, Aggarwal BB, Liu M. 2009. Acetyl-11-keto-â-boswellic acid inhibits prostate tumor growth by suppressing vascular endothelial growth factor receptor 2-mediated angiogenesis. *Cancer Research* 69(14): 5893-5900.

Park B, Prasad S, Yadav V, Sung B, Aggarwal BB. 2011. Boswellic acid suppresses growth and metastasis of human pancreatic tumors in an orthotopic nude mouse model through modulation of multiple targets. *PLoS One* 6(10): e26943.

Pathania AS, Wani ZA, Guru SK, Kumar S, Bhushan S. 2015. The anti-angiogenic and cytotoxic effects of the boswellic acid analogue BA145 are potentiated by autophagy inhibitors. *Molecular Cancer* 14: 6; DOI 10.1186/1476-4598-14-6.

Pungle P, Banavalikar M, Suthar A, Biyani M, Mengi S. 2003. Immunomodulatory activity of boswellic acids of *Boswellia serrata* Roxb. *Indian Journal of Experimental Biology* 41(12): 1460-1462.

Raja Af, Ali F, Khan IA, Shawl AS, Arora DS. 2011. Acetyl-11-keto-β-boswellic acid (AKBA); targeting oral cavity pathogens. *BMC Research Notes* 4: 406. DOI 10.1186/1756-0500-4-406.

Sadhasivam S, Palanivel S, Ghosh S. 2016. Synergistic antimicrobial activity of Boswellia serrata Roxb. ex Colebr. (Burseraceae) essential oil with various azoles against pathogens associated with skin, scalp and nail infections. Left appl.microbial.2016 Dec., 63(6): 495-501. Doi: 10.1111/1am.12683. E. pub. 2016 Nov. 6.

Shah BA, Qazi GN, Taneja SC. 2009. Boswellic acids: a group of medicinally important compounds. *Natural Product Reports* 26(1): 72-89.

Siddiqui MZ. 2011. *Boswellia Serrata*, a potential antiinflammatory agent: an overview. *Indian Journal of Pharmaceutical Sciences* 27(3): 255-261.

Singh P, Chacko KM, Aggarwal ML, Bhat B, Khandal RK, Sultana S, Kuruvilla B. 2012. A-90 day gavage safety assessment of *Boswellia serrata* in Rats. *Toxicology International* 19(3): 273-278.

Streffer JR, Bitzer M, Schabet M, Dichgans J, Weller M. 2001. Response of radiochemotherapy-associated cerebral edema to a phytotherapeutic agent, H15. *Neurology* 56(9): 1219-21.

Sunnichan VG, Mohan Ram HY, Shivanna KR. 2005. Reproductive biology of *Boswellia serrata*, the source of salai guggul, an important gum-resin. *Botanical Journal of the Linnean Society* 147(1): 73-82.

Takada Y, Ichikawa H, Badmaev V, Aggarwal BB. 2006. Acetyl-11-keto-β-boswellic-acid potentiates apoptosis, inhibits invasion, and abolishes osteoclastogenesis by supperessing NF-êB and NF-êB regulated gene expression. *The Journal of Immunology* 176(5): 3127-3140.

Xia L, Chen D, Han R, Fang Q, Waxman S, Jing Y. 2005. Boswellic acid acetate induces apoptosis through caspase-mediated pathways in myeloid leukemia cells. *Molecular Cancer Therapeutics* 4(3): 381-388.

Zaitone SA, Barakat BM, Bilasy SE, Fawzy MS, Abdelaziz EZ, Farag NE. 2015. Protective effect of boswellic acid versus pioglitazone in a rat model of diet induced non-alcoholic fatty liver disease: influence on insulin resistance and energy expenditure. *Naunyn Schmiedeberg's Archives of Pharmacology* 388(6):587-600.

Zutshi U, Rao PG, Kaur S, Singh GB, Atal CK. 1980. Mechanism of cholesterol lowering effect of salai guggul ex-*Boswellia serrata*. *Indian Journal of Pharmacology* 18: 182-183.

14

Phosphate Solubilizing Bacteria Isolated from Agricultural Soils of Delhi Shows Mineral Phosphate Solubilizing Ability

Rhituporna Saikia and Ratul Baishya*

*Department of Botany, University of Delhi, Delhi-110007, INDIA
*Email: rbaishyadu@gmail.com

ABSTRACT

Phosphorus is a macro nutrient, acts as a limiting factor in most ecosystems. Phosphorus immobilization is a major issue in most agricultural soils of India. Phosphate Solubilizing Bacteria (PSB) help in the solubilization of insoluble mineral phosphates by the production of organic acids. The aim of this research was to isolate, identify and characterize phosphate solubilizing bacteria from different agricultural crop soils of river banks of Yamuna, Delhi. Experiments were established in the laboratory to isolate phosphate solubilizing bacteria from various soil samples. Biochemical tests were performed to identify and characterize phosphate solubilizing bacteria. The phosphate solubilization efficiency of bacteria was measured. Organic acid production by different PSB for solubilization of insoluble phosphates were estimated. PSB from the soils of four agricultural crops such as Fennel, Spinach, Coriander and Mustard were isolated, identified and their ability to solubilize mineral phosphates by the production of organic acids were studied. Different PSB isolated were characterized as *Citrobacter* sp., *Pseudomonas* sp., *Staphylococcus* sp. and *Bacillus* sp. The PSB varied in phosphate solubilizing efficiency and production of organic acids. *Pseudomonas* sp. produced the maximum amount of organic acids and hence was the most efficient PSB. These PSB can be used as biofertilizers instead of chemical fertilizers and lead to sustainable agriculture.

Keywords: Agricultural Crops, Mineral Phosphates, Organic Acids, Phosphate Solubilizing Bacteria.

INTRODUCTION

Phosphorus (P) is the second most essential nutrient after nitrogen for growth and development of plants (Hameeda et al. 2008). Phosphorus in soluble ionic form (mainly $H_2PO_4^-$) at the plasma membrane of root cell is an effective nutrient for plants (Goldstein 1986). Phosphorus immobilization is a common problem faced in agricultural lands. *P immobilization* is due to the high reactivity of soluble P with Calcium, Iron or Aluminium (Yu et al. 2011). Much of P immobilization occurs due to the addition of chemical fertilizers as plants can utilize only a minute amount and rest is converted to insoluble complexes. This leads to the necessity of frequent addition of such chemical fertilizers (Azziz et al. 2012). Chemical fertilizers are expensive for farmers in developing countries (Bashan et al. 2012). Excessive application of chemical fertilizers also leads to environmental hazards such as water eutrophication (Zaidi et al. 2009, Correll 1998).

To meet the demands of the poor farmers as well as to maintain a healthy environment, use of Phosphate Solubilizing Bacteria (PSB) as biofertilizer has come up recently. PSB solubilize these insoluble phosphates which are then absorbed by plants and simultaneously maintain the soil ecosystem. Genera like *Pseudomonas, Bacillus* etc. are efficient phosphate solubilizers. The key mechanism behind mineral phosphate solubilization is the release of organic acids by PSB (Rodrýìguez and Fraga 1999). Organic acids such as Citric acid, Succinic acid, Oxalic acid *etc.* are produced by PSB (Sharma et al. 2013).

The objectives of the study were to isolate, identify and study the mineral phosphate solubilizing ability of different PSB from four agricultural crop soils of Delhi (Fennel, Spinach, Coriander and Mustard). PSB were identified with the help of biochemical tests. The phosphate solubilization efficiency of bacteria was estimated. Organic acids production by different PSB was estimated and the pH drop due to the production of organic acids was noted.

STUDY AREA

Agricultural field was selected on the banks of river Yamuna, Delhi. Maximum rainfall recorded were in the month of July (296 mm). June experiences maximum temperature up to 48°C, whereas, the lowest temperature recorded was 4°C at the end of December and early January. Average annual relative humidity is 54.3%. The collection sites for experimentation was located between 28°39′46.80″ and 28°39′48.96″N latitudes and 77°15′2.55″ and 77°15′5.15″E longitudes. The soil characterized were sandy and alkaline whose pH N varies from 7-8.

COLLECTION OF SOIL SAMPLES

Soils were collected from the rhizospheric zone of 4 different crops such as Fennel, Spinach, Coriander and Mustard. Samples were collected in triplicates from each individual of a crop species. For each crop species, there were three randomly selected individuals. Samples were analysed within 2 weeks.

ISOLATION OF PHOSPHATE SOLUBILIZING BACTERIA

Serial dilution method was used to isolate PSB (Motsara and Roy 2008). Pikovskaya Agar containing tricalcium phosphate was used for isolation of PSB (Pikovskaya 1948). Soil suspension was made with 10g of soil in 90ml of sterile water in a conical flask. The suspension was shaken for 5 minutes. The suspension was diluted serially as follows: 6 test tubes were arranged in a test tube stand containing 9 ml of sterile water. 1 ml of the suspension from the conical flask was taken and added to the first test-tube containing 9 ml of sterile water to make a total volume of 10ml. The test tube was shaken vigorously. 1 ml of the suspension was then pipetted out from this test-tube and added to the second test-tube and shaken. This process was repeated serially until the last test-tube. 1 ml was pipetted out from each test tube and poured onto Pikovskaya medium (containing insoluble phosphates) plates. The aliquot was spread and the plates closed with Para film and labelled. The plates were inverted and incubated at 30°C for 4 days. Microbial colonies with a transparent zone developed. Such colonies were picked and streaked on Pikovskaya plates and maintained. Colonies with halo zone were considered PSB.

IDENTIFICATION OF PHOSPHATE SOLUBILIZING BACTERIA

Standard biochemical tests as listed in Bergey's Manual of Determinative Bacteriology were used for identification of PSB (Holt et al. 1994). Various tests such as lactose fermentation, oxidase, citrate utilization, gelatin hydrolysis and catalase helped in the identification of the isolated PSB.

QUALITATIVE ANALYSIS OF PHOSPHATE SOLUBILIZING EFFICIENCY

Colony diameter and halozone diameter of colony inoculated at the centre of plate maintained at 30°C for 7 days were measured. Phosphate Solubilization Efficiency (PSE) was measured for each PSB isolate.

PSE= (Colony diameter + Halozone diameter)/Colony diameter (Premono et al. 1996).

CHANGE IN pH OF THE PIKOVSKAYA MEDIUM

pH change of the Pikovskaya broth inoculated with PSB was noted after 7 days. Initial pH of the medium was 7.

ESTIMATION OF ORGANIC ACID PRODUCTION

Organic acid was estimated following method outlined by Baliah et al. (2016). Pikovskaya broth inoculated with PSB was kept in rotary shaker for 7 days. The culture was then centrifuged at 10000 rpm for 15 min. In a flask, 2ml of the filtrate was taken and phenolphthalein added to it. 0.01N NaOH was used for titrating against the filtrate. Amount of 0.01N of NaOH consumed gave the estimate of organic acids produced.

RESULTS

Isolation of PSB

A total of 11 isolates (4 *Citrobacter*, 4 *Pseudomonas*, 2 *Staphylococcus* and 1 *Bacillus*) were obtained from all 4 soil samples (Fennel, Spinach, Coriander and Mustard). PSB were isolated based on the formation of halo zone on Pikovskaya solid medium (Fig. 1).

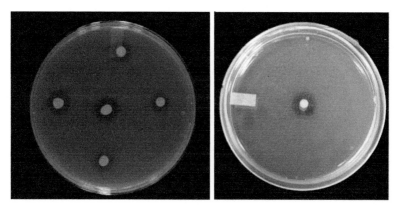

Fig. 1: Halo zone (transparent zone) formed by PSB in Pikovskaya medium

Identification of PSB

Various biochemical tests helped in the identification of 4 genera namely, *Citrobacter* sp., *Pseudomonas* sp., *Bacillus* sp. and *Staphylococcus* sp. (Table 1). Tests such as lactose, mannitol and sucrose fermentation, citrate utilization etc. were performed. The colony morphology and shape of bacteria were also noted.

Table 1: Morphological and biochemical characteristics of the isolated bacteria

Characteristics	*Citrobacter* sp.	*Pseudomonas* sp.	*Staphylococcus* sp.	*Bacillus* sp.
Colony morphology	White, round, Alliga smooth, shiny, entire margin	Off-white, round, entire margin, smooth, Shiny	Orange, round, entire margin, smooth	Dull white, dry, wavy margin
Shape	Straight rods	Straight or slightly curved rods	Spherical	Rods
Gram Reaction	-	-	+	+
Lactose fermentation	+	-	+	+
Mannitol fermentation	+	+	+	+
Sucrose fermentation	+	-	+	+
Citrate utilization test	+	+	+	+
Gelatin hydrolysis test	-	+	+	+
Indole test	-	-	-	-
Methyl red test	+	-	+	-
Vogues-Proskauer test	-	-	+	+
Oxidase test	-	-	-	-
Catalase test	+	+	+	+
Nitrate reduction test	+	+	+	+
Isolated from crop soils	Fennel, Spinach, Coriander, Mustard	Fennel, Spinach, Coriander, Mustard	Spinach, Mustard	Mustard

Phosphate Solubilizing Efficiency (PSE)

PSE significantly differed in different PSB (Table 2). It was highest in *Pseudomonas* sp. i.e. (3.9-4). Lowest PSE was shown by *Citrobacter* sp. (3.4-3.7). Halozone that was formed due to the solubilization of phosphates helped us to find out the phosphate solubilizing efficiency of the isolates.

Table 2: Phosphate Solubilizing Efficiency, pH Reduction and estimation of organic acids of PSB isolates

Crops	PSB	PSE	Reduction of pH	Organic acids (ml)
Fennel	*Citrobacter* sp.	3.4	5.35	1.5
	Pseudomonas sp.	3.9	3.9	2.7
Spinach	*Citrobacter* sp.	3.7	5.3	1.5
	Pseudomonas sp.	3.9	3.76	2.6
	Staphylococcus sp.	3.6	4.23	1.9
Coriander	*Citrobacter* sp.	3.6	5.21	1.3
	Pseudomonas sp.	4	3.82	2.7
Mustard	*Citrobacter* sp.	3.7	5.3	1.4
	Pseudomonas sp.	3.9	3.87	2.7
	Staphlococcus sp.	3.8	4.16	2.2
	Bacillus sp.	3.6	3.10	2.2

Reduction of pH and Organic Acid Estimation

The ability to produce organic acids varied significantly in different PSB (Table 2). Production of organic acids by PSB lowers the pH. Highest amount of organic acids production was shown by genus *Pseudomonas* sp. i.e. (2.6-2.7 ml). Maximum pH drop was shown by genus *Pseudomonas* sp. (3.76-3.9). Least drop of pH and least amount of organic acids production was noted in *Citrobacter* sp.

Discussion

Various kinds of PSB such as *Citrobacter* sp., *Pseudomonas* sp., *Staphylococcus* sp. and *Bacillus* sp. were isolated from rhizospheric soils of Fennel, Spinach, Coriander and Mustard. *Pseudomonas* sp. and *Citrobacter* sp. were the most frequently found which is depicted in our results that it was dominant in all the crop soils. Rodrýìguez & Fraga (1999) analyzed that bacterial strains from the genera *Pseudomonas* are the common phosphate solubilizers. According to Gulati et al. (2007), *Pseudomonas* is the dominating PSB in the rhizosphere. Our findings also showed that *Pseudomonas* as the common phosphate solubilizer. Bacillus was the least common PSB which was present only in Mustard. The PSB varied significantly in all the parameters such as PSE, pH reduction and amount of organic acids production. Our aim was to study the potential of different PSB to solubilize insoluble mineral phosphates. We could see that on inoculation in Pikovskaya broth, the PSB were able to solubilize good amount of mineral phosphates due to the production of organic acids which lowered the pH. The initial pH of the medium was 7 which decreased significantly during the period of 7 days. The more amounts of organic acids produced, it led to increase in acidity or lowering of pH significantly. Even in the case of Pikovskaya solid media, the production of halo is due to organic acids production. The organic acids solubilize the tricalcium phosphate present in the media leading to the production of halo or transparent zone. Eventhough *Citrobacter* sp. was also frequently found in all the crops, it was not an efficient mineral phosphate solubilizer as *Pseudomonas* sp. This species is an efficient solubilizer of insoluble phosphates by the production of organic acids which ultimately increase acidity of the medium or lowers the pH of the medium (Baliah et al. 2016, Tripti et al. 2012).

CONCLUSION

The present study was conducted on the soils from the flood banks of the river Yamuna. The bacteria isolated from the agricultural crop soils of Yamuna banks have the potential to solubilize mineral insoluble phosphates. Soils from different crop soils were collected and analysed. The primary objectives were to isolate, identify and characterize PSB from these soils. Various kind of PSB such as *Citrobacter, Pseudomonas, Bacillus* and *Staphylococcus* were isolated and characterized. The phosphate solubilizing efficiency was measured for all the PSB. The PSB were able to produce organic acids which solubilizes insoluble mineral phosphates. Thus, further field trials using these bacteria will help us to finalize the PSB which can be used as biofertilizers. Thus, these genera can be used as potent biofertilizers in agricultural crops facing phosphorus immobilization and thus reducing excessive use of chemical fertilizers. Excessive chemical fertilizers usage in soils disturbs the soil strata, degrades soil quality and affects agricultural productivity. As these biofertilizers are less costly and also eco-friendly, these can be used as alternative to chemical fertilizers and will benefit the poor farmers and also maintain the soil ecosystem.

Abbreviations

PSB-Phosphate Solubilizing Bacteria

PSE-Phosphate Solubilizing Efficiency

Conflict of Interest

The authors do not have any conflict of interest.

ACKNOWLEDGEMENT

Authors are thankful to UGC for providing Non-NET fellowship during the course of the study. Financial assistance provided by University of Delhi in the form of R&D projects to faculty members is also highly acknowledged.

LITERATURE CITED

Azziz G, Bajsa N, Haghjou T, Taulé C, Valverde Á, Igual J, Arias A. 2012. Abundance, diversity and prospecting of culturable phosphate solubilizing bacteria on soils under crop–pasture rotations in a no-tillage regime in Uruguay. *Applied Soil Ecology* 61: 320-326.

Baliah NT, Pandiarajan T, Kumar BM. 2016. Isolation, identification and characterization of phosphate solubilizing bacteria from different crop soils of SrivilliputturTaluk, Virudhunagar District, Tamil Nadu. *Tropical Ecology* 57(3): 465-474.

Bashan Y, Kamnev A, de-Bashan L. 2012. A proposal for isolating and testing phosphate- solubilizing bacteria that enhance plant growth. *Biology and Fertility of Soils* 49: 1-2.

Correll D. 1998. The Role of Phosphorus in the Eutrophication of Receiving Waters: A Review. *Journal of Environment Quality* 27: 261.

Goldstein A. 1986. Bacterial solubilization of mineral phosphates: Historical perspective and future prospects. *American Journal of Alternative Agriculture* 1: 51-57.

Gulati A, Rahi P, Vyas P. 2007. Characterization of Phosphate-Solubilizing Fluorescent Pseudomonads from the Rhizosphere of Seabuckthorn Growing in the Cold Deserts of Himalayas. *Current Microbiology* 56: 73-79.

Hameeda B, Harini G, Rupela O, Wani S, Reddy G. 2008. Growth promotion of maize by phosphate-solubilizing bacteria isolated from composts and macrofauna. *Microbiological Research* 163: 234-242.

Holt J, Krieg N, Sneath P, Staley J, Williams S. 1994. *Bergey's manual of determinative bacteriology*. 9th ed. Philadelphia: Lippincott.

PremonoME, Moawad AM, Vlek PLG.1996. Effect of phosphate-solubilizing *Pseudomonas putida* on the growth of maize and its survival in the rhizosphere. *Indonesian Journal of Crop Science* 11:13–23.

Motsara MR, Roy RN.2008. *Guide to laboratory establishment for plant nutrient analysis (Vol. 19)*. Rome: Food and Agriculture Organization of the United Nations.

Pikovskaya RI. 1948. Mobilization of phosphorus in soil in connection with the vital activity of some microbial species. *Mikrobiologiya* 17:362-370.

Rodrýiguez H, Fraga R. 1999. Phosphate solubilizing bacteria and their role in plant growth promotion. *Biotechnology Advances* 17: 319-339.

Sharma S, Sayyed R, Trivedi M, Gobi T. 2013. Phosphate solubilizing microbes: sustainable approach for managing phosphorus deficiency in agricultural soils. *Springer Plus* 2: 587.

Tripti, Kumar V, Anshumali 2012.Phosphate Solubilizing Activity of Some Bacterial Strains Isolated from Chemical Pesticide Exposed Agriculture Soil .*International Journalof Engineering Research and Development* 3:1-6.

Yu X, Liu X, Zhu T, Liu G, Mao C. 2011. Isolation and characterization of phosphate-solubilizing bacteria from walnut and their effect on growth and phosphorus mobilization. *Biology and Fertility of Soils* 47: 437-446.

Zaidi A, Khan M, Ahemad M, Oves M. 2009. Plant growth promotion by phosphate solubilizing bacteria. *Acta Microbiologicaet Immunologica Hungarica* 56: 263-284.

15

Investigation of Plants Utilization by Tribal Communities of Arunachal Himalayas in India

*Anup Kumar Das[1], Wishfully Mylliemngap[*1], Nako Laling[1]*
Om Prakash Arya[1] and R.C. Sundriyal[2]

[1]GB Pant National Institute of Himalayan Environment and Sustainable Development, North East Regional Center, Vivek Vihar, Itanagar - 791113 Arunachal Pradesh, INDIA
[2]GB Pant National Institute of Himalayan Environment and Sustainable Development, Kosi-Katarmal, Almora, Uttarakhand, INDIA
*Email: wishm2015@gmail.com

ABSTRACT

Arunachal Pradesh in Northeastern India covers a major portion of the Eastern Himalayas and is well-known for its rich biological and cultural diversity. The indigenous communities of the state living in close vicinity of forests depend upon wild plant-resources for meeting their various needs like medicine, food, fodder, fuel, etc. The present paper deals with documentation of plant species used by *Monpa* and *Adi* community residing in vast ecological range starting from subtropical to alpine ecosystems. A total of 346 species belonging to 215 genera and 102 families used by the two communities as wild vegetables (97 species), wild fruits (39), dye and color (13), spices and condiments (51), hunting and piscicide (40), paper pulp and fiber yielding plants (5) and medicinal use were reported (165). This study highlights the rich diversity of indigenous knowledge among the local communities in utilizing the diverse plant resources around them for their sustenance. Documentation of this knowledge is of utmost importance for identification of useful plant species for value addition, promotion as well as to formulate strategies for their sustainable utilization and conservation. This would also enhance the livelihood of the indigenous communities contribute to biodiversity conservation in these fragile mountain ecosystems.

Keywords: Traditional Knowledge, *Adi*, *Monpa*, Eastern Himalayas, Conservation

INTRODUCTION

The Indian Himalayan Region (IHR), with geographical coverage of over 5.3 lakh km^2, contributes greatly to richness at all levels of biodiversity components (i.e., genes, species and ecosystems). The region represents nearly 3.8% of total human population of the country and exhibits a great diversity of ethnic groups (171 out of a total 573 scheduled tribes in India) which inhabit remote inhospitable terrains (Samal et al. 2003). The eastern Himalaya supports one of the world's richest alpine flora, with high level of endemism. The temperate broadleaved forest type in the eastern Himalaya is among the most species-rich temperate forests in the world (Chettri et al. 2010).

Broadly the Indian East Himalayan region, which constitutes approximately 52% of total East Himalaya (total 524,190 sq. km), is often referred as North Eastern Region (NER). It consists of the contiguous, seven sister states namely Arunachal Pradesh, Assam, Manipur, Meghalaya, Mizoram, Nagaland and Tripura along with Sikkim. The NER covers an area of 2, 62,179 km^2 constituting 7.9% total geographical area of the country. Located at the tri-junction of Indo-Chinese, Indo-Malayan and Palaearctic biogeographic realms, the region exhibits diverse hilly terrain with wide ranging altitudinal range (GBPIHED 2011). It forms a unique biogeographic province encompassing major biomes recognized in the world. It has richest reservoir of plant diversity in India and is one of the 'biodiversity hotspots' of the world supporting 50% of India's biodiversity (Mao & Hynniewta 2000). The distinct tribes in the region have rich indigenous traditional knowledge system on the uses of components of biodiversity for their daily sustenance like food, fodder, shelter and healthcare. The knowledge and utilization of local plants depends on the ethnic group they belong to and also their remoteness from the modern world (Mao et al. 2009).

Arunachal Pradesh the largest state of northeast India with a geographical area of 83,743 sq. km covers a major portion of Eastern Himalaya. It opens pathway for the Palaeoarctic biota of Tibet and

Malayan elements from the South east, which resulted in enormous ecological and floristic diversity (Khongsai et al. 2011). It is home to over 26 major tribes and more than 100 subtribes, each having unique traditional and cultural practices. The different tribes are distributed in different climatic zones and thus have diverse tribe specific cultures. The indigenous communities inhabiting the remote areas of the state lives in close association with nature and possess immense knowledge on the use of wild plant species for food, shelter, medicine, hunting, fishing, art and craft, cultural and religious purposes (Singh et al. 2010). The biological diversity and natural resources of the state have sustained them since time immemorial. However, the scenarios are changing rapidly with accessibility of modern day facilities to every nook and corner of the state, which includes roads to inaccessible areas, modern medicines, food items, clothing, education, housing materials and so on. This has resulted in altered lifestyle of the people and thus a change in natural resource utilization pattern. Increase in natural resource use along with unsustainable developmental activity is primary concerns from ecological and environmental point of view. A number of researchers have documented ethnobotanically important species used by different indigenous communities of Arunachal Pradesh. However, such information are scattered in the form of isolated research papers or reports on different tribes or region. Therefore, the present paper are an attempt to consolidate the database on plant utilization pattern of two of the major tribes of the state, i.e., *Adi* and *Monpa* communities. Documentation of plant utilization pattern among tribal communities is important to highlight the rich plant resources of the region, identify important species for conservation and value addition for sustainable development of the communities as well as conservation of threatened plant species.

Geographic Location

Arunachal Pradesh with a geographical area of 83,743 sq.km located between 26°28' – 29°30' N latitudes and 91°30' – 97°30' E, lies in Eastern Himalaya Biodiversity Hot spot. The state shares international boundary with China in the north and northeast (1,080 km), Bhutan in west (160 Km), and Myanmar in east (440 Km) and in south with Indian states of Assam and Nagaland. The total forest cover of the state is 67,410 sq. km i.e., 80.5% of the total geographical area the forest in the state comprises of very dense forest (20,868 sq. km) moderately dense forest with 31,519 sq. km and open forest (15,023 sq. km) (FSI 2011). The forest vegetation comprises of many aromatic, medicinal, ethnobotanical, economically important species on which the people of the state are dependent. The state with population density of 17 per sq. km is predominantly a tribal state in this difficult and remote terrian of Eastern Himalaya (Census 2011, Doley et al. 2014).

The *Monpa* and *Adi* tribe of Arunachal Pradesh are two major tribes of the state with distinct culture customs and rich traditional knowledge base. *Monpa*s are settled in two districts, West Kameng and Tawang districts in the western & north western part of Arunachal Pradesh while *Adi* inhabit major parts of East Siang, Upper Siang, eastern part of West Siang and western part of Lower Dibang Valley districts in the eastern and north-eastern part of Arunachal Pradesh. Major economic activity of both the communities is agriculture and secondary occupation is animal husbandry. The *Monpas* practice settled agriculture, while in some pockets wet rice cultivation was also observed. *Adi* community practice shifting cultivation which is gradually changing towards settled cultivation. The major agriculture crop of *Monpas* are corn, millets and wheat, vegetable cash crop are potato, cabbage, capsicum, tomato and leafy vegetables, and fruits like orange, kiwi, walnut, pear, apple and plum. In case of *Adi* community major agriculture crops are rice and millets along with other vegetable and horticultural cash crops such as pineapple, orange, kiwi and large cardamom.

Data Collection

Data on wild plant species used by the *Adi* and *Monpa* communities were collected from various scientific publications including journals, books and other published literatures and consolidated plant resource tabulation have been done. The plant species classified into edible and non-edibles plant resources. Edible plant resource were further categorized into wild vegetables, wild fruits, medicinal plants and spices and condiments, while non-edible plant resource include dye and colour fixers, hunting and piscicide and paper, pulp and fiber producing plants.

PLANT WEALTH

A total of 346 species belonging to 215 genera and 102 families were reported to be used by *Adi* and *Monpa* communities. Out of these, 305 species were edible plant resources and 56 plant species were non-edible. Highest number of species were used as medicinal plants (165), followed by wild vegetables (97), spices and condiments (51), hunting and piscides (40), wild fruits (39), dye and colour fixer (13) and paper, pulp and fibre (5) (Fig. 1). Overall, Fabaceae was the most dominant family with 23 species being used for different purposes. Other dominant families were Zingiberaceae with 16 species followed by Asteraceae and Solanaceae (15 species each), Lauraceae (14), Polygonaceae (13 species), Euphorbiaceae and Rutaceae (12 species each), Poaceae (11), Moraceae, Piperaceae, Moraceae and Rubiaceae (10 each), Urticaceae (9), Amaryllidaceae, Rosaceae and Verbenaceae (7 species each), Musaceae (6), Apiaceae (5). The remaining 84 families were represented by 1-4 species (Table 1).

Among the two communities, the number of plant species reported used by the *Adi* community was maximum than that of the *Monpa* community, except for plants used as dye and colour fixers where more number of species was reportedly used by the *Monpa* community (Figure 1).

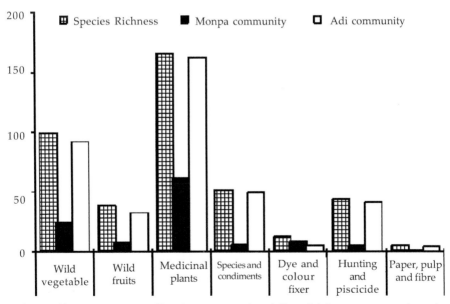

Fig. 1: Plant resource utilization pattern by *Adi* and *Monpa* community of Arunachal Pradesh

According to use pattern, 6 species were reported under 3-4 use categories (Table 1). Two species, viz., *Zanthoxylum rhetsa* and *Z. armatum* were reported to be used for 4 different purposes as wild vegetable, spices and condiments, piscicide and medicine. Tender

Table 1: Plant resources used by *Adi* and *Monpa* communities of Arunachal Pradesh

Species name	Family	Vernacular name*	Community	Use category**	Total no. of uses
Acacia pennata (L.) Wild.	Fabaceae	TatKung (A)	*Adi*	HP	1
Acacia rugata (Lam.) Voigt	Fabaceae	Riji/Pasoi tenga (A)	*Adi*	HP	1
Acanthopanax aculeatus (Aiton) Seem.	Araliaceae	Tako-laksin (A)	*Adi*	Med	1
Acmella oleracea (L.) R.K.Jansen	Asteraceae	Marshang (A)	*Adi*	HP	1
Acmella paniculata (Wall. ex DC.) R.K. Jansen	Asteraceae	*Adi* shena (A)	*Adi*	SC	1
Aconitum ferox Wall. ex Ser.	Ranunculaceae	Tsandu (M), Omi (A)	*Adi*	HP, Med	2
Aconitum heterophyllum Wall.	Rannanculaceae	Mran (A)	*Adi*	HP	1
Aconitum hookeri Stapf	Ranunculaceae	Zsa-tsandu (M)	*Monpa*	HP	1
Aegle marmelos Correa.	Rutaceae	Bel (As)	*Adi, Monpa*	Med	1
Aesculus assamica Griff.	Sapindaceae	Sarlok asing (A), Thretangshing (M)	*Adi, Monpa*	HP	1
Ageratum conyzoides L.	Asteraceae	Namsing Ing/Elee (A), Aieng-ying (*Adi-M*)	*Adi*	HP, Med	2
Albizia chinensis (Osbeck) Merr.	Fabaceae	Tat Kung asing (A)	*Adi*	HP	1
Albizia procera (Roxb.) Benth	Fabaceae	Sidak (A)	*Adi*	WV, HP	2
Albizzia lucida Benth.	Caesalpiniaceae	Tage (*Adi-M*)	*Adi*	Med	1
Allium chinense G.Don.	Amaryllidaceae	Dilab/mirrong (A)	*Adi*	Sc	1
Allium hookeri Thwaites	Amaryllidaceae	Disang-talap (A)	*Adi*	SC	1
Allium macranthum Baker	Amaryllidaceae	-	*Monpa*	WV, SC	2
Allium prattii C.H.Wright	Amaryllidaceae	-	*Monpa*	SC	1
Allium sativum L.	Amaryllidaceae	Jackok (A), Dilap (A)	*Adi*	SC, Med	2
Allium sp.	Amaryllidaceae	-	*Adi*	SC	1
Allium wallichii Kunth	Amaryllidaceae	-	*Monpa*	WV, SC	2
Alocasia macrorrhiza Schoff	Araceae	Engee (A)	*Adi*	Med	1
Alpinia malaccensis (Burm.f.) Rosc	Zingiberaceae	Puprere (A)	*Adi*	WV, WF	2

Contd.

Botanical name	Family	Local name	Tribe	Use	
Alpinia nigra Gaert.	Zingiberaceae	Bugbii Telli	*Monpa*	WF	1
Alpinia sp.	Zingiberaceae	-	*Adi*	SC	1
Alstonia scholaris (L.) Br.	Apocynaceae	Sing-gar changne (N)	*Adi, Monpa*	Med	1
Amaranthus gracilis Desf.	Amaranthaceae	Tai	*Adi*	WV	1
Amaranthus hybridus L.	Amaranthaceae	Gubor oying(A)	*Adi*	WV	1
Amaranthus spinosus L.	Amaranthaceae	Bundagmo (M), Gobrai (A)	*Monpa*	WV, Med	2
Amischotolype mollissima (Blume) Hassk	Commelinaceae	Tachar parin (A)	*Adi*	HP	1
Amomum aromaticum Roxb.	Zingiberaceae	Papia (A)	*Adi*	WV, SC	2
Amomum dealbatum Roxb.	Zingiberaceae	Papia (A)	*Adi*	WV, SC	2
Amomum subulatum Roxb.	Zingiberaceae	Jepo (A)	*Adi*	SC	1
Anacardium occidentale L.	Anacardiaceae	Kaju badam (As)	*Adi, Monpa*	Med	1
Ananas comosus Merr.	Bromeliaceae	Dibechengki/ Tako bela (A)	*Adi*	Med	1
Angiopteris evecta (Forst.) Hoffm.	Marattiaceae	Fadey (M), Taba (A)	*Adi, Monpa*	WV	1
Aporosa octandra (Buch.-Ham.ex D.Don) Vick.	Euphorbiaceae	-	*Adi, Monpa*	Med	1
Artemisia nilagirica (CB clarke) Pamp.	Asteraceae	-	*Adi*	Med	1
Artemisia vulgaris L.	Asteraceae	Khampa/Khanme (M)	*Monpa*	Med	1
Artocarpus chama Buch.Ham	Lauraceae	-	*Adi*	WV	1
Artocarpus lakoocha Roxb.	Moraceae	Belang (A)	*Adi*	WV, WF, SC	3
Artrocarpus heterophyllus Lam.	Moraceae	Belang (A)	*Adi*	WV, WF	2
Aspidopterys indica (Wild.) Hochreat	Malpighiaceae	-	*Adi*	HP	1
Averrhoa carambola L.	Oxalidaceae	Kadung (*Adi*-M)	*Adi*	Med	1
Axonopus compressus (SW.) P.Beauv	Poaceae	Bobosa (A)	*Adi*	Med	1
AzAdirachta indica A.Juss.	Meliaceae	Namsu (A)	*Adi*	Med	1
Balakata baccata (Roxb.) Esser.	Euphorbiaceae	Pukto asing (M), Samperai, Shigum(A)	*Adi, Monpa*	Med, WF	2
Baliospermum calycium Muell.-Arg	Euphorbiaceae	Gilgal (A)	*Adi*	WV	1
Bambusa balcooa Roxb.	Poaceae	Erbu (A)	*Adi*	WV	1
Bambusa pallida Munro	Poaceae	-	*Adi, Monpa*	Med	1

Contd.

Scientific name	Family	Local name	Tribe	Use	No.
Bambusa sp.	Poaceae	Shi (M)	*Monpa*	Med	1
Bambusa tulda Roxb.	Poaceae	-	*Adi*	WV	1
Bannaya reptans Sprengel	Scrophulariaceae	Kat-Buk usueng (A)	*Adi*	Med	1
Barringtonia acutangula (L.) Gaertn	Lecythidaceae	-	*Adi*	HP	1
Bauhinia acuminata L.	Caesalpiniaceae	Agok (Adi-M)	*Adi*	Med	1
Bauhinia purpurera L.	Fabaceae	-	*Adi, Monpa*	Med	1
Bauhinia variegata L.	Fabaceae	Kachnar, ogok (A)	*Adi, Monpa*	WV, Med	2
Begonia griffithiana (DC.) Warb.	Begoniaceae	Sudum Meku (A)	*Adi*	WV	1
Begonia josephii A.DC	Begoniaceae	Sisi baying (A)	*Adi*	WV, SC	2
Begonia palmata D.Don	Begoniaceae	Donpolapang (A)	*Adi*	WV	1
Begonia spp.	Begoniaceae	Abibying/Buk(A),	*Adi*	Med	1
Bidens biternata (Lour.) Merr. & Sherff	Asteraceae	-	*Adi*	Med	1
Bischofia javanica Blume.	Euphorbiaceae	-	*Adi, Monpa*	Med	1
Bixa orellana L.	Bixaceae	Hat-ranga (A)	*Adi, Monpa*	Med	1
Blumea fistulosa (Roxb.) Kurz.	Asteraceae	Rumdum (A)	*Adi*	WV, Med	2
Boehmeria macrophylla Don.	Urticaceae	So:bo Yapong (A)	*Adi*	WV	1
Brassica juncea (L.) Czern.	Brassicaceae	Tulang/tuka (A)	*Adi*	SC	1
Brassica spp.	Brassicaceae	-	*Adi*	SC	1
Bryophyllum pinnatum (Lam.) Oken	Crassulaceae	Yepe-tare (A), Nebi-Nelum (A)	*Adi*	Med	1
Calamus flagellum Griff. ex. Mart.	Arecaceae	-	*Adi*	WV	1
Calamus latifolius Roxb.	Arecaceae	Golar/Raiding (A)	*Adi*	WV	1
Callicarpa arborea Roxb.	Verbenaceae	Toti (A)	*Adi*	Med	1
Calotropis gigantea (L.) Br.	Asclepiadaceae	-	*Adi*	Med	1
Camellia sinensis L.	Theaceae	-	*Adi, Monpa*	Med	1
Canarium bengalense Roxb.	Burseraceae	-	*Adi, Monpa*	Med	1
Canarium strictum Roxb.	Burseraceae	Silum (A)	*Adi*	HP, Med	2
Capsicum annuum L.	Solanaceae	-	*Adi*	SC	1
Capsicum chinense Jacq.	Solanaceae	Sibol (A)	*Adi*	SC	1
Capsicum frutescens L.	Solanaceae	-	*Adi*	SC	1
Carex nubigena D.Don	Cyperaceae	Tabey (A)	*Adi*	Med	1

Contd.

Botanical name	Family	Local name (tribe)	Tribe	Use	No.
Carica papaya L.	Cariaceae	Omir (A)	*Adi*	Med	1
Cassia alata L.	Fabaceae	-	*Adi*	HP	1
Cassia fistula L.	Fabaceae		*Adi, Monpa*	Med	1
Cassia javanica L.	Fabaceae	Thedang (A)	*Adi*	HP	1
Cassia nodosa Buch.-Ham. ex Roxb.	Fabaceae	Thedang (A)	*Adi*	HP	1
Castanopsis echinocarpa King	Fagaceae	Hirang (Adi-M)	*Adi, Monpa*	Med	2
Castanopsis indica DC.	Fagaceae	Hi:rang (A), Kheshing (M)	*Adi*	WF, HP	1
Centella asiatica (L.) Urban	Apiaceae	Loram/Kipum Yayum (A)	*Adi, Monpa*	Med	1
Chenopodium album L.	Chenopodiaceae	Bathua, Machiosak (A)	*Adi, Monpa*	WV	1
Chlorophytum arundinaceum Baker	Asparagaceae		*Adi*	WV	2
Cinnamomum bejolgotha Buch.-Ham.	Lauraceae		*Adi, Monpa*	SC, Med	1
Cinnamomum camphora Nees. & Eber.	Lauraceae		*Adi, Monpa*	Med	1
Cinnamomum glandulifera Meissn.	Lauraceae		*Adi, Monpa*	Med	1
Cinnamomum glausence Drury.	Lauraceae		*Adi, Monpa*	Med	2
Cinnamomum tamala (Buch.-Ham.)	Lauraceae	Rapi-ising/Jongkengasing/Sepiriang (A)	*Adi, Monpa*	SC, Med	2
Cinnamomum verum J. Presl	Lauraceae	Hitipo-ri(A)	*Adi, Monpa*	SC, Med	2
Cinnamomum zeylanicum Blume	Lauraceae	Hitipo-ri (A), Solo sing (M)	*Adi, Monpa*	SC	1
Citrus limon L.	Rutaceae	Mori singkin (A)	*Adi*	WV	1
Citrus medica L.	Rutaceae	-	*Adi*	WF	1
Citrus paradisi MacF.	Rutaceae		*Adi, Monpa*	Med	1
Clacaria macrophylla Wall.	Rubiaceae	Pemi lagin (A)	*Adi*	Med	1
Clerodendrum colebrookianum Walp	Verbenaceae	Ongin (A)	*Adi*	WV, Med	2
Clerodendrum kaempferi (Jacq.)	Verbenaceae	Tapen (A)	*Adi*	WV	1
Clerodendrum spp.	Verbenaceae	Dumkar (A)	*Adi*	Med	1
Coffea bengalensis Roxb. ex. Schult.	Rubiaceae	Wanco (A)	*Adi*	Med	1
Colocasia esculenta (Linn.) Schott	Araceae	Enge (A), Sona (I)	*Adi*	WV, Med	2
Coptis teeta Wall.	Ranunculaceae	Mishmiteeta/Rinko (A)	*Adi*	Med	1
Coriandrum sativum L.	Apiaceae	Ori (A)	*Adi*	SC	1

Contd.

Crassocephalum crepidioides (Benth) S.Moore	Asteraceae	Hogegain (A), Gende (A/M)	Adi, Monpa	WV	1
Crepis japonica Benth	Asteraceae	Rum Dum (A)	Adi	WV	1
Curanga amara Juss.	Scrophulariaceae	Bon-ging (Adi-M)	Adi	Med	1
Curcuma aromatica Salisb.	Zingiberaceae	Jangali haladhi (A)	Adi	WV	1
Curcuma caesia Roxb.	Zingiberaceae	Kala haldi	Adi	Med	1
Curcuma longa L.	Zingiberaceae	Keloti (Adi-M)	Adi	SC, Med	2
Daphne papyracea Wall.	Thymelaeaceae	Shuksheng (M)	Monpa	DC, PPF	2
Debregeasia longifolia Burm.f. Wedd.	Urticaceae	Jirepole (A)	Adi	WV	1
Dendrobium nobile Lindl.	Orchidaceae	Hira-apum (Adi-M)	Adi	Med	1
Dendrocalamus hamiltonii Nees & Arnott ex Munro	Poaceae	Yayibyapu/Eni (A)	Adi	SC, PPF	2
Derris scandens (Roxb.) Benth.	Fabaceae	-	Adi	HP	1
Dicranopteris linearis Burm f.	Dicranopteridaceae	Tarang (A)	Adi	WV	1
Dillenia indica L.	Dilleniaceae	-	Adi	WV, Med	2
Dioscorea alata L.	Dioscoreaceae	Ogit/Mayong (A); Janghli alu (A)	Adi	WV, Med	2
Dioscorea bulbifera L.	Dioscoreaceae	Remet (A)	Adi	WV	1
Dioscorea hamiltonii Hook.f.	Dioscoreaceae	-	Adi	WV	1
Diospyros malabarica Kostel.	Ebenaceae	-	Adi, Monpa	Med	1
Diplazium esculentum (Retz.) Sw.	Athyriaceae	Takang/Dhekia saag (A)	Adi, Monpa	WV	1
Drymaria cordata Wild	Caryophyllaceae	Kaira (Adi-M)	Adi	Med	1
Echinocarpus assamicus Benth.	Elaeocarpaceae	-	Adi, Monpa	Med	1
Ehretia acuminata Roxb.	Boraginaceae	-	Adi, Monpa	Med	1
Elaeagnus latifolia L.	Elaeagnaceae	-	Monpa	WF	1
Elaeagnus parvifolia Wall. ex Royal	Elaeagnaceae	Damrep (M)	Monpa	WF	1
Elaeagnus umbellata	Elaegneaceae	-	Monpa	WF	1
Elettaria cardamomum Maton	Zingiberaceae	Khuisang (A)	Adi	SC	1
Embelia ribes Burm.f.	Myrsinaceae	Hinkong (Adi-M)	Adi	Med	1
Engelhardtia spicata Lesch. ex. Blume	Juglandaceae	Corcorshing (M)	Monpa	DC, PPF	1
Ensete glaucum (Roxb.) Cheesman	Musaceae	Ngakku (A)	Adi	WF	1

Contd.

Species	Family	Local name	Tribe	Use	No.
Ensete superbum Cheesman	Musaceae	Colon/Kopak (A)	*Adi*	WV	1
Entada scadens (L.) Benth.	Mimosaceae	Taboh (A)	*Adi*	Med	1
Equisetum sp.	Equisetaceae	Sisi dungki (A)	*Adi*	Med	1
Equisetum debile (Roxb. ex. Voucher	Equisetaceae	Hiru debung, Asi tabo (*Adi-M*)	*Adi*	Med	1
Eryngium foetidum L.	Apiaceae	Hariyo, Ori, Ori-ritak, migom ori (A)	*Adi*	SC	1
Erythrina stricta Roxb.	Fabaceae	-	*Adi, Monpa*	WV, Med	2
Euphorbia hirta L.	Euphorbiaceae	-	*Adi*	WV	1
Euphorbia wallichii Hook.f.	Euphorbiaceae	Tharnu (M)	*Monpa*	HP	1
Fagopyrum dibotrys (D.Don) H.Hara	Polygonaceae	-	*Adi*	WV	1
Fagopyrum esculentum Moench.	Polygonaceae	Amintatek (A)	*Adi*	WV, Med	2
Fagopyrum tataricum Gaertn.	Polygonaceae	Akka/Aabra (A)	*Adi*	Med	1
Ficus auriculata Lour.	Moraceae	Hote (A)	*Adi*	WF	1
Ficus hispida L.	Moraceae	-	*Adi*	WF	1
Ficus racemosa L.	Moraceae	Takuk (A)	*Adi*	WV	1
Ficus roxburghii Wall	Moraceae	Tatuk (A)	*Adi*	WF	1
Ficus semicordata Buch.-Ham. ex J.E.Sm.	Moraceae	Tukusen (A,M)	*Adi, Monpa*	Med	1
Ficus spp.	Moraceae	Takuk (A)	*Adi*	Med	1
Flemingia macrophylla (Willd.) Prain.	Fabaceae	-	*Adi*	DC	1
Foeniculum vulgare Mill		Ori (A)	*Adi*	SC	1
Garcinia pedunculata Roxb. ex Buch.-Ham.	Clusiaceae	Liba (A); Prejang bizi(A)	*Adi*	WF, SC, Med	3
Glycine max (L.) Merr.	Fabaceae	Rontung (A), Piyak	*Adi, Monpa*	WV, SC	2
Gmelina arborea L.	Verbenaceae	Gomari (As)	*Adi, Monpa*	Med	1
Gnaphalium affine D.Don	Asteraceae	Paput (A, M)	*Adi, Monpa*	WV	1
Grewia serrulata DC.	Tiliaceae	Hakobangi (A)	*Adi*	PPF	1
Grewia tiliaefolia Vahl.	Tiliaceae	Mekuri tai (A)	*Adi, Monpa*	Med	1
Gymnocladus assamicus U.N. Kanjilal ex P.C. Kanjilal	Mimosaceae	Dekang (*Adi-M*)	*Adi*	Med	1
Gymnocladus burmanicus C.E. Parkinson	Fabaceae	Dikang (A)	*Adi*	HPP	1
Gynocardia odorata R.Br.	Flacourtiaceae	Takuk-changne (A)	*Adi*	Med	1

Contd.

Species	Family	Local name	Use	Tribe	No.
Gynura cusimbua (D. Don) S.	Asteraceae	Nakling(A)	WV	Adi	1
Hedychium gardnerianum Sheppard ex Ker Gawl.	Zingiberaceae	Bivi (A)	WF	Adi	1
Hedychium spicatum Buch.-Ham. ex Sm.	Zingiberaceae	Royik (A)	Med	Adi	1
Hemidesmus indicus (L.) R.Br.	Apocynaceae	-	Med	Adi	1
Hibiscus sabdariffa L.	Malvaceae	Amta (A)	SC	Adi	1
Hibiscus tiliaceous L.	Malvaceae	-	Med	Adi, Monpa	1
Holboellia latifolia Wall	Lardizabalaceae	-	WF	Adi	1
Houttuynia cordata Thunb.	Saururaceae	Roram, Reram, Zizibaying (A)	SC	Adi	2
Hydnocarpus kurzii (King) Warb.	Flacourtiaceae	-	Med	Adi, Monpa	1
Hydrangea robusta Hook. f. & Thomson.	Hydrangeaceae	Takmi (A)	WV	Adi	1
Ilex embelioides Hook.f	Aquifoliaceae	-	DC	Monpa	1
Illicium griffithii Hook.f. & Thomson.	Schisandraceae	Munsheng (M); Lissi (M)	DC, SC	Monpa	2
Indigofera tinctoria L.	Fabaceae	Zia-shing (M)	DC	Monpa	1
Ipomoea batatas L.	Convolvulaceae	-	WV	Adi	1
Juglans regia L.	Juglandaceae	Kay (M)	DC	Adi	1
Lasianthus longicauda Hook. f.	Rubiaceae	-	HP	Adi	1
Leucosceptrum canum Sm.	Lamiaceae	Tote (A)	Med	Adi	1
Limacia oblonga Miers	Menispermaceae	Titmilie (A)	WF	Adi	1
Litsea citrata Blume	Lauraceae	Taier (A)	SC	Adi	1
Litsea cubeba (Lour.) Pers.	Lauraceae	Meinjin (A), Rajil/Taier (A)	SC, Med	Adi	2
Litsea monopetala (Roxb.) Pers.	Lauraceae	-	Med	Adi, Monpa	1
Magnolia oblonga Wall. ex Hook.f. & Thomson	Magnoliaceae	Scrio-changne (A)	Med	Adi, Monpa	1
Maianthemum purpureum (Wall.) La Frankie	Asparagaceae	-	WV	Adi, Monpa	1
Mallotus phillipinensis Muell.-Arg.	Euphorbiaceae	-	Med	Adi, Monpa	1
Mangifera sylvatica Roxb.	Anacardiaceae	Pen-lokar/Tabing/Bon-am (A)	Med	Adi	1
Mannihot esculenta Crantze	Euphorbiaceae	Shingjoktang (M)	WV	Monpa	1

Contd.

Scientific name	Family	Local name	Tribe	Use	No.
Melastoma malabathricum L.	Melastomataceae	Ke-Seng (A), Phutkala (A)	*Adi*	DC	1
Melastoma nomale D.Don	Melastomataceae	Padiraju (A)	*Adi*	WF	1
Melia azederach L.	Meliaceae	Jungli neem	*Adi*	Med	1
Melletia pachycarpa Benth.	Fabaceae	Tasmu (A)	*Adi*	Med	1
Melothria heterophylla (Lour) Cogn.	Cucurbitaceae	Kubumiku (A)	*Adi*	WV	1
Mentha arvensis L.	Lamiaceae	Pudinang (A)	*Adi*	SC	2
Mesua assamica (King ex Prain) Kosterm	Clusuaceae	Sia nahar	*Adi*	HP	1
Michelia champaca L.	Magnoliaceae	Tita sopa	*Adi, Monpa*	Med	1
Mikania micrantha Kunth.	Asteraceae	-	*Adi, Apatani*	Med	1
Milletia panchycarpa Benth	Fabaceae	Bokoa-bih/ Bokoabeh (A)	*Adi*	HP	1
Mimosa pudica L.	Mimosaceae	Anying ing (A)	*Adi*	Med	1
Mimusops elengi L.	Sapotaceae	Bokul	*Adi, Monpa*	Med	1
Moringa oleifera Lam.	Moringaceae	Drumstick tree, Sajana	*Adi, Monpa*	WV, Med	2
Morus alba L.	Moraceae	Hinsai (A)	*Adi*	Med	1
Murraya koenigii (L.) Spreng	Rutaceae	-	*Adi*	SC, Med	2
Murraya paniculata (L.) Jack	Rutaceae	Nyibumtarum (A)	*Adi*	WF	1
Musa acuminata Colla	Musaceae	Kulu (A)	*Adi*	WV	1
Musa balbisisana Colla	Musaceae	Kopak/Colon (A)	*Adi, Monpa*	WV	1
Musa velutina Wendl. & Drude.	Musaceae	Luro, Kopack, Kodum (A)	*Adi*	WV, Med	2
Mussaenda roxburghii Hook.f.	Rubiaceae	Tangmeng (A), Akshap (M)	*Adi, Monpa*	WV	1
Mycetia longifolia (Wall.) Kuntze	Rubiaceae	Tangmge (A)	*Adi*	WV	1
Nephelium lappaceum L.	Sapindaceae	-	*Adi*	WF	1
Nicotiana tabacum L.	Solanaceae	Dumv (G); Tampu (M)	*Monpa*	Med	1
Not Identified	Poaceae	Eeye epuk (A)	*Adi*	HP	1
Not Identified	Poaceae	Ekku (A)	*Adi*	HP	1
Not Identified	Poaceae	Porang (A)	*Adi*	HP	1
Not Identified	Poaceae	Kodong (A)	*Adi*	HP	1
Ocimum americanum L.	Lauraceae	Take-mareng/ Tasingoying/ Tare-mareng (A)	*Adi*	SC	1

Contd.

Scientific name	Family	Local name	Tribe	Use	No.
Oenanthe javanica (Blume) DC	Apiaceae	-	*Adi, Monpa*	WV	1
Ophiorrhiza nepalensis D.Don	Rubiaceae	Akhap (A)	*Adi*	WV	1
Ophiorrhiza fasciculate D.Don.	Rubiaceae	Akhap (A)	*Adi*	WF	1
Oroxylum indicum (L.) Vent.	Bignoniaceae	Tapatale (M), Paksum alang (A), Domir ettkung (A)	*Adi, Monpa*	Med	1
Oxalis corniculata L.	Oxalidaceae	Phakep (A)	*Adi*	WV, Med	2
Oxalis debilis Kunth.	Oxalidaceaea	Khui-hamang (Ap)	*Monpa*	WV	1
Paedaria foetida L	Rubiaceae	Yepe-tere (*Adi.M*)	*Adi*	WV, Med	2
Panax bipinnatifidus Seem.	Araliaceae	-	*Adi, Monpa*	WV	1
Pandanus nepaulensis H. St. John	Pandanaceae	Tako (A)	*Adi*	PPF	1
Paris polyphylla Sm.	Alismataceae	Dipogoiak (A); Apuk (*Adi.M*)	*Adi*	Med	1
Parkia timoriana (A.DC.) Merr.	Fabaceae	Bel (A)	*Adi, Monpa*	Med	1
Passiflora edules Sim.	Passifloraceae	Namdung (A), Tining (Ap)	*Adi*	WV	1
Perilla frutescens (L.) Britton	Lamiaceae		*Adi*	SC	1
Persea robusta (W.W.Sm.) Kosterm.	Lauraceae	-	*Monpa*	WF	1
Persicaria barbata H.Hara	Polygonaceae	Ruri (A)	*Adi*	HP	1
Persicaria nepalensis (Meisn.) Miyabe	Polygonaceae		*Adi*	SC	1
Phoebe cooperiana U.N Kanjilal ex A.Das	Lauraceae	Tapil/Mekahi/ Tambor (A)	*Adi, Monpa*	WV, Med	2
Phoebe paniculata Nees.	Lauraceae	-	*Adi, Monpa*	Med	1
Phrynium parviflorum Roxb.	Musaceae	Ko-pat/ Hakoda (A)	*Adi*	Med	1
Phyllanthus acidus (L.) Skeels.	Euphorbiaceae	Bhui amla (As)	*Adi, Monpa*	Med	1
Phyllanthus emblica L.	Euphorbiaceae	Amlakhi (As)	*Adi, Monpa*	Med	1
Phyllostachys assamica Gamble	Poaceae	Tempor (A)	*Adi*	Med	1
Physalis minima L.	Solanaceae	Bodopati (A)	*Adi*	WF	1
Pilea bracteosa Wedd.	Urticaceae	Guge(A)	*Adi*	WV	1
Pinus wallichiana A.B.Jackson	Pinaceae	Tongschi, Lamshing (M)	*Monpa*	DC	1
Piper attenuatum Buch.-Ham. ex. Miq.	Piperaceae	Dolopan (A)	*Adi*	Med	1
Piper betel L.	Piperaceae	Pan	*Adi*	Med	1
Piper longum L.	Piperaceae	Odor (A)	*Adi*	SC	1

Contd.

Species	Family	Local name	Tribe	Use	No.
Piper mullesua Don	Piperaceae	Odor (A)	*Adi*	SC	1
Piper nepalense Miq.	Piperaceae	Ene ro:ri (A)	*Adi*	WV	1
Piper nigrum L.	Piperaceae	Odor (A)	*Adi*	SC	1
Piper pedicellatum C. DC	Piperaceae	Namar (A)	*Adi, Monpa*	WV	1
Piper peepuloides Roxb.	Piperaceae	Rari (*Adi*.M)	*Adi*	Med	1
Piper sylvaticum Roxb.	Piperaceae	Rari (A)	*Adi*	WV	1
Piper wallichii (Miq.) Hand.-Mazz.	Piperaceae	Rari (*Adi*.M)	*Adi*	Med	1
Plantago erosa Wall.	Plantaginaceae	Doni-hankang (A)	*Monpa*	WV	1
Plantago major L.	Plantaginaceae	–	*Adi*	WV	1
Polygonum alatum Buch.-Ham. ex. Spreng.	Polygonaceae	Uyushayan (A)	*Adi*	WV	1
Polygonum barbatum L.	Polygonaceae	Rukji (*Adi*.M)	*Adi*	Med	1
Polygonum capitatum Ham.	Polygonaceae	Apu-mo (I)	*Adi*	Med	1
Polygonum chinensis var. *ovaliolia*	Polygonaceae	–	*Adi*	WF	1
Polygonum hydropiper L.	Polygonaceae	Chhum-gon (M)	*Adi, Monpa*	DC	2
Polygonum nepalense Meissn.	Polygonaceae	Ruri (A)	*Adi*	DC, HP	1
Polygonum pubescens Blume	Polygonaceae	Tamu (A)	*Adi*	HP	1
Polygonum runcinatum Buch.-Ham. ex. D.Don	Polygonaceae	Ruri (N)	*Adi*	WV	1
Portulaca oleracea L.	Portulacaceae	Pali echi, Gubar oying (A)	*Adi*	WV	1
Pothos scandens L.	Araceae	Loma losut (A)	*Adi*	Med	1
Pouzolzia bennettiana Wight.	Urticaceae	Oyin (A)	*Adi*	WV	1
Pouzolzia hirta Hassk.	Urticaceae	Oike/Oyik (A)	*Adi*	WV	1
Pouzolzia viminea Wedd.	Urticaceae	Oyek (A)	*Adi*	Med	1
Prasiola crispa Hook. f.	Prasiolaceae	Green algae	*Monpa*	WV	1
Premna bengalensis C.B.Clarke.	Verbenaceae	–	*Adi, Monpa*	Med	1
Prunus acuminata Hook f.	Rosaceae	–	*Adi*	Med	1
Prunus persica Batsch	Rosaceae	–	*Adi*	WF	1
Psidium guajava L.	Myrtaceae	Maduri (*Adi*.M)	*Adi*	Med	1
Pteridium aquilinium (L) Wild.	Pteridaceae	–	*Adi*	WV	1
Pueraria lobata (Willdenow) Ohwi	Fabaceae	Ridin (A)	*Adi*	Med	1
Pueraria thunbergia Benth.	Fabaceae	Ridin (*Adi*.M)	*Adi*	Med	1

Contd.

Scientific name	Family	Local name	Tribe	Use	
Punica granatum L.	Punicaceae	-	Adi, Monpa	Med	1
Pyrus pashia Buch.-Ham. ex. D.Don	Rosaceae	-	Adi, Monpa	WF, Med	2
Quercus serrata Thunb.	Fagaceae	Ruki (Adi.M)	Adi, Monpa	Med	1
Rauvolfia densiflora Benth.	Apocynaceae	Talo (A)	Adi	Med	1
Rhaphidophora decursiva Schott	Aracea	Amashi (I), Naga tenga (As)	Adi	HP	1
Rhus semialata Murr.	Anacardiaceae		Adi	Med	1
Riccinus communis L.	Euphorbiaceae	Toti/ Akirore (A), Akirokmi (A)	Adi	Med	1
Rubia cordifolia L.	Rubiaceae	Lining-Ru (M), Tamen (A)	Monpa, Adi	DC	1
Rubus ellipticus Sm.	Rosaceae	-	Monpa	WF	1
Rubus lineatus Reinw.	Rosaceae	-	Adi	WF	1
Rubus paniculata Sm.	Rosaceae	-	Adi	WF	1
Rubus rosifolius Sm.	Rosaceae		Adi	WF	1
Sarcochlamys pulcherrima Gaud.	Urticaceae	Ombe (A)	Adi, Monpa	Med	1
Saurauia armata Kurz.	Saurauiaceae	-	Adi	WF	1
Saurauia roxburghii Wall.	Saurauiaceeae		Adi	WF	1
Sauropus androgynous (L.) Merr.	Euphorbiaceae	Woein (Adi.M)	Adi	Med	1
Scoparia dulcis L.	Plantaginaceae	Peyong (A)	Adi	Med	1
Selaginella wallichii (Hook. & Grev.) Spreng.	Selaginellaceae	Hojum (A)	Adi	WV	1
Senna alata (L.) Roxb.	Caesalpiniaceae	Donyi gori (A)	Adi	Med	1
Sida acuta Burm.f.	Malvaceae	Holap (A)	Adi	WV	1
Silene heterophylla Freyn	Caryophyllaceae	-	Adi	WF	1
Smilax ovalifolia Roxb.	Smilacaceae	Yorit (Adi.M)	Adi	Med	1
Solanum xanthocarpum Schrad. & H.Wendl.	Solanaceae	Sengelathang (A)	Adi	Med	1
Solanum incanum L.	Solanaceae	Cheteka Kara (I)	Adi	Med	1
Solanum indicum L.	Solanaceae	Sotabayom (A)	Adi	WV	1
Solanum intregrifolium L.	Solanaceae	Cheteka (I)	Adi	Med	1

Contd.

Scientific name	Family	Local name	Tribe	Use	No.
Solanum khasianum C.B.Clarke	Solanaceae	Kopir (A)	Adi	WV	1
Solanum nigrum L.	Solanaceae	-	Adi	WV, Med	2
Solanum spirale Roxb.	Solanaceae	Bangko (A)	Adi, Monpa	WV, Med	2
Solanum stramonifolium Jack.	Solanaceae	Bangko Kopi (A)	Adi	WV	1
Solanum torvum Sw.	Solanaceae	Byakta (A), Kopi (A)	Adi	WV, Med	2
Solanum viarum Dunal	Solanaceae	Boromgbu (M), Bengela Tang (A)	Adi, Monpa	Med	1
Sonchus brachyotus DC.	Asteraceae	Paku Hadu Hamamng/ Kochi hama (A)	Adi	WV	1
Spilanthes acmella Murr.	Asteraceae	Marshang (A/M)	Adi, Monpa	WV	1
Spilanthes oleracea L.	Asteraceae	Marshang (A)	Adi	HP	1
Spilanthes paniculata C.B.Clarke	Asteraceae	Marsha/Marshang (A); Adi Marsang (Adi.M)	Adi	HP	2
Spiradiclis bifida Wall. ex Kurz.	Rubiaceae	Sokho (A)	Adi	WV	1
Spondias pinnata P.C. Kanjilal and Das	Anacardiaceae	-	Adi	WF, Med	2
Sterculia villosa Roxb.	Sterculiaceae	Hitum (A); Udal (A)	Adi	WF, PPF	2
Stereospermum chelonoides (L.f.) DC.	Bignoniaceae	Mano (M)	Adi, Monpa	Med	1
Stereospermum hypostictum Miq.	Bignoniaceae	-	Adi, Monpa	Med	1
Swertia chirata Ham.	Gentianaceae	-	Adi	Med	1
Syzygium cumini (L.) Skeels	Myrtaceae	-	Adi	WF, DC, Med	3
Syzygium megacarpum (Craib.) Rath.& Nair	Myrtaceae	-	Adi, Monpa	Med	1
Tacca intergrifolia Ker-Gawl.	Taccaceae	Tagon/babor (Adi.M)	Adi	Med	1
Tamarindus indica L.	Fabaceae	-	Adi, Monpa	Med	1
Tectona grandis L.	Verbenaceae	-	Adi, Monpa	Med	1
Terminalia arjuna W. & A.	Combretaceae	Arjuna	Adi, Monpa	Med	1
Terminalia bellerica (Gaertn.) Roxb.	Combretaceae	-	Adi, Monpa	Med	1
Terminalia chebula Retz.	Combretaceae	-	Adi, Monpa	WF, Med	2
Tetrameles nudiflora R. Br.	Datiscaceae	-	Adi, Monpa	Med	1
Thevetia peruviana Juss.	Apocynaceae	Parijat (As)	Adi, Monpa	Med	1

Contd.

		Cocoriang (Adi:M)	Adi	Med	1
Thunbergia grandiflora (Roxb.ex Roht.) Roxb.	Acanthaceae				
Trevesia palmata (Roxb.) Ves.	Araliaceae	Tago (A); Tang-gongs (A)	Adi	WV, HP, Med	3
Trichosanthes cordata Roxb.	Cucurbitaceae	Dongkyong riyong (A)	Adi	WV	1
Trigonella foenum-graecum L.	Fabaceae	-	Adi	SC	1
Urtica hirta Blume	Urticaceae	Oyik (A)	Adi	WV	1
Villebrunea integrifolia Gaudich.	Urticaceae	Tane (A)	Adi	HP	1
Vitis himalayana (Royle) Brandis	Vitaceae	Riying (A)	Adi	HP	1
Woodfordia fruticosa Kurz.	Lythraceace	Chot-tingba (M)	Monpa	DC	1
Zanthoxylum acanthopodium DC.	Rutaceae	Ombe/ombeng/Onger (A)	Adi	SC, Med	2
Zanthoxylum armatum DC.	Rutaceae	Onger(A), Khagi (M)	Adi, Monpa	WV, SC, HP, Med	4
Zanthoxylum burkillianum Babu	Rutaceae	-	Adi	SC	1
Zanthoxylum hamiltonianum Wall. ex Hook.f.	Rutaceae	Ombe (A), Ongar (A)	Adi	Med	1
Zanthoxylum nitidum (Roxb.) DC.	Rutaceae	Tejmai-bih (A)	Adi	HP	1
Zanthoxylum rhetsa (Roxb) DC.	Rutaceae	Onger (A)	Adi	WV, SC, HP, Med	4
Zingiber cassumunar Roxb.	Zingiberaceae	Kekir (A)	Adi	Med	1
Zingiber montanum (J. Koenig) Link ex A. Dietr	Zingiberaceae	Kekir (A)	Adi	SC	1
Zingiber officinale Rosc.	Zingiberaceae	Take/Kekir (A)	Adi	SC, Med	2
Zizyphus mauritiana Lam.	Rhamnaceae	-	Adi	WF	1

Note: * A-Adi; Adi:M- Adi Minyong; M-Monpa, As-Assamese; I- Idu-mishmi; Ap- Apatani; G- Galo; N- Nyishi
** WV-wild vegetable; WF-wild fruit; med-medicine; SC-spices & condiments; HP-hunting & piscicide; DC-dye & colour fixer; PPF-paper, pulp & fibre

Source: Ali & Ghosh (2006), Angami et al. (2006), Baruah et al. (2013), Bharali et al. (2017), Boko & Narsimhan (2014). Das et al. (2013), Devi et al. (2014), Doley et al. (2014), Dollo et al. (2005), Ghosh et al. (2014), Kagyung et al. (2010), Kar & Borthakur (2008), Kumar et al. (2015), Mahanta & Tiwari (2005), Namsa et al. (2011), Paul et al. (2006; 2013), Rethy et al. (2010), Saha et al. (2014), Singh & Srivastava (2010), Singh et al. (2010), Srivastava (2009), Srivastava et al. (2010), Tag & Das (2004), Tag et al. (2014, 2015), Yumnam & Tripathi (2013), Yumnam et al. (2011)

shoots were used as vegetables, while seeds and fruits were used as spices, medicine and fishing.

Fruits and seeds of *Zanthoxylum rhetsa* were reported for traditional use in treating fever and headache while *Z. armatum* were used for cold, cough, fever and toothache. Bark extract of plant is considered as general health tonic. Four species, viz., *Artocarpus lakoocha, Garcinia pedunculata, Syzygium cumini* and *Trevesia palmata* have three different uses. *Artocarpus lakoocha* was reported to be used as fruit, vegetable, and spice and condiment. *Garcinia pedunculata* was being used as wild fruit, spice and condiment and medicine for gastric, diarrhea, dysentery and stomach pain. *Syzygium cumini* was reportedly used as wild fruit and medicine, as stimulant. Bark and fruits of *S. cumini* were being used by *Adi* community to obtain black dye. *Trevesia palmata* leaves were used as wild vegetable, medicine for cold and cough while fruits were used as piscicidal plant. There were 51 species being used for two different purposes while 289 species have single use only.

Edible Plant Resources

Both the communities possess rich knowledge in utilization of many wild plant species as vegetable, fruit, medicine, spices and condiments which are collected from nearby forests. These plant species form inseparable part of their traditional food and medicine system. Use of wild plants as medicine, staple and supplementary food items, seasoning and flavouring ingredients to improve the taste of food has been a common practice among indigenous communities all over the world living in remote forest areas. Among *Monpa* community highest number of species were used as medicinal plants (61), followed by wild vegetables (24), wild fruits (8) and spices and condiments (6). In case of *Adi* community highest number of plant species were used as medicinal plants (162), followed by wild vegetable (92), spices & condiments (49) and wild fruits (32) (Figure 1).

Wild Vegetables and Fruits

Wild edible plants form an important part of diet for two communities. The common species used by both *Adi* and *Monpa* communities were *Chenopodium album, Piper pedicellatum, Solanum spirale, Spilanthes acmella, Zanthoxylum armatum, Zanthoxylum rhetusa, Maianthemum purpureum, Musa balbisisana, Panax bipinnatifidus,* and *Oenanthe javanica.* Many species of wild fruits and vegetables such as *Zanthoxylum* spp., *Spilanthes acmella, Solanum nigrum, Pouzolzia bennetiana, Phoebe cooperiana, Diplazium esculentum* were reported to be sold in local markets which also add to their income.

Many of the preferred species by the two communities such as *Zanthoxylum* spp., *Elaeagnus* spp., *Ficus* spp., *Castanopsis* spp., *Amaranthus* spp., *Solanum nigrum* were reported to have high nutritive values (Angami et al. 2006) may have traditional value. Other commonly available species which were abundant in the forests were not sold in the market. Therefore, study on the nutritive values of the wild edible plant species based on local preference of the communities would help in identifying potential highly nutritious species, introduction into agricultural systems as alternative to conventional cultivated crops which would also increase the food and nutritional diversity of local communities.

Spices and Condiments

Spices and condiments are used for improving the taste and aroma of the food items by both *Adi* and *Monpa* community. The plant species used are generally collected from wild while few of them were cultivated in agricultural systems.

Wild food plants not only form an important part of the diet but can have important cultural and social significance among the indigenous communities (Kuhnlein & Turner 1991, Power 2008). For some communities, wild plants are not just sources of nutrition or economic value, but are also regarded as spiritual beings having close relationship with people throughout their lives (Sylvester & García Segura 2016).

The importance of wild foods cannot be underestimated in the context of indigenous people living in the remote and rough terrain of Arunachal Himayas. Since the local communities living in these areas have poor access to transport and communication and proper markets, they are mostly dependent on subsistence agriculture and forest resources for their dietary needs. Wild edible plants can provide them with nutritional diversity, can alleviate food shortages during pre-harvest seasons and have been reported to contain important nutrients not often available in commonly consumed foods (Robsinson 2014). Nutritionists have found that wild plants provide key micronutrients and key sources of protein that can be lacking in people's regular diets (Grivetti & Ogle 2000, Johns & Maundu 2006, Powell et al. 2013). In modern days, the nutritional importance of wild foods is becoming more prominent as rural and agricultural diets increasingly rely on imported processed foods that are high in fat and refined sugar and low in fiber and micronutrients (Damman et al. 2008, Kuhnlein et al. 2013). Therefore, use of wild foods plays a great role in maintaining food and nutritional security of the indigenous communities living in the Arunachal Himalayas.

Medicinal Plants

Both *Adi* and *Monpa* communities have rich traditional knowledge in medicinal plant use which is evident from the highest number of recorded species of medicinal plants in the present investigation. Among *Adi* tribe, 162 plant species were used in traditional medicinal system whereas among *Monpa* community 61 plant species were used to cure various minor and major ailments such as gastro-intestinal problems, fever, liver disorder, bone fracture,skin problems, headache etc. Medicine preparation from the plants are either administered orally or applied externally on the skin in case of cuts and wounds, inflammation, skin diseases and related problems. Most of medicinal plants were collected from wild by knowledgable tribal people who know the distribution of the species.

With advancement of modern medicine and health care facilities such traditional practices are getting diminished gradually. However, there is still existence of traditional healers in remote villages where there is limited access to modern health care systems. In present time when people are increasingly inclined towards nutraceutical and pharmaceutical products of plant origin, preservation and utilization of the knowledge of traditional healers looks promising for further building up scientific research in the field. Training and capacitiy building of traditional healers would be beneficial to educate them about sustainable harvesting techniques, post harvest practices and the importance of maintaining hygienic conditions for medicine preparation, storage and packaging herbal medicines.

A number of medicinal plant species used by both the communities such as *Aconitum ferox, Illicium grifthii, Paris polyphylla, Coptis teeta* and *Swertia chirayita* are highly valued plants that are recognised as potental pharmaceutical species. For instance, *Paris polyphylla* is a well-known folk medicinal herb which has been used for a long time as an important herbal drug for the treatment of inflammation, fractures, parotitis, hemostatis, snake bite and abscess in folk medicine (Wu et al. 2012). Many species have been reported to be widely used in Ayurveda, Unani, Siddha, Homeopathy, Folk and Sowa-Rigpa systems of Indian Medicine (Anonymous 2018).

In recent times, focus on plant research has increased all over the world and a large body of evidence has been collected to show immense potential of medicinal plants used in various traditional systems (Devi et al. 2008). Therefore, ethnomedicinal plant diversity of the Arunachal Himalayas offers a promising scope for development of medicinal plant industry which can be a source of livelihood for the local communities as well as revenue for the state.

Non-Edible Plant Resources

Many non-edible plant resources were utilized by *Adi* and *Monpa* community for different purposes like dye and colour, hunting and piscicide, paper, pulp and fibre. Both *Adi* and *Monpa* communities are expert craftsmon, highly skilled in making handloom products using locally available plant species for making paper and fibres along with traditional dyes and colours. These communities also practice hunting and fishing using traditional techniques. Hunting and fishing are important activity of tribal people which fulfill the requirement of protiens in their diet. These activities are also deeply connected to their socio-cultural and religious practices. They have very good knowledge of using different species of plants for catching fishes and wild animals. As many as 56 plant species were used by *Adi* and *Monpa* community for dye & colour, hunting & piscicide and paper, fibre. Highest number of species were reported to be used for hunting and piscicide (40 species) followed by dye & colour (13) and paper & fibre (5). Out of the 40 species used in hunting and piscicide, maximum number of species (30) was used as piscicides while the rest of the species were used in making traps and poison arrows for hunting animals and birds.

Dye and Colour Fixers

Dyes are colouring agents which have profound utilization in various fields like textiles, handlooms, cosmetics and food. (Siva 2007). The use of natural dyes have been recorded even before 2600 BC by China. Earlier 1800 herbal dyes were extensively used for colouring cloths and from them onwards the synthetic colours started substituting the herbal dyes (Gokhale et al. 2004). Unlike synthetic dyes, natural dyes obtained from plant are environment friendly and have limited or no side effects (Mahanta & Tiwari 2005).

During the last few decades, there has been increase in demand of herbal dyes and as such documentation of dye-yielding plants is very necessary. Dye and colour yielding plants distributed in forests of the state are used by *Adi* and *Monpa* communities. Various parts of the plants like fruit, bark, roots and leaves were used to get distinct colours. These plant species can be used for obtaining total organic colours and may provide additional income for local people. Nine plant species were used by *Monpa* community while 5 plant species were used by *Adi* community for obtaining different colours. Organic colours like dark red, dark brown, black, yellow, indigo, purple black, blue black, red, reddish yellow were obtained from different parts of various plants. Among the recorded species, *Melastoma malabathricum* and *Rubia cordifolia* were reported to be used by the Meitei community of Manipur for dyeing different handloom and handicraft items (Potsangbam et al. 2008, Sharma et al. 2005).

Hunting and Piscicide

Hunting and fishing are important activities of tribal people (Yumnam & Tripathi 2013). The tools and techniques used for hunting and fishing including poisoning and harpooning fishing nets, fishing rods and indigenous equipments (Rao 2003, Yumnam & Tripathi 2013). Use of modern fishing technology which includes use of motor boat and synthetic gears has drastic effect on aquatic ecosytems. Besides, the modern fishing gears are very costly and indigenous communities are often unable to meet the expenses. On the other hand, traditional fishing techniques are eco-friendly, have very less impact on aquatic ecosystem and are very cost effective too (Rathakrishnan et al. 2009). The materials required for traditional fishing can be obtained from nearby forest.

The tribal people of Arunachal Pradesh are very fond of fishing and hunting and have rich knowledge of different plant species used for preparing the fishing gears and poisoning. The most commonly used plant species in traditional fishing and hunting by the two communities include *Aconitum* spp., *Cassia nodosa, Castanopsis indica, Derris scandens, Euphorbia wallichii, Gymnocladus burmanicus, Lasianthus longicauda, Mesua assamica* and *Milletia panchycarpa*. Hunting methods of tribal people in Arunchal Pradesh is also witnessing considerable changes as guns are steadily replacing the age-old practice of using bow and arrows. The shift from traditional hunting and fishing methods to using modern tools has adverse effects on wildlife and aquatic ecosytems as well. Therefore, it has become necessity of time to make tribal communities more aware of importance of their rich traditional practices. This will ensure conservation of traditional knowledge of tribal people as well as conservation of forest and aquatic ecosystem in particular and the whole environment in general.

Paper and Fibre

The demand for paper is increasing exponentially with the increase in human population. As wood of different tree species are conventional raw material for paper making, thus increasing demand for paper will increase pressure on the forest throughout the world (Smook 1992). Recently, non-wood fibres were increasingly used for paper production especially in developing countries (FAO 1994). To increase the non-wood pulp production documentation of potential plant species for pulp production is required. Paper and fiber extraction from plant species is an ancient art and in fact first true paper was made from inner bark of mulberry and bamboo fibres in 105 A.D. in China (Atchison & Mcgovern 1993). The forests of Arunachal Pradesh are rich in diversity of such important plant species. Both *Adi* and *Monpa* community have rich traditional knowledge of paper and fiber extraction from different plant species. *Monpa* community had been using *Daphne papyracea* for making paper (Paul et al. 2006). *Adi* community use species of *Dendrocalamus hamiltonii*, *Grewia serrulata*, *Pandanus nepaulensis* and *Sterculia villosa* for obtaining fibers to be used in different handicrafts. Such plant species are potential resource for pulp and fibre production and may be used as alternatives to wood for paper production.

CONCLUSION

The present study highlighted the rich traditional knowledge of plant utilization by *Adi* and *Monpa* communities residing in different parts of Arunachal Himalayas. The diverse utilization pattern of plants in food, medicine, dyes, hunting, piscicide, fiber and paper for their sustenance and livelihood shows their dependency on plant resources of the area. The study also revealed the potential of these species for maintaining food and nutritional security, and livelihood generation of the local communities. With gradual urbanization of villages and out-migration of younger generations, there is a change in lifestyle, food habits and traditional practices of indigenous communities. Therefore, further scientific research on plant resource utilization pattern of these communities is required to explore the potential of these plant species for value addition, promotion, and sustainable utilization towards livelihood development as well as conservation of the prioritised plant species.

ACKNOWLEDGEMENTS

The authors are grateful to Director G.B. Pant National Institute of Himalayan Environment and Sustainable Development for providing necessary facilities for carrying out research work. Financial support from Department of Science & Technology (DST), Govt. of India under National Mission for Sustaining the Himalayan Ecosystem (NMSHE) Task Force-5 Project is gratefully acknowledged.

LITERATURES CITED

Ali N, Ghosh B. 2006. Ethnomedicinal plants in Arunachal Pradesh: Some tacit prospects. *ENVIS Bulletin: Himalayan Ecology* 14(2): 19-24.

Angami A, Gajurel PR, Rethy P, Singh B, Kalita SK. 2006. Status and potential of wild edible plants of Arunachal Pradesh. *Indian Journal of Traditional Knowledge* 5(4): 541-550.

Anonymous 2018. Web page http://www.medicinalplants.in/ accessed on 8th May, 2018.

Atchison JE, McGovern JN. 1993. *History of paper and the importance of agricultural residues fiber*. Atlanta, GA.

Baruah S, Borthakur SK, Gogoi P, Ahmed M. 2013. Ethnomedicinal plants used by *Adi-Minyong* tribe of Arunachal Pradesh, eastern Himalaya. *Indian Journal of Traditional Knowledge* 4(3): 278-282.

Bharali P, Sharma M, Sharma CL, Singh B. 2017. Ethnobotanical survey of spices and condiments used by some tribes of Arunachal Pradesh. *Journal of Medicinal Plants* 5(1): 101-109.

Boko N, Narsimhan D. 2014. Rapid survey of plants used by *Adi* tribe of Bosing-Banggo, East Siang District, Arunachal Pradesh, India. *Pleione* 8(2): 271:282

Chettri N, Sharma E, Shakya B, Thapa R, Bajracharya B, Uddin K, Choudhury D, Oli KP. 2010. Biodiversity in the Eastern Himalayas: Status, trends and vulnerability to climate change. Technical report 2. International Centre for Integrated Mountain Development (ICIMOD), Kathmandu, Nepal.

Census 2011. *Census of India 2011*. Provisional Population Totals. New Delhi: Government of India.

Das M, Anju J, Hirendra SN. 2013. Traditional medicines of herbal origin practice by the *Adi* tribe of East Sang district of Arunachal Pradesh, India. *Global Journal of Research on Medicinal Plants & Indigenous Medicine* 2(5): 298

Damman S, Eide WB, Kuhnlein HV. 2008. Indigenous Peoples' Nutrition Transition in a Right to Food Perspective. *Food Policy* 33: 135–155.

Devi A, Rakshit K, Sarania B. 2014. Ethnobotanical notes on Allium species of Arunachal Pradesh, India. *Indian Journal of Traditional Knowledge* 13(3): 606-612.

Devi KN, Sarma HN, Kumar S. 2008. Estimation of essential and trace elements in some medicinal plants by PIXE and PIGE techniques. *Nuclear Instrumentation Methods in Physics Research* 266: 1605-1610.

Doley B, Gajurel P R, Rethy, P. and Buragohain, R. (2014). Uses of trees as medicine by the ethnic communities of Arunachal Pradesh, India. *Journal of Medicinal Plants Research* 8(24): 857-863.

Dollo M, Singh KI, Saha D, Chaudhry S, Sundriyal RC. 2005. Ethnically Diverse Area in Arunachal Pradesh. In: Bhatt BP, Bujarbaruah KM, eds. *Agroforestry in North East India: Opportunities and Challenges* 55-70.

FAO 1994. *Forest products*. Yearbook. Rome, Italy.

FSI 2011. *State of Forest Report 2011*. Forest Survey of India.

GBPIHED. 2011. Contribution towards developing a roadmanp for biodiversity and climate change, Indian part of East Himalaya. In: *Climate summit for a living Himalaya* - Bhutan 2011.

Ghosh G, Ghosh DC, Melkania U, Majumdar U. 2014. Traditional medicinal plants used by the *Adi*, Idu and *Khamba* tribes of Dehang-Debang biosphere reserve in Arunachal Pradesh. *International Journal of Agriculture, Environment and Biotechnology* 7(1): 165.

Gokhale SB, Tatiya AU, Bakliwal SR, Fursule RA. 2004. Natural dye yielding plants in India. *Natural Product Radiance* 3(4): 228-234.

Grivetti LE, Ogle BM. 2000. Value of Traditional Foods in Meeting Macro- and Micronutrient Need: The Wild Plant Connection. *Nutrition Research Reviews* 13: 31–46.

Johns T, Maundu P. 2006. Forest Biodiversity, Nutrition and Population Health in Market-Oriented Food Systems. *Unasylva* 224(57): 34–40

Kagyung R, Gajurel PR, Rethy P, Singh B. 2010. Ethnomedicinal plants used for gastrointestinal diseases by *Adi* tribes of Dehang-Debang Biosphere Reserve in Arunachal Pradesh. *Indian Journal of Traditional Knowledge* 9(3): 496-501.

Kar A, Borthakur SK. 2008. Medicinal plants used against dysentery, diarrhoea and cholera by the tribes of erstwhile Kameng district of Arunachal Pradesh. *Natural Product Radiance* 7(2): 176-181.

Khongsai M, Saikia SP, Kayang H. 2011. Ethnomedicinal plants used by different tribes of Arunachal Pradesh. *Indian Journal of Traditional Knowledge* 10(3): 541-546.

Kuhnlein HV, Erasmus B, Spigelski D, Burlingame B. 2013. *Indigineous Peoples' Food Systems and Well-Being: Interventions and Policies for Healthy Communities*. FAO, Rome, Italy.

Kuhnlein HV, Turner NJ. 1991. Traditional plant foods of Canadian Indigenous peoples: Nutrition, botany and use. In: Katz SH, ed. *Food and Nutrition in History and Anthropology*. Vol. 8, Taylor & Francis, Amsterdam, Netherlands.

Kumar N, Kumar S, Singh B, Mishra BP, Singh B, Singh V. 2015. Traditional practices of utilization and conservation of non-wood forest products by *Adi* tribes of Arunachal Pradesh. *Journal of Applied and Natural Science* 7(1): 111-118.

Mahanta D, Tiwari SC. 2005. Natural dye-yielding plants and indigenous knowledge on dye preparation in Arunachal Pradesh, northeast India. *Current Science* 88(9): 1474-1480.

Mao AA, Hynniewta TM. 2000. Floristic diversity of North East India. *Journal of Assam Science Society* 41(4): 255-66.

Mao AA, Hynniewta TM, Sanjappa M. 2009. Plant wealth of Northeast India with reference to ethnobotany. *Indian Journal of Traditional Knowledge* 8(1): 96-103.

Namsa ND, Mandal M, Tangjang S, Mandal SC. 2011. Ethnobotany of the *Monpa* ethnic group at Arunachal Pradesh, India. *Journal of Ethnobiology and Ethnomedicine* 7(1): 31.

Paul A, Arunachalam AA, Khan ML, Arunachalam K. 2006. *Daphne papyracea* Wall. ex Steud.–A traditional source of paper making in Arunachal Pradesh. *Natural Product Radiance* 5(2): 133-138.

Paul A, Kalita J, Khan ML, Tripathi OP. 2013. *Illicium griffithii* Hook.f. & Thomson-A potential source of natural off-farm income to the rural people of Arunachal Himalaya, India. *Indian Journal of Natural Products and Resources*. 4(2): 131-137.

Potsangbam L, Ningombam S, Laitonjam WS. 2008. Natural dye yielding plants and indigenous knowledge of dyeing in Manipur, Northeast India. *Indian Journal of Traditional Knowledge* 7(1): 141-147.

Powell B, Mandu P, Kuhnlein HV, Johns T. 2013. Wild Foods from Farm and Forest in the East Usambara Mountains, Tanzania. *Ecology of Food and Nutrition* 52: 451–478.

Power EM. 2008. Conceptualizing Food Security for Aboriginal People in Canada. *Canadian Journal of Public Health* 99: 95–97.

Rao VM. 2003. *Tribal women of Arunachal Pradesh: Socio economic status.* Mittal Publication, New Delhi.

Rathakrishnan T, Ramasubramanian M, Anandaraja N, Suganthi N, Anitha S. 2009. Traditional fishing practices followed by fisher folks of Tamil Nadu. *Indian Journal of Tradtitonal Knowledge* 8(4): 543-547.

Rethy P, Singh B, Kagyung R, Gajurel PR. 2010. Ethnobotanical studies of Dehang–Debang Biosphere Reserve of Arunachal Pradesh with special reference to Memba tribe. *Indian Jornal of Traditional Knowledge* 9(1): 61-67.

Robsinson J. 2014. *Eating on the Wild Side: The Missing Link to Optimum Health.* Little, Brown and Company. UK.

Saha D, Sundriyal M, Sundriyal RC. 2014. Diversity of food composition and nutritive analysis of edible wild plants in a multi-ethnic tribal land, Northeast India: an important facet for food supply. *Indian Journal of Traditional Knowledge* 13(4): 698-705.

Samal PK, Palni LMS, Agrawal DK. 2003. Ecology, ecological poverty and sustainable development in central Himalayan region of India. *International Journal of Sustainable Development & World Ecology* 10: 157-168.

Sharma HM, Devi AR, Sharma BM. 2005. Vegetable dyes used by the Meitei community of Manipur. *Indian Journal of Traditional Knowledge* 4(1): 39-46.

Singh RK, Pretty J, Pilgrim S. 2010. Traditional knowledge and biocultural diversity: learning from tribal communities for sustainable development in northeast India. *Journal of Environmental Planning and Management* 53(4): 511-533.

Singh RK, Srivastava RC. 2010. Bioculturally important plant diversity of Arunachal Pradesh: Learning from *Adi* and *Monpa* communities about 'Future crops of India. *Indian Journal of Traditional Knowledge* 9(4): 754-759.

Siva R. 2007. Status of natural dyes and dye-yielding plants in India. *Current Science* 10: 916-25.

Smook GA. 1992. *Handbook for pulp and paper technologists,* 2nd ed. Vancouver, Canada: Angus Wilde Publications.

Srivastava RC, Singh RK, Mukherjee TK. 2010. Bioculturally important rare new plant species of *Heteropanax* Seems (Araliaceae) from Eastern Himalaya, Arunachal Pradesh. *Indian Journal of Traditional Knowledge* 9(2): 242-244.

Srivastava RC. 2009. Traditional knowledge of *Adi* tribe of Arunachal Pradesh on plants. *Indian Journal of Traditional Knowledge* 8(2): 146-153.

Sylvester O, García Segura A. 2016. Landscape Ethnoecology of Forest Food Harvesting in the Talamanca Bribri Indigenous Territory, Costa Rica. *Journal of Ethnobiology* 36(1): 215–233.

Tag H, Das AK. 2004. Ethnobotanical notes on the Hill Miri tribe of Arunachal Pradesh. *Indian Journal of Traditional Knowledge* 3(1): 80-85.

Tag H, Tsering J, Gogoi BJ, Kalita B, Kalita P, Veer V. 2015. Ethnobotanical uses of poisonous plants in Arunachal Pradesh. *Journal of Bioresources* 2(2): 1-5.

Tag H, Tsering J, Hui PK, Gogoi BJ, Veer V. 2014. Nutritional potential and traditional uses of high altitude wild edible plants in Eastern Himalayas, India. *Waset* 8(2): 395-400.

Wu S, Gao W, Qiu F, Man S, Fu S, Liu C. 2012. Simultaneous quantification of *Paris Polyphyllin* D and Paris H, two potential antitumor active components in *Paris polyphylla* by liquid chromatography-tandem mass spectrometry and the application to pharmacokinetics in rats. *Journal of Chromatography B* 905: 54-60.

Yumnam JY, Bhuyan SI, Khan ML, Tripathi OP. 2011. Agro-diversity of East Siang-Arunachal Pradesh, Eastern Himalaya. *Asian Journal of Agricultural Sciences* 3(4): 317-326.

Yumnam JY, Tripathi OP. 2013. Ethnobotany: Plants use in fishing and hunting by *Adi* tribe of Arunachal Pradesh. *Indian Journal of Traditional Knowledge* 12(1): 157-161.

16

Withania somnifera (L.) Dunal, a Potential Indian Medicinal Plant with Multifarious Biological Activities

Gourav Paudwal and Prem N. Gupta

PK-PD-Toxicology and Formulation Division
CSIR-Indian Institute of Integrative Medicine, Jammu-180001
Jammu & Kashmir, INDIA
Email: gouravpaudwal@gmail.com; pngupta@iiim.ac.in

ABSTRACT

Herbal systems of medicine have been acceded by various drug regulatory authorities across the globe. In modern days, people prefer herbal medicines for the cure of various gentle to extreme and chronic disease, and another option is of allopathic system of medicine. In the present communication, the herbal formulations utilize extracted constituents of the various parts of *Withania somnifera* (L.) Dunal, a member of the famly Solanaceae, and commonly known as Ashwagandha, Indian ginseng, Asgand or winter cherry. This is a principal medicinal plant that has been used in various Ayurvedic formulations from more than 3000 years. Wide distribuion of this species have been reported from the various Indian states like are Gujarat, Madhya Pradesh, Maharashtra, Rajasthan, Uttar Pradesh, Punjab and mountainous regions of Himachal Pradesh and Jammu. This plant is commonly used, either alone or in combination, with other herbs for the cure of various conditions e.g stress and inflammation. The major phytochemicals of this species are alkaloids and steroidal lactones, and due to presence of various bioactive compounds, this plant species possess various multitudinous biological activities such as adaptogenic, antibiotic, aboritifacient, aphrodisiac, astringent, anti-inflammatory, diuretic, sedative, deobstruent, anti-stress, anti-oxidant, anti-carcinogenic, anti-aging, cardio protective, hypothyroid activity, immunomodulatory. Keeping these aspects into consideration, this herb has enormous potential for the management of a number of diseases. In the market, *W. somnifera* is available in various forms including decoctions, infusions, ointments, powder, tablet, capsule and syrup. This communication provides an account of various chemical constituents and their associated various biological functions.

Keywords: Medicinal Plant, *Withania somnifera*, Herbal Botanical, Phytochemical constituents, Biological Activities, Drug Discovery.

INTRODUCTION

Medicinal plants exhibit a food wellspring of normally available drugs in Indian literature having therapeutic significance. These therapeutic plants are in charge of delivering different phytochemicals which giving insusceptibility to battle in opposing plant illness. The various chemical substences investigation of therapeutic plants is vital to recognize the nearness of dynamic components present in them. Various or different phytochemicals likewise named as auxiliary substances include alkaloids, flavonoids, steroidal lactones, saponins, tannins, terpenoids, and so forth which show medicinal and biological activities, for example, anti-bacterial activity, anti inflammatory effects. Because of rising wellbeing records of natural medicines in contrast with synthetic medications, restorative plants-based medications have regained prominence for treating even the unending human ailment. The most seasoned composed confirmation of home grown drug utilization has been found on a Sumerian earth piece (around 5000 years of age) from Nagpur, India, including 12 home grown arrangements by utilizing around 250 distinct plants (Kelly et al. 2009, Petrovska 2012). Because of new disclosures expressing present day remedial uses of bioactive mixes from therapeutic plants, analysts are giving tremendous consideration regarding the therapeutic plant science. Out of all over 18,000 blooming plants on our blue planet nearly about 50% are of therapeutic significance and are the good wellspring of phytochemicals (Handa et al. 2006).

Withania somnifera (L.) Dunal is a little hard bush ordinarily known as "Indian Ginseng" (Fig. 1). It is familiar as 'Ashwagandha' in sanskrit and in Urdu as 'Asgand' (Dhuley 1998 and Ziauddin et al. 1996). It has a place with the family Solanaceae and achieves a stature of 0.5– 2 m. It is broadly disseminated in the dry regions of tropical and subtropical zones starting from the Canary Islands, South Africa, Middle East, Sri Lanka, India and to China. It is grown in not and dry climate of Europe and has turned into a unwanted plant in a few sections of Australia (Hepper, 1991 and Purdic et al. 1982). In India it is developed as a restorative product (Van et al. 2004). The entire plant or its distinctive parts are generally utilized as a part of Ayurvedic and as well as Unani frameworks of prescription (native frameworks of pharmaceutical in India) for its medicinal value properties and has been utilized since relic. The plant is said as an official medication in Indian Pharmacopeia-1985 (Uddin et al. 2012, Singh et al. 2011). In Ayurvedic system it is a conspicuous natural Rasayana and recognized as "Sattvic Kapha Rasayana". It is a natural or on the other hand metallic preparation that is utilized for biological properties, for example, in tonic, opiate, diuretic, hostile to helminthic, astringent, stimulant, anti-stress, anti-carbuncle, anti-ulcer, debility from maturity, ailment, vitiated states of vata, leucoderma, obstruction, sleep deprivation, mental meltdown, goitre, leucorrhoea, bubbles, pimples, worms, heaps, and oligospermia adaptogenic and anti-inflammation (Changhadi1938, Sharma 1999, Bhandari 1970, Basu 1935, Mishra 2004, Sharma et al. 1985). Furthermore, it is endorsed for snake bite and antidote for scorpionftin. Unani system of medicine. (Machiah et al. 2006, Machiah and Gowd 2006, Aggarwal et al. 1999). In the Unani arrangement of solution, the plant has been specified in an former confirmation "Kitab-ul-Hashaish" by Dioscorides in 78 AD. Asgand has different remedial employments. It has been prescribed for the cure of different sicknesses, incorporate joint pain, backpain carbuncle, spermatorrhoea, asthma, leucoderma, sexual debility, tension, despondency, scabies, ulcers, and leucorrhoea (Uddin et al. 2012, Ali et al. 1997, Ghani 1920, Kabiruddin 1955, Nadkarni 1982, Tiwari et al. 2014). Inferable from its articulated pressure busting characteristics. Its species name 'somnifera', has been drived from a Latin word signifying 'rest inducer' (Ven et al. 2010, Seenivasagam et al. 2011).

Pharmacologic impacts and traditional employments of W. somnifera are much the same as that of Korean Ginseng tea, which outfits an unobtrusive clarification for calling W. somnifera as Indian Ginseng (Grandhi et al. 1994). In Unani and Ayurvedic frameworks of prescription, for the most part foundations of W. somnifera are utilized as restorative purposes. Its pharmacologic activity loses following 2 years, subsequently newly dried roots are favored for better results (Uddin et al. 2012, Singh et al. 2011). The leaves of the plant are mainly used in fever and difficult swelling. The blooms are styptc, depurative, diuretic, and aphrodisiac. The seeds are against helminthic; used for

Fig. 1: Habit of *Withania somnifera*.

spots of the cornea, increment sperm check, and additionally testicular development. The natural products have been customarily utilized as a topical treatment for tumours and tubercular organs, carbuncles, and skin ulcers (Singh et al. 2011, Kaur et al. 2004, Chopra et al. 2004).

Phytochemical constituents

The study of various chemicals substances of plant have demonstrated the availability of various synthetic components in different parts of *W. somnifera*. In excess of 12 alkaloids, 40 withanolides and a few sitoindosides have been secluded and announced from the *Withania somnifera* (Mirjalili 2009). *Withania somnifera* is massively rich store of an extensive variety of optional metabolites. The number of inhabitants in naturally dynamic mixes incorporates around 40 withanolides. Out of different withanolides the few withanolides which are present in abundance are withanolide A, withanolide D, withanone and withafewin A. and all of these have different activities (Fig. 2, Table 1)

Table 1: Various bioactive withanolides

Bioactive compounds	Bioactivities	References
Withaferin A, Withanolide D, Withalongolide A, Withalongolide A-4,19,27-b) triacetate, Withalongolide B-4,19-diacetate	Anti-cancer	Mandal et al. (2008) Subramanian et al. (2014) Mondal et al. (2010, 2012a, b)
3β-hydroxy-2,3-dihydrowithanolide F	Hepatoprotective	Budhiraja et al. (1986)
4 β-hydroxywithanolide E, Withaferin A	Hypoglycemic	Takimoto et al. (2014) Gorelick et al. (2015)
3β-hydroxy-2,3-dihydrowithanolide F, Withaferin A	Cardiovascular	Budhiraja et al. (1984) Ravindran et al. (2015)
17β-hydroxywithanolide K, Withaferin A, 14,15β-epoxywithanolide I, Withanolide F, Withanolide D	Anti-microbial	Choudhary et al. (2010) Subramanian and Sethi (1969), Choudhary et al. (2010)
Withanolide A	Alzheimer's disease	Sehgal et al. (2012)
Withaferin A	Anti-inflammatory	Noh et al. (2016)
Withaferin A	Antioxidant	Bhattacharya et al. (1997)
Withanolide A	Neuroprotection	Baitharu et al. (2014)
3β-hydroxy-2,3-dihydrowithanolide F	CNS related	Budhiraja et al. (1984)
Withaferin A, Withanolide E, 5,20 α(R)-dihydroxy-6α,7 α-epoxy-1-oxo-(5α)-with a-2,24dienolide	Immunosuppressive	Shohat et al. (1978) Shohat et al. (1978) Bähr and Hänsel (1982)

All withanolides are having C-28 steroidal lactone triterpenoids with backbone ergostane structure. The structure have six-membered lactone ring and oxidation takes place at C-22 and C-26 carbon atom. The nomenleture for most of structures were done as 22-hydroxy ergostane-26 oic, 26, 22 lactones. Most of the witharnolides contain six or five membered lactone ring as basic moiety with side thain at C-8 or C-9 carbon atom. Various other compounds are also found in withania somnifera products e.g. 27-hydroxy withanolide A, iso-withanone, 6,7-epoxy-1,3,5-trihydroxy-witha-24-enolide (Chatteriee et al. 2010, Dhar et al. 2015, Lal et al. 2006).

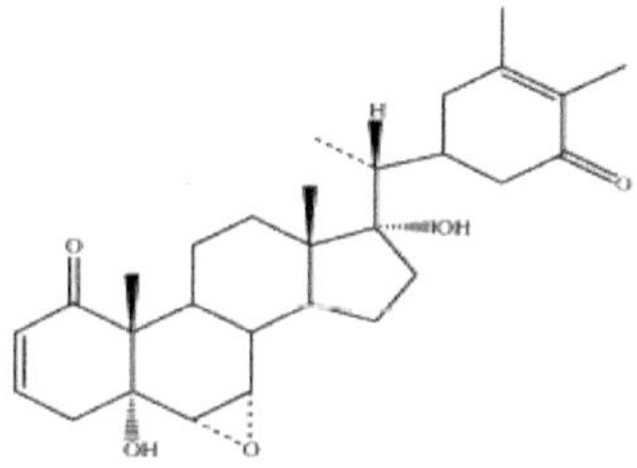

Withanone

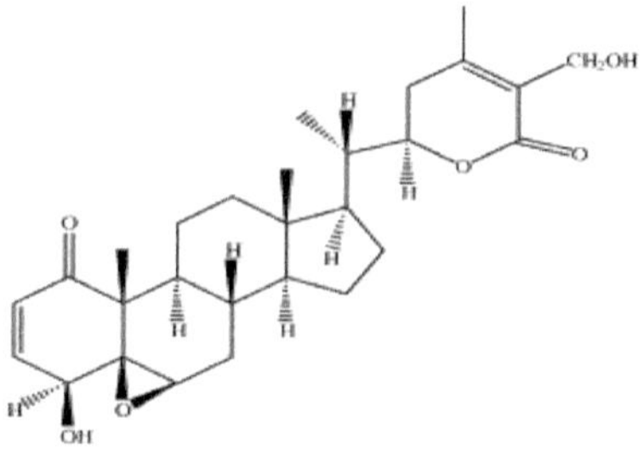

Withaferom A

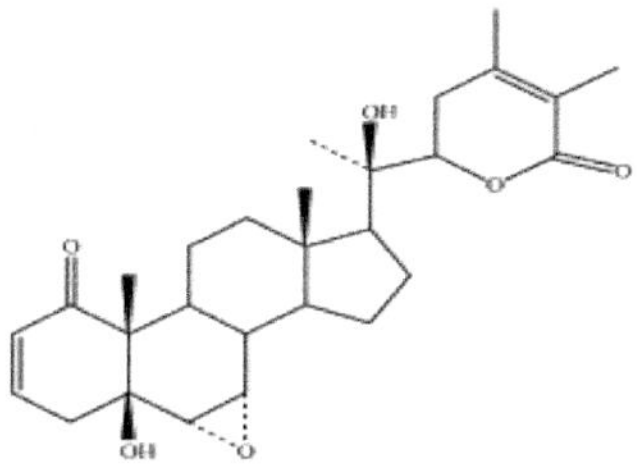

Withanolide A

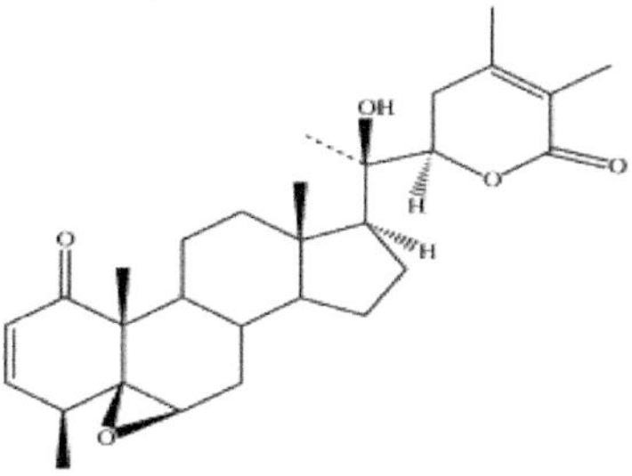

Withanolide D

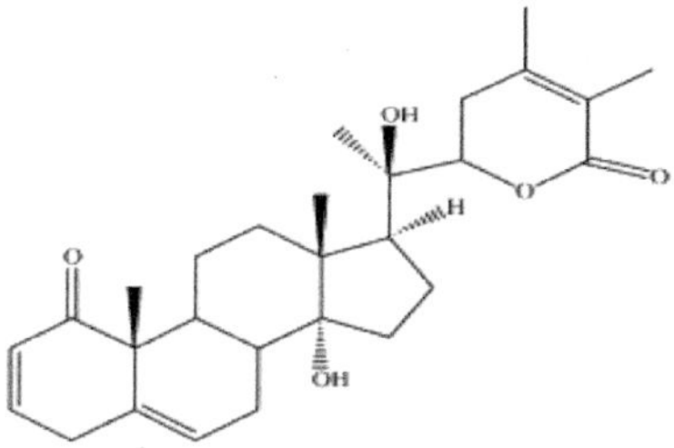

Withanolide G

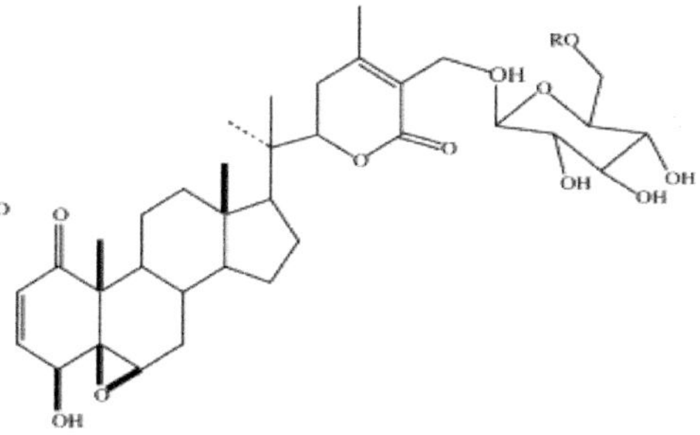

Sitoindoside IX (withaferin-A-C27-O-b-D-glucoside
Sitoindoside X (6'-O-palmitoyl-A-C27-O-b-D-glucoside

Fig. 2: Major active constituents of *Withania somnifera* (Mishra et al. 2000).

Different alkaloids, a few steroids, ashwagandhine, cuscohygrine, anahygrine, salts, flavonoids, nitrogen containing mixes, steroids, steroidal lactones are also observed to be available in various products by Withania somnifera. The various parts of plant contains differents others phytochemicals like tnopane alualoids, withanosides Steroidal saponins, lignanamides, flavonoids, coayulins (C,H,L) and withanolide glycosides. (Mirjalili et al. 2009, Chatterjee et al. 2010, Singh et al. 2015, Jadaun et al. 2016, Kushwaha et al. 2013a and b). The steroidal lactones on withandides are responsible for the gritty smell of ashwagandha. The various parts of plants contain different amount of withaferin and its amount in various parts are in decreasing order as shoot >leaves>internodes>flowers>seeds (Praveen and Murthy 2010).

Biological activity of *Withania somnifera*

Withania sominfera is the major medicinal plant and centre of fascination for research due to its biological activities including antimicrobial, activity in mitigating against joint, anti- diabetic, cardiodefensive, neuroprotective, anti-leishmaniasis (Fig. 3) (Tripathi and Verma 2014,Dhar et al.; 2015). It has anticancer activity and act by various mechanisms like cell cycle arrest apoptosis outophagy adn somoothering different oncogenic pathways. Therefore, it has great potential for the various tumors like breast cervical, lung, colon, prostate and medullary thyroid malignancy as an anticancer agent (Palliyaguru et al. 2016).

The leaves of Withania somnifera have been appeared to possess antimicrobial activity against both gram-positive (methicillin-safe Staphylococcus aureus and Enterococcus spp) (Bisht and Rawat 2014) and gram-negative microbes (Salmonella typhi, Escherichia coli, Pseudomonas aeruginosa, Proteus mirabilis, Klebsiella pneumonia and Citrobacter freundii) (Singh et al. 2006, Alam et al. 2012).

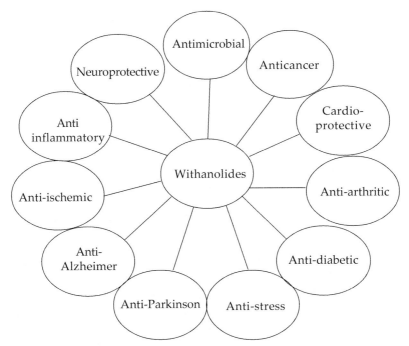

Fig. 3: Various biological activities of Withanolides

The oral bacterial contamination caused by Streptococcus mutans, and Streptococcus sobrinus can be stopped by the use of leaves of withania somnifera (Pandit et al., 2013). The roots have neuroprotective activity, immunomodulatory action and activity against tumor. It is observed to be successful against treatment of Alzheimer's and Parkinson's sickness. The underlying foundations of Withania somnifera have appeared to hinder a few markers of irritation, for example, cytokines (Interleukin-6, Tumor corruption factor-α), responsive oxygen species (ROS) and Nitric oxide (NO). The detail of plant part of Withania somnifera and its pharmacological activity is summarized in Table 2.

Table 2: Pharmacological actions of bioactive compounds of *Withania somnifera*

Part of plant	Extract type	Action	References
Leaf	Methanolic	Antimicrobial	Widodo et al (2008), Corelick et al. (2015)
		Anticancer against cancer cells (TIG1, U2OS, and HT1080) by activating p53, apoptosis pathway & cell cycle arrest	Konar et al. (2011) Kuboyama et al. (2014) Bisht and Rawat (2014) Alfaifi et al. (2016)
		Anti-diabetic (Alloxan-induced diabetes mellitus in rats) Hypoglycaemic (Increased uptake of glucose in myotubes and adipocytes) Neuroprotection through activation of neuronal proteins, oxidative stress and DNA damage Anticancer Neuroblastoma Antimicrobial (against methicillin resistant Stephylococcus aureus and Enterococcus sps.) Antiproliferative against MCF-7,HCT116 and HepH2 cell lines	
	Ethanolic	Antimicrobial activity	Dhiman et al. (2016)
	Hydroalcoholic	Anti-cancer (Breast Cancer)	Nema et al. (2013)
	Aqueous	Anti- cancer	Wadhwa et al. (2013)
Root	Chloroform	Anti-cancer (liver, breast, colon) Prostate cancer	Siddique et al. (2014)
	Alcoholic	Anti-tumor	Khazal et al. (2013)

Contd.

	Ethanolic	Anti-cancer (Cervical) Anti-diabetic (Stabilized blood glucose levels) Anti-diabetic (Non-insulin-dependent Diabetes mellitus in rats)	Jha et al. (2014)
	Aqueous	Anti-stress	Jain and Saxena (2009)
	Ethanolic	Antidiabetic Alloxan induced diabetes illustrates, normalised the urine sugar, blood glucose, glucose-6-phosphate and tissue glycogen levels	Jha et al. (2014)
Stem	Methanolic	Hepatoprotective role in acetoaminophen-intoxicated rats	Devkar et al.; (2016)
Fruit		Alzhemizer's disease	Jayaparkasham et al. (2010)

Antimicrobial activity

Predictable with the folkloric utilization of *Withania somnifera* against diseases, methanolic leaf concentrate of *W. somnifera* has demonstrated anti-bacterial activity against Gram-positive clinical segregates of ·methicillin-safe *Staphylococcus aureus* and *Enterococcus* spp (Bisht et al. 2014). Moreover, W. somnifera exhibited intense anti-microbial activity against Gram-Negative species, for example, *Escherichia coli, Salmonella typhi, Proteus mirabilis, Citrobacter freundii, Pseudomonas aeruginosa,* and *Klebsiella pneumonia* (Singh et al. 2011, Alam et al. 2012). The strength of *W. somnifera* has been seen to fluctuate in various examinations against various life forms. The anti-microbial activity was credited to cytotoxicity, quality quieting, and immunopotentiation (Mwitari et al. 2013). W. somnifera has strong activity against *Salmonella typhimurium* (Alam et al. 2012). Moreover, improved survival rate, lessened bacterial heap of different fundamental organs of mice with salmonellosis has been accounted for after organization of W. somnifera (Owais et al. 2005). W. somnifera removes synergized increment the counter bacterial impact of Tibrim (rifampicin and isoniazid) against *Salmonella typhimurium* and *E. coli* (Arora et al., 2004). Further, W. somnifera restrained biofilm development of oral microorganisms, *Streptococcus mutans* and *Streptococcus sobrinus* at even sub-minimum inhibitory concentration (MIC) levels. There was additionally a dosage related increment in multiplying times of *Streptococcus mutans* and *Streptococcus sobrinus* up to 258 and 400%, respectively (Pandit et al. 2013).

Withanolides instigates apoptosis-like passing in Leishmania donovani in vitro by inciting DNA scratches, cell cycle capture at the sub G_0/G_1 stage, and externalization of phosphatidyl serine in a measurements and time dependent way through an expansion in receptive oxygen species (ROS) and a diminishing in mitochondrial potential (Chandrasekaran et al. 2013) by hindering the protein kinase-C flagging pathway (Grover et al. 2012). Vitally, against leishmanial action was displayed by *W. somnifera* against free-living promastigotes and intracellular amastigotes of Leishmania major with a greatest inhibitory impact of 50 % (El et al. 2009). *W. somnifera* synergized assurance in cisplatin-treated L. donovani-tainted mice when contrasted with just *W. somnifera*-treated L. donovani-contaminated mice by upgrading the level of T cells and normal executioner cell-related marker, NK1 (Sachdeva et al. 2013). Moreover, flavonoids extracted from *W. somnifera* have been accounted for to be successful against Candida albicans with MIC of 0.039 and least fungicidal focus (MFC) of 0.039. In addition, it was illustrated that *A. flavus* and *Aspergillus niger* were impervious to *W. somnifera* (Singh et al. 2011).

Anti-inflammatory Activity

Withania somnifera has shown stamped calming impacts in different ailment models. Its root extract shown mitigating and muco-helpful action by settling rot, edema, neutrophil invasion in trinitro-benzyl-sulfonic (TNBS) prompted provocative gut model (Pawar et al. 2011). Powder of its foundations was found to have a strong inhibitory impact on proteinuria, nephritis, and other markers, for example, cytokines including interleukin (IL)- 6 and tumor nacrosis factor (TNF), nitric oxide (NO), and ROS in a mouse model of lupus (Minhas et al. 2011; Minhas et al. 2012). In human umbilical vein endothelial cells (HUVECs), withaferin-A was appeared to hinder phorbol-12-myristate-13-acetic acid derivation (PMA)-actuated shedding of endothelial cell protein-C- receptor (EPCR) by hindering TNF and interleukin (IL)-1b.

Additionally, in mouse withaferin-A attenuated cecal ligation and cut (CLP)-incited EPCR shedding by decreasing the articulation and movement of tumor necrosis factor (TNF). Also, withaferin-A weakened PMA-invigorated phosphorylation of p38, extracellular managed kinases (ERK) and c-Jun N-terminal kinase (Ku et al. 2014). Withaferin-A secures vascular boundary trustworthiness in HUVECs and in mice, actuated by high versatility gathering box-1-protein (HMGB1) by hindering hyper permeability, articulation of cell bond particles (CAM), attachment also, relocation of leukocytes, creation of interleukin-6 and TNF (Lee et al., 2012).

Anti-arthritic activity

Adequate point of reference recommends a noteworthy part for W. somnifera in joint pain. Aqueous concentrates of *W. somnifera* root powder demonstrated a transient chondroprotective impact on harmed human osteoarthritic ligament by huge and reproducible hindrance of the gelatinase action of collagenase type-2 protein in vitro (Sumantran et al. 2007) and by fundamentally diminished NO discharge (Sumantran et al. 2008). Furthermore, the unrefined ethanol concentrate of *W. somnifera* fundamentally stifled lipopolysaccharide (LPS)- initiated generation of genius incendiary cytokines TNF, IL-1b, and IL-12 in fringe blood mononuclear cells from typical people and synovial liquid mononuclear cells from rheumatoid joint pain patients potentially by repressing atomic translocation of the interpretation factors NFj-b and activator protein-1 (AP-1) and phosphorylation of Ij-b. Furthermore, it normalised LPS-induced NO creation in RAW 264.7 cells (Singh et al. 2007).

In a rodent model of adjuvant-instigated joint inflammation, *W. somnifera* root powder weakened ligament debasement as surveyed by estimation of bone collagen (Rasool et al. 2007). Fluid concentrate of W. somnifera root forestalled expanded joint list, autoantibodies, and C-responsive protein-P in collagen-initiated joint rats (Khan et al. 2015). Organization of W. somnifera root powder to the ligament rats altogether diminished the seriousness of joint pain by viably enhancing the practical recuperation of engine movement and radiological score (Gupta et al. 2014). Moreover, W.somnifera as a constituent in a polyherbal plan (BV- 9238) decreased TNF-an and NO creation, with no cytotoxic impacts in Freund's total adjuvant-prompted joint pain in rats and a mouse macrophage cell line (Dey et al. 2014). More critically, W. somnifera helps collagen adjustment by repressing collagenase (Ganesan et al. 2011). A few investigations have detailed conflictual reports with respect to Withaferin-A. In rabbit articular chondrocytes, Withaferin- A actuated loss of sort collagen articulation and fiery reactions intervened up-direction of cyclooxygenase-2 (COX-2) articulation through enactment of microRNA-25 (Kim et al. 2014, Yu et al. 2013). In addition, checked intensification in the generation of intracellular ROS joined by apoptosis and expanded p53 articulation were observed, and these impacts were subject to PI3 K/AKT and JNK pathways (Yu et al. 2013, Yu et al. 2014).

Anticancer activity

Different kinds of diseases or malignancy related changes in cell lines have been observed by *W. somnifera* or its synthetic constituents. Molecular docking investigation showed the utilization of withaferin-A and withanone for drug development (Vaishnavi et al. 2012). Leaf concentrate of *W. somnifera* and its segments reduce malignancy cells by no less than five unique pathways—p53 flagging, granulocyte–macrophage state animating element (GM-CFS) flagging, demise receptor flagging, apoptosis flagging and by G2-M DNA harm direction pathway (Widodo et al. 2008).

Withaferin-A showed effect against malignancy by inciting ROS-mediated apoptosis in melanoma cells by slamming Bcl-2/Bax and Bcl-2/Bim proportions. This apoptotic course utilized the mitochondrial pathway and was related with Bcl-2 down-direction, translocation of Bax to the mitochondrial film, arrival of cytochromec into the cytosol, repeal of transmembrane potential, enactment of caspases-9 and 3. The withanolide-incited early ROS loss and mitochondrial film potential unsettling influences taken after by the arrival of cytochrome c, translocation of Bax to mitochondria, and apoptosis-instigating factor to cell cores. These occasions paralleled initiation of caspases-9 and 3, Poly-(ADP-Ribose) Polymerase (PARP) fracture of DNA (Mayola et al. 2011). Withaferin-A likewise prompted the overexpression of TNFR. These examinations propose that withaferin-A executes harmful cells by apoptosis that can be reliant or potentially free of mitochondrial components (Malik et al. 2007). In a kidney malignancy cell line, Withaferin-An actuated dose dependent apoptotic cell demise and PARP cleavage through down-control of the STAT-3 pathway (Choi et al. 2011, Um et al. 2012).

Cardio-protective activity

Withania somnifera has cardio-protective activity (Das et al. 1964). *W. somnifera* exhibited cardiotropic and cardioprotective properties in experimental models (Ojha et al. 2009, Prince et al. 2008). Polyherbal formulation which had W. somnifera as a part demonstrated cardioprotection in animal models (Thirunavukkarasu et al. 2006, Mohan et al. 2006) by actuating nuclear factor-erythroid-2-related interpretation factor (Nrf)- 2, fortifying stage II detoxification chemicals, repealing apoptosis in a Nrf-2-subordinate way (Reuland et al. 2013).

Anti-diabetic activity

Different polyherbal formulation (Dianix, Trasina) of Indian Systems of Medicine demonstrated strong anti diabetic activity in human subjects (Gautam et al. 2013, Mutalik et al. 2005, Bhattacharya et al. 1997). In patients, W. somnifera root powder balanced out blood glucose that was practically identical to that of an oral hypoglycemic medication daonil, at the point when treated orally for 30 days (Andallu et al. 2000). Also, W. somnifera treatment essentially enhanced insulin sensitivity index and hindered the ascent in homeostasis display appraisal of insulin opposition in non-insulin-subordinate diabetes mellitus in rats (Anwer et al. 2008). In concurrence with these contemplates, W. somnifera leaf and root extract progressed glucose take-up in skeletal myotubes and adipocytes in a measurement subordinate way, with the leaf extract illustrating more articulated impacts than the root extract (Gorelick et al. 2015).

Anti-stress activity

Withania somnifera brought about better pressure resilience in animals (Bhattacharya et al. 1995, Kaur et al. 2001, Singh et al. 2001). The aqueos conent of *W. somnifera* roots mitigated incessant pressure and initiated decrease of T cell populace and up-regulated Th1 cytokines in mice (Khan et al. 2006). In a clinical report for the wellbeing and adequacy of a high concentration full-range concentrate of W. somnifera roots in human subjects, serum cortisol levels were reduced, without causing any significant symptoms (Chandrashekhar et al. 2012). Moreover, EuMil, a poly natural formulation, uniquely improved cerebral monoamine (nor-adrenaline, dopamine, and 5-hydro-xytryptamine) levels actuated by ceaseless electroshock stress (Bhattacharya et al. 2002).

Neuro-protective activity

Numerous investigations have recorded the neuroprotective impacts of *W. somnifera* (Ven et al. 2010, Durg et al. 2015, Wollen 2010, Singh et al. 2008, Kuboyama et al. 2014). The leaf concentrate and its part withanone secure scopolamine-instigated dangerous changes in both neuronal and glial cells. Scopolamine-initiated inactivation of neuronal cell markers, for example, NF-H, MAP-2, PSD-95, GAP-43, and glial cell marker glial fibrillary acidic protein (GFAP) and with DNA harm and oxidative pressure markers was particularly lessened by *W. somnifera* (Konar et al. 2011). *W. somnifera* extract lessened lead-actuated lethality in glial cells by adjusting the outflow of GFAP and warmth stun protein (HSP70), mortalin, and neural cell attachment atom (NCAM) (Kumar et al. 2014). Also, extract of *W. somnifera* prevent streptozotocin-actuated oxidative harm in treated mice by alleviating oxidative pressure (Parihar et al. 2004).

Anti-Parkinson activity

Point of reference exists in literature for a noteworthy part for *W. somnifera* in Parkinson's sickness. *W. somnifera* has been appeared to constrict Parkinson indications and pathology in a 6-hydroxydopamine (6-OHDA) rodent model for the ailment. The investigation showed the reclamation of the substance of striatal dopamine and its metabolites doubtlessly through its articulated antioxidant activity as confirm by the lessening of LPO, decreased glutathione (GSH) content, and exercises of glutathione-S-transferase (GST), glutathione reductase (GR), GPX, SOD, and CAT. Change of striatal catecholamine content because of *W. somnifera* may have turned around the functional impairments like locomotor action and solid coordination and medication actuated rotational conduct. This investigation likewise showed up-direction of dopaminergic D2 receptor populaces in striatum, which goes about as a compensatory mechanism after enlistment of Parkinsonism to get each accessible dopamine molecule. Moreover, *W. somnifera* has prompted an increment in the quantity of surviving dopaminergic neurons as evaluated by tyrosine hydroxylase marking (Ahmad et al. 2005). *W. somnifera* root extract re-established antioxidant status, lessened oxidant stress, and subsequently standardized catecholamine content in mid mind of 1-methyl-4-phenyl-1,2,3,6-tetrahydropyridine (MPTP)-inebriated parkinsonian mice. These biochemical changes went with the improvement in utilitarian action of the model (Sankar et al. 2007, Rajasankar et al. 2009, Rajasankar et al. 2009).

Anti-Alzheimer activity

Literature proposes a noticeable part of *W. somnifera* in major improvement against Alzheimer's disease. Standardized fluid concentrate of *W. somnifera* enhanced intellectual and psychomotor execution in healthy human volunteers (Pingali et al. 2014). *W. somnifera* root extract turned around the behavioural changes, neurotic pieces of information and A β-clearance in Alzheimer's ailment models by up-controlling lipoprotein receptor-related protein in liver (Sehgal et al. 2012). Simulation studies have indicated that withanamides-A and -C exceptionally tie to the dynamic theme of (Aβ) and recommend that withanamides have the capacity to keep the fibril development and in this way ensure cells from Aβ toxicity (Jayaparkasam et al. 2010). Besides, docking simulation studies have anticipated restraint of human acetyl cholinesterase by withanolide-A for Alzheimer's treatment (Grover et al. 2012).

Withania somnifera bears a useful impact on intellectual shortfall by improving oxidative harm incited by streptozotocin in a model of intellectual impairment (Ahmed et al. 2013). *W. somnifera* re-established cell morphology in Aβ-treated SK-N-MC cell line by upgrading cell suitability and the peroxisome proliferator-enacted receptor-γ (PPAR-γ) levels (Kurapati et al. 2013). Further, it prompted restraint of acetylcholinesterase action (Kurapati et al. 2014).

Anti-ischemic and anti-hypoxic activity

Withania somnifera constricted middle cerebral supply induced improvement of the oxidative stress marker malondialdehyde, diminishment in sore zone, and rebuilding of neurological deficiencies (Chaudhary et al. 2003). *W. somnifera* granted practical rebuilding and lessening of infarct volume in mice subjected to permanent distal middle cerebral artery occlusion (pMCAO). It prompted recovery of hemeoxygenase-1 (HO-) articulation and subsided the upregulation of the proapoptotic protein PARP-1 by means of the PARP-1 apoptosis-inciting factor (AIF) pathway that was adjusted by pMCAO in mouse cortex. This phenomena prompted blockage of the apoptotic cascade by preventing nuclear translocation of AIF. Moreover, semaphorin-3A articulation was expanded by pMCAO and it starts inhibitory signs that frustrate repair. *W. somnifera* altogether lessened the outflow of Semaphorin-3A and accordingly started repair systems (Raghavan et al. 2014 and Raghavan et al. 2015).

Toxicological studies of *Withania somnifera*

Withania somnifera has been utilized for different pharmacological activities for long time for all age gatherings and both genders and not withstanding amid pregnancy without toxic effects (Sharma et al. 1985). Prabu et al. (2013)have assessed hydro-alcoholic root concentrate of *W. somnifera* against acute and sub-acute oral toxicities in Wistar rats and discovered it non-toxic even at 2000 mg/kg body weight. The extract was given at 2000 mg/kg and watched for 14 days for acute toxicity at 500, 1000 and 2000 mg/kg and watched for 28 days for sub-acute toxicity, anyway there was no critical change in body weight, organ weight and hemato-biochemical parameters.

Pharmacokinetic Studies of *Withania somnifera*

Various examinations have been done in various biological models to illustrate the pharmacokinetics of W. somnifera. Two noteworthy constituents withaferin-A and withanolide-A have been seen after oral administration of standardised *W. somnifera* aqueous extract in mice. A dose of 1000 mg/kg extract (proportional to 0.4585 mg/kg of withaferin-A and 0.4785 mg/kg of withanolide-A) showed nearly comparable pharmacokinetic designs for both of these withanolides with mean plasma concentrations (C_{max}) of 16.69 ± 4.02 and 26.59 ± 4.47 ng/ml for withaferin-A and withanolide-A, with T_{max} (time taken to achieve C_{max}) of 10 and 20 min, respectively, demonstrating their fast absorption. The area under the plasma concentration– time curve from 0 to 4 h (AUC_{0-4h}) was 1572.27±57.80 and 2458.47 ± 212.72 min ng/ml, respectively. The $T_{1/2}$ of 59.92 ± 15.90 min and 45.22 ± 9.95 min and distribution of 274.10±9.10 and 191.10±16.74 ml/min/kg for withaferin-A and withanolide-A, respectively, were observed.

Generally relative oral bioavailability has been observed to be 1.44 times more noteworthy for withaferin-A than withanolide-A (Patil et al. 2013). Furthermore, Thaiparambil et al. have demonstrated that withaferin-A achieves crest fixations up to 2 µM in plasma with a half-life of 1.36 h following a single 4 mg/kg measurements in 7–8-week-old female Balb/C mice, whereas the clearance from plasma was quick (0.151 ng/ml/ min).

Marketed Formulation of Ashwagandha

Ashwagandha is pushed as a defensive medication against atherosclerosis, hypertension, and coronary heart ailments (Mehra et al. (2009). It diminishes the affectability of the heart to adrenergic incitement and consequently ensures the heart against thoughtful upheavals. Moharana et al. (2008) announced that the roots and leaves of Ashwagandha are customarily utilized as a part of the type of powder, decoction, or oil. These have been utilized as a part of people drug against general incapacity, hypertension, aggravation and wounds.

Thirunavukkarasu et al. (2006) discovered Ashwagandha to have vitality boosting properties and suggested its use as a dietary supplement for cardioprotection. The impact of Ashwagandha pull was assessed for lipid peroxidation in stress. The herb was found to have exceptionally great cell reinforcement action, which may halfway clarify the antistress, clog encouraging, mitigating and antiageing effects of this herb (Moharana 2008, Table 3).

Table 3: List of marketed formulation of *Withania somnifera*

Dosage form	Marketed Product	Manufacturer	Usefulness
Capsule	Stresswin	Baidynath Ayurved Bhawan	Combating exertion, reduction in anxiety, strain, and stress, improvement ofstamina, relief from disturbed sleep, mental alertness
Capsule	Stresscom	Dabur India Ltd.	Relieves anxiety neurosis, physical and mental stress, and relieves general debility and depression
Tablet	Brento	Zandu Pharmaceutical Works Ltd.	Nerve tonic
Capsule	Ashvagandha	Morpheme Remedies	Combating stress
Powder	Dabur Ashwagandha Churna	Dabur	Combating stress
Capsule	Ashwagandha	Ayurceutics	Stress reliever
Oil	Himalaya Massageoil	The Himalaya Drug Co.	Stress relief and relief from insomnia

Recent patents on Ashwagandha

Ghosal et al. (2004) US patent distribution 2004/0166184, uncovers preparation of *Withania somnifera* from roots and leaves containing 8-25% withanolide glycosides and sitoindosides, around 25-75% oligosaccharides and under 2% of free withanferin A (aglycone). Ghosal et al US patent No. 6,153,198 uncovers a high purity Withania Somnifera extract from root of Ashwagandha containing at slightest 3% withanolide glycoside and sitoindoside, no less than 3% oligosaccarides and under 0.5% of cytotoxic withferin A (aglycones) as a high, purity stable powder for delivering an upgraded cognizance impact for the utilization and to improve the learning ability. U.S. Pat. No. 7,108,870, Sangwan et al. reported an improved procedure of analytical and quantitative isolation of withaferin A from *Withnia somnifera*. Patent US 20140087009, McNeary et al. reveals an arrangement including mixes of β-glucan and Withania Somnifera for improving the invulnerable action of certain target cytokines and diminishing cortisol or corticosterone. Sumithradevi et al reported a straightforward strategy to purify withanolide A from the roots of Withania somnifera. Uddin et at revealed the phytochemical and biological activity of *Withania somnifera*, WO2012160569, Jayesh Panalal et al uncovers a process for extraction of Withanoside IV and V from Ashwagandha roots and its composition.

REFERENCES CITED

Agarwal R, Diwanay S, Patki P, Patwardhan B. 1999. Studies on immunomodulatory activity of *Withania somnifera* (Ashwagandha) extracts in experimental immune inflammation. *Journal of Ethnopharmacology* 67(1): 27–35.

Ahmad M, Saleem S, Ahmad AS, Ansari MA, Yousuf S 2005 Neuroprotective effects of *Withania somnifera* on 6-hydroxydopamine induced Parkinsonism in rats. *Human and Experimental Toxicology* 24: 137–147.

Ahmed ME, Javed H, Khan MM, Vaibhav K, Ahmad A .2013. Attenuation of oxidative damage-associated cognitive decline by *Withania somnifera* in rat model of streptozotocin induced cognitive impairment. *Protoplasma* 250: 1067–1078.

Alam N, Hossain M, Mottalib MA, Sulaiman SA, Gan SH 2012. Methanolic extracts of Withania somnifera leaves, fruits and roots possess antioxidant properties and antibacterial activities. *BMC Complementary and Alternative Medicine* 12: 175.

Alfaifi MY, Saleh KA, El-Boushnak MA, Elbehairi SEI, Alshehri MA, Shati AA 2016 Antiproliferative activity of the Methanolic extract of *Withania somnifera* leaves from Faifa Mountains, Southwest Saudi Arabia, against several human cancer cell lines. *Asian Pacific Journal of Cancer Prevention* 17(5): 2723–2726.

Ali M, Shuaib M, Ansari SH 1997 Withanolides from the stem bark of *Withania somnifera*. *Phytochemistry* 44:1163–1168.

Andallu B, Radhika B. 2000. Hypoglycemic, diuretic and hypocholesterolemic effect of winter cherry (*Withania somnifera*, Dunal) root. *Indian Journal of Experimental Biology* 38: 607–609.

Anwer T, Sharma M, Pillai KK, Iqbal M. 2008. Effect of *Withania somnifera* on insulin sensitivity in non-insulin-dependent diabetes mellitus rats. *Basic Clinical Pharmacology and Toxicology research* 102: 498–503.

Arora S, Dhillon S, Rani G, Nagpal A. 2004. The in vitro antibacterial/synergistic activities of *Withania somnifera* extracts. *Fitoterapia* 75: 385–388.

Bähr V, Hänsel R 1982 Immunomodulating Properties of 5, 20α (R)-Dihydroxy-6α, 7 α-epoxy-1- oxo-(5α)-witha-2, 24-dienolide and Solasodine. *Planta Medica* 44(01): 32–33.

Basu KA 1935 *Withania somnifera*, Indian medicinal plants, 2nd edn. IIIrd Lalit Mohan Basu, Allahabad, pp 1774–1776.

Bhandari CR. 1970. Ashwagandha (*Withania somnifera*) Vanaushadhi Chandroday (*An Encyclopedia of Indian Herbs*), CS Series, Varanasi Vidyavilas Press, Varanasi, India, pp 96–97

Bhattacharya SK, Bhattacharya A, Sairam K, Ghosal S. 2000. Anxiolytic-antidepressant activity of *Withania somnifera* glycowithanolides: an experimental study. *Phytomedicine* 7: 463–469

Bhattacharya SK, Kumar A, Ghosal S. 1995. Effects of glycowithanolides from *Withania somnifera* on an animal model of Alzheimer's disease and perturbed central cholinergic markers of cognition in rats. *Phytotherapy Research* 9:110–113.

Bhattacharya SK, Satyan KS. 1997. Experimental methods for evaluation of psychotropic agents in rodents: I-Anti-anxiety agents. Indian *Journal Experimental Biology* 35: 565–575

Bhattacharya SK, Satyan KS, Chakrabarti A. 1997. Effect of Trasina, an Ayurvedic herbal formulation, on pancreatic islet superoxide dismutase activity in hyperglycaemic rats. *Indian Journal Experimental Biology* 35: 297–299.

Bisht P, Rawat V 2014 Antibacterial activity of *Withania somnifera* against Gram-positive isolates from pus samples. *Journal of Ayurveda and integrative medicine* 35: 330

Budhiraja RD, Garg KN, Sudhir S, Arora B. 1986. Protective effect of 3-ss-hydroxy-2, 3-dihydrowithanolide F against CCl4-induced hepatotoxicity. *Planta Medica* 52(01): 28–29

Budhiraja RD, Sudhir S, Garg KN. 1984. Antiinflammatory activity of 3 β -Hydroxy-2, 3-dihydrowithanolide F. *Planta Medica* 50(02): 134–136.

Chandrasekaran S, Dayakar A, Veronica J, Sundar S, Maurya R. 2013. An in vitro study of apoptotic like death in Leishmania donovani promastigotes by withanolides. *Parasitology International* 62: 253–261

Chandrasekhar K, Kapoor J, Anishetty S. 2012. A prospective, randomized double-blind, placebo-controlled study of safety and efficacy of a high-concentration full-spectrum extract of Ashwagandha root in reducing stress and anxiety in adults. Indian *Journal of Psychology and Medicine* 34: 255–262.

Changhadi GS. 1938 .Ashwagandharishta—*Rastantra* Sar Evam Sidhyaprayog Sangrah. Krishna-Gopal Ayurveda Bhawan (Dharmarth Trust), Nagpur, pp 743–774.

Chatterjee S, Srivastava S, Khalid A, Singh N, Sangwan RS, Sidhu OP, Roy R, Khetrapal CL, Tuli R. 2010. Comprehensive metabolic fingerprinting of *Withania somnifera* leaf and roots. *Phytochemistry* 71:1085–1094

Chaudhary G, Sharma U, Jagannathan NR, Gupta YK. 2003. Evaluation of *Withania somnifera* in a middle cerebral artery occlusion model of stroke in rats. *Clinical and Experimental Pharmacology and Physiology* 30: 399–404

Chopra A, Lavin P, Patwardhan B, Chitre D. 2004. A 32-week randomized, placebo-controlled clinical evaluation of RA-11, an Ayurvedic drug, on osteoarthritis of the knees. *JCR J Clinical Rheumatology* 10: 236–245

Choudhary MI, Hussain S, Yousuf S, Dar A, Mudassar A-u-R. 2010. Chlorinated and diepoxy withanolides from *Withania somnifera* and their cytotoxic effects against human lung cancer cell line. *Phytochemistry* 71:2205–2209

Das PK, Malhotra CL, Prasad K. 1964. Cardiotonic activity of Ashwagandhine and Ashwagandhinine, two alkaloids from Withania ashwagandha, Kaul *International Pharmacodynamics Therapy* 150:356–362

Devkar ST, Kandhare AD, Zanwar AA, Jagtap SD, Katyare SS, Bodhankar SL, Hegde MV. 2016. Hepatoprotective effect of withanolide-rich fraction in acetaminophen-intoxicated rat: decisive role of TNF-α, IL-1β , COX-II and iNOS. *Pharmaceutical Biology* 54(11): 2394–2403

Dey D, Chaskar S, Athavale N, Chitre D. 2014. Inhibition of LPS-induced TNF-alpha and no production in mouse macrophage and inflammatory response in rat animal models by a novel Ayurvedic formulation, BV-9238. *Phytotherapy Research* 28: 1479–1485

Dhar N, Rana S, Bhat WW, Razdan S, Pandith SA, Khan S DUHP, Dhar RS, Vaishnavi S, Vishwakumar R, Latto Sk. 2013. Dynamics of withanolide biosynthesis in relation to temporal expression pattern of metabolic genes in *Withania somnifera* (L.) Dunal: a comparative study in two morpho-chemovariants. *Molecular Biology Reports* 40(12): 7007–7016

Dhar N, Rana S, Razdan S, Bhat WW, Hussain A, Dhar RS et al. 2014. Cloning and functional characterization of three branch point oxidosqualene cyclases from *Withania somnifera* (L.) dunal. *Journal of Biological Chemistry* 289(24): 17249–17267

Dhar N, Razdan S, Rana S, Bhat WW, Vishwakarma R, Lattoo SK. 2015. A decade of molecular understanding of withanolide biosynthesis and in vitro studies in *Withania somnifera* (L) Dunal: prospects and perspectives for pathway engineering. *Frontiers in Plant Science* 6: 1031 doi : 10.3389/fpls.2015.01031.

Dhiman R, Aggarwal N, Aneja KR, Kaur M. 2016. In vitro antimicrobial activity of spices and medicinal herbs against selected microbes associated with juices. *International Journal of Microbiology* 2016: 9015802 doi : 10.1155/2016/9015802

Dhuley JN. 1998. Effect of ashwagandha on lipid peroxidationin stress-induced animals. *Journal of Ethnopharmacology* 60: 173–178

Durg S, Dhadde SB, Vandal R, Shivakumar BS, Charan CS. 2015. *Withania somnifera* (Ashwagandha) in neurobehavioural disorders induced by brain oxidative stress in rodents: a systematic review and meta-analysis. *Journal of Pharmacy and Pharmacology* 67(7) : 879-899.

El-On J, Ozer L, Gopas J, Sneir R, Enav H et al. 2009. Antileishmanial activity in Israeli plants. *Annals of Tropical Medicine and Parasitology* 103:297–306

Ganesan K, Sehgal PK, Mandal AB, Sayeed S. 2011. Protective effect of *Withania somnifera* and Cardiospermum halicacabum extracts against collagenolytic degradation of collagen. *Applied Biochemistry and Biotechnology* 165: 1075–1091

Gauttam VK, Kalia AN. 2013. Development of polyherbal antidiabetic formulation encapsulated in the phospholipids vesicle system. *Journal of Advanced Pharmaceutical Technology and Research* 4:108–117

Ghani N. 1920. *Khazain-ul-Adviyah*, vol I. Munshi Nawal Kishore, Lucknow, pp 230–231

Ghosal. S. 2000. *Withania somnifera* composition. US Patent No. 6 : 153-158.

Gorelick J, Rosenberg R, Smotrich A, Hanuš L, Bernstein N. 2015. Hypoglycemic activity of withanolides and elicitated *Withania somnifera*. *Phytochemistry* 116: 283–289

Grandhi A, Mujumdar AM, Patwardhan B. 1994. A comparative pharmacological investigation of Ashwagandha and Ginseng. *Journal of Ethnopharmacology* 44: 131–135

Grover A, Katiyar SP, Jeyakanthan J, Dubey VK, Sundar D. 2012. Blocking Protein kinase C signaling pathway: mechanistic insights into the anti-leishmanial activity of prospective from *Withania somnifera*. 13 : doi : 10.1186/1471-2146-13-57-520.

Grover A, Shandilya A, Agrawal V, Bisaria VS, Sundar D. 2012. Computational evidence to inhibition of human acetyl cholinesterase by withanolide a for Alzheimer treatment. *Journal of Biomolecular Structure and Dynamics* 29: 651–662

Gupta A, Singh S. 2014. Evaluation of anti-inflammatory effect of *Withania somnifera* root on collagen-induced arthritis in rats. *Pharmaceutical Biology* 52: 308–320

Hahm ER, Singh SV. 2013. Autophagy fails to alter Withaferin A-mediated lethality in human breast cancer cells. *Current Cancer Drug Targets* 13:640–650

Handa SS, Rakesh DD, Vasisht K 2006 Compendium of medicinal and aromatic plants ASIA.

Hepper FN. 1991. Old World Withania (Solanaceae): a taxonomicreview and key to the species. In: Hawkes JG, Lester RN,Nee M, Estrada N (eds) Solanaceae III: taxonomy, chemistry,evolution. Royal Botanic Gardens Kew and Linnean Society of London, Londonherbal drugs from Withania somnifera. *BMC Genomics* 13:S20

Heyninck K, Lahtela-Kakkonen M, Van der Veken P, Haegeman G, Vanden Berghe W. 2014. Withaferin A inhibits NFkappaB activation by targeting cysteine 179 in IKKbeta. *Biochemical Pharmacology* 91:501–509

Jadaun JS, Sangwan NS, Narnoliya LK, Tripathi S, Sangwan RS. 2016. Withania coagulans tryptophan decarboxylase gene cloning heterologous expression and catalytic characteristics of the recombinant enzyme. *Protoplasma* 254:181–192

Jain SM, Saxena P. 2009. Protocols for in vitro cultures and secondary metabolite analysis of aromatic and medicinal plants. In: Methods in Molecular Biology, vol 1391, 2nd edn. Humana Press, New York, pp 303–315

Jayaprakasam B, Padmanabhan K, Nair MG. 2010. Withanamides in Withania somnifera fruit protect PC-12 cells from β -amyloid responsible for Alzheimer's disease. *Phytotherapy Research* 24(6):859–863

Jha AK, Nikbakht M, Capalash N, Kaur J. 2014. Demethylation of RARβ 2 gene promoter by Withania somnifera in HeLa cell line. *European Journal of Medicinal Plants* 4(5): 503–510

Kabiruddin M. 1955. Makhzan-ul-Mufradat. Nadeem University Printers, Lahore, pp 75–76

Kaur K, Rani G, Widodo N, Nagpal A, Taira K et al. 2004. Evaluation of the anti-proliferative and anti-oxidative activities of leaf extract from in vivo and in vitro raised Ashwagandha. *Food and Chemical Toxicology* 42: 2015–2020

Kaur P, Mathur S, Sharma M, Tiwari M, Srivastava KK et al. 2001. A biologically active constituent of *Withania somnifera* (ashwagandha) with antistress activity. *Indian Journal of Clinical Biochemistry* 16: 195–198

Kelly-Pieper K, Patil SP, Busse P et al. 2009. Safety and tolerability of an Antiasthma HerbalFormula (ASHMI™) in adult subjects with asthma: a randomized, double-blinded, placebocontrolled,dose-escalation Phase I study. *Journal of Alternateive and Complementary Medicine* 15(7): 735–743

Khan B, Ahmad SF, Bani S, Kaul A, Suri KA et al. 2006. Augmentation and proliferation of T lymphocytes and Th-1 cytokines by *Withania somnifera* in stressed mice. *International Immunopharmacolology* 6: 1394–1403

Khan MA, Subramaneyaan M, Arora VK, Banerjee BD, Ahmed RS. 2015. Effect of *Withania somnifera* (Ashwagandha) root extract on amelioration of oxidative stress and autoantibodies production in collagen-induced arthritic rats. *Journal of Complementary and Integrative Medicine* 12: 117–125

Khazal KF, Samuel T, Hill DL, Grubbs CJ. 2013. Effect of an extract of Withania somnifera root on estrogen receptor-positive mammary carcinomas. *Anticancer Research* 33(4): 1519–1523

Kim SH, Singh SV. 2014. Mammary cancer chemoprevention by withaferin A is accompanied by in vivo suppression of selfrenewal of cancer stem cells. *Cancer Prevention Research* 7: 738–747

Konar A, Shah N, Singh R, Saxena N, Kaul SC, Wadhwa R, Thakur MK. 2011. Protective role of Ashwagandha leaf extract and its component withanone on scopolamine-induced changes in the brain and brain-derived cells. *PLoS One* 6(11):e27265 doi : 10.1371/journal.pone.0027265.

Ku SK, Han MS, Bae JS. 2014. Withaferin A is an inhibitor of endothelial protein C receptor shedding in vitro and in vivo. *Food and Chemical Toxicology* 68: 23–29. : e77624 doi : 10.1371/journal .pone.0077264.e collection 2013.

Kuboyama T, Tohda C, Komatsu K. 2014 .Effects of Ashwagandha (roots of Withania somnifera) on neurodegenerative diseases. *Biological and Pharmaceutical Bulletin* 37(6): 892–897

Kuboyama T, Tohda C, Zhao J, Nakamura N, Hattori M et al. 2002. Axon- or dendrite-predominant outgrowth induced by constituents from Ashwagandha. *NeuroReport* 13:1715–1720

Kumar P, Singh R, Nazmi A, Lakhanpal D, Kataria H. 2014. Glioprotective effects of Ashwagandha leaf extract against lead induced toxicity. *Biomedical Research International* 2014: 182029 doi : 10.1155/2014/182029.

Kurapati KR, Atluri VS, Samikkannu T, Nair MP. 2013. Ashwagandha (Withania somnifera) reverses beta-amyloid1-42 induced toxicity in human neuronal cells: implications in HIV associated neurocognitive disorders (HAND). *PLoS One* 8: (10)

Kurapati KR, Samikkannu T, Atluri VS, Kaftanovskaya E, Yndart A l. 2014. beta-Amyloid1-42, HIV-1Ba-L (clade B) infection and drugs of abuse induced degeneration in human neuronal cells and protective effects of ashwagandha (*Withania somnifera*) and its constituent Withanolide A. *PLoS One* 9: e112818

Kushwaha AK, Sangwan NS, Tripathi S, Sangwan RS. 2013a. Molecular cloning and catalytic characterization of a recombinant tropine biosynthetic tropinone reductase from Withania coagulans leaf. *Gene* 516: 238–247

Kushwaha AK, Sangwan NS, Trivedi PK, Negi AS, Misra L, Sangwan RS. 2013b. Tropine forming tropinone reductase gene from *Withania somnifera* (Ashwagandha): biochemical characteristics of the recombinant enzyme and novel physiological overtones of tissue-wide gene expression patterns. *PLoS One* 8:e74777 doi : 10.1371/journal.pone.0074777

Kushwaha S, Roy S, Maity R, Mallick A, Soni VK, Singh PK et al. 2013c. Chemotypical variations in *Withania somnifera* lead to differentially modulated immune response in BALB/c mice. *Vaccine* 30(6): 1083–1093

Kushwaha S, Soni VK, Singh PK, Bano N, Kumar A, Sangwan RS, Misra-Bhattacharya S. 2012b.*Withania somnifera* chemotypes NMITLI 101R, NMITLI 118R, NMITLI 128R and Withaferin A protect Mastomys coucha from Brugia malayi infection. *Parasite Immunology* 34: 199–209

Lal P, Misra L, Sangwan RS, Tuli R. 2006. New withanolides from fresh berries of Withania somnifera. *Zeitschrift für Naturforschung* B 61:1143–1147

Lee W, Kim TH, Ku SK, Min KJ, Lee HS et al. 2012. Barrier protective effects of withaferin A in HMGB1-induced inflammatory responses in both cellular and animal models. *Toxicology and Applied Pharmacology* 262:91–98

Machiah DK, Girish K, Gowda TV. 2006. A glycoprotein from a folk medicinal plant, Withania somnifera, inhibits hyaluronidase activity of snake venoms. *Comparative Biochemistry and Physiology* 143:158–161

Machiah DK, Gowda TV. 2006. Purification of a post-synaptic neurotoxic phospholipase A 2 from Naja naja venom and its inhibition by a glycoprotein from Withania somnifera. *Biochimie* 88:701–710

Malik F, Kumar A, Bhushan S, Khan S, Bhatia A et al. 2007. Reactive oxygen species generation and mitochondrial dysfunction in the apoptotic cell death of human myeloid leukemia HL-60 cells by a dietary compound withaferin A with concomitant protection by N-acetyl cysteine. *Apoptosis* 12:2115–2133

Mandal C, Dutta A, Mallick A, Chandra S, Misra L, Sangwan RS, Mandal C. 2008. Withaferin A induces apoptosis by activating p38 mitogen-activated protein kinase signaling cascade in leukemic cells of lymphoid and myeloid origin through mitochondrial death cascade. *Apoptosis* 13: 1450–1464

Mayola E, Gallerne C, Esposti DD, Martel C, Pervaiz S et al. 2011. Withaferin A induces apoptosis in human melanoma cells through generation of reactive oxygen species and downregulation of Bcl-2. *Apoptosis* 16:1014–1027

McNeary P S. 2013. Composition of Beta- glucan and Ashwagandha. US Patent No. 8,597,697 B2

Mehra, R., Prasad, M., Lavekar, G.S. 2009. An approach ofAshwagandha + Guggulu in atheromatous CHD associated with obesity. *Journal of Research In Ayurveda* 30 (2): 121–125.

Minhas U, Minz R, Bhatnagar A. 2011. Prophylactic effect of *Withania somnifera* on inflammation in a non-autoimmuneprone murine model of lupus. *Journal of Drug Discovery and Therapeutics* 5:195–201

Minhas U, Minz R, Das P, Bhatnagar A. 2012 .Therapeutic effect of *Withania somnifera* on pristane-induced model of SLE. *Inflammopharmacology* 20: 195–205

Mirjalili MH, Moyano E, Bonfill M, Cusido RM, Palazon J. 2009. Steroidal lactones from *Withania somnifera*, an ancient plant for novel medicine. *Molecules* 14:2373–2393.

Mishra LC, Singh BB, Dagenais S. Scientific basis for the therapeutic use of *Withania somnifera* (ashwagandha) a review.2000 Alternate Medicine Review; 5: 334–46.

Mishra B. 2004. Ashwagandha—Bhavprakash Nigantu (*Indian Materia Medica*). Varanasi, Chaukhambha Bharti Academy,pp 393–394

Mohan IK, Kumar KV, Naidu MU, Khan M, Sundaram C. 2006. Protective effect of CardiPro against doxorubicin-induced cardiotoxicity in mice. *Phytomedicine* 13: 222–229

Mohan R, Hammers HJ, Bargagna-Mohan P, Zhan XH, Herbstritt CJ. 2004 .Withaferin A is a potent inhibitor of angiogenesis. *Angiogenesis* 7:115–122

Moharana, D. 2008. Shatavari, Jastimadhu and Aswagandha the Ayurvedic Therapy. Orissa Rev., pp. 72–77.

Mondal S, Bhattacharya K, Mallick A, Sangwan R, Mandal C. 2012a. Bak compensated for Bax in p53-null cells to release cytochrome c for the initiation in Withania somnifera. *PLoS One* 11:e0149691 doi : 10.1371/journal.pone.0034277

Mondal S, Mandal C, Sangwan R, Chandra S, Mandal C. 2010. Withanolide D induces apoptosis in leukemia by targeting the activation of neutral sphingomyelinase-ceramide cascade mediated by synergistic activation of c-Jun N-terminal kinase and p38 mitogen-activated protein kinase. *Molecular Cancer* 9:239 doi : 10.1186/1476-4598-9-239

Mondal S, Roy S, Maity R, Mallick A, Sangwan R, Misra-Bhattacharya S, Mandal C. 2012b. Withanolide D, carrying the baton of Indian rasayana herb as a lead candidate of antileukemic agent in modern medicine. In: Biochem roles Eukar cell surface Macromolecular research Springer, New York, pp 295–312

Mutalik S, Chetana M, Sulochana B, Devi PU, Udupa N. 2005. Effect of Dianex, a herbal formulation on experimentally induced diabetes mellitus. *Phytotherapy Research* 19:409–415

Mwitari PG, Ayeka PA, Ondicho J, Matu EN, Bii CC. 2013. Antimicrobial activity and probable mechanisms of action of medicinal plants of Kenya: Withania somnifera, Warbugia ugandensis, Prunus africana and Plectrunthus barbatus. *PLoS One* 8(6):e65619 doi : 10.1371/journal.pone.00655619

Nadkarni KM. 1982. *Indian Materia Medica*, 3rd edn, vol I. Popular Prakashan Pvt Ltd, Bombay, pp 1292–1294

Nagalingam A, Kuppusamy P, Singh SV, Sharma D, Saxena NK. 2014. Mechanistic elucidation of the antitumor properties of withaferin a in breast cancer. *Cancer Research* 74:2617–2629

Nema R, Khare S, Jain P, Pradhan A. 2013. Anticancer activity of Withania somnifera (leaves) flavonoids compound. *International Journal of Pharmaceutical Sciences Review and Research* 19(1):103–106

Noh EJ, Kang MJ, Jeong YJ, Lee JY, Park JH, Choi HJ. 2016. Withaferin A inhibits inflammatory responses induced by Fusobacterium nucleatum and Aggregatibacter actinomycetemcomitans in macrophages. *Molecular Medicine Reports* 14(1):983–988

Ojha SK, Arya DS. 2009. *Withania somnifera* Dunal (Ashwagandha): a promising remedy for cardiovascular diseases. *World Journal of Medical Sciences* 4:156–158

Owais M, Sharad K, Shehbaz A, Saleemuddin M. 2005. Antibacterial efficacy of *Withania somnifera* (ashwagandha) an indigenous medicinal plant against experimental murine salmonellosis. *Phytomedicine* 12: 229–235

Palliyaguru DL, Singh SV, Kensler TW. 2016. *Withania somnifera*: from prevention to treatment of cancer. Molecular Nutrition and Food Res 60(6): 1342–1353

Pandit S, Chang K-W, Jeon J-G. 2013. Effects of *Withania somnifera* on the growth and virulence properties of *Streptococcus mutans* and *Streptococcus sobrinus* at sub-MIC levels. *Anaerobe* 19: 1–8

Parihar MS, Hemnani T. 2004. Alzheimer's disease pathogenesis and therapeutic interventions. *Journal of Clinical Neuroscience* 11:456–467

Patil D, Gautam M, Mishra S, Karupothula S, Gairola S. 2013. Determination of withaferin A and withanolide A in miceplasma using high-performance liquid chromatography-tandem mass spectrometry: application to pharmacokinetics after oral administration of *Withania somnifera* aqueous extract. *Journal Pharmaceutical Biomedical Analysis* 80:203–212

Pawar P, Gilda S, Sharma S, Jagtap S, Paradkar A. 2011. Rectal gel application of *Withania somnifera* root extract expounds anti-inflammatory and muco-restorative activity in TNBS-induced inflammatory bowel disease. *BMC Complementary Alternative Medicine* 11:34 doi : 10.1186/1472-6882-11-34

Petrovska BB. 2012. Historical review of medicinal plants? Usage. *Pharmacognosy Reviews* 6(11):1-5

Pingali U, Pilli R, Fatima N. 2014. Effect of standardized aqueous extract of *Withania somnifera* on tests of cognitive and psychomotor performance in healthy human participants. *Pharmacognosy Research* 6:12–18

Prabu PC, Panchapakesan S, Raj CD. 2013. Acute and subacute oral toxicity assessment of the hydroalcoholic extract of *Withania somnifera* roots in Wistar rats. *Phytotherapy Research* 27: 1169–1178

Prakash J, Chouhan S, Yadav SK, Westfall S, Rai SN e 2014 *Withania somnifera* alleviates parkinsonian phenotypes by inhibiting apoptotic pathways in dopaminergic neurons. *Neurochemical Research* 39: 2527–2536

Prakash J, Yadav SK, Chouhan S, Singh SP. 2013. Neuroprotective role of Withania somnifera root extract in manebparaquat induced mouse model of parkinsonism. *Neurochemical Research* 38:972–980

Praveen N, Murthy HN. 2010. Production of withanolide-A from adventitious root cultures of *Withania somnifera*. *Acta Physiologiae Plantarum* 32: 1017–1022

Praveen N, Murthy HN. 2012. Synthesis of withanolide a depends on carbon source and medium pH in hairy root cultures of Withania somnifera. *Industrial Crops and Products* 35:241–243 Med 67:432–436

Prince PSM, Suman S, Devika PT, Vaithianathan M. 2008. Cardioprotective effect of 'Marutham'a polyherbal formulation on isoproterenol induced myocardial infarction in Wistar rats. *Fitoterapia* 79:433–438

Purdie RW, Symon DE, Haegi L. 1982. Solanaceae. *Flora Aust* 29:184

Raghavan A, Shah ZA 2014 Withania somnifera improvesischemic stroke outcomes by attenuating PARP1-AIF-mediated caspase-independent Apoptosis. *Molecular Neurobiology*. doi:10.1007/ s12035-014-8907-2

Raghavan A, Shah ZA. 2015 .*Withania somnifera*: a pre-clinical study on neuroregenerative therapy for stroke. *Neural Regeneration Research* 10:183–185

RajaSankar S, Manivasagam T, Sankar V, Prakash S, Muthusamy R. 2009. *Withania somnifera* root extract improves catecholamines and physiological abnormalities seen in a Parkinson's disease model mouse. *Journal of Ethnopharmacology* 125:369–373

Rajasankar S, Manivasagam T, Surendran S. 2009. Ashwagandha leaf extract: a potential agent in treating oxidative damage and physiological abnormalities seen in a mouse model of Parkinson's disease. *Neuroscience Letters* 454:11–15

Ramanathan M, Balaji B, Justin A. 2011. Behavioural and neurochemical evaluation of Perment an herbal formulation in chronic unpredictable mild stress induced depressive model. *Indian Journal of Experimental biology* 49:269–275

Rasool M, Varalakshmi P. 2007. Protective effect of *Withania somnifera* root powder in relation to lipid peroxidation, antioxidant status, glycoproteins and bone collagen on adjuvantinduced arthritis in rats. *Fundamental and Clinical Pharmacology* 21:157–164

Reuland DJ, Khademi S, Castle CJ, Irwin DC, McCord JM .2013. Upregulation of phase II enzymes through phytochemical activation of Nrf2 protects cardiomyocytes against oxidant stress. *Free Radial Biology and Medicine* 56: 102–111

Sangwan R, Chaurasiya N A, Mishra L N , Lal P, Uniyal G C, Sangwan N S, Srivastav A K, Suri K A, Qazi G N and Tuli TR. 2006. Process for isolation of Withaferin- A from plant materials and products therefrom. US Patent No. 7,108,870 B2

Sankar SR, Manivasagam T, Krishnamurti A, Ramanathan M. 2007. The neuroprotective effect of root extract in MPTP-intoxicated mice: an analysis of behavioral and biochemical variables. *Cellular Molecular Biology Letters* 12: 473–481

Seenivasagam R, Sathiyamoorthy S, Hemavathi K. 2011. Therapeutic impacts of Indian and Korean ginseng on human beings—a review. *International Journal of Immunoogicall Studies* 1:297–317

Sehgal N, Gupta A, Valli RK, Joshi SD, Mills JT. 2012. *Withania somnifera* reverses Alzheimer's disease pathology by enhancing low-density lipoprotein receptor-related protein in liver. *Proceedings National Academy of Sciences of the united states of America* 109: 3510–3515

Sharada A, Solomon FE, Devi PU. 1993. Toxicity of Withania somnifera root extract in rats and mice. *Pharmaceutical Biology* 31:205–212.

Sharma PV. 1999. Ashwagandha. Dravyaguna Vijana, Chaukhambha Viashwabha rti Varanasi, pp 763–765

Sharma S, Dahanukar S, Karandikar S. 1985. Effects of long term administration of the roots of ashwagandha and shatavari in rats. *Indian Drugs* 29: 133-139

Shohat B, Kirson I, Lavie D. 1978. Immunosuppressive activity of two plant steroidal lactones withaferin A and withanolide E. *Biomedicine* 28(1):18–24

Siddique AA, Joshi P, Misra L, Sangwan NS, Darokar MP. 2014. 5 6-De-epoxy-5-en-7-one-17-hydroxy withaferin A a new cytotoxic steroid from Withania somnifera L Dunal leaves. *Natural Product Research* 28:392–398

Singh AK, Varshney R, Sharma M. 2006. Regeneration of plants from alginate-encapsulated shoot tips of *Withania somnifera* (L) Dunal a medicinally important plant species. *Journal of Plant Physiology* 163:220–223

Singh B, Saxena AK, Chandan BK, Gupta DK, Bhutani K.K et al. 2001. Adaptogenic activity of a novel, withanolide-free aqueous fraction from the roots of Withania somnifera Dun. *Phytotherapy Research* 15:311–318

Singh D, Aggarwal A, Maurya R, Naik S. 2007. Withania somnifera inhibits NF-kappaB and AP-1 transcription factors in human peripheral blood and synovial fluid mononuclear cells. *Phytotherapy Research* 21:905–913

Singh G, Kumar P. 2011. Evaluation of antimicrobial efficacy of flavonoids of *Withania somnifera* L. *Indian Journal of Pharmaceutical Sciences* 73:473

Singh G, Tiwari M, Singh SP, Singh S, Trivedi PK, Misra P. 2016. Silencing of sterol glycosyltransferases modulates the withanolide biosynthesis and leads to compromised basal immunity of *Withania somnifera*. *Scientific Reports* 6:25562

Singh N, Bhalla M, de Jager P, Gilca M . 2011. An overview on ashwagandha: a Rasayana (rejuvenator) of Ayurveda. *African Journal of Traditional, Complementary and Alternate Medicine* 8(5):208–213.

Singh P, Guleri R, Singh V, Kaur G, Kataria H, Singh B, Kaur G, Kaul SC, Wadhwa R, Pati PK. 2015. Biotechnological interventions in *Withania somnifera* (L) Dunal. *Biotechnology and Genetic Engineering Reviews* 31:1–20.

Singh RH, Narsimhamurthy K, Singh G. 2008. Neuronutrient impact of Ayurvedic Rasayana therapy in brain aging. *Biogerontology* 9:369–374

Siriwardane AS, Dharmadasa RM, Samarasinghe K. 2013.Varieties of Withania sommfera (L.) Dunal. Grown in Sri Lanka. *Pakistan journal of Biological Sciences* 16(3): 141–144 .

Sivanandhan G, Dev GK, Jeyaraj M. 2013. A promising approach on biomass accumulation and withanolides production in cell suspension culture of Withania somnifera (L) Dunal. *Protoplasma* 250: 885–898.

Sivanandhan G, Selvaraj N, Ganapathi A. 2014. Enhanced biosynthesis of withanolides by elicitation and precursor feeding in cell suspension culture of Withania somnifera (L) Dunal in shake-flask culture and bioreactor. *PLoS ONE* 9:e104005 doi : 10.1371/journal.pone.010-4005.

Srivastava S, Sangwan RS, Tripathi S, Mishra B, Narnoliya LK, Misra LN, Sangwan NS. 2015. Light and auxin responsive cytochrome P450s from Withania somnifera Dunal: cloning, expression and molecular modelling of two pairs of homologue genes with differential regulation. *Protoplasma* 252(6): 1421–1437.

Subramanian C, Zhang H, Gallagher R, Hammer G, Timmermann B, Cohen M. 2014. Withanolides are potent novel targeted therapeutic agents against adrenocortical carcinomas. *World Journal of Surgery* 38(6): 1343-1352.

Subramanian SS, Sethi PD. 1969. Withaferin–A from Withania somnifera coagulants roots. *Current Science* (India) 38:267–268.

Sumanth M, Nedunuri S. 2014. Comparison of bioavailability and bioequivalence of herbal anxiolytic drugs with marketed drug alprazolam. *World Journal of Pharmaceutical Research* 3: 1358–1366.

Sumantran VN, Chandwaskar R, Joshi AK, Boddul S, Patwardhan B. 2008. The relationship between chondroprotective and antiinflammatory effects of Withania somnifera root and glucosamine sulphate on human osteoarthritic cartilage in vitro. *Phytotherapy Research* 22: 1342–1348.

Sumantran VN, Kulkarni A, Boddul S, Chinchwade T, Koppikar SJ et al. 2007. Chondroprotective potential of root extracts of *Withania somnifera* in osteoarthritis. *Journal of Biosciences* 32: 299–307.

Takimoto T, Kanbayashi Y, Toyoda T, Adachi Y, Furuta C, Suzuki K et al. 20144. β-Hydroxywithanolide E isolated from Physalis pruinosa calyx decreases inflammatory responses by inhibiting the NF-êB signaling in diabetic mouse adipose tissue. *International Journal of Obesity* 38(11): 1432–1439.

Thaiparambil JT, Bender L, Ganesh T, Kline E, Patel P. 2011. Withaferin A inhibits breast cancer invasion and metastasis at sub-cytotoxic doses by inducing vimentin disassembly and serine 56 phosphorylation. *International Journal of Cancer* 129:2744–2755.

Thirunavukkarasu M, Penumathsa S, Juhasz B, Zhan L, BagchiM et al . 2006 . Enhanced cardiovascular function and energy level by a novel chromium (III)-supplement. *BioFactors* 27:53–67.

Tiwari R, Chakraborty S, Saminathan M, Dhama K, Singh SV. 2014. Ashwagandha (*Withania somnifera*): role in safeguarding health, immunomodulatory effects, combating infections and therapeutic applications: a review. International *Journal of Biological Sciences* 14(2): 77–94.

Tripathi V, Verma J. 2014. Current updates of indian antidiabetic medicinal plants. *International Journal of Research in Pharmacy and Chemistry* 4(1): 114–118.

Uddin Q, Samiulla L, Singh V, Jamil S. 2012. Phytochemical and pharmacological profile of Withania somnifera dunal: areview. *Journal of Applied Pharmaceutical Sciences* 02(01): 170–175.

Um HJ, Min KJ, Kim DE, Kwon TK. 2012. Withaferin A inhibits JAK/STAT3 signaling and induces apoptosis of human renal carcinoma Caki cells. *Biochemical and Biophysical Research Communications* 427: 24–29.

Vaishnavi K, Saxena N, Shah N, Singh R, Manjunath K. 2012. Differential activities of the two closely related withanolides, Withaferin A and Withanone: bioinformatics and experimental evidences. *PLoS One* 7: e44419 doi : 10.1371/ journal.pone.0044419 .

Van Wyk B-E, Wink M. 2004. Medicinal plants of the world.Briza Publications, Pretoria

Ven Murthy M, Ranjekar PK, Ramassamy C, Deshpande M. 2010. Scientific basis for the use of Indian Ayurvedic medicinal plants in the treatment of neurodegenerative disorders: 1. Ashwagandha. *Central Nervous System Agents in Medicinal Chemistry* 10: 238–246.

Wadhwa R, Singh R, Gao R, Shah N, Widodo N, Nakamoto T, Kaul S. 2013. Water extract of Ashwagandha leaves has anticancer activity: identification of an active component and its mechanism of action. *PLoS One* 8(10): e77189 doi : 10.1371/annototion/6705/27-5970-4734-8601-9913-adcce984.

Widodo N, Takagi Y, Shrestha BG, Ishii T, Kaul SC et al. 2008. Selective killing of cancer cells by leaf extract of Ashwagandha: components, activity and pathway analyses. *Cancer Letters* 262 (1): 37–47.

Wollen KA. 2010 Alzheimer's disease: the pros and cons of pharmaceutical, nutritional, botanical, and stimulatory therapies, with a discussion of treatment strategies from the perspective of patients and practitioners. *Alternative Medicine Review* 15(3): 223–244.

Yu SM, Kim SJ. 2013. Production of reactive oxygen species by withaferin A causes loss of type collagen expression and COX-2 expression through the PI3 K/ Akt, p38, and JNK pathways in rabbit articular chondrocytes. *Experimental Cell Research* 319: 2822–2834.

Yu SM, Kim SJ. 2014. Withaferin A-caused production of intracellular reactive oxygen species modulates apoptosis via PI3K/Akt and JNKinase in rabbit articular chondrocytes. *Journal of Korean Medical Sciences* 29: 1042–1053.

Ziauddin M, Phansalkar N, Patki P, Diwanay S, Patwardhan B. 1996. Studies on the immunomodulatory effects of Ashwagandha *Journal of Ethnopharmacology* 50: 69–76.

17

Volatiles Profiling and Agronomic Practice of *Cymbopogon khasianus* [IIIM (J) CK-10 Himrosa] for Commercial Cultivation and Value Addition

Rajendra Bhanwaria[1], Bikarma Singh[2] and Rajendra Gochar[1]*

[1]*Genetic Resources and Agrotechnology Division,* [2]*Plant Sciences (Biodiversity and Applied Botany Division), CSIR-Indian Institute of Integrative Medicine Jammu 180001, Jammu & Kashmir, INDIA*
Email: drbikarma@iiim.ac.in*, rbhanwaria@iiim.ac.in

ABSTRACT

Cymbopogon khasianus (Hack.) Stapf. ex Bor. variety [IIIM (J) CK-10 Himrosa] is a commercially important aromatic grass of the family Poacae and contains high valued volatile constituents which has high demand in pharmaceutical, flavour, fragrance and cosmatic industries. This plant variety is one of the rich source of two monoterpenoids, geraniol and ocimene. Geranoil is an alcohol frequently used as terpenoid fragrance material, and ocimene is a group of isomeric hydrocarbons. The present communication deals with the herb yields, volatile constituents and agrotechnology of CK-10 Himrosa. The volatile constituents vary from season to season and from geographic locations of cultivation. Data presented were collected from field trials and experiment works conducted on farmer field at Balesar area of district Jodhpur (Rajasthan). It has been observed that [IIIM (J) CK-10 Himrosa] is rich in geraniol whose percentage varies from 70-80%, cisocimene (10-11%) trans-cimene (5-6%) geranyl acetate (2-3%), and various others constituents present in minor quantities. For analysis of volatile constituents, Gas Chromatography–Flame Ionization Detector (GC-FID) and GC-Mass Spectrometry (GC–MS) methods were developed. Agronomic data indicate substantial variations in the essential oil compositions, which varies due to season of harvest and place of cultivation. This crop is hardy in nature and from future perspectives; extension of these aromatic crops in rainfed and saline areas could brought un-utilized barenlands and wastelands under cultivation. Biomass and essential oil obtained from this crop would be helpful in development of new value added products. This may serves as economic crop for marginal farmers and helpful in gaining international recognition after development of products in the form of perfumes, soaps, cosmetics and other products for human use.

Keywords: Aromatic Plants, Essential Oils, Agro-technology, Value Addition.

INTRODUCTION

Aromatic grass *Cymbopogon* Spreng is one of the most important aroma bearing genus of the family Poaceae (or Gramineae) widely adapted to various agroclimatic zones, and few of them grows as wild in natural vegetation (Singh et al. 2000). Currently 52 species recommended as accepted names under the genus *Cymbopogon* (TPL 2108), reported to have wide distribution in tropical to temperate regions of the world (Bertea and Maffei 2010). Several species under this genus have been reported to be used in traditional herbal medicine, while those known to be rich in volatile constituents have usefulness in the cosmetics, pharmaceuticals and perfumery industries (Jeong et al. 2009, Avoseh et al. 2015, Verma et al. 2015). The volatile oils of these grasses shows various biological functions such as antimcrobial, immunomodulatory and antioxidant properties (Lagouri et al. 1993, Lee et al. 2008, Kalemba et al. 2012). *Cymbopogon* plants are perennial grasses, with narrow and long leaves that are mostly characterized by the presence of silica thorns aligned on the leaf edges. Leaves bear glandular hairs, usually each with a basal cell that is wider than the distal cell (Spies et al. 1994, Mathews et al. 2002, Bertea and Maffei 2010).

Cymbopogon khasianus (Hack.) Stapf. ex Bar. variety [IIIM (J) CK-10 Himrosa] commonly called 'Himrosa' is a commercial valuable aromatic crop suitable for hardy environment in rainfed and saline tropical regions of India and adjoining areas of Southeast Asian countries. Volatile oil of this species possess anti-helmintic, anti-inflammatory, anti-ageing, anti-microbial, mosquito repellant, pesticidal and larvicidal activities (Singh et al. 2000). The oil is pale yellowish in color and rich in geraniol (70–80%), a monoterpenols which has high demand in pharmaceutical and industries sectors looking for flavor and fragrance value added products for human welfare. Fungal endophyte enhances biomass production and essential oil yield of East Indian lemongrass (Ahmad et al. 2001). Essential oils are produced as secondary metabolites by many plants and can be distilled from all parts of plant (O'Bryan et al. 2015). The volatile essential oil has wide application in industries and extensively used for making mosquito repellent products, soaps, rose-like perfumes, cosmetics preparations, and many similar value-added products for human use.

In the present communication, volatile constituents and agrotechnology of Himrosa CK-10 were studied in details and recommended as an important plant for future perspective of grasses in value addition sector and product development sector and growth in flavour, fragrance pharmaceuticals and cosmetic industries.

History and Distribution

Cymbopogon khasianus (Hack.) Stapf. ex Bor is one of the best known species of the genus *Cymbopogon* with the synonyms *Andropogon khasianus* Munro ex Hack., *Andropogon nardus* var. *khasianus* Hack., *Cymbopogon auritus* B.S.Sun and *Cymbopogon khasianus* var. *khasianus*. This species was entomology to a popular tribe called *khasis*, residing in the khasi mountains of Meghalaya in Northeast India. It is native to Asia (Bhutan, China, India, Myanmar and Thailand). *C. khasianus* is known by numerous common names, such as CK-10, Himrosa, and Geraniole Ghash.

Taxonomic Enumeration

Cymbopogon khasianus is a tall perennial fast growing grass with tuft of geraniol scented leaves from the annulate and sparingly branched roots. It grows to a height upto 2 m, and has a distinct bluish-green leaflet which does not produce seed, and instead it has inflorescence in the form of spikelets. This species has many bulbous stems that increases the clump size as the plant grows and matures. The leaves of *C. khasianus* is long, glaucus green, glabrous, linear, with short ligule, tapering upward and along the margins, and tightly clasp sheaths at the base, narrow and separating at the distal end.

Field Trials and Extension for Experimentation

Initial *C. khasianus* first introduced in CSIR-IIIM campus as aromatic crops, later on experimentation were conducted related to agronomy, chemistry and essential oil contents and variety was developed and released. For the presented research, field trials and experiment works

Fig. 1: Experimetal sites of CK-10 Himrosa in Jodhpur, Rajasthan.

were conducted on farmers field at Balesar area (26.3974°N, 72.4791°E) of district Jodhpur (Rajasthan state). The experimental sites experiences arid climate having cold winters and hot summers. The minimum temperature recorded was 3.0°C in winter, and the maximum temperature of 47.0°C in summer. A pictorial representation of the experimental area is given in Figure 1.

Essential Oils Constituents

Experimental trials and investigation of essential oils reveals that the chemical composition of these volatile oils and extracts of *Cymbopogon khasianus* varies according to the geographical origin and seasonal growth of herbage. The chemical classes of isolated bioactives from this species include flavonoids, phenols, ketones, sterols, sugars, tannins and terpenoids. *Cymbopogon khasianus* variety [IIIM (J) CK-10] is having geraniol rich essential oil whose composition varies from 70-80% and percentage of major constituents depends on climatic factors, geographical location, soil conditions and seasonal growth of this crop in the area of cultivation. Geraniol is a monoterpenoid pharmaceutically important compound, and by dehydration and isomerization of geraniol, myrcene and ocimene are formed which has application in flavour, cosmetics and fragrance industries. Other essential oil constituents of the variety are cis-ocimene (10-11%), trans-ocimene (5-6%), geranyl acetate (2-3%), and various others in minor quantities. The experiments were conducted at Jodhpur and oils were extracted, and GC-MS analysis result is given in Table 1.

Table 1: Volatile constituents of *Cymbopogon khasianus* [IIIM (J) CK-10 Himrosa] of Jodhpur, Rajasthan

Name of compound	Retention time	(%)
p-Cymene	6.44	1.11
α- Phellandrene	7.27	1.01
cis-Ocimene	7.97	10.87
trans-Ocimene	8.24	5.36
Linalool	9.70	2.09
p-Menth-2-en-l-ol	10.42	2.16
l-Terpineol	10.91	1.22
Geraniol	14.13	73.01
Geraniol acetate	17.20	2.09
Caryophyllene	18.36	1.07

The presence of geraniol, ocimene and geranyl acetate are the main characteristic of the essential oil of CK-10 Himrosa. This plant variety cultivated is superior in terms of geraniol content as compared to other grass variety cultivated in India. Figure 2 provide the structure of major chemical constituents of IIIM(J) CK-10 Himrosa.

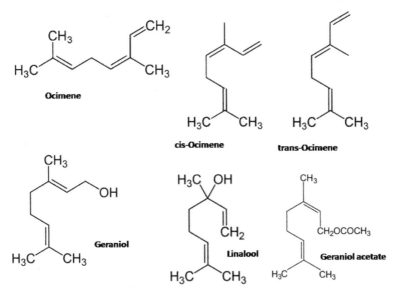

Fig. 2: Structure of major essential oil constituents of CK-10 Himrosa

AGROTECHNOLOGY ASPECTS

Soil and Climate

Cymbopogon khasianus is resistance to drought, stress and prefer hardy environment. It can easily grow, in different types of soils such as sandy loam, medium black and alluvial soil slightly acidic to alkaline in nature. The pH usually ranges from 6.5 to 9.5. The standing water or water logged condition is harmful for this variety. This crop can also be cultivated in deserts having slightly loamy soil of Rajasthan and elsewhere where it could be beneficial to prevent desertification and wind soil erosion. The average annual rainfall 700-750 mm and temperature varies between 3.0°C in winter to 47.0°C in summer. The tropical and sub tropical climatic condition and low hill elevation of 1200 to 1400 m above mean sea level would be good for this crop.

Land Preparation and Plantation

Prior to plantation of crops, land preparation is an important requirement for plantation and establishment of this crop in the field. First time field should be ploughed three–four times with disc plough and cross harrowing and leveled to bring the field to a good tilth condition. Properly leveled beds of 10m to 10m usually bed size depends on availability of irrigation facilities, and land area available. The slips need to be transplanted in the month of February-March or August- September. Rainy season is ideal for first time plantation of this crop area in rainfed areas and Kandy belts. The slips can be obtained by splitting of well grown clumps and it is estimated that 45 to 60 slips can be obtained per clump from one year old planted seedling. Slip is planted with sufficient spacing 40 x 40 or 40 x 50 cm apart from plant to plant and from row to row. Average 50,000 to 62,500 slips would be required for planting of this crop in one hectare area. New leaves start sprouting from slips within 15 to 20 days after plantation and establishment of this crop in field. Tillering would be required after 30 to 40 days and this would be helpful for 90% to 95% survivable rate.

Manure and Fertilizers

Cymbopogon khasianus requires high nitrogen, phosphorus and potassium element for its luxurious vegetative growth. Total nitrogen is a basic requirement to produce optimum oil yield. Organic manure is good sources of essential elementals nutrients for the crop application of 10-12 tonne/hectare well decomposed FYM mixed in soil would be required before 15 days of transplantation of slips and basic application of N: P: K 200: 80: 60 kg/hectare ratio during first year and 255-250 kg nitrogen per hectare during the second and subsequent years, using urea, Diammonium Phosphate (DAP) and Muriate of Potash (MOP) is suitable for proper growth of this crop. Nitrogen in the form of urea can be applied through top-dressings at each harvest or cutting would be beneficial for vegetative luxuriant growth. Nitrogen application increases the herbage yield which leads to more oil production. Excessive application of nitrogen beyond recommended level may adversely affects the oil quality and the quantity. Besides, the application of micronutrient fertilizer @ 20-25 kg/ha can produce better crop yield.

Irrigation

Aromatic grasses are generally raised as rainfed crops (Singh et al. 2000), but crop required irrigation immediately after the plantation of slips in field. The crop needs five or six irrigation during summer, while three to four irrigations is sufficient in winter season. The crop requires immediate irrigation within two days of each harvest. Frequent irrigations is required to this commercial crop to obtain good growth and essential oil yields.

Weed Control and Management

Hoeing and hand-weeding required for CK-10 Himrosa crop because the weeds affects the growth, oil quantity and chemical constituents of essential oils. Two-three weeding would be necessary during the first year of luxuriant weeds and crop. Competition would be seen during the early stages of crop growth. One-two hand weeding required within 45-50 days of plantation after successful establishment of slips in field, hoeing can be applied as weed control technique. Application of 2,4-D @1.5 kg ai/ha can be used pre and post emergence for the broad leaved weeds in early stages of crop growth. For satisfactory results, one has to adopt an integrated approach of using chemical and mechanical weed control methods for this crop.

Harvesting

Cymbopogon khasianus is important commercial crop and timely harvesting is important for obtaining the higher yield of volatile oils with superior quality and for maintaining the bioactive constituents at good percentage in the oil. First harvests of this crop could be done after four-five months after planting the slips. There-after, this crop can be harvested at an interval of 50 to 60 days. The vegetative parts of the plant materials can be chopped 8-10 inches above the soil surface at the time of maturity for the better growth of next cutting. During rainy and winter season, the moisture content in the leaves used to be high. Therefore, before distillation, it is desirable to allow the crop material wilt for 24 to 48 hours as it reduces the moisture content in herbage. The essential oil can be obtained by the steam distillation methods of herbage which takes about 3 to 4 hours to complete the distillation process. Nearly 75% to 82% of oil can be obtained during the first hour and 15 to 20% during the second hour, and heavy molecular weight compound takes slightly longer time to recover from this method.

Herbage and Essential Oil Yield

First year crop gives an average of 50 to 60 tonnes of fresh herbage and 160 to 180 kg essential oil yield per hectare per annum. After one to two harvest/cutting increases the diameter of clump, results increase in fresh herbage yield. Second year, 65 to 70 tonnes of fresh herbage yield can be obtained and 180 to 200 kg essential oil yield per hectare per annum. Essential oil content varies from 0.4 to 0.6 % depending upon the season's climatic condition and geographical area of plantation. An average oil recovery of 0.40% is obtained at pilot scale. It is observed that under water stress and well management conditions, the oil recovery increases upto 1% and oil yield has been observed 200 to 220 kg oil/ha during the first year, while in second and subsequent year's 250 to 280 kg oil/ha can be distilled out from fresh material.

Economics

Geranoil rich essential oils has high demand in the international markets and it's price is increasing every year. The prevailing price of the essential oil of this crop in the Indian market is Rs. 1,500-2000 per kg. Under well management condition and using of all package and practices, crop will give approximately net profit of Rs. 1 to 1.25 lakh per annum in first year and subsequently in the next year onwards.

CONCLUSION

The volatile oils have a long history of use for medicine, in perfumes, cosmetics, as botanicals and spices for foods. *Cymbopogons* widely adapted to the different agroclimatic zones of the country, however, they are sensitive to environmental conditions such as rainfall, humidity, temperature and soil fertility. There can be wide variation in the yield and the quality of volatile produced at different locations. CK-10

Himrosa can be best suited for cultivation in wastelands, particularly saline soils, alkaline soils, hill slopes and marginal lands. Besides fulfilling the indigenous requirement, they can earn a sizeable amount of foreign exchange, for the poorer section of the socity. Agrotechnology provided are cost effective and this condition can improves the soil yields. This crop is hardy in nature and from future perspective, extension of these aromatic crops in rainfed and saline areas could brought un-utilized barrenlands and wastelands under crop cultivation, Biomass and essential oil obtained from this crop would be helpful in development of new value added products. This may serves as economics for marginal farmers and helpful in gaining international recognition after development of products in the form of perfumes, soaps, cosmetics and other products for human use.

ACKNOWLEDGEMENT

Authors would like to thank Director CSIR-IIIM Jammu and Head of the Division for the necessary facilities and encouragement. It bears institutional publication number IIIM/2237/2018.

LITERATURES CITED

Ahmad A, Alam M, Janardhanan KK. 2001. Fungal endophyte enhances biomass production and essential oil yield of East Indian lemongrass. *Symbiosis* 30: 275-285.

Avoseh O, Oyedeji O, Rungqu P, Nkeh-Chungag B, Oyedeji A. 2015. Cymbopogon species: ethnopharmacology, phytochemistry and the pharmacological importance. *Molecules* 20: 7438-7453.

Bertea CM, Maffei ME. 2010. The Genus Cymbopogon Botany, Including Anatomy, Physiology, Biochemistry, and Molecular Biology, pp. 1-24. In Edit. Book Akhila A. Essential Oil bearing Grasses. Published by CRC Press Taylor & Francis Group Raton, United State of America.

Jeong M, Park PB, Kim D, Jang Y, Jeong HS, Choi S. 2009. Essential oil prepared from Cymbopogon citrates exerted an antimicrobial activity against plant pathogenic and medical microorganisms. *Mycobiology* 37: 48–52.

Kalemba D, Matla M, Smetek A. 2012. Antimicrobial activities of essential oils. *Dietary Phytochemicals and Microbes* Berlin: Springer p. 157-183.

Lagouri V, Blekas G, Tsimidou M, Kokkini S, Boskou D. 1993. Composition and antioxidant activity of essential oils from oregano plants grown wild in Greece. *European Food Research and Technology* 197 (1): 20-23.

Lee SH, Lillehoj HS, Lillehoj EP, Cho SM, Park DW, Hong YH, Chun HK, Park HJ. 2008. Immunomodulatory properties of dietary plum on coccidiosis. *Comparative, Immunology, Microbiology Infectious Diseases* 1: 389-402.

Mathews S, Spangler RE, Mason-Gamer RJ, Kellogg EA. 2002. Phylogeny of Andropogoneae inferred from phytochrome B, GBSSI, and NDHF. *International Journal of Plant Sciences* 163: 441-450.

O'Bryan CA, Pendelton SJ, Crandella PG, Ricke SC. 2015. Potential of Plant Essential Oils and Their Components in Animal Agriculture – *in vitro* Studies on Antibacterial Mode of Action. *Frontiers in Veterinary Science* 2: 35. doi: 10.3389/fvets.2015.00035.

Singh K, Kothari SK, Singh DV, Singh VP, Singh PP. 2000. Agronomic studies in cymbopogons - a review. *Journal of Spices and Aromatic Crops* 9(1): 13-22.

Spies JJ, Troskie TH, Vandervyver E, Vanwyk SMC. 1994. Chromosome studies on African plants-II. The tribe Andropogoneae (Poaccae, Panicoideae). *Bothalia* 24: 241-246.

TPL. 2018. The plant lis 2013, version 1-1 published on the internet: http://www.theplantlist.org (Accessed: 10 August 2018).

18

In-vitro Investigation of Anti-diabetic, Phytochemical and Silver Nanoparticles Synthesis of *Dendrophthoe falcata* (L.f.) Ettingsh: A Hemiparasitic Taxa

Ayan Kumar Naskar, Souradut Ray and Amal Kumar Mondal*

Plant Taxonomy, Biosystematics and Molecular Taxonomy Laboratory
UGC-DRS-SAP, DBT-BOOST-WB Funded Department
Department of Botany & Forestry, Vidyasagar University, Midnapore-721102
West Bengal, INDIA
*Email: naskarayan8@gmail.com, souradutray@gmail.com
amalcaebotvu@gmail.com; akmondal@mail.vidyasagar.ae.in

ABSTRACT

Dendrophthoe falcata (L.f.) Ettingsh, a unique hemiparasitic taxa belongs to the family Loranthaceae, spreads over the entire forest in the South West Bengal. It is generally stem hemiparasitic taxa, as it is found mostly on aerial part of the host plant due to lack of root system. It depend on the host plant for mainly water and to some extend carbon. It synthesizes its own food. This plant has haustoria instead of root, they penetrate their haustoria into the vascular bundle mainly in the xylem tissue of the host plants. This species has numerous medicinal values with a long history of its use in Chinese traditional medicines. The present work deals with the anti-diabetic, phytochemicals and silver nanoparticles properties of *D. falcata*. The responsible enzyme i.e. Alpha-amylase breaks the large starch molecules which produces free glucose and simultaneously increases the blood sugar level and as a result hyperglycemia occurs. This study also reveals that the plant's leaves extracts have inhibitory activities on this key alpha amylase enzyme and apart from that aqueous extract of this plant contains some secondary metabolites which is very crucial data for drug preparation and any other research purposes. The maximum amount of silver nanoparticles is present in this plant which is valuable as a silver line in near future implications for different drugs preparations

Keywords: *Dendrophthoe falcata*, Loranthaceae, Medicinal value, Alpha Amylase enzyme, Phytochemistry, Silver nanoparticles.

INTRODUCTION

Hemiparasitic taxa represents only 1% in world vegetations. Among them *Dendrophthoe falcata* (L.f.) Ettingsh holds the major portion. This plant taxa is distributed throughout the natural forest vegetation of South West Bengal, mainly on the aged plants. As per investigation, it was observed that, it is a rare plant in the forest of West Bengal. As this plant born on the rough surface bark of the host trees, dispersal of its seeds mainly occurred by sunbirds, as a result less than 50% successful rate of germination take place. Their abundance is very less in nature for this reason they shows camouflage for survival. The recognition or finding out of this plant is very laborious work because they look like a branch of the host trees. The leaf of this plant is green, wavy, thick, exstipulate and margin of the leaf reddish in color. The bark of stem is brown in color, uniformly distributed lenticels; slightly rough in texture.

There are many different types of human diseases found in our social life and diabetes mellitus is one of them. This disease is chronic, severely characterized by hyperglycemia, due to improper regulation of insulin hormone (Aguwa, 2004). To prevent hyperglycemia or diabetes there is one way, to keep glucose level near to normal as much as possible in the blood. The non-pharmacological and pharmacological both approaches can manage diabetes. The non-pharmacological approaches to resist diabetes are exercise; diet control, surgery etc. alongside pharmacological approaches includes the use of drugs (insulin) or oral hypoglycemic agents. These remarkable anti diabetic drugs are not only costly but also associated with lots of side effects (Aguwa 2004, Adeneye and Agbaje 2008). Therefore devoid of side effect and low-cost drugs are searched by the scientist from a long while. Interestingly, some plants are now identified to cure this lifelong disorder. One of them is *Dendrophthoe falcata*, hemi-parasitic taxa (Fig. 1) which is available in the natural forest vegetation of South west-Bengal on a different host plants (Table 1). However, the responsible active molecules to cure diabetes from this plant have not been isolated till now. Many works had also been reported on the pharmacological activities such as antihypertensive, anti-epileptic and effects of immune stimulants (Pernossian and Kocharan 2004, Imafidon and Igbinaduwa 2007).

Alpha-amylase is the major form of amylase found in humans and other mammals. This enzyme cuts off alpha-bonds of large sugar molecules (Fig. 2). This enzyme is found in plants (barley), Ascomycotina or Basidiomycotina fungus and bacteria (*Bacillus* sp.). Our human saliva also contains amylase enzyme which is digests the food initially. Foods that contain high amount of starch (rice, potato), or slightly sweet in test, when they are chewed in the mouth, amylase secretion is starts and breaks down some of starch into sugar in our mouth. The pancreas also synthesizes alpha amylase to hydrolyze starch from dietary materials into mono, di and tri-saccharides which supply the body energy. The objectives of this study was to determine the inhibitory effect of *Dendrophthoe falcata* on α-amylase enzyme.

Some natural occurring substances obtained from animals, plants and minerals are used for preparing drug. These organic substances could be obtained from both primary and secondary metabolic processes and they also provide a good source in medicinal science at the earliest time.

Table 1: Host plants of *Dendrophthoe falcata*

Sr. No.	Botanical name	Common name	Family
1.	*Careya arborea* Roxb.	Wild guava	Lacythidaceae
2.	*Bombax ceiba* L	Cotton tree	Malvaceae
3.	*Shorea robusta* Roth.	Sal	Dipterocarpaceae
4.	*Terminalia catappa* L.	Indian Almond	Combretaceae
5.	*Lagerstroemia speciosa* (L.) Pers.	Jarul	Lythraceae
6.	*Toona serrata* M.Roem.	Chinese Mahogany	Meliaceae
7.	*Swietenia macrophylla* King.	Mahagoni	Meliaceae
8.	*Tectona grandis* L.f.	Teak	Lamiaceae
9.	*Facocia indica*		
10.	*Aegle marmelosa* L.	Beal, golden apple	Rutaceae
11.	*Madhuca longifolia* (Konig) J.F.Macbr.	Mahua	Sapotaceae
12.	*Ficus religion* L.	Bo-tree. Peepal	Moraceae
13.	*Albizia lebbeck* (L.) Benth.	Siris	Leguminaceae
14.	*Terminalia arjuna* (Roxb.) Wight & Arn.	Aurjun	Cmbretaceae
15.	*Ficus hispida* L.	Dumur	Moraceae
16.	*Limonia acidissima* L.	Wood apple	Rutaceae
17.	*Magnifera indica* L.	Mango	Anacardiaceae
18.	*Melia azedarach* L.	Mahaneem	Meliaceae

Fig. 1: Habitat of *Dendrophata falcata* (Flowering stage, Flower, Fruits and bark)

Phytochemicals are the chemical compounds that occur naturally in plants and responsible for color and have organoleptic properties. Scientists estimated that there may be as many as 10,000 different phytochemicals with potential to cure diseases such as cancer, stroke or metabolic syndromes (Ganatra et al. 2012). The antibiotics or antimicrobial properties of these plant are due to the presence of phytochemicals such as saponins, glycosides, flavonoids and alkaloids. Although these active compounds are not well established due to the lack of knowledge and proper techniques.

Now a days the nano-technology is very unique technique in science by which we synthesize nanoparticles such as silver and gold. which is actually have a various applications including medicine, mechanical and bio-medical electronics (Kalimuthu et al. 2008, Smitha et al. 2008). Nanoparticles are usually a cluster of atoms ranging between 1-100nm in size and they exhibit new properties based on their size, distribution and morphology (Satyavani et al. 2011). The materials are synthesized in nano size for metals are commonly used for synthesis of nanoparticles by chemical and biological methods. Though biological method is commonly adopted for the synthesis of silver nanoparticles, use of plant extracts is widely studied due to its advantages over others. In the field of nanotechnology researcher are finding that metal nanoparticles have all kinds of previously unexpected benefits. They are usually prepared from noble metals, that is, silver, gold, platinum and palladium while silver nanoparticles (AgNPs) are the most exploited (Roy 2010), because of its wider range of application in medicine, electronics, energy saving, environment, textile, cosmetics, biomedical and many more.

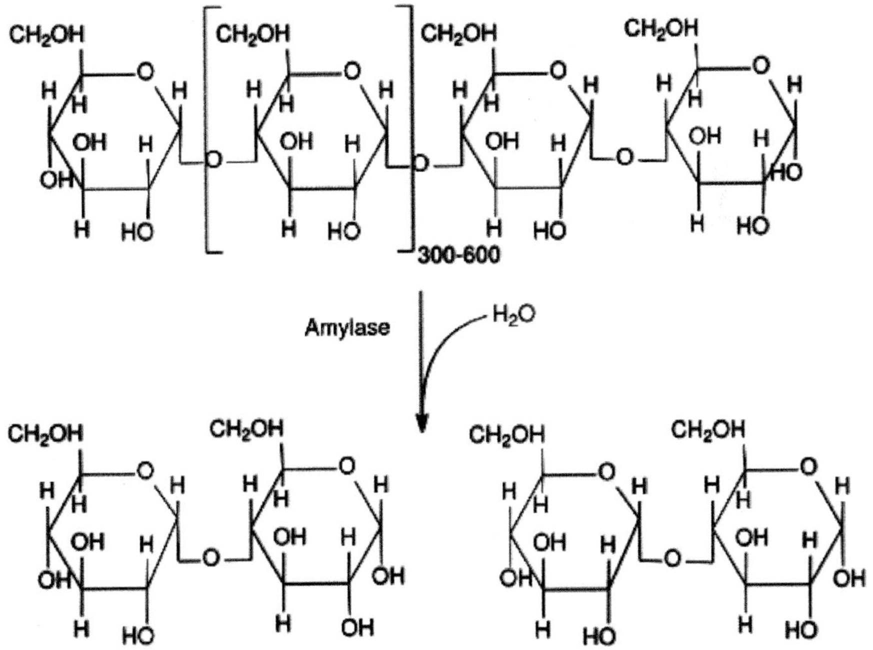

Fig. 2: Mechanism of Alpha-amylase

ETHONOBOTANICAL APPLICATIONS

Dendrophthoe falcata is known as "Vanda" in the Indian Ayurvedic system of medicine. It is a perennial woody parasitic plant, widespread throughout India (Fig. 3), including Andaman & Nicobar Islands, Srilanka, Nepal, Bhutan, Indo-china, Thailand, Tropical Australia, Bangladesh, Malaysia and Myanmar (Balakrishnan 2012).

Fig. 3: Distribution of *D. falcata* in India (Source: envis.frlht.org/maps)

Dendrophthoe falcata is a hemiparasitic plant whose whole plant used in indigenous system of medicine as a potential medicinal agents such as cooling, bitter, astringent, aphrodiasic, narcotic and diuretic, and is useful in pulmonary tuberculosis, asthma, menstrual disorders, swelling wounds, ulcers, renal and vesical calculi (Moshrafuddin 2010). Ashes of bark used to wash clothes; decoction given in swellings and juice poured in ears to reduce ear-ashes, wood is useful to in tannery (Balakrishnan 2012). Decoction of plant is used by women as antifertility agents and also possess anticancer activity (Nadkarni 1976, Pattanayak et al. 2008). Leaf paste is used in skin diseases and also applied on boils and setting dislocated bones. The plant has been scientifically proved to have antilithiatic, diuretic, cytotoxic and immune-modulatory activities (The Wealth of India 1969, Pattanayak 2009, Allekutty 1993, Mary et al. 1993).

TECHNIQUES

Dendrophthoe falcata leaves are selected for present study. The sample were collected from the different natural, undisturbed forests area and rural area of South West Bengal of Medinapore (Salboni forest, Arabari forest, Hoomgarh forest), Purulia (Ajodhya Hills), Bankura (Sarenga jungle) during January to March 2017.

Plant fresh material were dried under sunlight up to 4-5 days then grinded with the help of motor and pestel to make fine powder and after that homogenized with water and boiled for 10 minutes. The extract was filtered and then centrifuged. The supernatant was used for determination of enzyme inhibitory property. Assay of α-amylase inhibitory property: Plant extract (0.1ml) was incubated with 0.2 ml of properly diluted enzyme for 20 minutes at 37⁰C temperature. Then 0.1 ml of starch solution was added to the reaction mixture and incubated for 3 minutes at 37⁰C temperature. The enzyme reaction mixture was interrupted by the addition of 0.2 ml Dinitrosalicylic (DNS) acid and heated for 5 minutes in boiling water. Then the tube containing the mixture was cooled under running tap water. Then 4 ml water was added and optical density of the solution was determined by spectrophotometer at 540 nm. A blank mixture was prepared in same manner without adding enzyme in the mixture, percentage inhibition of enzyme activity was measured.

PHYTOCHEMICAL SCREENING

Test for Tannins

With 5 ml plant aqueous extract, few drops (2-3) of 0.1% ferric chloride was added. Carefully observed that any precipitation or any colour changes happen or not. A bluish-black or brownish-green precipitate indicated the presence of tannins (Ejikeme et al. 2014).

Test for Saponins

Demonstration of frothing -2.5ml of filtrate was diluted to 10ml with distilled water and shaken vigorously for 2 minutes (frothing indicated the presence of saponin in the filtrate).

Demonstration of emulsifying properties - 2 drops of olive oil was added to the solution and shaken vigorously for a few minutes, formation of a fairly stable emulsion indicated the presence of saponins (Ejikeme et al. 2014).

Test for Terpenoids

5 ml of extract was mixed in 2 ml of chloroform. Then 3 ml of concentrated H_2SO_4 was added to form a layer. A reddish brown precipitate colorations at the interface formed indicated the presence of terpenoids (Ejikeme et al. 2014).

Test for Flavonoids

5 ml of extract solution from stalk solution taken and then few drops of 20% sodium hydroxide solution was added. A change to yellow color which on addition of acid changed to colorless solution depicted the presence of flavonoids (Harborne, 1973 & Sofowora, 1993).

Test for Cardiac Glycosides

5 ml of extract was treated with 2ml of glacial acetic acid containing one drop of ferric chloride solution. This was underplayed with 1ml of concentrated sulphuric acid. A brown ring at the interface indicated the deoxysugar characteristics of cardenolides. A violet ring may appear below the ring while in the acetic acid layer, a greenish ring may be formed (Harborne 1973, Sofowora, 1993).

Test for Free Anthraquinones

5 ml of chloroform was added to 0.5 g of the powdered specimen. The resulting mixture was shaken or 5 mins after which it was filtered. The filtrate was then shaken with equal volume of 10% ammonia solution. The presence of a bright pink colour in the aqueous layer indicated the presence of free anthraquinones (Harborne 1973, Sofowora 1993).

Test for Carotenoids

1 g of specimen sample was extracted with 10 ml of chloroform in a test tube with vigorous shaking. The resulting mixture was filtered and 85% sulphuric acid was added. A blue colour at the interface showed the presence of carotenoids (Harborne 1973, Sofowora 1993).

Test for Reducing Compounds

The filtrate solution make alkaline with the help of 20% sodium hydroxide solution. The resulting solution was boiled with an equal volume of Benedict qualitative solution on a water bath. The formation of a brick red precipitate depicted the presence of reducing compound (Harborne 1973, Sofowora 1993).

Test for Alkaloids

1g of powdered sample of the specimen was separately boiled with water and 10ml hydrochloric acid on a water bath and filtered. The pH of the filtrate was adjusted with ammonia to about 6-7. A very small quantity of the following reagent was added separately to about 0.5ml of the filtrate in a different test tube and observed (Harborne 1973, Sofowora 1993).

- Picric acid solution.

- 10% tannic solution.

- Mayer's reagent (Potassium mercuric iodide solution).

The test tubes were observed for colored precipitates or turbidity.

For Silver Nanoparticls

90ml 1mM $AgNO_3$ taken into a conical flasks and 1ml of plant extract mixed drop by drop and the solution are homogenized about 15 minutes The color change of the solution was checked periodically and the conical flasks were incubated at room temperature for 3 hours. After incubation, the color of the solution is checked.

RESULTS

The plant extract shows anti-diabetic activity has an important role in inhibiting the glucose level thus providing protection to human against hyperglycemia. This study is carried out to evaluate the anti-diabetic activity of aqueous extract of the leaves of *D. falcata*. In our experiment different conc. of plant extracts are applied on the solution mixture which contain alpha-amylase enzyme, starch and DNS. Measure the Optical Density of the mixture by UV Spectrophotometer (UV-1800 SHIMADZU) and a significant result we found (Table 2). The result of the experiment is given in following tabulated form and graphical representation of the data present in Fig. 4.

Table 2: O.D values of the DNS in respect to Different plant concentration

Concentration of plant extract (mg/ml)	OD value
100	1.895
200	1.621
300	1.425
400	1.329
500	1.286

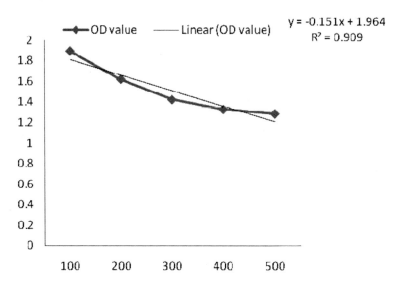

Fig. 4: Amylase inhibitory activity of different concentration of plant extract

The data presented here indicate that aqueous extract of *D. falcata* possesses significant amylase inhibition activity in in-vitro condition. DNS or dinitrisalisilic acid is a chemical compound that binds to the glucose (Reducing sugar), it cannot bind with the non-reducing sugar.

In this experiment, when plant extract, enzyme and starch solution were incubated for 3 min at 37⁰C, enzyme break down the starch and produced glucose and side by side DNS react with this free glucose and give ANS which shows maximum absorbance at 540 nm but when we increase the plant extract concentration, the enzyme amylase inhibited slowly as a result production of free glucose is decreased and ANS production is also decreased slowly in the reaction mixture that's why the OD value decreased slowly. It represent that the plant extract having a alpha amylase inhibitory properties.

In phytochemical investigation, the aqueous extract of the leaves of *D. falcata* were prepared and different tests for phytochemical constituent was performed using generally accepted laboratory technique for qualitative determinations. The result shows that the presence of phytochemicals such as alkaloids, flavonoids, tannins, terpenoids, reducing sugars, carbohydrates and cardiac glycosides in aqueous extract of the plant. When a new drug to be discovered, qualitative phytochemical analysis is a very important step as it gives information about the presence of any particular primary or secondary metabolite in the extracts of the plant which is having a clinical significance.

The synthesis of silver nanoparticles using *D. falcata* leaves extract was found to be significant in boiled extract method compared to other methods. The colour of the reaction medium gradually changed to dark brown because of the surface plasmon resonance. An absorption peak between 430-460nm under UV-Visible spectroscopy confirms the presence of silver nanoparticles (Fig. 5).

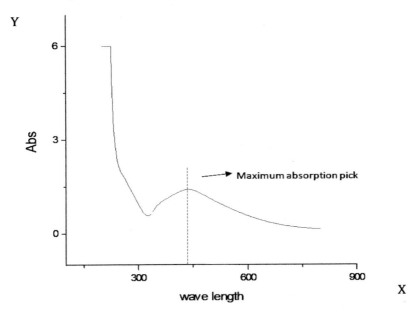

Fig. 5: UV-Visible spectroscopy showing the presence of Nano-particles

A similar pattern was observed by (Zarchi et al. 2011), where the synthesis of silver nanoparticles was done using ethanol extract of *Andrachnea chordifolia*, by sunlight irradiation. In our experiment we plotted Absorption Spectrum according to the "Y" axis and wave length plotted according to "X" axis. The absorption pick increase from 692μm, and maximum increase at 440μm. then the pic decrease. The absorption pick at 440m indicate the presence of Silver nano particle in the *D. falcata* leaves powder.

PHARMACOLOGY ASPECTS

Wound Healing

Chaitanya et al. (2010) investigated that the ethanolic extract of leaves and stem had a very good efficiency on excision and incision wound models in rats. The extract of *D. falcata* showed the potent wound healing capacity as evident from the wound contraction and increased tensile strength (Chaitanya 2010).

Antimicrobial activity

Antimicrobial activity was investigated with aqueous, acetone and methanol extract of this plant on four pathogenic bacterial strains like *Bacillus subtilis* (Cohn 1872), *Escherichia coli* (Castellani and Chalmers 1979), *Klebsiella pneumonia* (Trevisan 1887) and *Aeromonas hydrophila* (Stainer 1943). In this experiment, it compatible with four bacterial strain about their inhibition capability. We found that, in case of aquous extract after 24 hour incubation result is *K. pneumoniae>A. hydrophila>E. coli>B. subtilis* where after 48 hours of incubation it shows, *A. hydrophila>K. pneumoniae>E. coli>B. subtilis*. In case of Methanolic extract- after 24 hours incubation result is *A. hydrophila>E. coli = K.pneumoniae>B. subtilis* but after 48 hours incubation there is no significant changes occurred. In case of acetone extract, after 24 hours incubation it shows *A. hydrophila >K. pneumoniae>E. coli>B. subtilis*, where after 48 hours incubation the order is *A. hydrophila = K. pneumoniae>E. coli = B. subtilis* (Ray et al. 2015).

Anthelmintic activity

Ethyl acetate and methanolic extracts of the *D. falcata* leaves showed significant activity at 40 mg/ml (Dipak 2009).

Tablet binder

Tablets were prepared with *D. falcata* mucilage and evaluated for tablet characteristics. Wet granulation technique was used for the preparation paracetamol granules. The tablet binder concentrations used in formulations were 2, 4, 6 and 8% w/w. Tablets were compressed to hardness at 6.6 to 6.9 kg/cm2. The evaluation of tablet showed 0.98 to 0.53% friability, 10 to 17 min disintegration time and more than 90% dissolution in 70 min. Tablets at 6% w/w binder concentration showed more optimum results as tablet binder. The *Dendrophthoe falcata* mucilage was found to be useful for preparation of uncoated tablet dosage form (Kothawade 2010).

Antihyperlipidaemic Activity

Hyperlipidaemia was induced by administration of High fat diet (HFD) for 42 days which showed marked elevated levels of serum TC, TG, LDL VLDL, and reduction in level of HDL as compared to control group fed with normal diet. Administration of *Dendrophthoe falcata* leaves ethanolic extracts with daily dose of 300 mg/kg, significantly altered the levels of serum TC, TG, LDL, VLDL and serum HDL level at different degrees. Ethanolic extract (70%) at 300mg/kg showed significant (p<0.01) antihyperlipidaemic activity in HFD induced hyperlipidaemia (Tenpe 2008).

CONCLUSION

For complementary or synergistic effects of herbal products, it may be prepare with the help of a single herb or combination of multiple herbs. Some time animal's product and minerals are included to the herbal products for formulating the traditional medicine. It will be sold in market either in raw plants or extraction of the plant's part. Although raw extraction of the various plant's parts have medicinal importance but modern drugs come in market after comprehensive investigation on bioactivity, pharmaco-therapeutics, mechanism of action, and toxicity and after proper standardization and clinical trials. Present study shows that the plants *Dendrophthoe falcata* have inhibitory activity of alpha amylase. Since alpha amylase is the one of the most important enzyme which mainly is responsible for hyperglycemia, by inhibiting its enzymatic activity with the above mentioned plant are having anti diabetic properties. From above plants, the drug-development programmer may be take response to develop modern drugs with the help of mastered chemical compound for that disease. Now time has to be changing, the use of that non-toxic plant that's having traditional medicinal value, development of modern drugs from these plants should be emphasized for the control of various diseases. In fact, time has come to make good use of centuries- old knowledge on plants through modern approaches of drug development. An extensive research and development work should be undertaken on these plants and its products for their better economic and therapeutic utilization.

The phytochemical tests performed on the aqueous extracts of *D. falcata* leaves shows the presence of alkaloids, flavonoids, tannins, terpenoids, reducing sugar, carbohydrates and cardiac glycosides. These results may help to discover new chemical classes of antibiotic substances that could serve as selective agents for infectious disease chemotherapy and control. *D. falcata* leaves can be further analysed for qualitative and quantitative extraction of reported phytochemical to explore the possibilities of using it as an herb medicine on scientific ground. The effect of these plants on pathogenic organism and toxicological investigations and further purification however, needs to be carried out.

In the nanoparticle study it can be concluded that nanoparticle research is currently an area of scientific research, due to a wide variety of potential application in biomedical and optical fields. Nanoparticles are of great scientific interest as they are effectively a bridge between bulk materials and atomic or molecular structures. It is clear that from the present study that the sliver nanoparticles synthesized from leaves of *D. falcata* in a high amount can be used effectively in the treating bacterial disease and diabetes management.

Abbreviation

Abbreviation	Full Name
D. falcata	*Dendrophthoe falcata*
OD	Optical density
DNS	3,5-Dinitrisalicylic acid
ANS	3-amino 5-nitrosalicylic acid
HFD	High fat diet
TC	Total cholesterol
TG	Triglycerides
LDL	Low density lipoproteins
VLDL	Very low density lipoproteins
HDL	High density lipoproteins

ACKNOWLEDGEMENTS

We highly acknowledge to UGC-DRS-SAP-I for their financial support (2012-2017). The work based on M. Sc. project work including DRS-SAP-I under the Co-ordinator of Prof A. K. Mondal FLS, FIAAT, Professor of Botany and proper discussion with Dr. S. Mondal Parui, HOD, Co-ordinator, P.G. Department of Zoology, Section Biochemistry, Lady Brabourne Collage, Kolkata. We are thankful to Dr. Sougata Sarkar, Research Associate CSIR-IIIM, Jammu, for providing the opportunity to undertake the curriculum and also help me every time. We would like to express our deep sense of gratitude to Dr. Bikarma Singh, Scientist-Plant Taxonomy and Ecology, Value Addition laboratory section Herbarium & Crud Drug Repository Division, Indian Institute of Integrative Medicine, Jammu, for giving me such type of innovative opportunity. I want to give thanks to Debjani Singha for supporting me all the time. I am also thankful to other research scholars of Plant Taxonomy, Biosystematics and Molecular Taxonomy Laboratory for their valuable guidance and help throughout the study specially field survey ad laboratory work.

LITERATURES CITED

Adeneye AA, Agbaje EO. 2008. Pharmacological evaluation of oral hypoglycemic and antidiabetic effects of fresh leaves of ethanol extract of *Morinda lucida* benth in normal and alloxan-induce diabetic rats. *African Journal of Biomedic* 11(1): 65-71.

Aguwa CN. 2004. *Therapeutic basis for clinical pharmacy in the tropics, (3rd ed.)*. Enugu.

Alkutti NA, Srinivasan KK, Gundu RP, Udupa AC, Keshavamurthy KR. 1993. Diuretic and antilithiatic activity of *D. falcata. Fitoterapia* 64: 325-331.

Allekutty NA, Sriniwasan KK, Gundu RP, Udupa AC. and Keshawamurthy KR. 1993. Diuretic and Antilithiatic activity of *Dendrophthoe falcata. Fitotherapia* 64(5): 325–331.

Balakrishnan NP, Chakrabarty T, Sanjappa M, Lakshminarsimhan P, Singh P. 2012. *Flora of India*: Botanical Survery of India, Kolkata, 23: 3.

Chaitanya Sravanthi K, Sarvani Manthri, Srilakshmi S, Ashajyothi V. 2010. Wound Healing Herbs – A review. *International Journal of Pharmacy & Technology* 2(4): 603- 624.

Dipak RN, Subodh PC, Subhash MC. 2009. Anthelmintic potential of *Dendrophthoe falcata* Etting. (l.f.) Leaf. *International Journal of Pharmaceutical Research and Development* 6: 002.

Ejikeme CM, Ezeonu CS, and Eboatu AN. 2014. Determination of physical and phytochemical constituents of some tropical timbers indigenous to Niger Delta Area of Nigeria. *European Scientific Journal* 10(18): 247–270.

Ganatra H, Durge P. and Patil SU. 2012. Preliminary Phytochemicals Investigation and TLC Analysis of *Ficus racemosa* Leaves. *Journal of Chemical and Pharmaceutical Research* 4(5): 2380-2384.

Gupta RS, Kachhawaa JBS. 2007. Evaluation of contraceptive activity of methanol extract of *Dendrophthoe falcata* stem in male albino rats. *Journal of Ethnopharmacology* 112(30): 215-218.

Harborne JB. 1973. *Phytochemical Methods*. London.

Imafidon KE. and Igbinaduwa P. 2007. Effects of dried powdered leaves of *Loranthus bengwensis* L. (African mistletoe) on blood pressure and electrolyte leves of normal and hypertensive rats. *Global Journal Biotechnology Biochemistry* 2(2): 51-53.

Kothawade SN, Shinde PB, Agrawal MR, Aragade PD, Kamble HV. 2010. Preliminary Evaluation of *Dendropthoe falcata* Mucilage as Tablet Binder. *International Journal of PharmTech Re*search 2(2): 1474.

Mary KT, Kuttan R. and Kuttan G. 1993. Cytotoxicity and Immuno modulatory activity of *Loranthes* extract. *Amala Research Bulletin* 13: 53–58.

Moshrafuddin A. 2010. *Medicinal Plants*. MJP Publishers, Chennai.

Nadkarni's KM. 1976. *Indian Materia Medica*. Bomby Popular Prakashan.

Pattanayak SP and Mazumder PM. 2010. Therapeutic potential of *Dendrophthoe falcata* (L.f) Ettingsh on 7, 12-dimethylbenz(a) anthracene-induced mammary tumorigenesis in female rats: effect on antioxidant system, lipid peroxidation, and hepatic marker enzymes. *ResearchGate* 20(4): 381-392.

Pattanayak SP, Mazumder PM. 2009. Effect of *Dendrophthoe falcata* (L.f.) Ettingsh on female reproductive system in Wistar rats: a focus on antifertility efficacy. Contraception: *An International Reproductive Health Journal* 80(3): 314-20.

Pattanayak SP, Mitra Mazumder P. 2009. Assessment of neurobehavioral toxicity of Dendrophthoe falcata (L.f.) Ettingsh in rat by functional observational battery after subacute exposure. *Pharmacognocy Magazine* 5:98–105.

Pattanayak SP, Sunita P and Muzumder PM. 2008. *Dendrophthoe falcata* (L.f.) Ettingsh: A consensus review. *Pharmacognosy Review* 2(4): 359-368.

Pernossian A, Kocharan A. 2004. Pharmacological activities of phenyl propanoids of mistletoe (*Loranthus parasiticus*) sourced from different host trees, *Bioresearch Journal* 2(1):18-23.

Ray S, Mondal AK, Verma NK. 2015. Preliminary phytochemical analysis of leaf extract of *Loranthus parasiticus* merr. with reference to their antibacterial activity. *International Journal for Current Research* 7(3): 13661-13666.

Roy N., Barik A. 2010. Green synthesis of silver nanoparticles from the unexploited weed resources. *Interntional Journal of Nanotechnology* 4: 95.

Satyavani K, Gurudeeban S, Balasubramanian T R. 2011. Biomedical potential of silver nanoparticles synthesized from calli cells of *Citrullus colocynthis* (L) Schrad. *Journal of Nanobiotech* 9:43.

Sofowora A. 1993. *Medicinal Plants and Traditional Medicine in Africa*. Ibadan, Nigeria

Tenpe CR, Upaganlawar AB, Khairnar AU, Yeole PG. 2008. Antioxidant, Antihyperlipidaemic and Antidiabetic Activity of *Dendrophthoe falcata* Leaves-A Preliminary study. *Pharmacognocy Magazine* 4(16).

The Wealth of India. 1969. *Raw Materials*. New Delhi: CSIR, NISCOM. 3: 34–36.

Zarchi AAK, Mokhtari N, Rehman M, Ali T. and Amini M. 2011. A sunlight-induced method for rapid biosynthesis of silver nanoparticles using an *Andrachnea chordifolia* ethanol extract. *Applied Physics* 103(2): 349-353.

19

Crop-Weather Interactions, Phytochemistry, Pharmacology and Evaluation of the Phenological Models for *Echinacea purpurea* (L.) Moench under Subtropical and Temperate Environments

Kiran Koul[1†], Bikarma Singh[1,†,6], Mahendra Kumar Verma[2] Sumit Singh[1], Pooja Goyal[5], Govind Yadav[3], Bishander Singh[6], Surinder Kitchlu[1] and Rajneesh Anand[4]*

[1]Plant Sciences (Biodiversity and Applied Botany Division), [2]Medicinal Chemistry Division, [3]Mutagenesis lab (Animal House Division),[4]Instrumentation Division [5]Plant Biotechnology CSIR-Indian Institute of Integrative Medicine Jammu-180001, Jammu & Kashmir, INDIA
[6]Academy of Scientific and Innovative Research, Anusandhan Bhawan New Delhi - 110001, INDIA
[†]First Author as equal contributor
[6]Department of Botany, Veer Kunwar Singh University, Ara, Bihar, INDIA
*Corresponding author email: drbikarma@iiim.res.in; drbikarama@iiim.ac.in

ABSTRACT

Echinacea purpurea (L.) Moench. is a medicinal and nutraceutically important herb of American origin, extended its distribution to other parts of the globe in subtropical and temperate environment. This species is known for various biological properties such as anti-microbial, immuno-stimulatory and anti-inflammatory properties due to the presence to several bioactive compounds loaded in the form of phenylpropanoids, flavonoids, terpenoids, lipids, nitrogenous compounds, carbohydrates and others such as ascorbic acid, sitosterol, linoleic acid, cyanidin glycosides and sesquiterpene esters in different quantities. Published data reveals that its major chemical constituents possess anti-anxiety, anti-depression, cytotoxicity, anti-infective and anti-mutagenicity properties. Besides, plants respond to abiotic and climatic signals that are responsible for various adaptive responses at the cellular level and vary with ecosystem in which the species grows. It may behave differently in different ecological climates, thereby, modifying their gene expressions, and this reveals that the plant growth and the adaptability are mainly affected by microclimatic factors. Experimental trials conducted to collect data on growth phenology in relation to crop-weather interaction and investigated result is presented in this research. Studies on chemical evaluation of *E. purpurea* under two constrasting environment, subtropical and temperate climate, with reference to cichoric acid and chlorogenic acid undertaken for this plant is also presented. Future studies in relation to the application of biotechnological interventions and chemical evaluation at different time scale would be a great contribution to the science for this species as it has the tremendous potentials which may lead to development of various nutraceuticals and value added products in future for human health and medicines.

Keywords: Medicinal Plant, *Echinacea purpurea*, Crop Weather Interaction, Phyochemistry, Pharmacology, Nutraceuticals.

INTRODUCTION

Echinacea purpurea (L.) Moench is a medicinal and nutraceutically important herb of American origin, extended its distribution to other parts of the globe in subtropical and temperate environment. It is a herbaceous perennial plant of Asteraceae (or Compositae) members native to America and the wild habitat of this species includes prairies and rocky edaphic conditions. Although the natural range of *E. purpurea* is quite broad, but natural growth truly represents a peculier ecotone (McKeown 1999) preferring the shade edges with partial sun exposure. *Echinacea* phytopharmaceuticals are the most popular botanicals as immunostimulants in the Europe and USA which is evident from the fact that more than 800 drugs are currently in German market having constituents of this species (Rawls 1996). *Echinacea* products rank in top ten best selling drugs in USA. All parts of the plant are useful and can be harvested for polysaccharides (Bauer and Wagner 1991, Bonadeo et al. 1971), polyacetylenes (Bauer et al. 1988, Schulte et al. 1967), caffic acid (Bauer and Foster 1991), cichoric acid (Hobbs 1989) and alkyamides (Schulthess et al. 1991). These bioactive principles are implicated in the immunostimulatory effect of this species (Bauer and Wagner 1991). Besides, it helps in curing cold, flu and other infections particularly of respiratory disorders (Hobbs 1989, 1994; McKeown 1999). This can also be used as medicine externally for curing burns, snake and insect bites (Busing 1952, Hill et al. 1996, Selvanayahgam et al. 1994). Hot water extract of the dried leaves can be taken orally for inflammations (Mascolo 1987). Essential oils present in the achenes of *E. purpurea* have an important role as phytochemical marker in differentiating various species (Lienert et al. 1998, Schulthess et al. 1989). The volatile components present in the essential oils the extracted from the species is specific to climatic factors, and variation in temperature, rainfall and other climatic parameters influence their concentration and quality (Thappa et al. 2004).

Plants respond to abiotic and climatic signals may leads to various adaptive responses at the cellular level, change in habitat and climatic variables may causes species to behave differently with respect to the ecological zones, there by, modifying their gene expressions (Fernandez et al. 2004). This proves that the plant growth and the adaptability were mainly affected by climatic factors. Therefore, an attempt has been to study the phenological adaptations of *E. purpurea* and presented in this communication.

Growth models have been developed to compare and predict the number of days to flowering and physiological maturity in terms of thermal units under temperate and sub-tropical environments as degree-day. Such model are based on the agro-meterological application of temperature effect on the plant growth (Dwyer and Steward 1986) which invariably applied to correlate the phase-wise development in crops to predict sowing and maturity date (Wurr et al. 2002). Hence, one of the parameters of the present investigation is based on these concepts and experimental trials were conducted at two different environment, the temperate and the sub-tropical climatic conditions in Western Himalaya of Jammu and Kashmir State. The study expresses the thermal requirement and crop-weather interactions at various phenological events and also predicting the occurrence of phenophases at two climatic conditions with respect to *E. purpurea*.

STUDY AREA, TECHNIQUES AND ANALYSIS

Experimental Site

The experimentation trials were conducted at two different climatic zones of Western Himalaya in Jammu and Kashmir State of India (Jammu: 33.44°N latitude 74.50°E longitude representing subtropical climate, Kashmir: 34.07°N latitude/74.51°E longitude representing temperate climate). The seedlings were raised in nursery and allowed to develop their leaves before transplanting them in experimental plots following randomized block design model (Fig. 1). Three replicates of plots having distance of 40cm x 40cm from plant to plant and from row to row 100 cm x 100 cm were planted. Under the sub-tropical conditions, the seedlings were transplanted on 25th February 2014, where as, same species were planted in the first week of May i.e. 5th May 2014 under temperate climate. Plant growth parameters were monitored continuously for two growing seasons and the data were recorded related to the development of phenophases based on twenty replicates of plants per treatment.

Fig. 1: Experimental trials for Value-Added Products from *Echinacea purpurea*

Phenology

Phenological time scale for enumerating the phenophases have been investigated based on the appearance of vegetative and reproductive phases. Data on seedling emergence, leaf initiation and elongation, leaf induction and its continuation, thereafter flowering initiation and flower induction (25%, 50%, 100%), leaf senescence and seed setting were recorded during the period of experimentation.

Energy Summation Indices

Energy summation indices representing radiations, thermal unit and transpiration could be used for identification of phenological events and maturity dates in most of the cultivated crops. Growing degree days are a tool in phenology which is used to measure the heat accumulated to predict the plant development rate such as the date that a flower would be blooming and time of crop to reach maturity. It is a way assigning a heat value to degree days calculated in each day using the maximum daily temperature, the minimum daily temperature and the base temperature. It can be computed by using the following formula (Shahi et al. 2005, Parthasarthi et al. 2013).

$$\text{Growing Degree Days} = \sum\nolimits_{th}^{tp}(TA - TB)$$

where, TA=average of the daily maximum and the minimum ambient temperature(C), TB=base/threshold temperature (5°C) below which the plant development is usually bound to cease, tp/th = time of transplanting, th = time of harvest. The threshold temperatures used to be higher for tropical crops and lower for temperate crops.

Phenothermal Index

Phenothermal (PTI) index is the ratio of degree days to the number of days between the two phenological stages. Influence of an ambient temperature on phenological events can be studied using the accumulated degree-day and growth day (Chakravarty and Sastry 1983).

It can be calculated by as PTI equals to the degree days consumed between two phenological streams divided by the number of days between two the phenological stages.

DISCUSSION

Crop-Weather Interactions

It has been quantified as phenothermal index which exhibited a linear response to phasic growth development indicates that the appropriate application of heat unit concept in subtropical and temperate environment, has different responses to the adaptation of plants. The data obtained while evaluating the growth models of *E. purpurea* in two different contrasting climatic conditions to develop phasic model or phenology module for prediction of calendar days as dependent variable by using degree days (independent variable) in linear regression equations is provided in Table 1.

Table 1: Growth performance of *Echinacea purpurea* under sub-tropical and temperate environments

Morphological traits	1st Year Environment			2nd Year Environment		
	Temperate	Sub-tropical	t-test	Temperate	Sub-tropical	test
Plant height (cm)	59.55	49.15	2.522	81.5	52.1	9.161
Tillers/plant (nos)	1.9	17.25	N/S	8	12.3	N/S
Flowers/plant (nos)	21.45	27.25	N/S	60.95	27.25	6.301
Flower diameter (cm)	8.635	6.245	5.480	8.605	6.245	6.028
Fresh herb/plant (g)	276.75	155.95	6,834	633	243	8.100
Fresh root/plant (g)	28.65	44	N/S	131	60.25	9.801
Dry herb /plant (g)	55.4	39.72	3.965	169.6	61.999	8.683

It has been experimentally proven that approximately 220-230 calendar days are required for getting better quality of crop produce which corresponds to about 3000 degree days for temperate, and 3800 degree-days for subtropical climatic conditions. Phenothermal index indicate that *E. purpurea* exhibits 50% flowering at 12.98 optimal values for the temperate and 16.50 optimal value for the subtropical environment. Further delay in harvest results percentage decrease in quality contents of different secondary metabolites. It helps in scheduling optimal harvest period in accordance to phenological time scale. It can be assumed that the phenothermal index not only quantify the crop-weather interaction, but also acts as a indicator for obtaining the better quality and quantity of produce for *E. purpurea* and similar crops growing in tropical to temperate environment (Table 2).

Further, it has been observed that the coefficient of variation (CV%) is comparatively low 5.86 (subtropical) and 10.16 (temperate) at vegetative phase, while CV% was 39.60 and 11.10 during reproductive phase under sub-tropical and temperate environments respectively. When the entire growth cycle taken together as a single unit, consistency can be observed having 12.85 mean value of CV% for the temperate climate, and 10.45 for the sub-tropical environment. Therefore the index seems to be the most effective under temperate condition in taking the account and expressing the effect of varying ambient temperature on the duration between the phenological events for comparing the crop response in respect to the ambient temperature between the different phenological stages.

Table 2: Phenological time scale in relation to summation of energy indices for *Echinacea purpurea* at varied environments

Phenophase (s)	Sub-tropical Climate Energy Indices			Temperate Climate Energy Indices		
	Calen-dar days	Degree-days	Phenother-mal Index	Calen-dar days	Degree-days	Phenothermal Index
Emergence (100%)	11	71.00	6.45	11	134.50	12.22
VEGETATIVE PHASE						
i. Leaf initiation and elongation	36	221.20	6.14	26	276.91	10.65
ii. Leaf induction	46	297.70	6.47	31	321.41	10.36
iii. Leaf induction continue	79	568.70	7.19	112	1485.26	13.26
iv. Coefficient of variation (CV%)			5.86			10.16
REPRODUCTIVE PHASE						
i. Flower initiation	110	940.70	8.55	117	1576.26	13.47
ii. Flower induction	135	1679.70	12.44	127	1758.26	13.84
iii. Flowering (25%)	175	2391.70	13.66	138	1944.76	14.09
iv. Flowering (50%)	226	3730.90	16.50	218	2831.41	12.98
v. Flowering, leafing, senescence and initiation of seeding (100%)	231	3852.15	16.67	220	3108.91	14.80
vi. Mean	-	-	10.45	-	-	12.85
vii. Coefficient of variation (CV%)	-	-	39.60	-	-	11.10

Phenophasic Model

A linear regression models based on thermal units requirement have been developed for the first time of *E. purpurea* at sub-tropical and temperate climatic conditions as mentioned below:

Y_{st} =34.556 + 0.0536 AGDD (R^2.97)

Y_t =6.058 + 0.069 AGDD (R^2.99)

where, Y_{st} and Y_t are the number of calendar days predicted at sub-tropical and temperate environments, respectively. AGDD is the accumulated degree-days for the particular phenophase. The models indicated that accumulated thermal units accounted for nearly 94% and 98% for sub-tropical and temperate climatic conditions and verification of occurrence, respectively.

PHYOTOCHEMICALS

Genus *Echinacea* species contain several varieties of chemical components that contribute to their activity. Few chemical constituents were reported to be unique to one species, while others occur in two or more of the same species. Different classes of secondary compounds were considered as important to the therapeutic actions of this genus, and these has received the most intensive study (Bauer and Wagner 1991, Bauer et al. 1988). Various compounds belonging to different classes of secondary metabolites have been isolated and identified from the extracts and some of the major chemical constituents of *E. purpurea* are given in Fig. 2. Phenyl propanoids, flavanoids, terpenoid, lipids, nitrogenous and carbohydrate compounds are the main constituents of *Echinacea purpurea*, which is presented below briefly in subheads:

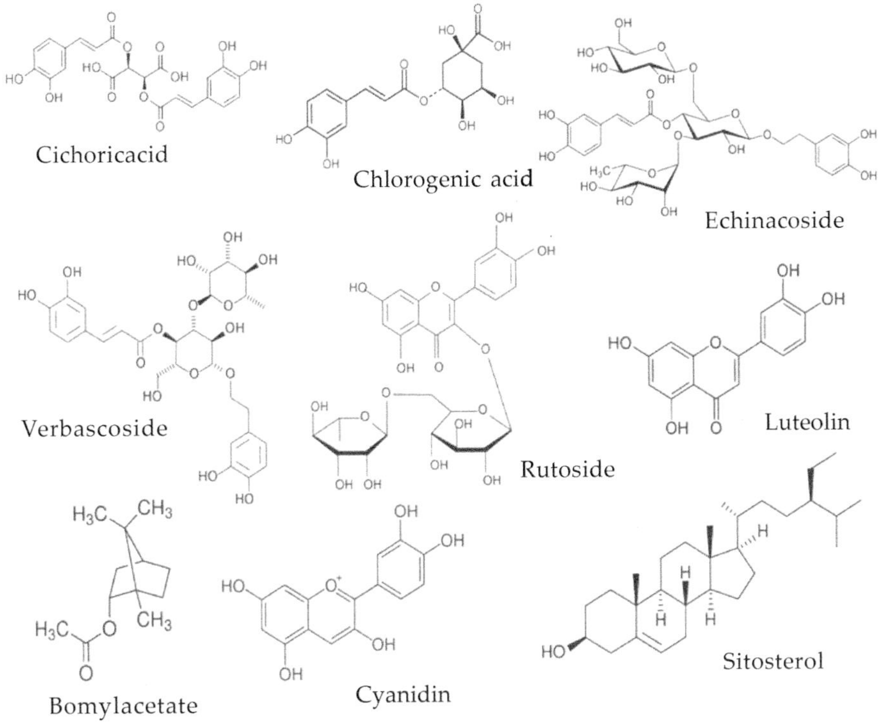

Fig. 2: Chemical structure of major compounds in *Echinacea purpurea*

Phenylpropanoids

Caffeic acid isolated from *E. purpurea* includes echinacoside, verbascoside, des-rhamnosylverbascoside, caftaric, 6-O-caffeoy-lechinacoside, cynarin, cichoric acid, chlorogenic and isochlorogenic acids (Bauer and Wagner 1991, Wagner et al. 1989).

Flavonoids

Rutoside, luteolin, kaempferol, quercetin, quercetagetin, apigenin, isorhamnetin were the main flavonoids isolated and characterized from *Echinacea* species. Usually flavonoids content in leaves of *E. purpurea* reported to be 0.48% (Bauer and Wagner 1991).

Terpenoid Compounds

Essential oils were reported from roots, leaves, flowers and other aerial parts of *Echinacea purpurea*. The concentration of the volatile oils varies for the fresh materials from 0.05-0.48%, where as 0.1-1.25% in dried plant materials (Bauer and Wagner 1998). The major constituents are borneol, bornylacetate, pentadeca-8-ene-2-one, germacrene D, caryophyllene, caryophyllene epoxide and palmitic acid (Bauer and Wagner 1998).

Lipid Compounds

Polyacetylenes such as Trideca-1-en-3,5,7,9,10-pentayne and pontica-epoxide reported in *Echinacea purpurea* (https://www.mdidea.com/products/herbextract/ echinacea).

Nitrogenous Compounds

Alkylamides such as echinacein (dodeca-(2E,6Z,8E,10E)-tetraenoic acid) in the roots (Bauer et al. 1988). There is a report that 1-1.5 years old plant contain the major amount of alkylamide *ca* 14.1 mg/g in the vegetative stems, *ca* 5.7 mg/g in rhizomes, *ca* 2.7 mg/g in flowers, and *ca* 1.7 mg/g in roots. *E. purpurea* also contain alkaloids in the form of glycine betaine, traces of pyrrolizidine, tussilagine and isotussilagine.

Carbohydrates

Two immunostimulatory polysaccharides (carbohydrates), PS I and PS II, were isolated from the aerial parts of *Echinacea purpurea*. Chemical structure suggested that PS I to be a 4-O-methylglucuronoarabinoxylan with an average molecular weight (MW) of 35 kDa, while PS II is an acidic arabinorhamnogalactan of 50 kDa. A xyloglucan (79 kDa) was isolated from the leaves and stems of *E. purpurea* (Bauer and Wagner 1991). Fructose and fructan polymers are also reported in this species (Giger et al. 1989).

Besides, other chemical constituents of *E. purpurea* were ascorbic acid, sitosterol, linoleic acid, cyanidin glycosides, sesquiterpene esters, echinadiol-, echinaxanthol-and many more in minor quantities.

Evaluation of Cichoric acid and Chlorogenic acid in *Echinacea purpurea*

Experimental trials were conducted between 2015 and 2018 by CSIR-Indian Institute of Integrative Medicine Jammu under two adverse climates (temperate and subtropical) for evaluation of cichoric acid and chlorogenic acid content of *Echinacea purpurea*. During analysis, it was found that the leaves are the main source of cichoric acid and chlorogenic acid. It has also been recorded that a marked decrease in Cichoric acid (43.62%) and Chlorogenic acid (73%) along with extraction value (37.14%) have been observed under sub-tropical environment (Table 3). Therefore, the temperate environment is more conducive for development and accumulation of these chemical compounds in *Echinacea purpurea*.

Table 3: Chemical evaluation of *Echinacea purpurea* under temperate and subtropical environments Cichoric acid (%)Chlorogenic acid

Plant part(s)	Cichoric acid (%)		Chlorogenic acid (%)		Extraction Value (%)	
	Temperate	Subtropical	Temperate	Subtropical	Temperate	Subtropical
Leaves	1.080	0.608	0.510	0.154	28.0	17.6
Stem	0.360	0.348	0.147	0.055	26.0	17.9
Flowers	0.294	0.214	0.620	0.380	18.4	14.6
Roots	0.443	0.562	0.172	0.046	14.5	09.8

BIOLOGICAL STUDIES

According to Manayi et al. (2015), alkamides, polysaccharides, caffeic acid and its derivatives considered as an important chemical constituents of *Echinacea purpurea*. A number of studies revealed that alkamides are involved in the immunomodulatory properties of *Echinacea* extracts *in vitro* and *in vivo* (Goel et al. 2002, Gertsch et al. 2004). Caffeic acid or its derivatives applied toward authentication and quality control of the plant extracts, where as polysaccharides has crucial role in anti-inflammatory activity of *Echinacea* preparations (Laasonen et al. 2002).

Immunimodulatory effects

As per Manayi et al. (2015) and Anonymous (1999), immunostimulant activity of *Echinacea purpurea* preparations is caused by three mechanisms: Phagocytosis activation, fibroblast stimulation and the enhancement of respiratory activity which results in augmentation of leukocyte mobility. Numerous published *in vivo* studies related to immunomodulatory and anti-inflammatory effects of *E. purpurea* indicate that innate immunity is enhanced by administration of the plant compounds, which strengthened immune system against pathogenic infections through activation of the neutrophils, macrophages, polymorphonuclear leukocytes (PMN), and natural killer (NK) cells (Manayi et al. 2015, Barnes et al. 2005). Akamides effective on cannabinoid receptor type 2 (CB2), is considered as a possible mechanism of immunomodulatory properties (Woelkart and Bauer 2007). Molecular mechanism could be increased of cyclic adenosine monophosphate (cAMP), p38 mitogen-activated protein kinases (p38/MAPK), and c-Jun N-terminal kinases (JNK) signaling, as well as nuclear factor kappa-light-chain-enhancer of activated B cells (NF-κB), activating transcription factor 2/cAMP responsive element binding protein 1 (ATF-2/CREB-1) in primary human monocytes and macrophages (Gertsch et al. 2004, Woelkart and Bauer 2007, Chicca et al. 2009). Sketch representing biological and pharmacological activities of alkamides isolated from *E. purpurea* is presented in Fig. 3.

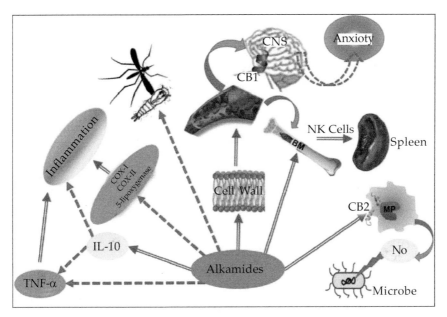

Fig. 3: Sketch representing biological and pharmacological activities of alkamides isolated from *Echinacea purpurea*. BM = Bone marrow, CB = Cannabinoid receptor, CNS = Central nervous system, COX = Cyclooxygenase enzyme, IL-10 = Interleukin-10, MP = Macrophage, NK = Natural killer cells, NO = Nitric oxide, TNF-α = Tumor necrosis Factor-α (*Courtesy*: Mayani et al. 2015, Pharmacognosy review)

Anti-infective Activity

Goel et al. (2004) reported that *Echinacea purpurea* could be effective in preventing respiratory associated diseases. As reported in various literatures (Gwaltney 2002, See and Wark 2008, Nichols et al. 2008), acute respiratory infections in humans usually caused by one or more of a group of well-known viruses, which includes over 100 rhinoviruses such as influenza viruses A and B, para-influenza viruses, corona viruses, respiratory syncytial virus, and certain adeno-viruses. Studies on *E. purpurea* indicate multiple actions of the herbal preparation, resulting either from the individual activities of several compounds or the synergistic effect of different compounds. It includes direct virucidal activity against several viruses involved in respiratory infections at concentrations which are not cytotoxic, direct bacteriocidal actions against certain potentially pathogenic respiratory bacteria, inactivation of microbial pathogens pertinent to humans and their domesticated animals, pro-inflammatory response of epithelial cells and tissues to viruses and bacteria, modulation of certain immune cell functions and reversal of the excessive mucin secretion induced by rhinovirus, which results due to gene expression. Combination of these beneficial activities could reduce the amount of prevailing viable pathogens, and their transmission and lead to amelioration of symptoms of the infection (Pleschka et al. 2009, Hudson 2009, Sharma et al. 2010, Schneider et al. 2010). Besides, the extracts of *E. purpurea* can affect various signaling pathways of epithelial cells and inhibit pathogens or bacterium-induced secretion of cytokines or chemokines and other inflammatory mediators that were responsible for the pulmonary symptoms (Altamirano-Dimas et al. 2007, 2009; Wang et al. 2008).

Veterinary Applications

Echinacea herb reported to have modern tradition of veterinary applications in North America and Europe, and few reports have described analogous for human diseases, or even controlled trials in animals, invariably the treatments were concluded to be safe and free of significant side effects (Lans et al. 2007a,b). Domestic animals such as pets, livestock and fish requires treatment at some point in their lives for viral and microbial diseases, and the causative organisms are usually analogous to the corresponding human counterparts as for example, avian influenza viruses, animal herpes viruses, various respiratory viruses and bacteria, and many fungal and parasitic infections. Consequently some of them should be responsive to *Echinacea* treatment, either as direct antivirals, antimicrobials, or as an anti-inflammatory agent. In addition some of these organisms, especially bacteria such as *Salmonella* and *Campylobacter* species, are important sources of contaminated foods. Furthermore some commentators have pointed out the need to evaluate herbal preparations as replacements for at least part of the antibiotic onslaught that farmed animals often receive (Hudson 2012). Allen (2003) investigation in chicks infected with the protozoan parasite Coccidia concluded that dietary supplementation with *Echinacea purpurea* root extract significantly decreased lesion scores and improved the health of the animals, in comparison with animals raised on a normal diet, although immune parameters were not measured. On the other hand, Hermann et al. (2003) studies in young pigs indicate that dietary *E. purpurea* was found to offer no protection against the porcine reproductive and respiratory syndrome virus (PRRS virus). Since this virus is a member of the arteri-virus family (related to corona-viruses) and possesses a membrane, it would be expected to be susceptible to direct contact with *E. purpurea*. Studies carried out in uninfected horses by O'Neill et al. (2002) and on fish by (Aly and Mohamed 2010) suggested possibilities for *Echinacea* preparations. Again safety was not considered a problem for the animals. Fish, like other farmed animals, are always potentially vulnerable to viral and microbial infections, especially under conditions of stress; consequently alternative treatments to synthetic antimicrobials.

CONCLUSION

Echinacea purpurea is one of the important medicinal plants that stimulate overall activity of the cells responsible for fighting all kinds of infection. Its immunomodulatory and anti-inflammatory properties are known in drug discovery which is capable of modulating of various immune system pathways. Published data reveals that this species has long history of traditional usages for wide range diseases. While studying crop weather interaction, it was observed that the combined effect of growing conditions of *E. purpurea* of heat growing degree days was significant at all phenological stages. Future studies in relation to biotechnological intervention and chemical evaluation at different time scale is required as this is one of the plants which has potentials in development of nutraceuticals and value added products in time to come.

ACKNOWLEDGEMENT

Authors are thankful to Director IIIM, Dr Ram A. Vishwakarma for facilities. This article bears institutional publication number IIIM/2245/2008.

REFERENCES CITED

Allen PC. 2003. Dietary supplementation with Echinacea and development of immunity to challenge infection with coccidia. *Parasitology Research* 91(1): 74-78.

Altamirano-Dimas M, Hudson JB, Cochrane D, Nelson C, Arnason JT. 2007. Modulation of immune response gene expression by *Echinacea* extracts: results of a gene array analysis. *Canadian Journal of Physiology and Pharmacology* 85(11): 1091-1098.

Altamirano-Dimas M, Sharma M, Hudson JB. 2009. *Echinacea* and anti-inflammatory cytokine responses: results of a gene and protein array analysis. *Pharmaceutical Biology* 47(6): 500-508.

Aly SM, Mohamed MF. 2010. *Echinacea purpurea* and *Allium sativum* as immunostimulants in fish culture using Nile tilapia (Oreochromis niloticus). *Journal of Animal Physiology and Animal Nutrition* 94(5): e31-e39.

Anonymous. 1999. Geneva: World Health Organization. WHO monographs on selected medicinal plants; pp. 136-145.

Barnes J, Anderson LA, Gibbons S, Phillipson JD. 2005. *Echinacea* species (*E. angustifolia* (DC.). Hell, *E. pallida* (Nutt.) Nutt, *E. purpurea* (L.) Moench): A review of their chemistry, pharmacology and clinical properties. *Journal of Pharmacy and Pharmacology* 57: 929-954.

Bauer R, Khan IA, Wagner H. 1988. TLC and HPLC analysis of *Echinacea pallida* and *E. angustifolia* roots. *Planta Medica* 54: 426-430.

Bauer R, Wagner H. 1991. *Echinacea* species as potential immunostimulatory drugs. *Medicinal Plant Research* 5: 253-321.

Bauer V, Remiger P, Wagner H . 1998. New Alkamides from *Echinacea angustifolia* and *Echinacea purpurea* Roots. *Planta Medica* 54: 563-564.

Baur R, Foster S. 1991. Analysis of alkamides and caffeic acid derivatives from *Echinacea simulata* and *E.paradoxa* roots. *Planta Medica* 57: 447-449.

Busing K. 1952. Hyaluronidasechummung durch echinacin. *Arzneimittel Forschung* 2: 467-469.

Chakravarty NVK, Sastry 1983. Phenology and accumulated heat unit relationships in wheat planting dates in the Delhi region. *Agriculture Sciences Proceedings* 1: 32-42.

Chicca A, Raduner S, Pellati F, Strompen T, Altmann KH, Schoop R, et al. 2009. Synergistic immunomopharmacological effects of N-alkylamides in *Echinacea purpurea* herbal extracts. *International Immunopharmacology* 9: 850–858.

Dwyer LM, Stewart DW. 1986. Leaf area development in field grown maize. *Agronomy Journal* 18: 334-338.

Fernandez GCJ, Chen HK. 1989. Temperature and photoperiod influences on reproductive development of reduced photoperiod sensitive mugbean genotype. *Journal of American Society for Horticultural Science* 114(2): 204-209.

Gertsch J, Schoop R, Kuenzle U, Suter A. 2004. Echinacea alkylamides modulate TNF-alpha gene expression via cannabinoid receptor CB2 and multiple signal transduction pathways. *FEBS Letters* 577: 563-569.

Giger E, Keller F, Baumann TW. 1989. Fructans phytotherapeutic preparation. *Planta Medica* 55: 638-639

Goel V, Chang C, Slama JV, Barton R, Bauer R, Gahler R, Basu TK. 2002. Alkylamides of *Echinacea purpurea* stimulate alveolar macrophage function in normal rats. *International Immunopharmacology* 2: 381-387.

Goel V, Lovlin R, Barton R, Lyon MR, Bauer R, Lee TD, Basu TK. 2004. Efficacy of a standardized echinacea preparation (Echinilin) for the treatment of the common cold: a randomized, double-blind, placebo-controlled trial. *Journal of Clinical Pharmacy and Therapeutics* 29(1): 75-83.

Gwaltney JM. 2002. Clinical significance and pathogenesis of viral respiratory infections. *American Journal of Medicine* 112(6): 13-18.

Hermann JR, Honeyman MS, Zimmerman JJ, Thacker BJ, Holden PJ, Chang CC. 2003. Effect of dietary *Echinacea purpurea* on viremia and performance in porcine reproductive and respiratory syndrome virus-infected nursery pigs. *Journal of Animal Science* 81(9): 2139-2144.

Hill N, Stam C, Hanselen RA. 1996. The efficacy of prikweg R. gel in the treatment of insects bites: A double-blind, placebo-controlled clinical trial. Pharmacy. *World and Science* 17: 35-41.

Hobbs CR. 1994. *Echinacea*- A literature review. *Herbalgram* 30: 33-49.

Hobbs CR. 1989. The *Echinacea* handbook. Botanica Press, Capitola

Hudson JB. 2009. The use of herbal extracts in the control of influenza. *Journal of Medicinal Plant Research* 3(13): 1189-1194.

Hudson JB. 2012. Applications of the phytomedicine *Echinacea purpurea* (Purple Coneflower) in Infectious Diseases. *Journal of Biomedicine and Biotechnology* 1-16. doi:10.1155/2012/769896.

Laasonen M, Wennberg T, Harmia-Pulkkinen T, Vuorela H. 2002. Simultaneous analysis of alkamides and caffeic acid derivatives for the identification of *Echinacea purpurea, E. angustifolia, E. pallida* and *Parthenium integrifolium* roots. *Planta Medica* 68: 572-574.

Lans C, Turner N, Khan T, Brauer G, Boepple W. 2007a. Ethnoveterinary medicines used for ruminants in British Columbia, Canada. *Journal of Ethnobiology and Ethnomedicine* 3(1) 11. DOI: 10.1186/1746-4269-3-11.

Lans C, Turner N, Khan T, Brauer G. 2007b. Ethnoveterinary medicines used to treat endoparasites and stomach problems in pigs and pets in British Columbia, Canada. *Veterinary Parasitology* 148(3-4): 325-340.

Lienert D, Anklam E, Panne U. 1998. Gas chromatography-mass spectral analysis of roots of *Echinacea* species and classification by multivariate data analysis. *Phytochemical Analysis* 88-98.

Makeown KA. 1998. A review of the taxonomy of the genus *Echinacea*. *Perspectives on New Crops and New Uses* 482-489.

Mascolo N, Autore F, Capasso A, Menghini, Fasulo MP. 1987. Biological screening of Italian medicinal plants for antiinflammatory activity. *Phytotherapy Research* 1(1): 28-31.

Mayani A, Vazinan M, Saridnia S. 2015. *Echinacea purpurea*, pharmacology, phytochemistry and analysis methods. *Pharmacognosy Review* 9(17): 63-72.

Nichols WG, Peck Campbell AJ, Boeckh M. 2008. Respiratory viruses other than influenza virus: impact and therapeutic advances. *Clinical Microbiology Reviews* 21(2): 274-290.

O'Neill W, McKee S, Clarke AF. 2003. Immunologic and hematinic consequences of feeding a standardized Echinacea (*Echinacea angustifolia*) extract to healthy horses. *Equine Veterinary Journal* 34(3): 222-227.

Parthasarthi T, Velv G, Jeyakumar P. 2013. Impact of crop heat unit on growth and developmental physiology of future crop production: a review. *Journal of Crop Science and Technology* 2(1): 1-11.

Pleschka S, Stein M, Schoop R, Hudson JB. 2009. Anti-viral properties and mode of action of standardized *Echinacea purpurea* extract against highly pathogenic avian Influenza virus (H5N1, H7N7) and swine-origin H1N1 (S-OIV). *Virology Journal* 13(6): 197.

Rawls R. 1996. Europe's strong herbal brew. *Chemical and Engineering News* 53-60.

Schneider S, Reichling J, Stintzing FC, Messerschmidt S, Meyer U, Schnitzler P. 2010. Anti-herpetic properties of hydroalcoholic extracts and pressed juice from *Echinacea* pallida. *Planta Medica* 76(3): 265-272.

Schulte KE, Ruecker G, Perlick J. 1967. The presence of polyacetylene compounds in *Echinacea purpurea* and *Echinacea angustifolia*. *Arzeimittel Forschung* 17: 825-829.

See H, Wark P. 2008. Innate immune response to viral infection of the lungs. *Paediatric Respiratory Reviews* 9(4): 243-250.

Selvanayahgam ZE, Gnanevendhan SG, Balakrishna K, Rao RB. 1994. Antisnake venom botanicals from ethnomedicine. *Journal of Herbs, Spices and Medicinal Plants* 2(4): 45-100.

Shahi AK, Kaul MK, Gupta R, Dutt P, Chandra S, Qazi GN. 2005. Determination of essential oil quality index by using energy summation indices in a elite strain of *Cymbopogon citratus* (DC) Stapf [RRL(J)CCA12]. *Flavour and Fragrance Journal* 20: 118-121.

Shahi AK, Singh A. 1987. The significance of heat unit system for crop management practices in aromatic plants-An introductory study on Mentha arvensis. *Indian Perfumer* 31: 100-108.

Sharma SM, Anderson M, Schoop SR, Hudson JB. 2010. Bactericidal and anti-inflammatory properties of a standardized *Echinacea* extract (Echinaforce): dual actions against respiratory bacteria. *Phytomedicine* 17(8-9): 563-568.

Suchulthess B, Giger E, Baurmann TW. 1989. May achene analysis serve for species diagnosis in *Echinacea*. *Plant Medica* 55: 213-367.

Suchulthess BH, Giger E, Bumann TW. 1991. *Echinacea*: anatomy, phytochemical pattern, and germination of the achene. *Planta Medica* 57: 384-388.

Thappa RK, Bakshi SK, Dhar PL, Agarwal SG, Kitchlu S, Kaul MK, Suri KA. 2004. Significance of changed climatic factors on essential oil composition of *Echinacea purpurea* under subtropical conditions. *Flavour and Fragrances Journal* 19: 452-454.

Wagner H , Breu W, Willer F , Wierer M, Remiger P, Chwenker G. 1989. Invitro inhibition of arachidonate metabolism by some alkamides and preylated phenols. *Planta Medica* 55.

Wang CY, Staniforth V, Chiao MT, Hou CC, Wu HM, Yeh KC, Chen CH, Hwang PI, Wen TN, Shyur LF, Yang NS. 2008. Genomics and proteomics of immune modulatory effects of a butanol fraction of *Echinacea purpurea* in human dendritic cells. *BMC Genomics* doi: 10.1186/1471-2164-9-479. 9: 479.

Woelkart K, Bauer R. 2007. The role of alkamides as an active principle of *Echinacea*. *Planta Medica* 73: 615-623.

Wurr DCE, Fellow JR, Phelps K. 2002. Crop scheduling and prediction-Principles and Opportunities with field vegetables. *Advances in Agronomy* 76: 201-234.

20

Note on Commercial Valuable Medicinal Wealth in Hamirpur District of Himachal Pradesh in Western Himalaya

Anjna Sharma and Gurdarshan Singh

PK-PD Toxicology and Formulation Division
CSIR-Indian Institute of Integrative Medicine, Canal Road, Jammu - 180001
Jammu & Kashmir, INDIA
*Email: sharmaanjna87@gmail.com

ABSTRACT

The popular use of local plant wealth among the rural people in Hamirpur district of Himachal Pradesh reflects their greater interest in the traditional medicine and other traditional and commercial purposes. Hamirpur, one of the pioneer Himalayan district is a rich repository of medicinal flora. It is the smallest district of Himachal Pradesh due to its area, surrounded by thick forest area which is rich in diverse vegetation. Medicinal plants are used in the Ayurvedic, Unani and other traditional systems of medicine and in plant-based pharmaceutical industries. The patterns of the use of the local medicinal herbs for the treatment of various diseases and ailments have been an old practice. But the people of the district still depends upon the uses of local plants for their primary health problems treatment. This paper provides the information on the traditional uses of some of the plants such as *Asparagus racemosus, Acacia catechu, Albizzia lebbek, Berberis lyceum, Bryonia laciniosa, Azadirachta indica, Adhatoda vasica, Andographis peniculata, Aegle marmelos, Celastrus peniculata, Cymbopogon flexsosus, Cuscuta reflexa* and *Holarrheana antidysentric* of Hamirpur district for the treatment of various disorders such as anthelmintic, anti-fungal, diabetic, antibacterial, cancer, diarrhea, bronchitis, tuberculosis etc. Hence the objective of the study was to document ethnobotanical knowledge primarily of notable herbs employed by the different backward people, whether tribal or rural, in the area under study. Therefore the documentation and recording of traditional knowledge associated is necessary step for the preservation of traditional knowledge about the use of local plant wealth and conservation of plant species of study area. The data would be useful for further scientific exploration.

Keywords: Western Himalaya, Medicinal uses, Chemistry, Pharmacology and Conservation

INTRODUCTION

Medicinal plants contains substances that can be used for therapeutic purpose or which is a precursor for synthesis of useful drugs. Man has been using various plant species since the time immemorial for well-being. The plants used for curing various diseases in human have been mentioned in ancient literatures like Rig-Veda, Bible and Quran. Plants are the basic source of knowledge of modern medicine, out of the total 422,000 flowering plants reported from the world, more then 50,000 are used for medicinal purposes. In India, more than 43% of the total flowering plants are reported to be of medicinal importance. Utilization of plants for medicinal purposes in India has been documented long back in ancient literatures. However, organized studies in this direction were initiated in 1956. Majority of the people living in the developing world is struggling to increase the standard of living and to improve the health care delivery in the face of increasing poverty and growing population.

It has been estimated that 70-80% of population in the developing countries relis only on their traditional herbal remedies for their ailments as the costly pharmaceuticals are out of their reach. Keeping in view the above fact it can be inferred that by careful collection of data and experimentation, medicine of much higher value and low cost can be isolated from the plants, to fulfill the requirement of the major portion of the world population especially that of developing world. Therefore, importance, necessity and potentiality of medicinal plants cannot be overlooked. In India, medicinal plants form the backbone of several indigenous traditional systems of medicine. The first mention of medicinal use of plant has been documented in Rig-Veda which mentioned 67 plants having therapeutic effects. The properties of traditionally used drugs have been documented in details in "Ayurvedic" one of the sacred texts of the Hindu philosophy.

WHO report depicts that more than 80% of world's population rely on plants based products to meet their health care needs (Ramya et al. 2008 (Pandey et al. 2016). Nearly, 25 to 45% of modern prescriptions contain plant derived lead molecules as a basic source in drug formulations. Furthermore, about 42% of 25 top selling drugs marketed worldwide are either directly obtained from natural sources or entities derived from plant products. There has been a revival of interest in herbal medicines. This is due to increased awareness of the limited ability of synthetic pharmaceutical products to control major diseases and the need to discover new molecular structures as lead compounds from the plant kingdom.

Plants are the basic source of knowledge of modern medicine. The basic molecular and active structures for synthetic fields are provided by rich natural sources. This burgeoning worldwide interest in medicinal plants reflects recognition of the validity of many traditional claims regarding the value of natural products in health care. The relatively lower incidence of adverse reactions to plant preparations compared to modern conventional pharmaceuticals, coupled with their reduced cost, is encouraging both the consuming public and national health care institutions to consider plant medicines as alternatives to synthetic drugs.

Being a hilly State, Himachal Pradesh has rich plant diversity due to varying degree of agro-climatic zone from subtropical to extreme cold. The state is a bucket of large variety of medicinal herbs. There are about 3500 known plant species recorded in the state, and out of these about 500 are reported as of medicinal value. This plants diversity is used by the local people for various activities like, traditional healthcare. Hamirpur district of Himachal Pradesh is a hilly track and has rich diversity of flora. So this is a suitable area for study related to medicinal plants.

GEOGRAPHIC LOCATION

Himachal Pradesh experiences diverse climatic conditions due to the wide variations in altitude ranging from 350-6500 m above mean sea level. Himachal Pradesh is prosperous in plant diversity with large numbers of plant resources confined to different climatic zones on the basis of temperature, precipitation, latitude and altitude.

Intensive ethno botanical exploration field works were undertaken in selected places of Hamirpur district to find out medicinal plants used for various ailments. Data on medicinal plants were collected by making exploration trips to different areas of Hamirpur district. Local inhabitants of the selected study areas were interviewed for making sustainable use of plant resources. An attempt was made to get the useful information on medicinal plants. Data related to each ethno botanical aspect were collected from local people of that area. After collecting the probable information on ethno- medicinal plants the data were analyzed and compiled with related literature and then the report was documented.

STUDY AREA

The Hamirpur (Himachal Pradesh) district is situated between 76°-17'50" to 76°43'42" east longitudes and 31°24'48" to 31°53'35" north latitudes. It is located in the south western part of Himachal Pradesh. It is a part of Lower Himalayas, the elevation varies from the 400 meters to 1,100 meters. The town of Hamirpur lies to the east of Jakh range where the country is undulating, but in the north and north east bare and rugged hills, deep ravines with precipitous sides transform the landscape into what has been described as an agitated sea suddenly arrested and fixed stones. The main hill ranges of the district are known as Jakh dhar and Sola Singhi Dhar (Piplu ki dhar). The Jakh dhar runs in continuation of Kali dhar range in the Kangra district. It enters in Hamirpur district near Nadaun and transverses it into southeastern direction. The rivers are drained by a number of perennial streams which are tributaries of either river Beas or river Satluj. Bakar Khad, Kunha Khad and Man Khad drain into river Beas, while Sukar Khad and Mundkhar Khad drain into Seer Khad which ultimately mingles into the river Satluj.

MEDICINAL PLANTS AND THEIR USAGES

While data collection of medicinal plants, various species were documented from Hamirpur district of Himachal Pradesh. The details are given below and in table 1:

1. Asparagus racemosus

Asparagus racemosus belongs to the family Asparagaceae *Asparagus racemosus* is a plant used in traditional Indian medicine (Ayurveda). The root is used to make medicine. Asparagamine A, a polycyclic alkaloid was isolated from the dried roots and subsequently synthesized to allow for the construction of analogs. Two new steroidal saponins, shatavaroside A and shatavaroside B together with a known saponin, filiasparoside C, were isolated from the roots of Asparagus racemosus. Five steroidal saponins, shatavarins VI-X, together with five known saponins, shatavarin I (or asparoside B), shatavarin IV (or asparinin B), shatavarin V, immunoside and schidigerasaponin D5 (or asparanin A), have been isolated from the roots of Asparagus racemosus. Also known is the isoflavone 8-methoxy-5,6,4'-trihydroxyisoflavone 7-O-beta-D-glucopyranoside (Lakhsmi et al. 2011).

Medicinal uses

It is used in case of upset stomach (dyspepsia), constipation, stomach spasms, and stomach ulcers. It is also used for fluid retention, pain, anxiety, cancer, diarrhea, bronchitis, tuberculosis, dementia, and diabetes.

2. Acacia catechu

Acacia catechu is a deciduous tree with a light feathery crown and dark brown, glabrous, slender, thorny, shining branchlets, usually crooked. Bark dark brown or dark grey, brown or red inside, nearly 12-15 mm in thickness, rough, exfoliating in long narrow rectangular flakes which often remain hanging. Branchlets armed with pseudo-stipular spines in pairs below the petioles. Pod, 10-15 cm long and 2-3 cm thin, straight, flat, glabrous dark-brown and shining when mature.

Medicinal Uses

The different parts of the tree have a variety of medicinal uses, which in haemoptysis (spitting of blood). A paste of the bark is useful in conjunctivitis. The bark is reported to be useful in the treatment of snake bites. Flowers: A mixture of flower tops, cumic, milk and sugar is useful in gonorrhea. Wood: Cutch and katha obtained from the heartwood have great medicinal value. It is cooling, digestive and a very valuable astringent, specially in chronic diarrhea and dysentery, bleeding piles, uterine haemorrhages, leucorrhoea, gleet, atonic dyspepsia, chronic bronchitis, etc. It is also useful in cases of mercurial salivation, bleeding or ulcerated or spongy gums, hypertrophy of the tonsils, relaxation of the uvula, aphthous ulceration of the month, etc. A mixture of catechu and myrrh (Kathol) is usually prescribed as a tonic and as a galactagogue to women after confinement. Kheersal is used as a remedy for chest diseases, especially for the treatment of asthma, cough and sore throat (Lakhsmi et al. 2011, Singh et al. 2006).

3. Albizzia lebbek

It is a medicinal tree native to India which is found throughout the country. In Ayurveda, it is used to prepare various medicines. This tree contains alkaloids, tannins, saponins and flavonoids which has medicinal action. It is a nitrogen fixing tree. This species is used in treatment of bites and stings from poisonous animals such as snake.

Medicinal Uses

Bark of the plant is used in the treatment of leucoderma, itching, skin diseases, piles, exercise perspiration, inflammation, erysipelas and bronchitis, Bark of the plant is used in the treatment of asthma and allergic disorders, Leaves of the plant are used in night blindness and strengthen gums and teeth. The seeds are useful as aphrodisiac and tonic to the brain. Used for gonorrhoea and tuberculosis glands; oil is applied topically in leucoderma. Flowers are given in case of asthma and snake-bite.

4. Berberis lyceum

Berberis lycium is an evergreen shrub belongs to family Berberidaceae. It is also known as Indian berberry in English, kashmal or kasmal in Hindi and Ishkeen in Urdu. It is a suberect, rigid, spiny shrub 2.7- 3.6 ml in height. The genus *Berberis* is widely distributed in America, Europe and Asia (Agharkar et al., 1991)

Medicinal Uses

The extract of the root bark of Berberis lycium is known to produce antidiabetic effects on the rabbits and helps in reducing blood glucose level. The crude powder of *Berberis lycium* root has anti-hyperlipidemic effect in alloxanized rabbits. Hypercholesterolemia and hypertriglyceridemia have been reported to occur in alloxan induced diabetic rabbits (Wojtowich et al. 2004). *Berberis lycium* root bark powder significantly reduces the total cholesterol and triglyceride and LDL of treated rabbits as compared to untreated diabetic rabbits (http://natureconservation.in/description-and-medicinal-uses-of-azadirachta-indica-neem/). *Berberis lycium* have shown hepatoprotective effect also.

5. Bryonia laciniosa

Bryonia laciniosa is an Ayurvedic herb used in traditional medicine as an aphrodisiac and pro-fertility compound, touted to increase masculinity and enhance youthfulness during aging. It belongs to the category of *Vrishya rasayana* alongside *Anacyclus pyrethrum*. The plant belongs to Cucurbitaceae and *Bryonia alba* reported has anti-tumor properties and *Bryonia diocia* (brony root, a diuretic).

Medicinal Uses

Byronia laciniosa is touted as an anti-pyretic (against fever) and was found to exert anti-pyretic effects (at 500mg/kg) at an efficacy similar to the control drug, paracetamol (150mg/kg). This study also noted analgesic (painkilling) actions in a dose-dependent manner, although even 500mg/kg *Byronia Laciniosa* was less effective (52.61% inhibition relative to control) than 100mg/kg Aspirin (68.87% inhibition) (Sood et al. 2012).

6. Azadirachta indica

It is a tree in the mahogany family Meliaceae. It is one of two species in the genus *Azadirachta*, and is native to India, Myanmar, Bangladesh, Sri Lanka, Malaysia and Pakistan, growing in tropical and semi-tropical regions. Neem is a fast-growing tree that can reach to a height of 15-20 m, rarely to 35–40 m. It is evergreen, but in severe drought it may shed most or nearly all of its leaves. The branches are wide spread.

Medicinal Uses

Azadirachta indica is variously known as Sacred Tree. Products made from neem tree have been used in India for over two millennia for their medicinal properties. All parts of neem tree used as anthelmintic, anti-fungal, anti-diabetic, antibacterial, antiviral, contraceptive and sedative. Neem tree is used in many medicinal treatment like skin diseases, healthy hair, improve liver function, detoxify the blood, Pest and disease control, fever reduction, dental treatments, cough, asthma, ulcers, piles, intestinal worms, urinary diseases etc. Oil of neem used in soap, shampoo, balms and creams as well as toothpaste. Small branches of neem used as toothbrush. Neem gum is used as a bulking agent and for the preparation of special purpose food (for diabetics). Neem leaf paste is applied to the skin to treat acne, and in a similar vein is used for measles and chicken sufferers. Practitioners of traditional Indian medicine recommend that patients suffering from chickenpox sleep on neem leaves (Agharkar et al. 1991).

7. Adhatoda vasica

Adhatoda vasica is a small evergreen plant of the Acanthaceae family, with broad, lanceolate (sharp and pointed like a lance) leaves measuring 10 to 16 centimeters in length and 5 centimeters wide. They become greenish-brown when dried and have a bitter taste. They have a smell similar to strong tea.

Medicinal Uses

The leaves, roots and flowers of *Adhatoda vasica* also called vasa or vasaka were used extensively in traditional Indian medicine for thousands of years to treat respiratory disorders such as asthma. *Adhatoda vasica* is considered useful in treating bronchitis, tuberculosis and other lung and bronchiole disorders. Vasicine stimulates the respiratory system. It induces bronchodilation and relaxes the tracheal muscles. It provides significant protection against histamine induced bronchospasm. It reduces the elasticity and viscosity of tracheal mucus. Vasicine also exhibites moderate hypotensive activity and cardiac-depressant effect. It also shows anti-oxidation and hyaluronidase inhibiting activities, with improved moisture holding property. Alkaloids showes both bronchodilatory and bronchoconstrictory activity. It stimulates the respiratory system. It shows potent anti-inflammatory effect and also possesses abortifacient activities.

8. Andographis paniculata

The genus *Andrographis* belongs to Acanthaceae family, comprised of about 40 species. Only a few are popular for their use in folk medicine for assorted health concerns. Of these few, *A. paniculata* is the most important. *A. paniculata*, commonly known as King of Bitters or kalmegh, is an annual, branched, erect handsome herb running half to one meter in height.

Medicinal Uses

It has been used by traditional medical practitioners for stomachaches, inflammation, pyrexia, and intermittent fevers. The whole plant has been used for several applications such as anti-dote for snake-bite and poisonous stings of some insects, and to treat dyspepsia, influenza, dysentery, malaria and respiratory infections. The leaf extract is a traditional remedy for the treatment of infectious disease, fever-causing diseases, colic pain, loss of appetite, irregular stools and diarrhea (Chopra et al. 1980, Jarukamjorn et al. 2010, Chaturvedi et al. 1983, Balu et al., 1993)..

9. Aegle marmelos

It is a spinous deciduous and aromatic tree with long, strong and axillary spines of family Rutaceae. This tree grows up to 18 mt in height and thickness of tree is about 1-2.5 m. leaves are 3-5 foliate, leaflets are ovate and have typical aroma. Flowers are greenish white and sweet scented. Fruits are large, woody, greyish yellow, 8-15 celled and have sweet gummy orange colored pulp. Seeds are compressed, oblong and numerous found in aromatic pulp. Main chemical components are marmelosin, alloimperatorin, marmelide, tannic acid, marmin, umbelliferone, isoimperatorin, isopimpinellin, skimmin, marmesin, marmesinin, fatty acids, beta-sitosterol (http://www. motherherbs. com/aegle-marmelos-extract.html).

Medicinal Uses

Extract of *Aegle marmelos* fruits shows hypoglycaemic activities. Problems of the female reproductive system like *leucorrhea*, menstrual irregularities, vaginal hemorrhages etc. are also relieved with the use of bilva along with other herbs.

10. Celastrus paniculata

Celastrus paniculatus of the family Celastraceae is a large woody climbing shrub with a yellowish bark which is referred to as *Jyothismati* in Ayurveda. It is a deciduous vine plant which grows natively throughout the Indian subcontinent. It has been used in various Indian cultural and medical traditions, such as Ayurveda and Indian Unani. It contain sequesterpene alkaloids celastrine and celastrol (0.15% of arils) as well as celapanin, celapanigin, celapagin, paniculatin and paniculatadiol (seed oil) malkanguniol (a major consituent of the seed oil), pristimerin, zeylasterone and zeylasteral (bark extract) α-dihydroagarofuranoid sesquiterpenes, polyalcohol A-D and the polyalcohol dulcito.

Medicinal Uses

Celastrus Paniculatus also called Jyotishmati in Sanskrit or Malkangani in Hindi improves memory and cognitive functions. It is beneficial to neurological diseases and pain disorders including muscle cramps, backache, sciatica, osteoarthritis, facial paralysis and paralysis.The seeds of *Celastrus paniculatus* have very hot potency, which means seed powder produces heat in the body and gives feeling of warmness after its consumption. This effect generally appears with larger dosage of Malkangn.

11. Cymbopogon flexuosus

Indian Lemongrass is an aromatic, evergreen, clump-forming perennial grass producing numerous, erect stems up to 3 metres tall from a short, thick rhizome.

Indian lemongrass is one of the main sources of lemongrass essential oil. The plant is often cultivated, both in gardens and commercially, for this oil which has a multitude of uses as a food flavouring, in perfumery, medicinal etc. Rhizomes of *C. citratea* were reported to contain about 0.52% alkaloids from 300 g plant material. Some of the flavonoids isolated from *Cymbopogon* species are Isoorientin, tricin, luteolin, luteolin 7-O-glucoside (cynaroside), isoscoparin and 2''-O-rhamnosyl isoorientin. Catechol, chlorogenic acid, caffeic acid and hydroquinone are important constituents.

Medicinal Use

Lemongrass is used for treating digestive tract spasms, stomachache, high blood pressure, convulsions, pain, vomiting, cough, achy joints (rheumatism), fever, common cold and exhaustion. It has potent in vivo activity against Ehrlich and Sarcoma-180 tumors (Sharma et al. 2009).

12. Cuscuta reflexa

Cuscuta reflexa is commonly called as dodder plant, and also known as devil's hair, witch's hair, love vine, amarbel and akshabela etc. *Cuscuta reflexa* is a parasitic weed plant and also an extensive climber. *Cuscuta* grows as homoparasite, it has very low level of chlorophyll and photosynthesis activity, completely depends over the host plant for its survival. *Cuscuta reflexa* varies in the colour of flowers produced from white to pink. Flowers generally produced in the early summer and autumn but also depend on the species. Seeds are produced in the large quantities. Kaempferol-3-o- glucoside, quercetin, quercetin-3-o-glucoside cuscutin, stearic acid, palmitic acid, dulcitol, luteolin, coumarina and cuscutamine.

Medicinal Uses

Hydroalcoholic extract of *C. reflexa* showed hepatoprotectic activity in albino rats against paracetamol induced hepatic damage and act as hepatoprotective agent. Antitumor activity chloroform and ethanol extracts of *C. reflexa* showed antitumor activity against Ehrlich ascites carcinoma tumor in mice at doses of 200 and 400 mg/kg body weight orally. Acute oral toxicity studies were also performed to determine the safety of the extracts antioxidant activity In vitro antioxidant activity of *Cuscuta reflexa* stems were investigated by estimating the degree of non enzymatic haemoglobin glycosylation (Saini et al. 2015).

13. Holarrhena antidysentria

Holarrhena antidysenterica plant grows as a deciduous shrub up to 3ms high. The leaves are opposite, ovate and 10-cm long. The stem has several branches. The flowers are corymb-like cymes. The flowers have five white petals and grow up to 2-3cm high. The fruits are cylindrical and paired and the seeds are light brown. It is a medicinal plant in Ayurveda.

Medicinal Uses

It is used for the treatment of dysentery caused due to amoeba. The seeds are antibilious and promote conception. It is also used for toning up vaginal tissue after delivery in women. The plant is used for the treatment of skin diseases such as scabies, ringworm, itching and other infections. The plant is used as the rejuvenating agent for the immune system in the body. It also cures rheumatoid arthritis and osteoarthritis. In treatment of bleeding piles, kurai checks the secretion of mucus and blood. It is also used for mal-absorption. It is also used for the treatment of colic and also used for the treatment of urinary tract infection. It possesses many biological activities which are attributed to their antimicrobial activity. These compounds also demonstrate antiviral, anti-inflammatory and anticancer properties.

CONCLUSION

Ethnobotanical and traditional knowledge contributes to the conservation of biodiversity and provides resource of economic and ecological interest. It has been realized that medicinal herbs are going to play an important role in future for drug programme. Herbal drugs provide strength to the body organs and stimulate normal functioning. The herbal drugs act selectively without disturbing other system. These herbal products are the symbol of safety as compare to the synthetic drugs, which is regarded as unsafe to human being and environment. Modern medicine affects several metabolic activity in the human system and has side effects which makes body more susceptible to other diseases. Our attempts for this research work will not only provide recognition to this treasure, but also help in the conservation of these medicinal plants for further researchers in India and elsewhere worldwide. Although herbs had been priced for their medicinal, flavouring and aromatic qualities for centuries, the synthetic products of the modern age surpassed their importance, for a while. However, dependence on synthetics is over and people are returning to the

Table 1: Major plants of commercial values of Hamirpur district of Himachal Pradesh

Sr.No.	Botanical names/Family	Local name	Part used	Disease/ailment	Treatment
1.	*Asparagus racemosus* Wild./ Asparagaceae	Shatavar	Roots	Aphrodisiac	The dry roots are powdered and taken orally with warm water once a day for 10-15 days.
2.	*Azadirachta indica* A.Juss/ Meliaceae	Neem	Roots	Joints pain	Roots paste is applied on the af fected area.
3.	*Adhatoda vasica* L./Acantheceae	Adusha	Leaves	Antiseptic	Leaves paste is used
4.	*Andrographis paniculata* Nees/ Acanthaceae	Kalmegh	Whole plants	Fever	Entire plant is ground, boiled in water and filtered, taken regularly till fever breakdown
5.	*Aegle marmelos* (L.) Corrêa/ Rutaceae	Bil	Fruits	Gastric problems	Fruit juice is taken for a few days
6.	*Celastrus paniculata*/ Celastraceae	Malkangni	Seeds	Joint pain, rheumatism	Seed oil is applied on the affected part.
7.	*Cymbopogon flexosus* (Nees ex Steud.) W.Watso/ Poaceae	Nimbu	ghass Leaves	Back ache	Leaves paste is applied to affected parts
8.	*Cuscuta reflexa* Roxb/ Convolvulaceae	Akahash bel	Leaves	Cold	Leaves extract is used.
9.	*Albizzia lebbek* (L.) Benth / Fabaceae	Arjun	Bark	Blood pressure	The dry bark is powdered and taken orally with lukewarm water twice a day for 3-4 days
10.	*Holorrhena antidysentric*/ Apocynaceae	Vatsaka	seeds	rheumatoid arthritis	Rheumatoid arthritis,osteoarthris
11.	*Acacia catechu* (L.f.) Willd/ Mimosaceae	Katha	Bark	Throat infection	A paste of the bark is useful in con junctivitis
12.	*Berberis lyceum* Royle/ Berberidaceae	Kaimblu	Flowers	Antiprolific, anti-psoriatic	Used and powder form
13.	*Bryonia laciniosa*/ Cucurbitaceae	Climber	Seeds	Aphrodisiac and pro-fertility	Flower and seeds

naturals with hope of safety and security. It's time to promote them globally.

Competing Interest

Author declares no conflict of interest.

ACKNOWLEDGEMENTS

Authors are grateful to Director, CSIR-Indian Institute of Integrative Medicine, Jammu for support in this research work. This research was supported by Council of Scientific and Industrial Research, New Delhi. Author acknowledge AcSIR for their enrollment in Ph.D. program.

REFERENCES CITED

Agharkar SP. 1991. Medicinal plants of Bombay presidency. Scientific Publishers.India. 44-48.

Balu S, Alagesaboopathi C. 1993. Anti-inflammatory activities of some species of *Andrographis* Wall. *Ancient Science of Life* 13: 180–184.

Chaturvedi GN, Tomar GS, Tiwari SK, Singh KP. 1983. Clinical studies on Kalmegh (*Andrographis paniculata* Nees) in infective hepatitis. *International Journal of Ayurveda Research* 2: 208–211.

Chopra RN. 1980. Glossary of Indian medicinal plants. Council for Scientific and Industrial Research, New Delhi

http://natureconservation.in/description-and-medicinal-uses-of-azadirachta-indica-neem/

http://www.motherherbs.com/aegle-marmelos-extract.html.

Jarukamjorn K, Kondo S, Chatuphonprasert W, Sakuma T, Kawasaki Y, Emito N. 2010. Gender-associated modulation of inducible CYP1A1 expression by andrographolide in mouse liver. *Eurorean Journal of Pharmaceutical Science* 39: 394-401.

Lakshmi T, Anitha R, Geetha RV. 2011. Acacia catechu willd -A gift from ayurveda to mankind – A Review. 5: 273-93.

Pandey A, Singh S. 2016. Traditional phytotherapy for various diseases by the local rural people of bharai village in the kullu district of himachal pradesh (india). *International Journal of Pharmaceutical Science and Research* 7: 1263- 1270

Ramya S. 2008. Phytochemical screening and antibacterial activity of Leaf Extracts of *Pterocarpus marsupium. Ethnobotanical Leaflet* 12: 1029-1034.

Saini P, Mithal R, Menghani E. 2015. A parasitic Medicinal plant *Cuscuta reflexa*: An Overview. *International Journal of Scientific & Engineering Research* 951: 2229-5518.

Sharma P, Mondhe DM, Muthiah S, Harish C. Pal, Ashok K. Shahi, Ajit K. Saxena, Ghulam N. Qazi. 2009. *In -vivo* activity against Ehrlich and Sarcoma-180 tumors. 15: 160-168.

Singh KN, Lal B. 2006. Notes on Traditional Uses of Khair (Acacia catechu Willd.) by Inhabitants of Shivalik Range in Western Himalaya. *Ethnobotanical Leaflets* 10: 109-12.

Sood P, Modgil R, Sood M. 2012. *Berberis lycium* a medicinal plant with immense value. *Journal of Pharmaceutical and Biological Research* 1: 27-37.

21

Revisiting Himalayan High Altitude Plants for Skin Care and Disease

Javaid Fayaz Lone[1] and Bikarma Singh[1,2]*

[1]*Plant Sciences (Biodiversity and Applied Botany), CSIR-Indian Institute of Integrative Medicine, Jammu - 180001, Jammu and Kashmir, INDIA*
[2]*Academy of Scientific and Innovative Research, New Delhi-110001, INDIA*
*Corresponding email: drbikarma@iiim.res.in, drbikarma@iiim.ac.in

ABSTRACT

Himalaya represents an important focal point of botanical rich diversity, and home of many medicinal plants known for their various therapeutic properties. Since ancient times, several plant species have been used as ethnomedicine, and only a few of these species have been fully explored and validated scientifically. Today, cosmetic and pharmaceutical industries are the fastest growing sectors that use botanical extracts to maintain the health and support integrity of the skin and other parts of human body. People are demanding cosmeceutical additives from market made from plants as herbal botanicals as these causes fewer or no side effects, and therefore, ethnomedicine are gaining immence importance for cosmetic and value added products development of natural origin. Human and animal skin is the largest covering of the body as it provide barrier to the internal cells or tissues, and protect the body from infections and toxic chemicals. Keeping in mind the importance of skin, investigations and literatures were consulted to document plants protecting skin infections from external environment. The potential high altitude plants of Himalaya were screened out to be used in skin care and treatment of various skin pathogens. In the present communication, total 81 plants of 72 genera belonging to 40 families are presented that have ethnobotanical application in skin care and usages in curing skin diseases. This data will provide a base-line information for cosmetic and pharmaceutical companies for development of various new value-added products for human and animal health care.

Keywords: Medicinal Plant, High Altitude, Unexplored, Skin Care, Himalayas.

INTRODUCTION

In modern world, cosmetic and pharmaceutical industries are the fastest growing sector that uses the botanical extracts to maintain human health care, and support integrity of skin and other parts of human body (Stallings and Lupo 2009). People is demanding the cosmeceutical additives from market as herbal products for their skin care, fragrance and cosmotic as they cause fewer or no side effects (Lal and Kishore 2014). Skin represents the outer covering of the body that provides protection to internal tissues from injury and infection (Mir et al. 2014). Besides preventing the harmful substances from entering the body, skin prevents the loss of nutrients and protects the internal cells and tissues from extreme temperature. It has been scientifically validated that a large number of endogenous and exogenous variables are known to affect skin such as age, body part, skin type, sweat, soap, and other human-use products (Yosipovitch and Maibach 1996, Panther and Jacob 2015).

Tribal population in hilly regions are living in harmony with nature and depend on plants for their daily needs as food and medicine since human civilization (Chopra 1933, Kirtikar and Basu 1981, Chopra et al. 1986, Singh et al. 2016, Singh and Bedi 2017). Herbal medicines or ethnobotany represents useful component of the natural wealth and is recognized across the world for less or no side effects (Singh et al. 2014). Ethno medicinal plants are the main locally available resources for curing various ailments as evident from published literatures. Indian Himalaya is a hub and rich repository of medicinal plants distributed in different altitude, exposed to varying environmental conditions, and each of them has a folk knowledge associated. Lots of ethnobotanical investigations have been carried out across the world with reference to plants used as skin care and other diseases (Kirtikar KR 1935, Jain 1981, Dastur 1970, Wahid and Siddiqui 1961, Padal et al. 2013, Pushpangaden 1986, Raghunathan 1976, Rout et al. 2009, Sebastian and Bhandari 1984, Sinhababu and Banerjee 2013, Tabbassum and Hamdani 2014). The present investigation were carried out to explore and to document the plant species used to cure skin diseases and associated problems in Himalaya regions of India such as Jammu and Kashmir, Himachal Pradesh, Uttarakhand, Sikkim and Arunachal Pradesh.

Himalayan High Altitude Ethnomedicinal Plants Used for Skin Care

Several papers pertaining to the ethnobotanical studies curing dermal disorders and diseases have been published worldwide (Khan and Chaghtai 1982, Manishayadav 2012, Policepatel and Manikrao 2013). Hence, an attempt have been made to record some of these high value medicinal plants distributed throughout Indian Himalaya used for skin care. In present communication, total 81 plants of 72 genera belonging to 40 families have been mentioned which are commonly used for skin associated problems and health care. Botanical name, family, vernicular name, altitudinal distribution range and method of usage are presented in Table 1. Family such as Asclepiadaceae, Asteraceae, Berberidaceae, Caprifoliaceae, Caryophyllaceae, Convolvlaceae, Euphorbiaceae, Fabaceae, Gentianaceae, Lamiaceae, Liliaceae, Ranunculaceae and Zingiberaceae were the most dominant family used in skin care and associated diseases frequently used by tribal people of Himalaya as medicine. These medicinal plants were not only used for skin care, but can be used for other treatments due to high medicinal demand by virtue of diverse chemical constituents.

CONCLUSION

Himalayan hotspot is a repository of medicinal plants and recognized as an important focal point of botanical rich diversity that have therapeutic properties. Since ancient times, several plant species have been used in folk medicine and only a few species have been fully explored and validated scientifically. Now a day, ethnomedicinal plants are gaining immence importance as people are demanding cosmetic products of natural origin especially plant-based products. The plants which has different application in skin care has great economic values and can be used on daily basis as medicine or as plants for skin care. The medicinal plants described in this communication cover a number of skin diseases, and this investigated data can serve as the baseline information for development of value added medicine and products for human and animal skin care. Realizing the importance of medicinal plants in skin care, there is also need to conserve these valuable plant for future. Upcoming researches should focused relates to intervention of molecular and biotechnolgical sciences which will definetely helped in development of new varieties and can be brought under captive cultivation as a continuous source of plant materials for product development.

Table 1: Medicinal plants distribution range and mode of usages in skin from high altitude Himalaya regions of India

Botanical name/Family	Vernicular name	Distribution range (m ASL)	Mode of Usages
Achyranthes aspera L./ Amaranthaceae	Latijra	300- 2500	Leaf extract used to cure inflammation.
Aconitum heterophyllum Wall. ex Royal/ Ranunculaceae	Patris	2400-4500	Paste prepared from whole plant or roots used to cure smallpox and boils.
Actaea spicata L./ Ranunculaceae	Baneberry	2000-3300	Berries crushed and applied to cure boils and cuts.
Ailanthus excelsa Roxb./ Simarubaceae	Maharukha	2500 -3000	Leaf juice used as anti-septic agent for wounds and skin eruptions.
Ajuga bracteosa Wall. ex Benth./ Lamiaceae	Kauributi	1200-2400	Root decoction used as blood purifier.
Albizia lebbeck Benth./ Mimosaceae	Siris	800-1600	Bark crushed and used in curing cuts or drunk as a blood tonic.
Allium cepa L./Liliaceae	Gunda	1000-3000	Crushed bulb ties on boils, which helps them to ripen and evacuate the pus in short time, and provide relief from pain.
Anemone rapicola Camb. /Ranunculaceae	Batkul.	3000-4000	Bulb powder mixed with ghee and applied on burnt skin and on part cuts.
Argemone mexicana L./ Papaveraceae	Shialkanta	300-1600	Leaf juice used to cure skin diseases, and crushed seeds used to applies on sores.
Arisaema jacquemontiana Blume/ Araceae	Hapat makei	2000-3200	Paste of rhizomes mixed with mustard cooking oil as massage to regain muscular strength, and also applied to cure blisters and pimples of face skin.
Arisaema flavum (Forssk.) Schott /Araceae	Bang	2000-2500	Dried rhizome paste is used to cure skin allergies.
Asparagus racemosus Willd. /Asparagaceae	Satvar	1000-1900	Leaves and roots used in treatment of burns.
Cannabis sativa L./ Cannabaceae	Bang/ chars	500-3000	Fresh paste of leaves and stems applied externally to cure boils and wounds.

Contd.

Plant name / Family	Local name	Altitude	Uses
Cassine glauca (Rottb.) Kuntze / Celastraceae	Bakra	500-1500	Decoction of leaves given to patients suffering from eczema.
Catunaregam spinosa (Thunb.) Tirveng. / Rubiaceae	Maniphal	500-1000	Juice of the plant applied to cure skin inflammation arise from allergic reactions.
Centaurium centaurioides (Roxb.) SR Rao & Hemadri / Gentianaceae	Belgaum	1000-1500	Seeds used as blood purifier.
Albizia chinensis (Osbeck) Merr./Mimosaceae	Khanujera, Siras	600-1450	Infusion of bark useful in scabies.
Cichorium intybus L./ Asteraceae	Kasni	1200-2000	Root juice used as blood purifier.
Crotalaria sericea Retz./ Fabaceae	Jhunjhunia	1300-2500	Plant used in scabies.
Cullen corylifolium (L.) Medik./ Fabaceae	Babchi	300-800	Powdered seeds administered orally with warm water.
Curcuma aromatic Roxb./ Zingiberaceae	Jangli haldi	1200-2600	Powdered dried rhizomes applied as a poultice on skin eruption and infections.
Curcuma longa L. / Zingiberaceae	Nisa	1300-1600	Root decoction used to cure scabies.
Cuscuta reflexa Roxb./ Convolvulaceae	Akasbel	1200-2000	Whole plant boiled in water and then cooled, used as bath to cure dermatitis and body itching.
Dalbergia sisoo Roxb. / Fabaceae	Sisu	600-1600	Powdered roots taken with hot water as blood purifier.
Datura metel L./ Solanaceae	Sadahdhatura	300-1200	Fresh fruits crushed and applied to to cure boils and cuts.
Desmodium gangeticum (L.) DC./ Fabaceae	Sarivan	1600-2000	Seeds powder and paste applied on infected area to cure boils.
Desmodium triflorum (L.) DC. / Fabaceae	Janglimethi	1300-2100	Leaf juice used to cure wounds and boils.
Drosera peltata Thunb./ Droseraceae	Mukhajal	1300-2600	Juice from leaves mixed with salt applied to cure blisters.
Eclipta prostrata (L.) L. / Asteraceae	Bhring raj	500-1500	Root juice help to cure skin boils and cuts.
Euphorbia caducifolia Haines/ Euphorbiaceae	Danda Thor	1200-2000	Fresh plant latex applied on skin to cure blisters and wounds.
Euphorbia thymifola L. / Euphorbiaceae	Chotidudi	1300	Juice of plant used for treatment of snake bite.
Euphorbia parviflora L./ Euphorbiaceae	Dudhi	800-2000	Whole plant taken as blood purifier and tonic.

Contd.

Scientific name / Family	Common name	Altitude	Uses
Evolvulus alsinoides L./ Convolvulaceae	Sankhapushpi	1000-1800	Decoction of whole plant drunk as blood purifier.
Ficus racemosa L. / Moraceae	Gular	300-1200	Leaves and bark applied externally on skin as poultice in case of curing eczema.
Flemingia lineata L. (Aiton) / Fabaceae	Randa	800- 1750	Ash of whole plants mixed with coconut oil and used to cure fresh skin cuts and boils.
Fritillaria cirrhosa D.Don / Liliaceae).	Sheethkar.	2000-3200.	Bulbs used to cure wounds and burns.
Gentiana dahurica Fisch./ Gentianaceae	Gentian	2000 -2500	Whole plant used as tonic or energy giving substance.
Hemidesmus indicus (L.) R.Br. ex Schult. / Asclepiadaceae	Sugandhi-Pala	700-1500	Root decoction used for curing smallpox.
Hiptage benghalensis Kurz./ Malpighiaceae	Madhavilata	2000-3000	Leaf juice useful for curing boils and inflammation.
Hydnocarpus wightiana Blume / Flacourtiaceae	Kowti	1000-1500	Crushed powdered seeds with honey or milk cures boils, sores and ulcers.
Hydrocotyle javanica Thunb./ Apiaceae		700-2500	Plant used as tonic and help to cure leucoderma.
Indigofera tinctoria L./ Fabaceae	Neel	500-1350	Leaf paste applied on skin to induce hair growth.
Ipomoea nil (L.) Roth / Convolvlaceae		800-2000	Decoction of plant used as blood purifier.
Jasminum grandiflorum L./ Oleaceae	Chamba	1000-2000	Flower juice helpful in curing corn of toes and hands.
Jasminum arborescens Roxb. / Oleaceae	Chameli	1000	Leaves applied in the form of poultice to occur eczema.
Jasminum officinale L./ Oleaceae	Chamba	1000-1500	Oil extracted from flowers used to cure ringworms.
Juglans regia L. /Juglandaceae	Akhrot	2000-3300	Oil from kernel used to cure wounds and used to cure bone.
Lactuca serriola L. /Asteraceae	Kahu ,salad	2000-3000	Whole plant used as tonic and as blood purifier.
Lavandula angustifolis L./ Lamiaceae	Dharu	1700-2000	Plant extract used for blood purification.
Lawsonia inermis L. / Lamiaceae	Hena, Mehndi	300-1800	Decoction of leaves and tender twigs are effective in case of eczema and leprosy.
Leonotis neptafolia R.Br. / Lamiaceae	Hejurchei	800-1500	Burlt flower ash used externally to cure skin boils

Contd.

Plant name / Family	Local name	Altitude	Uses
			and wounds.
Lepidium ruderale L./ Brassicaceae	Towdri	2500-4000	Leaves used to cure skin fistula eruption appear on face and nostril.
Lindenbergia indica Vatke/Scrophulariaceae	Gazdar	2000-2500	Juice of leaves applied with coriander to cure skin eruption.
Mallotus philippensis (Lam.) Muell. Arg. / Euphorbiaceae	Kamila	500-1500	Powderd seeds and roots mixed to form paste and used to cure dermatitis.
Melia azedarach L./ Meliaceae	Bakiain	1600-200	Stem bark decoction used for healing ulcers and abrasions of skin.
Meriandra strobilifera Benth/Lamiaceae		1600-2000	Decoction of leaves useful for healing ulcers and abrasions.
Mimosa pudica L./ Mimosaceae	Lajwanti	500-1500	Whole plant crushed and used to cure itching.
Neolitsea umbrosa (Nees) Gamble/ Lauraceae	Chirindi	1200-2500	Fruits used as application for skin diseases.
Nepeta catarialL. /Lamiaceae	Gandsoi	2000-3000	Paste made from leaves useful for wound healing.
Ocinum americanum L./Lamiaceae	Kali tulsi	500-1500	Paste made from the leaves applied to cure parasitical skin infections.
Plumbago zeylanica L./Plumbaginaceae	Lal chitra	5300-1600	Root paste applied to cure boils and wounds.
Portulaca quadrifida L./Portulacaceae	Chhotaluniya	1500 -2000	Juice of whole plant with oil used to cure skin inflammation.
Pterocarpus santalinus L.f. /Fabaceae	Lalchnadan	300-1500	Plant extract used as blood purifier.
Rheum australe D.Don/Polygonaceae	Pumbchaln	2500-3500	Rhizome as in paste used to cure skin burns and other skin infections.
Rubia cordifolia L./Rubiaceae	Manjith	2600	Root and fruit used to cure paralysis, ulcers and boils.
Salvia moorcroftiana Wall. ex Benth. /	Shollari	1500-3200	All parts are used in treating of skin boils.
Sapindus trifoliatus L./ Sapindaceae	Ritha	500-1200	Fruits powdered and used to cure leucoderma.
Saussurea obvallata (DC.) Edgew./ Asteraceae	Brahamakamal	3500-5000	Juice of roots applied on cuts and bruises.
Sigesbeckia orientalis L./Asteraceae	Katampam	1000-2000	Plant extract used to cure ulcers.
Sinopodophyllum hexandrum (Royle) T.S. Ying / Berberidaceae	Banwangun	1800 -3700	Rhizome used to cure cuts, wounds and small pox.

Contd.

Botanical name / Family	Local name	Altitude (m)	Uses
Smilax china L./ Liliaceae	Chobchini	1600-2700	Whole plant used as tonic and blood purifier.
Solanum americanum Mil. / Solanaceae	Makoh	1400-2500	Crushed leaves applied to cure boils and sores.
Sphaeranthus indicus L./ Asteracease	Mundi	800-1600	Whole plant boiled and drunk as tonic.
Stellaria media (L.) Vill. / Caryophyllaceae	Losdi	1900-4000	Leaf is used to cure skin allergy and boils.
Tamarix gallica L. / Tamaricaceae	Jhau	1000-4500	Rhizome boils and drunk as tonic.
Taraxacum campylodes G.E.Haglund/ Asteraceae	Dudhi		Root juice applied to cure boils and smallpox.
Tephrosia purpurea (L.) Pers./ Fabaceae	Sarphonka	1000-2000	Cooked tubers used as blood purifier and to cure skin infections.
Thalictrum foliolosum DC./ Ranunculaceae	Mamirapinjari	1500-2500	Fresh roots crushed and juice applied to cure wounds.
Thespesia lampas (Cav.) Dalzell./ Malvaceae	Bankapas	300-1300	Bark boiled with water and bath taken to cure scabies and allergy.
Thespesia populnea (L.) Sol. ex Correa / Malvaceae	Parasipal	1000-1500	Leaves crushed and used to cure scabies and psoriasis.
Viburnum grandiflorum Wall. ex DC./ Caprifoliaceae	Kalmach	2500-3500	Poultice made from dried powdered stems used to heal fractures.

REFERENCES CITED

Chopra RN, Nayar SL, Chopra IC. 1986. Glossary of Indian Medicinal plants. Council of Scientific and Industrial Research, New Delhi, India.

Chopra RN. 1933. Indigenous Drugs of India. Council of Scientific and Industrial Research, Calcutta, India.

Dastur JI. 1970. Medicinal plants of India and Pakistan Taraporevala DB Sons and Co. Pvt. Bombay, India.

Jain SK. 1981. Glimpses of India Ethanobotany. Oxford and IBH Publishing Co., Delhi, India.

Khan SS Chaghtai SA. 1982. Ethnobotanical study of some plants used for curing skin affliction. *Ancient Science of Life* 1(4): 236 - 238

Kirtikar KR, Basu BD. 1981. Indian medicinal plants. Bishan Singh Mahendra Pal Singh, Dehradun, India.

Kirtikar KR. 1935. Indian medicinal plants. Lalit Mohan Basu, Allahabad.

Lal N, Kishore N. 2014. Are plants used for skin care in South Africa fully explored?. *Journal of Ethnopharmacology* 153(1): 61-84.

Manishayadav K, Khan K, Baeg MZ. 2012. Ethnobotanical plants used for curing skin diseases by tribal's of Rewa Districts (Madhya Pradesh). *Indian Journal of Life Science* 2(1): 123-126.

Mir MY. 2014. Indigenous knowledge of using medicinal plants in treating skin disease by tribal's of Kupwara, J&K, India. *International Journal of Herbal Medicine* 1(6): 62-68.

Padal SB, Chandrasekhar P, Satyavathi K. 2013. Ethnomedicinal investigation of medicinal plants used by the tribes of Pedabayalu Mandalam, Visakhapatnam district, Andhra Pradesh, India. *International Journal of Computational Engineering Research* 3(4): 8-13.

Panther DJ, Jacob SE. 2015. The importance of acidification in atopic eczema: an underexplored avenue for treatment. *Journal of Clinical Medicine* 4(5): 970-978.

Policepatel SS, Manikrao VG. 2013. Ethnomedicinal plants used in the treatment of skin diseases in Hyderabad Karnataka region, Karnataka, India. *Asian Pacific Journal of Tropical Biomedicine* 3(11): 882-886.

Pushpangaden CKA. 1986. Ethnomedical and ethnobotanical investigations among some scheduled caste communities of Travancore, Kerala, India. *Journal of Ethanopharmacology* 16(2-3): 175-186.

Raghunathan K. 1976. Preliminary Techno-economical Survey of Natural Resources and Herbal Wealth of Ladakh. Published by CCRIMH, New Delhi, India.

Rout SD, Panda T, Mishra N. 2009. Ethno-medicinal plants used to cure different diseases by tribals of Mayurbhanj district of North Orissa. *Ethno-Medicine* 3(1): 27-32.

Sebastian MK, Bhandari MM. 1984. Medico-ethno botany of Mount Abu, Rajasthan, India. *Journal of Ethanopharmacology* 12: 223-230.

Sinhababu A, Banerjee A. 2013. Ethno botanical study of medicinal plants used by tribal's of Bankura Districts West Bengal. *Indian Journal of Medicinal Plant Studies* 98-104.

Stallings AF, Lupo MP. 2009. Practical uses of botanicals in skin care. *The Journal of Clinical and Aesthetic Dermatology* 2(1): 36-40.

Tabassum N, Hamdani M. 2014. Plants used to treat skin diseases. *Pharmacognosy Review* 8(15): 52-60.

Wahid A, Siddiqui HH. 1961. A survey of drugs. Institute of the History of Medicine and Medical Research, Delhi, India.

Yosipovitch G, Maibach HI. 1996. Skin surface pH: A protective acid mantle. *Cosmetics and Toiletries* 111: 101-102.

Singh B, Borthakur SK, Phukan SJ. 2014. A survey on ethnomedicinal plants utilized by the indigenous people of Garo Hills with special reference to the Nokrek Biosphere Reserve (Meghalaya), India. *Journal of Herbs, Spices & Medicinal Plants* 20(1): 1-30.

Singh B, Sultan P, Hassan QP, Gairola S, Bedi YS. 2016. Ethnobotany, traditional knowledge, and diversity of wild edible plants and fungi: A case study in the Bandipora district of Kashmir Himalaya, India. *Journal of Herbs, Spices & Medicinal Plants* 22(3): 247-278.

Singh B, Bedi YS. 2017. Eating from raw wild plants in Himalaya: traditional knowledge documentary on Sheena tribes along LoC Border in Kashmir. *Indian Journal of Natural Products and Resources* 8(3): 269-275.

22

Forest Resources, Village Livelihood and Investigation of Relic Places in the Fringes of Pangolakha Wildlife Sanctuary in Sikkim Himalaya of India

Arun Chettri

Department of Botany, Sikkim University, 6th mile, Samdur, Tadong-737102
Gangtok, Sikkim, INDIA
Corresponding author: sentinchettri@gmail.com

ABSTRACT

The present communication deals with the impact and contribution on livelihood sustainability based on forest resources and sacred places in nine fringe villages of Pangolakha Wild Life Sanctuary (PWLS), located in Sikkim Himalaya. Populations density in villages varies from 23 at Nayabusty and 274 at Padamchen. Populations were categorised under three age groups *viz.* minor (15-25 yrs), major (26-36 yrs) and elder (over 36 years). Approximate by 27.6% populations had minimal primary education. Agriculture is the main activity in Suvaneydara (86.7%), Agamlok (85.3%) and in Nayabustyy (82.3%). Villagers in Padamchen (12.5%) and in Nathang (12.5%) are either dependent on Government job (38.2%) or seasonal flow of tourist (25%). The educational qualification in these villages is classes 5 to 10 standards. 91.1% of people rely on forest resources. About 70% of the villagers depicts that forest cover of the region is improving. The availability of fodders (70%) and fuel woods (60%) are still in abundance. People depend on *tirja* (private*), khashmal* and *goucharan* areas for grazing and fuel wood collection. Forty nine economical economically by valued plant species belonging to 46 genera and 35 families were documented. Subtropical ecosystems had high number of plant species. Poaceae was the dominant family (17%). Tree life form is abundant (51%). Total 24 species are harvested for edible purpose, 24% are freely sold in the local *haats* (market) for their livelihood sustenance. Principal Component Analysis (PCA) of nine villages with respect to forest resources indicated the village specific use pattern of plant resources. Indigenous inherent knowledge of local plants is decreasing in Lingtam. PWLS and its adjoining villages have important places with religious importance, heritage sites managed by the people. It was observed that exploitation of plant resources, lack of maintenance and fast changing concept of traditional ethics among youths, plant resources were degraded and in regenerating stage.

Keywords: Forest Resources, Pangolakha Wild Life Sanctuary, Principal Component Analysis, Sacred Places, Sikkim Himalaya.

INTRODUCTION

Forest is defined as all lands bearing vegetative association dominated by trees of any size, exploited or not, capable of producing wood or other forest products, or extending an influence on the climate or water regime or providing shelter for livestock and wildlife. People have used forests for thousand of years. It is an important renewable resource and its sustainable utilization can contribute significantly in economic growth of the nation. Records tell us that India ideally have 33% of forest covers, but has decreased to about 11% in the last century (www.angrau.net). The north-eastern Indian states have more than 60% of the total geographical area under forest cover (FSI 1997). Sikkim state has 45.9% of the total geographical under forest cover[3]. The state of Sikkim comprised of 0.2% of the country areas is rich in biodiversity and is one of the hot-spot in the world (Tambe et al. 2012, Das et al. 2012). It accounts for 4500 species of flowering plants and over 424 medicinal plant species (Watanabe 2010).

The contribution of forest wealth to the economy of Sikkim is significant. People of this region utilize the forest resources as fodder, timber, medicines, fuel wood, wild edible and building materials. Local people have always been at the forefront of conservation measures. Forest of reverence, sacred forest, and government controlled reserved and protected forests are well documented in this part of the region. Protected areas like PWLS provide ecosystem services such as water, clean air, and protection from soil erosion, conservation of field gene bank, rich biodiversity and livelihood supplements from various activities such as tourism. However, natural and anthropogenic disturbances like frequent landslides, windthrow, forest fire, road construction, mega hydroelectric projects, agricultural expansion, cattle grazing, logging for sale as timber or pulp, illegal settlements are some of the common factors of deforestation of the region (Chettri 2006). As a result it causes the destruction of habitat, biodiversity loss and aridity. Further, it causes extinction of endangered species, shift in climatic conditions and desertification. Due to growing impact of deforestation, it is important to protect the endangered species of wildlife as well as those that are on the verge of extinction in their natural habitat. To protect the wildlife, many areas rich in diversity have been converted into reserved forests or sanctuaries. These sanctuaries not only protect wild life and provide natural habitat to them, but also brings us close to nature. But alienating access of local people to forest resources may affect their livelihood, living condition, ethical and aesthetic value of biodiversity.

The Pangolakha range separates Sikkim from Bhutan. Pangolakha wildlife sanctuary was notified in the year 2002, with an area of 128 km^2 and is a trans-boundary protected area bordering with China and Bhutan in its East and the Neora Valley National Park in the South. It is famous for Red Panda besides being a corridor for Tiger and Takin. Two broad types of ecosystems are present (i) Grasslands, and (ii) Woodlands in the PWLS. The former are represented by the sub-tropical grassland, the temperate pastureland and alpine meadow, while the later comprise sub-tropical, temperate forests and alpine ecosystem. The diverse ecosystem types have originated due to complex landscape structure wide-ranging in the altitude. The buffer zone of the PWLS comprises the sub-tropical riverine forest and grassland ecosystems. The sub-tropical forests stretch along the rivers and tributaries in PWLS. The temperate forest occupies the mid-altitude in the form of a belt extending from the buffer to the alpine forest at the core zone. The alpine scrubs are found upto about 4000 m above the tree-line. The vast stretches of alpine meadows are found between tree line and snow line (recent observation). However, Champion and Seth (1968) and Grierson and Long (1983-1991) had classified the forests of these region consisting of sub-tropical, temperate, mixed coniferous, dry temperate coniferous and alpine meadows which are ideal habitat for great diversity of floral and faunal elements.

Hathitshirey (elephant corridor) forms the tri-junction between Bhutan, Sikkim and West Bengal, which continuous with Neora Valley National Park in Gorkha land Territorial Administration (GTA) region of Darjeeling, West Bengal. It is a point of convergence location during the interstate migration of animal species. The PWLS area have diverse plant species such as *Rhododendron* spp., *Abies* spp., *Juniperus* spp., *Santalum* spp., *Juglans* spp., Oak species with dense bamboo brakes and provides a natural habitat for wild faunal species such as the Himalayan Monal (*Lophophorus impejanus*), Red Panda (*Ailurus fulgens,* Sikkim's State animal) and Himalayan Black Beer. These diverse natural resource needs of the people in the region are collected from the forests. The different community in the study area gather about 193 plant species as resources (Subba 2009).

With increasing human interference, particularly at lower elevation, several of these species now faces the danger of extinction. On the other hand, areas in the region with difficult terrain have never been explored. Thus our understanding on ecology of plant richness of the region is extremely poor. Therefore, the present study in a limited time period was carried out with following broad objectives: a) to document livelihood dependence of people on forest resources near Pangolakha wildlife sanctuary. b) to document the economically important plants in the forests, and c) to document the sacred groves vis- a-vis conservation of forests, traditional belief and practices among community.

Study Site

The study was conducted in Pangolakha Wildlife Sanctuary (PWLS 128 Km^2) and adjoining forest of surrounding villages (27°20'28" N and 88°46'42"E) (Fig. 1) in the Eastern Himalayan state of Sikkim in north-east India. The important tree species in this sanctuary are *Alnus nepalensis* D.Don, *Castanopsis tribuloides* DC., *Engelhardtia spicata* Blume, *Ficus semicordata* Sm., and *Lyonia ovalifolia* Drude, *Betula alnoides* D.Don, *Lithocarpus pachyphylla* Rehder, *Magnolia campbellii* Hook.f. & Thomson, *Abies densa* Griffith ex Parker, *Rhododendron* spp., and *Tsuga dumosa* Eichler. The climate of the study area is monsoonal. Three seasons are distinguishable in a year: winter (October–March), summer (March–May) and monsoon (June–September).

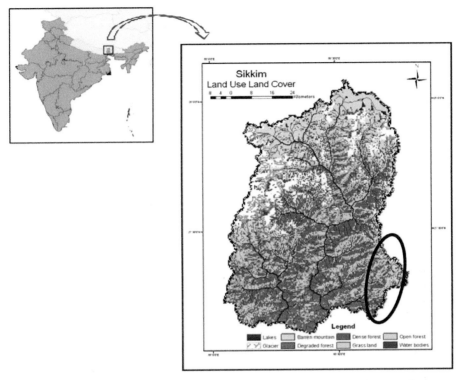

Fig. 1: Location of Pangolakha Wildlife Sanctuary in Sikkim, India

Questionnaire Development and Data Collection

The study relies on both primary and secondary data. For generating primary data, nine villages proximal to the forests were selected. From each village, households were randomly selected for studying the problems of the forest. The primary data were collected using a structured questionnaire. Besides, 10 elder respondents from the above mentioned villages were selected for collecting historical data regarding sacred places, heritage sites, migration, socio-cultural tradition of the people and destruction of forests, of the region. Interviews and focus group discussions with local people were conducted to acquire information on whether a given area in the forest was disturbed or not. Information from group discussions and from open-ended questions of personal interviews is presented in graphical and ordinal form after statistical analysis.

Statistical Analyses

Principal Component Analyses (PCA): was carried out to explore the local people perspective on use of economically important plant species that are useful for livelihood sustenance in nine villages along the PWLS. The obtained data sets from 9 villagers were analyzed for Principal Component Analysis (PCA) using CAP 4, Version 4.1.4 (Pisces Conservation Ltd. 2007). The plot ordination was created using the variance-correlation matrix calculated with all variables $\log_{(x+1)}$ transformed. In this ordination type eigenvectors show a number of tight pairs and it is possible to place the respondent and the species eigenvectors on the same ordination plot. Total 100 interviews were selected representing different groups of peoples having disproportionate age category. The question which plant is of economical importance or preferred species' were asked from the 50 categories of species. Each informant/respondent was asked the question separately. In the event of positive reply ('Yes'), the value 1 was allotted; and if the reply was 'No', 0 value was assigned (Hoft et al. 1999).

Human Population Distribution

The nine villages along the periphery of PWLS had variable number of populations. From these nine villages, 870 individuals were thoroughly interviewed during the survey. In the study, we encountered sporadic number of individuals, which ranged from 23 individuals (2.6%) at Nayabusty (lowest) to 274 (31.4%) individuals at Padamchen (highest). These people come under three age group categories *viz.* minor (15-25 yrs), major (26-36) and > than 37 yrs of age. Overall populations in the villages had greater proportion of peoples in major age group (59.7%), followed by minor one (26.4%) and 13.9% of individuals were of older age group. Populations of minor group were more in Dakline (38.75%) and least at Dokching (9.6%). Dokching had highest proportion of individuals (74.2%) in the intermediate age group and Suvaneydara had the lowest population's count of about 52.8% in this group. Populations under older age group were more in Nayabusty (17.3%) and least at Dakline (9.59%) (Fig. 2).

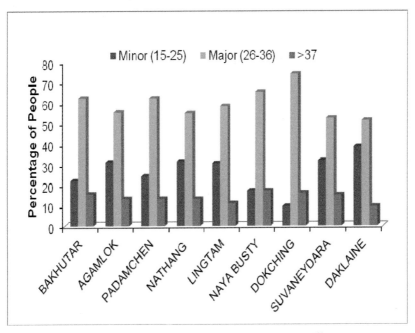

Fig. 2: Population categorisation in nine villages

Occupations and Educations in Villages

It is evident from the Fig. 3, that most of the people in all the villages near PWLS are dependent on agriculture or forest resources. The villages like Agamlok (85.3%), Suvaneydara (86.7%) and Nayabusty (82.3%) had highest number of people involved in agriculture. Their livelihood and sustenance is solely dependent on agriculture only. But it is interesting to find that villagers in Padamchen (12.5%) and Nathang (12.5%) are least involved in agriculture. Most of the people in these two villages are either dependent on Government job (38.2% in Padamchen and 25% in Nathang) or dependent on seasonal in flow of tourist. People in these villages are also involved as labourers in daily wage system under scheme like NREGA, road construction, tourist guide, etc., which accounts for about 12.6% of the population.

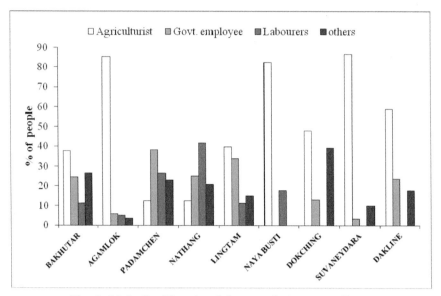

Fig. 3: Main livelihoods of the people in nine villages

(46.7%) number of people having studied upto class 10 standards. People in the study area had only about 5.6% of the populations who had been able to study upto class 12 standard and graduations respectively. Dakline people are educationally sound because they accounts for about 12.9% populations who have studied upto 12 standards or graduate level respectively.

Community Dependence on Forest Resources

Forest resources improve the livelihoods opportunity for forest dependent community and also improve the ecological conditions of forests. Forest-derived product increases the income for households, household needs and provide food resources to the community. However, the data collected from surrounding nine villages' (70%) depicts that forest cover of the PWLS and its adjoining reserved forests is improving. But it was interesting to learn that average 30% of the people were of the notion that forest cover is declining. People mostly from Bakhutar (94%) are of the notion that forest cover is not decreasing. Contrastingly, Nathang village people insisted that forest cover is declining (56%).

People still depend on forest resources for their livelihood sustenance. 91.1% of the people are of the opinion that villagers still rely on forest resources for their daily requirements. However, 8.9% of the villagers (Bakhutar, Agamlok and Padamchen) are of the opinion that they do not depend on forest resources for livelihood sustenance.

Forest department has restricted the open grazing of cattle in nearby reserved forests. People have to depend on surrounding plant resources in order to rear the cattle. Availability of fodder is still abundant and 70% of the people in the region are of the opinion that fodder availability is in abundance. Similarly, availability of fuel wood is also in abundance as stated by 66% of the people in the region. Only people from the Dakline area are of the notions that there has been decline in the availability of resources like fodder, fuelwood, and overall decrease in livestock populations.

Diversity and Distribution of Economic Plants in PWLS

Forty-nine plant species belonging to 46 genera and 35 families were recorded from the forests of PWLS and adjoining villages. The number of species was highest in the subtropical forest (21) followed by subtropical/temperate species (14), temperate (10) and temperate/subalpine (04) forests (Fig. 4).

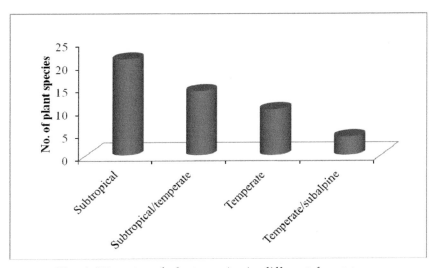

Fig. 4: Diversity of plant species in different forest types

Family Poaceae (grasses) was the dominant (17%) followed by Rutaceae (8%) in the study sites. The uses, vernacular name, and forest types, were identified across the PWLS, which are used by the different communities for various purposes (Table 1).

Economically important plant species under different life forms are widely available. About 25 (51%) are trees, 12 (24.5%) are herbaceous species followed by 6 (12%) species among the lianas and shrubby

Table 1: Community dependence on economically important plant species of Pangolakha wildlife sanctuary.

Vernacular names	Botanical names	Family	Uses	Forest types
Amliso	*Thysaenolaena maxima* (Roxb.) Kuntze	Poaceae	Cash crop/fodder	Subtropical/Temperate
Bantey	*Lithocarpus pachyphylla* Rehder	Fagaceae	Timber/edible/fuel wood	Temperate
Bara Alainche	*Amomum subulatum* Roxb.	Zingiberaceae	Cash crop	Subtropical forest
Bhadrasey	*Elaeocarpus lanceaefolius* Roxb.	Elaeocarpaceae	Timber/fuel wood	Temperate
Bhui Aishelu	*Rubus calycinoides* Kuntze	Rosaceae	Edible	Temperate
Bimiro	*Citrus* spp.	Rutaceae	Edible	Subtropical
Bokey Timbur	*Zanthoxylum acanthopodium* DC.	Rutaceae	Edible	Subtropical/Temperate
Budo okhati	*Astilbe rivularis* Buch.-Ham. ex D.Don	Saxifragaceae	Medicinal	Temperate
Chanp	*Michelia* spp.	Magnoliaceae	Timber	Subtropical/Temperate
Charchare Lahara	*Tetrastigma serrulatum* Planch.	Vitaceae	Fodder/edible	Subtropical
Chewri	*Bassia butyracea* Roxb.	Sapotaceae	Edible	Subtropical/Temperate
Chimphing	*Heracleum wallichii* DC.	Apiaceae	Medicinal	Temperate/Alpine
Chinde	*Pentapanax leschenaultii* Seem.	Araliaceae	Young twigs edible	Temperate
Chirawto	*Swertia chirayita* Karsten	Gentianaceae	Medicinal	Subtropical/Temperate
Choya/Tama Bans	*Dendrocalamus hamiltonii* Nees	Poaceae	NTFP	Subtropical
Dhupi	*Cryptomeria japonica* Don.	Pinaceae	Avenue tree/plantation	Subtropical/ Temperate
Gantey	*Gynocardia odorata* R.Br.	Flacourtiaceae	Medicinal	Subtropical
Ginger	*Zingiber officinale* Roscoe.	Zingiberaceae	Cash crop	Subtropical/Temperate
Gobre Salla	*Abies* spp.	Taxodiaceae	Timber/fuel wood	Subtropical/Temperate
Gogun	*Saurauia napaulensis* DC.	Saurauiaceae	Fodder/fuel wood	Subtropical
Golkakra	*Solena* spp.	Cucurbitacee	Edible	Subtropical
Gope Bans	*Cephalostachyum capitatum* Munro	Poaceae	NTFP	Subtropical
Gufla	*Holboellia latifolia* Wall.	Lardizabalaceae	Edible/medicinal	Temperate
Gurans	*Rhododendron* spp.	Ericaceae	Ornamental/fuel wood	Temperate/Sub-alpine
Indreni	*Trichosanthes dioica* Roxb.	Cucurbitaceae	Edible	Subtropical
Jaringo	*Phytolacca acinosa* Roxb.	Phytolaccaceae	Medicinal	Subtropical
Kabra	*Ficus infectoria* Roxb.	Moraceae	Leaf buds edible	Subtropical

Local name	Scientific name	Family	Use	Climate
Kaphal	*Myrica* spp.	Myricaceae	Fuel wood	Subtropical
Katus	*Castanopsis* spp.	Fagaceae	Edible/timber/fuel wood	Subtropical/Temperate
Khanakpa	*Euodia fraxinifolia* Hook.f.	Rutaceae	Medicinal	Subtropical/Temperate
Lapche Kawla	*Machilus edulis* King	Lauraceae	Edible	Temperate
Lapsi	*Spondias axillaris* Roxb.	Anacardiaceae	Edible/fuel wood	Subtropical
Lekh Pangra	*Entada phaseoloides* (L.) Merr.,	Fabaceae	Medicinal	Subtropical
Mahua	*Engelhardtia spicata* Blume.	Juglandaceae	Fuel wood	Subtropical
Maize	*Zea mays* L.	Poaceae	Food crop	Subtropical/Temperate
Maling	*Arundinaria maling* Gamble	Poaceae	NTFP	Temperate/Sub-alpine
Nakima	*Campylandra aurantiaca* Baker	Liliaceae	NTFP/Medicinal	Temperate
Nebaro	*Ficus roxburghii* Hook.f.	Moraceae	Ornamental/fuel wood	Subtropical
Okhar	*Juglans regia* L.	Juglandaceae	Timber/fuel wood/edible	Subtropical
Pakhanbed	*Bergenia ciliata* (Haw.) Stenb.	Saxifragaceae	Medicinal/ornamental	Subtropical/Temperate
Potato	*Solanum tuberosum* L.	Solanaceae	Food crop	Subtropical/Temperate
Rukh Aishelu	*Rubus ellipticus* Smith	Rosaceae	Edible	Subtropical
Singhane Baans	*Arundinaria hookeriana* Munro.	Poaceae	NTFP	Subtropical
Singhatta lahara	*Schisandra grandiflora* Hook f. & Thomson	Schisandraceae	Edible	Temperate
Singjango	*Gaultheria fragrantissima* Wall.	Ericaceae	Medicinal	Sub-alpine/Temperate
Tekhiphal	*Actinidia callosa* Lindl.	Actinidiaceae	Edible	Temperate
Titey	*Campylandra aurantiaca* Baker	Liliaceae	NTFP	Temperate
Totala	*Oroxylum indicum* Vent	Bignoniaceae	Medicinal/Religious	Subtropical
Tothney	*Aconogonum molle* Don.	Polygonaceae	Young twigs edible	Subtropical/Temperate
Uttis	*Alnus nepalensis* Don	Betulaceae	fuel wood	Subtropical

plant species. As per the plant used by the local villagers, the maximum numbers of species are harvested for edible purpose (24 species), followed by medicine (12 species), fuel wood (11 species) and timber (7 species) (Fig. 5). 12(24%) of the cultivated species such as *Campylandra aurantiaca* Baker, *Citrus* spp., *Juglans regia* Linn., *Spondias axillaris* Roxb., *Amomum subulatum* Roxb., *Zanthoxylum acanthopodium* DC., *Bassia butyracea* Roxb., *Thysaenolaena maxima* (Roxb.) Kuntze, *Zingiber officinale* Roscoe, *Zea mays* L., *Euodia fraxinifolia* Hook.f. are freely sold in the local markets for their income generation.

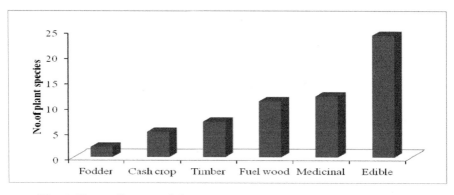

Fig. 5: Forest flora used for various purposes in the PWLS region.

Local People Perspectives on Forest Resources

In the ordination biplot, positions of the villagers with respect to plant resources are plotted in the reduced space. This allowed us to visualize the information from respondents from 9 villages regarding the use of plant resources along the buffer zone of PWLS. Further the eigen vector plots showed that economically important plants species in the knowledge of local people at Lingtam are decreasing. Similarly, villagers from Nathang, Agamlok and Bakhutar also had decreasing notions of forest resources like medicinal plant species and fuel wood plants. Further, informations of these species were characterized by high number of respondents as evident from strong loading pattern (Fig. 6).

The result of the PCA showed that the percentage of variance explained by the first two axes is 65.5%. Axis 1 and 2 contributed 36.8% and 29.2% of the variation respectively, therefore axis 1 is important in influencing the peoples perspective on the knowledge of plant resources (Fig. 6 and Table 2). Eigen-vector matrices with the loading of each species is presented in Table 3.

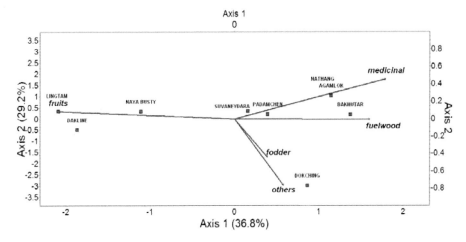

Fig. 6: Projection of villagers in the space defined by the first and second principal axis on people perspective about forest plant resources availability.

Table 2: Eigen-values for the principal axis for the data on people's perspective of forest resources.

Axes	Eigen values	Cumulative total	% of Total variance
1	1.8	1.8	36.4
2	1.5	3.3	29.1
3	1.1	4.4	21.6
4	0.6	4.9	11.5
5	0.1	5.0	1.4

Table 3. Eigen-vector matrix with the loading of each resource components in each principal axis for the data on people's perspective of forest resources.

Forest resources	Principal Axis 1	Principal Axis 2
Fuel wood	0.49	-0.01
Fodder	0.12	-0.45
Fruits	-0.64	0.09
Medicinal plants	0.55	0.46
Others	0.18	-0.77

Sacred places in PWLS and Adjoining areas

Pangolakha areas and its adjoining villages harbour some important areas having religious or historical importance often managed by the village people themselves. The important places of the area are Pangolakha top, Deorali, Gumpas-Also known as Agamlok buddhist Gumpa or Tashi Choekar Ling Gumpa. It was constructed around 20 years ago. In this Gumpa people used to celebrate, Dukpa tsechu, Buddha jayanti, and Dashami. During the celebration of all these important events Chandra *samaj* plays the leading role. This gumpa is considered as the guardian deity for the entire villages. *Pokhari* (lake), *Chyandara, Dakline (*heritage post office*), Ghatta tar* (heritage water mill), *War Crematorium-* Queens regiment army who died during the Indo-Tibet war (1889-1895) at Nathang (Appendix S1) are some important documented places. The forests of groves are dominated by the trees species like *Machilus edulis, Campylandra aurantiaca, Rubus calycinoides, Lithocarpus* spp., *Pentapanax leschenaultia, Holboellia latifolia* etc. Most of the visited sacred groves were in the regenerating phase due to over exploitation of plant resources from the area, lack of maintenance and fast changing concept of groves among the young generation.

DISCUSSION

The rich deposits of coal in different parts of the country confirm the earlier existence of gigantic and novel forests in India. It was the time (250 million year back) of the golden age of the forest in Indian history (Anonymous 1971). However, the migration of Aryans led to systematic deforestation of forests for their settlements. As the population increased more and more forest were cleared for cultivation, habitation and grazing (Chatterjee 1965). Common property resources in India once included vast stretches of forests, grazing lands and aquatic ecosystems. But due to ever increasing encroachment on forest resources these areas has been converted into Government 'Reserved Forests' for in situ conservation of natural resources in an undisturbed ecosystems. PWLS is one of the protected reserved in east Sikkim district. But this has alienated local people from having a stake in preserving bio-resources in this protected habitat. This in turn leads to large-scale losses in forest cover and the creation of wasteland.

Mortality and natality of species is a continuous phenomenon and it continues in its own pace. But there has been unpredictable rise in the human population around the globe and more prominently in India. This has adversely created great pressure on the natural resources. Abundant availability of natural resources and various developmental schemes in Sikkim has led to disproportionate distribution of population. Human population distribution varied across the villages in east district of Sikkim. The village like Nayabusty had lowest number of populations (2.6%). This may be attributed to its locational disadvantage and also insignificant investment on public infrastructure. Moreover, the tendency of villagers to migrate to the urban areas like

Lingtam, Rongli, Rangpo, Gangtok areas for better education, employment and also had impact of modernisation in the traditional life style has contributed a lot in the disproportionate distribution of populations. On ther hand, higher number of populations in Padamchen village (31.4%) is because of heavy public investment on infrastructure facilities and presence of army, SSB battalion and state police outpost.

The majority of population of the nine villages are in intermediate age group (59.7%) demographic profile with only 13.9% of the population in the older age group. This result corroborate the findings of lama's in Sikkim (Lama 2001), where 60 plus age category population were very few in Sikkim.Thus, implementaion of state policy in this part of the region should give emphasis to elder people. The existence of many traditional forest laws indicates that forests are important source of livelihoods for the Sikkimese people (Lama 2001). It is also evident from the study that people in the study area are agriculturist or depend solely on forest resources. The different communities in the area gather about 193 of plant species as resources (Subba 2009). The maximum number of villagers in Agamlok, Suvaneydara and Nayabusty are still dependent on forest resources. As depicted, 70% of the people residing adjoining to reserve forests are of the notions that forest resources is not declining. The reason could be that restrictions in the collection of fodder from the forest improved the forest ecosystem, but on the other hand affected poorer people in rearing livestock. People in the remaining villages are either government employee, or dependent on in flow of tourist. Futher, people in these villages are also involved as labourers in daily wage system under government scheme like NREGA, road construction and tourist guide, which accounts for about 12.6% of the population.

Employment of people in the low profile jobs indicates that majority of the people in all the villages are educationally backward. Most of the students are school dropout (30.6%). This indicates that literacy rates among the masses in the remote corner of the state are not impressive. In spite of highest priority given to education sector in the Sikkim. People access to the forest resources and their involvement in the decision making directly impact their livelihoods. In the similar study, it has been reported that the improvement in forest condition has not improved the community access to forest products such as timber, firewood and other non-wood forest products (ICIMOD 2004). Therefore, participation of poorer people, illiterate and undereducated people of the community in decision making and distribution of forest product is equally important to uplift their economic condition. Therefore, present study sites needs more number of higher levels of educational institutes and technical colleges.

Primary forests not only conserve the biodiversity of the area but provides with enormous amount of the economically important plants for sustenance of village livelihood. 49 plants species of economical importance have sustained the livelihood of the villagers in the nine villages. However, this requires, in addition to the appropriate policy instruments, a strong scientific basis for determining harvesting and extraction levels, value addition, marketing and benefit sharing. The diversity of these plant species to be more in the subtropical forest is evident in the present study. A difference in plant diversity in different forest types is due to the elevational gradient, topography and climatic variability (Chettri et al. 2010). Total 12(24%) of the cultivated species such as *Campylandra aurantiaca* Baker, *Citrus* spp., *Juglans regia* L., *Spondias axillaris* Roxb., *Amomum subulatum* Roxb., *Zanthoxylum acanthopodium* DC., *Bassia butyracea* Roxb., *Thysaenolaena maxima* (Roxb.) Kuntze, *Zingiber officinale* Roscoe., *Zea mays* L., *Euodia fraxinifolia* Hook.f. are freely sold in the local markets for their livelihood sustenance. It is also estimated that about 140 tonnes of wild foods is sold in the rural market of Sikkim annually (Sundriyal 1999). Collection and commercial exploitation of economically important plants have provided a great deal of livelihood sustainability among the villagers. This also depends on use value and use pattern of plant resources in the mind sets of the peoples using it for various purposes. The relationship between the use value of plant resources and villagers was evident, and it varied among the villages. This has led to the segregation of villager use of plant resources in different type. The extraction of plant resources for its edible purpose in the villages like Lingtam, Naya Busty and Dakline seemed to have wider use implication, which was evident from their strong loading with Principal Axis 1 in PCA ordination. Some of the plant resources in the form of medicinal values among the peoples of Suvaneydara, Phadamchen, Nathang, and Agamlok, which had strong loading with Principal Axis 2. Wider knowledge on use value of tree species among the people of Bakhutar as fuel wood may be due to human needs for daily energy requirement. The people of Sikkim, in the Sikkim Himalayas, have been living in the valley areas of this state since long. The aspirations of these people are unique because of their close integration with the forest environment (Ramakrishnan 1993). Worship of nature and its various components form an important part of the religious practices. Existence of sacred places, heritage sites in PWLS and nearby by places in the form of *Pangolakha top, Deorali, Gumpas-*Also known as *Agamlok buddhist Gumpa* or *Tashi Choekar Ling Gumpa,* constructed around 20 years ago. In this Gumpa people celebrate, *Dukpa tsechu (festival), Buddha jayanti, Dashami.* This gumpa is considered as the guardian deity for the entire villages. *Pokhari* (lake), *Chyandara, Dakline* (heritage post office), *Ghatta tar* (heritage water mill), *War Crematorium* are indicative of the widespread respect with this sacred region and are worshipped and preserved by the people. The uniqueness of these sacred groves, heritage sites is that the value system here is interpreted in a more holistic sense, *i.e.,* soil, water, biota, visible water bodies, river and less obvious notional lakes, all are to be taken together with the physical monuments. This supports the notion that sacred places as protected small patches of forest vegetation and area of special religious importance to a particular culture and community and is associated with festivals, religious rites, and it serve as an example of habitat preservation areas (Avasthe et al. 2004, Gadgil and Vartak 1975). Annually, ceremonies used to be performed regularly in these sacred places and others to propitiate the ruling deity. Propitiating these deities through various religious ceremonies is considered important for the welfare of the Sikkimese people and removal of plants or plant parts is considered to offend the ruling deity, leading to local calamities. The traditional religious belief is that the gods and the spirits of the ancestors live in these groves. Most of the visited groves were in the succession stage due to over exploitation of plant resources from the area, lack of maintenance and fast changing concept of groves among the young generation.

There is a great potential for developing and enhancing forest-based livelihoods in many parts of the region. Specific options need to be studied throughout the life cycle from harvesting to benefit sharing in order to develop mechanisms that can enable forest based livelihoods to play a role in economic development as well as an incentive for conservation.The rural communities inhabiting in forests respects the floral and faunal elements that give them sustenance. We must recognise the role of these people in restoring and conserving forests. The modern knowledge and skills of the forest department should be integrated with the traditional knowledge and experience of the local communities. The strategies for the joint management of forests should be evolved in a well-planned way. Their local traditional knowledge systems can be taken as a base on which modern concepts can be built, rather than by fostering concepts that are completely alien to their own knowledge systems. These rural people in the villages have a deep insight on the need for sustainable use of natural resources and know about methods of conservation, there are however several newer environmental concerns that are frequently outside their sphere of life experiences. Therefore, the study facilitates further research with active participation of local peoples and it also helps in divulging the rural people knowledge in the understanding of newer concerns and challenges.

FUTURE IMPLICATIONS

PWLS is relatively unexplored protected ecosystem compared to other protected areas in Sikkim. Therefore, further research in this area can help to strengthen our understanding of natural resources and their utilization. More engaging paleontological research to reconstruct the natural, oral history, documenting heritage sites of the area can be carried out in collaboration with World Heritage Conservation Society, Archaeological survey of India, Botanical survey of India etc. The vegetation characteristics of the sanctuary have under gone change ever since grazing activity in the area was banned. High diversity and density of bamboo species indicates that bamboo colonize quickly after disturbance, given a proximate seed source. Their importance in early successional communities suggests an important role in establishment of forest ecosystem functioning. Therefore, extensive study can be carried out on bamboo ecology as keystone species, habitat for threatened animal species, linkages and history with people (e.g. Singhane baans-*Arundinaria hookeriana* Munro-relates to Singhane dara village, death, birth), livelihood generation etc. Considering the extremely fragile nature of the Eastern Himalayan ecosystems and a growing threat from climate change, the future role of bamboo species as indicator species in ecosystem monitoring cannot be underestimated (field observation). Conflict between man-animal in the transition ecosystem needs more engaging study in future.

ACKNOWLEDGEMENTS

Author is thankful to Forest, Environment and Wildlife Management Department, Gangtok, Government of Sikkim for allowing to conduct study in the Pangolakha Wildlife Sanctuary and also thankful to the local people, Panchayat secretary members, for providing valuable information and wonderful hospitality. Thankful to the local faith healers and the local people living alongside the boundary of PWLS for their valuable information and cooperation.

REFERENCES CITED

Anonymous, 1971. *Report Central Forestry Commission, Ministry of Agriculture*, New Delhi, India.

Anonymous 1970. *Year book of the forest products*, F.A.O., 1970, pp. 1969-70.

Avasthe RK., Rai PC, Rai LK. 2004. *Sacred groves as repositories of genetic diversity-*A case study from Kabi-Longchuk, North Sikkim,.ENVIS Bulletin 12 (1): Himalayan Ecology.

Champion HG, Seth SK. 1968. *A revised survey of forest types of India*, Government of India, Delhi.

Chatterjee CD. 1965. Forestry in ancient India, West Bengal Forests, Centenery Commemoration Volume, page 2.

Chettri A, Barik SK, Pandey HN, Lyngdoh, MK. 2010. Liana diversity and abundance as related to microenvironment in three forest types located in different elevational ranges of the Eastern Himalayas. *Plant Ecology and Diversity* 3(2). 175–185.

Chettri N, Sharma E. 2006. Assessment of natural resources uses patterns: A case study along a trekking corridor of Sikkim Himalaya, India. *Resources, Energy and Development* 3(1): 21-34.

Das T, Mishra SB, Saha D, Agarwal S. 2012. Ethnobotanical Survey of Medicinal Plants Used by Ethnic and Rural People in Eastern Sikkim Himalayan Region. *African Journal of Basic & Applied Sciences* 4 (1): 16-20.

FSI. 1997. *State of Forest Report.* Forest Survey of India (Ministry of Environment and Forests), Dehra Dun.

Gadgil M, Vartak VD. 1975. Sacred groves of India – a plea for continued conservation. *Journal of Bombay Natural History Society*, 73 : 623-647.

Grierson AJC, Long DG. 1983-1991. *Flora of Bhutan*, vols 1-3. Royal Botanical Garden, Edinburgh, UK.

Hoft M, Barik SK, Lykke AM. 1999. *Quantitative Ethnobotany. Applications of multivariate sand statistical analyses in ethnobotany.* People and Plant Working Paper, pp. 46.http://www.angrau.net/StudyMaterial/EnviSci/BIRM301.pdf

ICIMOD. 2004. *Biodiversity and livelihood in the Hindu-Kush Himallayan Region.* International Center for Integrated Mountain Development (ICIMOD), Newsletter No. 45. ICIMOD, Kathmandu, Nepal.

Lama MP. 2001. *Sikkim Human Development Report*, Government of Sikkim, Social Science Press, Delhi.

Pisces Conservation Ltd. 2007. IRC House, the Square, Pennington, Lymington. Hampshire, SO41 8GN, UK.

Ramakrishnan PS. 1992. *Shifting agriculture and sustainable development: an interdisciplinary study from north-eastern India.* Paris, UNESCO-MAB Series, Carnforth, UK, Parthenon, 424 pp. (Republished by Oxford University Press, New Delhi, 1993).

Singh KK, Rai LK, Gurung B. 2009. Conservation of Rhododendrons in Sikkim Himalaya: An Overview. *World Journal of Agricultural Sciences* 5(3): 284-296.

Subba JR. 2009. Indigenous knowledge on bio-resources management for livelihood of the people of Sikkim. *Indial Journal of Traditional Knowledge* **8**(1):56-64.

Sundriyal M. 1999. Distribution, propagation and nutritive value of some wild edible plants in the Sikkim Himalaya, Ph D thesis submitted to Garhwal University, Srinagar (Garhwal).

Tambe S, Kharel G, Arrawatia, ML, Kulkarni, H, Mahamuni, K, Ganeriwala, AK. 2012. Reviving Dying Springs: Climate Change Adaptation Experiments from the Sikkim Himalaya. *Mountain Research and Development* 32(1):62-72.

Watanabe K. 2010. *Study report for the first session.* Study of Eco-Tourism for the Sikkim Biodiversity Conservation and Forest Management Project.

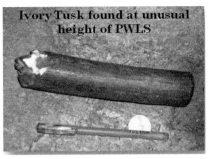

Appendix S1: Sacred places, heritage sites and account of unusual ivory tusk in PWLS and in its adjacent villages.

23

Himalayan *Saussurea costus* (Falc.) Lipsch.: Traditional uses, Phytochemistry, Therapeutic Potential and Conservation Perspective of Critically Endangered Medicinal Plant

Bikarma Singh[1,2] and Sneha[3]

[1]*Plant Sciences (Biodiversity and Applied Botany), CSIR-Indian Institute of Integrative Medicine, Jammu-180001, Jammu and Kashmir, INDIA*
[2]*Academy of Scientific and Innovative Research, Anusandhan Bhawan New Delhi-110001, INDIA*
[3]*CSIR-Central Institute of Medicinal and Aromatic Plant, Lucknow-226015 Uttar Pradesh, INDIA*
Email: drbikarma@iiim.res.in; drbikarma@iiim.ac.in

ABSTRACT

Saussurea costus (Falc.) Lipsch. is a perennial rhizomatous herbaceous plant, commonly called as *kuth* had a wide application in drug discovery programme and indigenously known for its medicinal usage in Indian system of medicine. Wide distribution of this species recorded from Himalayas and Indo-Mynamar hotspot biodiversity regions. This species is traditionally used for treatment of several seasonal diseases happen to occurs in hilly and mountaneous regions. It has been scientifically validated that this species possess various biological activities such as anti-microbial, anti-cancer, anti-arthritic, anti-convulsant, anti-oxidant, angiogenesis effect, hepatoprotective, gastro-protective, anti-obesity, and anti-inflammatory due to the presence of various bioactive principles such as costunolide, germacrene, lappadilactone, isodihydrocostunolide, cynaropicrin, linoleic acid, cyclocostunolide, alantolactone, isoalantolactone, sesquiterpene-saussureamines and several others in minor quantities. This species naturally grows in wild and mostly collected from the wild for its herbal usages, and this wild collection activities put S. *costus* under threatened and critically endangered categories, hence there is need for the development of agrotechnology for cultivation in large scale. This species has the potential components for developing new molecules in future for drug development. The present communication provides traditional usages, phytochemistry, therapeutic potentials, and future perspectives of *S. costus*. This baseline information provided in this article will be helpful in future for research to be carried forward in drug discovery programme using this plant species.

Keywords: Himalayan medicinal plant, *Saussurea costus*, Traditional knowledge, Chemistry, Pharmacology, Drug discovery.

INTRODUCTION

Medicinal plants are known for their bioactives chemical constituents synthesized in different parts as a secondary metabolites. *Saussurea* genus named after Horace Benedict de Saussure (1740-1799) and currently comprised of 433 species (TPL 2018). This species globally recognized as one of the largest genus under the family Asteraceae (or Compositae). Asteraceae represented by 32,913 species belonging to 1,911 genera (TPL 2018), has the ability to grow and distribute in all types of climates ranging from tropical to alpine, and several species of these are also cultivated for their economic demands at international markets for medicine and nutraceuticals. The maximum diversity of *Saussurea* DC. reported from Sino-Himalayan region, however, this genus extended its distribution to Europe, Asia and North America (Chopra and Vishwakarma 2018). Indian boundary encompassed with 61 species and most of these species restricts their distribution to Alpine Himalaya (Pusalkar and Singh 2009) and also evident that 37 species have been reported from the Himalayas (Chopra and Vishwakarma 2018).

Saussurea costus (Falc.) Lipsch. is a perennial rhizomatous herbaceous plant, commonly called *kuth* had a wide application in drug discovery programme and traditionally known for its medicinal usages in Indian system of medicine. Wide distribution of *S. costus* recorded from Himalayas and Indo-Myanmar hotspot biodiversity region, however, majority of the population of *S. costus* is restricted to Western Himalayas in India, China and Pakistan (Shah 2006). This species is found growing in close association with *Betula* forests on hill slopes of Himalayas (FRLHT 2009), and grows well in the moist slopes between the elevation ranges of 2000-4000 m MSL (Hajra et al. 1995). According to Chopra and Vishwakarma (2018), this species reported naturally growing in

Rajouri of Jammu, Bhaderwah of Doda, Pir Panjal Mountain range of Shopian, Kishanganga valley of Kashmir, Chenab valley and Suru valley of Ladakh, Nanda Devi National Park and Valley of Flowers National Park in Uttarakhand and Churah in Himachal Pradesh. In India, Butola and Samant (2010) reported the cultivation of *S. costus* from Himachal Pradesh and Uttarakhand, and in addition to these, Kokate et al. (2002) reported this species cultivated in Tamil Nadu and Uttar Pradesh to meet the commercial demand of the market. Kuniyal et al. (2005) indicated Indo-China and Vietnam are also the place for commercial multiplication of this species in large scale for industries and pharmaceutical sectors.

Saussurea costus has wide application as medicine because its constituents possess various pharmacological activities such as angiogenesis effect, anti-bacterial, anti-arthritic, anti-cancer, anti-convulsant, anti-ulcer, gastric function, hepatoprotective, anti-inflammatory and anti-viral (Cheng 1995, Kim et al. 2012, Chopra and Vishwkarma 2018). Its bioactive constituents include costunolide, germacrene, lappadilactone, isodihydrocostunolide, alantolactone, cynaropicrin, linoleic acid, cyclocostunolide, isoalantolactone, sesquiterpene-saussureamines and several others chemical constituents (Madhuri et al. 2012, Zahara et al. 2014). Several biological activities of this species are well proved, and well established which gave a rationale scientific approach to the traditional claims (Madhuri et al. 2012). Therefore, the present communication provides traditional uses, phytochemistry, therapeutic approaches and conservation perspectives of *S. costus*.

Taxonomy of Saussurea costus

Saussurea costus is an upright, robust, perennial herb growing to a height of 1-5 m. Root has strong aromatic odour. Leaves are triangular to ovate, upper surface scaberulous, lower surface glabrous. Petioles of leaves are 50-90 cm long, 30-40 cm broad and winged. Lamina is thin and margin of terminal lobe irregularly lobed. Head is formed terminal. Involucral bract is usually tetra-seriate, ovate-lanceolate, purple, brown inside. Receptacular bristles is 1.2-1.5 cm long, and corolla tube dark purple. Achene is oblong, slightly curved, and compressed, 0.7-0.8 cm long, 0.3-0.4 cm wide, lower part narrow, brown and tetrangular. Pappus is brown, plumose, setae 0.8-1 cm long and unequal. Chopra and Vishwakarma (2018) provided the microscopic characters and reported transverse section of the fresh root show uniform periderm followed by broad zone of phloem and xylem. Cork is 3 to 5 layered, and phloem fibre is exhibited. Medullary ray is multiseriate and wider in phloem region than in xylem. Oleoresin cavities and inulin is also present. Under microscope, irregular bits of yellow, brown or orange-red fragments of resin and oil associated with thin-walled parenchymatous cells, broken bit of xylem vessel with scalariform, reticulate thickening and horizontal end wall is present (Anonymous 2001).

Phenology and Reproductive Aspects

The phenological period of *Saussurea costus* varies across the place of occurrence of this species in Himalaya regions and elsewhere across the globe. As per Chopra and Vishwakarma (2018), flowering of this species starts in June and extends upto the middle of August. Fruiting starts in August and can last upto the end of September. In winter, plants in wild habitat were not visible and dies off due to harsh climate and resume their growth after mild showers in April-May. While, studying cytology and chromosomes of *S. costus*, VirJee and Kachroo (1985), and Siddique and Wafai (1993) observed the two types of cytotypes i.e. 2n=26 and 2n=36. When flowering starts in the month of June, and after deposition of pollen to the stigmas, the seed development usually starts from first week of August and lasts upto the last week of September. The senescence initiates in October and is completed at the end of November. *S. costus* produces a large number of pollen grains than the ovules. Each floret possesses only one ovule. The peripheral floret gets modified into petal like structure to attract pollinators. The coloured heads of *S. costus* and the numerous scented and smelly pollen grains that they produce attracts the insects from a long distance. The insects insert their mouth parts into the anther collar in search of nectors, and get heavily loaded with pollen grains and when pollinators visited another flowers to a far flung areas, pollination takes place. On anthesis, pollen masses discharges out through the tips of floret in the form of white dust. The stigma still remains concealed within the floret till all the pollen grains get completely thrown out of the anther column. As the anther dehiscence completes, the rod like style elongates and emerges out, and fertilization takes place. According to Wani et al. (2006), studying the reproductive biology is fundamental to developing productive protocols for elite, endangered and threatened plants. Flowers essentially act as a food source for pollinators and other visitors, thereby, giving the study of flowering phenology both ecological and evolutionary significance (Wani et al. 2006). As per Chopra and Vishwakarma (2018), few literatures are available regarding the reproductive biology of *S. costus*. The study reveals that the low temperature due to ice and snow at the high altitude topography responsible for the late reproductive cycles of this species as compared to the other species of Asteraceae growing at the lower elevations. The undergrowth roots of *S. costus* remain dormant throughout the winter months, and the sprouting of the undergrowth roots varies in time and change of weather.

Phytochemistry

Saussurea costus is a repository of various bioactive constituents belongs to different classes of compounds which were mostly isolated from the roots. By using petroleum, hexane and methanolic extracts, various chemical constituents have been isolated and characterized from this species (Fig. 1).

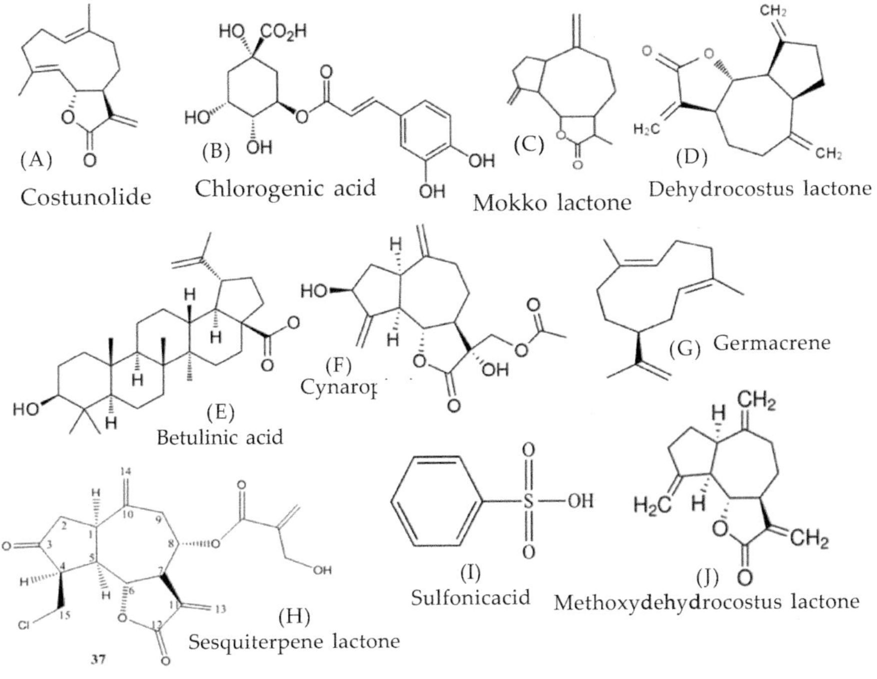

Fig. 1: Major phytochemical constituents and their structures isolated from *Saussurea costus*

Costunolide (cyclocostunolide, dihydro costunolide), chlorogenic acid, mokko lactone, dehydrocostus lactone (4α, 4β methoxydehydrocostus lactone), betulinic acid, linoleic acids, cynaropicrin, sesquiterpene lactones (sulfocostunolide A, sulfocostunolide B), germacrenes [(+)-germacrene A, germacra-1(10),4,11(13)-trien-12-ol, germacra 1(10),4,11(13)-trien-12 al and germacra-1 (10),4,11(13)-trien-12-oic acid], lappadilactone, and saussuramines (A,B,C,D,E) were isolated, studied and characterized from *S. costus* (Salooja et al. 1950, Rao et al. 1960, Govindan and Bhattacharaya 1977, Dhillon et al. 1987, Singh et al. 1992, Talwar et al. 1991, Talwar et al. 1992, Yoshikawa et al. 1993, Taniguchi et al. 1995, Kalsi et al. 1995, Yang et al. 1997, Matsuda et al. 2000, de Kraker et al. 2001, Pandey et al. 2004, Robinson et al. 2008, Wang et al. 2008, Choi et al. 2009).

Non-cytotoxic compounds, shikokiols, sesquiterpenes, guaianolidetype with a C_{17} skeleton and lappalone were isolated (Jung et al. 1998, Sun et al. 2003, Chhabra et al. 1998). Bedi et al. (2011) reported the chemical analysis of *S. costus* and discussed the presence of reducing sugars, tannins, resins, essential oils and alkaloids. Madhuri et al. (2012) investigated that from the acidic fraction of the *S. costus* oil, α-amorphenic acid can be isolated, and these have been studied for its spectral studies (Ruke et al. 1978). Under the group of guainolides, isodehydrocostus lactone and isozaluzanin C were isolated from the plant and confirmed by correlation with dehydrocostus lactone (Kalsi et al. 1984, Choi et al. 2009). The sesquiterpene lactones belonging to guaiane type, sulfocostunolide A and sulfocostunolide B (Wang et al. 2008) and with sulfonic acid group named 13-sulfodihydrosantamarine were also studied from this medicinal plant species.

Drugs and Medicines

It has been of the record that the traditional system of medicine in India such as ayurveda, amchi, siddha and unani uses *S. costus* as an ingredient for herbal formulation. Whole plant or sometime roots of this species used in asthma, cough, cholera, skin infection and rheumatism (Pandey et al. 2007, Chopra and Vishwakarma 2018). In traditional unani system, this species is used as tonic for germinal organ, brain, liver and heart (Rawat and Pangtey 1987). Besides, this species has wide application in treatment of paralysis and local herbal medicine men used this species as diuretic and anthelminthic. Ayurvedic doctors used medicine of this plant as bitter, phlegm, fever, phthisis, cough, dyspepsia, pain relief, dropsy, skin infections and jaundice (Chopra and Vishwakarma 2018). There is a report that the local hakims and vaidyas use roots of *S. costus* in curing malaria, leprosy and rheumatism. The chinese apply the root of this species in toothache (Nadkarni 1982). Several traditional tibetan herbal medicines using this plant were made for curing chronic inflammation of lungs, cough and chest congestion (Tsarong 1994). Dried root mixed with cooking oil applied on head as tonic to improve the condition of hairfall and scalp. Root of this species is also reported to be smoked as stimulant in China and Indian Himalaya as a substitute for the opium in winter. According to Bedi et al. (2011), this species used as a perfume and also sprayed in house for protecting the clothes from insects, moths and vermins. In Ayurvedic medicine, 0.2-1.0 gram of root drug in the form of powder is recommended for skin ailments (Anonymous 2001). Kim et al. (2012) reported that *S. costus* used in Chinese traditional medicine for the treatment of abdominal pain, nausea and cancer.

Volatile Constituents

Several studies focused on the different composition of volatile oils, and their content varies according to season (Gwari et al. 2013, Benedetto et al. 2018). Singh et al. (1957) reported that the volatile oils can be obtained from the fresh roots by a number of methods. It is of general practice that volatile oil can be extracted from the fresh roots of the plant using Hydro-Distillation Method in a Clevenger Type Apparatus for 3.5 to 4 hours at room temperature (25-30°C). Volatile fractions can be separated by using sodium sulphate anhydrous powder, and its constituents can be analyzed using an Agilent Technologies Gas Chromatography Device connected to Mass System (Agilent USA) with DB-5 fused silica column (30 m×0.25 mm i.d., film thickness 0.25 µm). The oven temperature requires being at 50°C for 5 min and raised to 280 °C at a rate of 10°C/min. Helium gas used as the carrier gas at a flow rate of 1 mL/min, and injector and detector temperatures to be kept at 280°C. Ion source temperature to be maintained at 150°C and scan mass range of m/z can be maintained at 50-550. The compounds can be identified by comparison of their mass spectra with the Wiley Libraries and Retention Indices with those reported in the literature. Analysis of essential oil components indicates thirty-one different volatile compounds, and the predominant compounds include monoterpenes and sesquiterpenes classes of compounds. Dehydrocostus lactone (17.73%), 1,3-cyclooctadiene (16.1%), elema-1,3,11(13)-trien-12-ol (11.56%), beta-elemen (5.9%), valerenol (5.28%) and various others in minor percentage (Bagheri et al. 2018, Table 1).

Table 1: Chemical constituents of essential oils from roots of *Saussurea costus*

Volatile constituents	Percentage of Oil
Thymol	1.07
Carvacrol	0.70
beta-Elemen	5.90
trans-Caryophyllene	4.37
alpha-Ionone	1.67
trans-α-Bergamotene	0.53
alpha-Humulene	0.45
Geranylacetone	0.54
beta-Selinene	3.15
alpha-Curcumene	2.93
alpha-Selinene	1.43
Cetene	0.38
cis-gamma-Bisabolene	0.48
Elemol	2.70
gamma-Eudesmol	1.01
beta-Eudesmol	1.14
alpha-Eudesmol	1.44
Elema-1,3,11(13)-trien-12-ol	11.56
Tetradecyne	1.61
1,3-Cyclooctadiene	16.10
Cyclododecene, 12-methyl-1-(1-propynyl)	0.36
(3E,5E,8Z)-3,7,11-Trimethyl-1,3,5,8,10-dodecapentanene	0.17
Z-alpha-trans-Bergamotol	0.30
(+)gamma-Costol	1.18
Valerenol	5.28
(-)alpha-Costol	3.74
2(3H)-Benzofuranone	0.37
Germacra-1(10),4,11(13)-trien	0.21
(-)Isodiospyrin	0.27
Dehydrocostus lactone	17.73
Costunolide	0.34
Monoterpenes	4.28
Sesquiterpenes	65.57
Hydrocarbons	20.30
Other compounds	47.90

Biological Applications

Saussurea costus have been studied in detail by several workers across the globe for biological functions, and its potential has been scientifically validated as antimicrobial, angiogenesis, anti-cancer, anti-convulsant, anti-arthritic, anti-inflammatory, anti-oxidant, hepatoprotective, anti-cytotoxic, anti-septic and anti-ulcer activities using different in-vitro and in-vivo models (Madhuri et al. 2012, Chopra and Vishwakarma 2018). The pharmacological activities investigated from the *S. costus* extracts and its isolated phytochemicals have been summarised below in subhead.

Anti-microbial Activity

Root extracts of *Saussurea costus* were evaluated for the action against Hepatitis B Virus (HBV). Costunolide and dehydrocostus lactone suppressed the expression of Hepatitis B surface antigen (HBsAg) in human hepatoma Hep3B cells in a dose-dependent manner (IC50 values 1.0 and 2.0 mM) following the method of northern blotting analysis and the suppression observed in the human hepatoma cell line HepA2 derived from the HepG2 cells (Chen et al. 1995, Madhuri et al. 2012, Chang et al. 2011). It was proved that the tested compounds showed significant activity against Hepatitis B Virus. Yu et al. (2007) studied the inhibitory effects of *S. costus* ethanolic extract on the growth, acid production, adhesion and water-insoluble glucan synthesis of *Streptococcus mutans*, and the inhibition of caries-inducing properties of *Streptococcus mutans* has been validated. Kaewsomboon et al. (2012) studied the antibacterial properties of Konjac Glucomannan (KGM) film containing extract of two herbs (*Atractylodes lancea* and *Saussurea costus*) at different concentration and results when compaired against two pathogenic bacterias, *Bacillus subtilis* and *Staphylococcus aureus* shows the higher bacterial activity. Volatile oils extracted from roots is suitable in food products as a novel food preservative because this inhibit the growth of *Staphylococcus aureus*, and represses the production of exotoxins, particularly *Staphylococcal enterotoxins* (Qui et al. 2011, Chopra and Vishwakarma 2018). This investigation suggested that the anti-bacterial activity of the plant is mainly due to the volatile oils present (Li et al. 2005).

Angiogenesis Effect

Chemical constituents costunolide isolated from *Saussurea costus* investigated to be able to inhibit the endothelial cell proliferation which is induced by Vascular Endothelial Growth Factor (VEGF). In-vitro study conducted by Madhuri et al. (2012) indicate that the chemotaxis induced by VEGF of human umbilical vein endothelial cells (HUVECs) can significantly inhibits at IC_{50} of 3.4 μM. When costunolide tested for angiogenesis in in-vivo method by mouse corneal micro pocket assay, the neo vascularisation of mouse corneal induced by VEGF had significantly inhibited at the dose of 100 mg/kg/day, and this demonstrated that this species can produce angiogenesis effect. Betulinic acid, methyl ester, mokko lactone and dehydrocostus lactone showed potent activity for protein tyrosine phosphatase 1B (PTP1B) inhibitors and Choi et al. (2009) proves this by *in vitro* protein tyrosine phosphatase 1B (PTP1B) inhibition assay.

Anti-cancer Activity

Kim et al. (2008) attempted to investigate the hexane extract of *Saussurea costus* and proves that it induces apoptosis of DU145 cells. The active principle dehydrocostus lactone isolated through column chromatography inhibits the cell growth and induced apoptosis in DU145 human prostate cancer cell lines (Choi and Gun-Hee 2010). In traditional system of medicine, the dried roots of *S. costus* are used for the treatment of cancer (Chopra and Vishwakarma 2018). Sesquiterpene, costunolide, β-cyclocostunolide, dihydro costunolide and dehydro costuslactone compounds tested for their *in vitro* cytotoxic activity, Robinson et al. (2008) proves that these compounds exhibit potent cytotoxic activity. Cynaropicrin was tested for its immunomodulatory effects on cytokine release, nitric oxide production and immunosuppressive effects. The compound inhibited the cell lines such as U937, Eol-1 and Jurkat T cells with IC50 values of 3.11, 10.9 and 2.36 ìM respectively in a dose-dependent manner, but Chang liver and human fibroblast cell lines does not show any such effect and it was found that the effect may be due to apoptosis. Cho et al. 2004 investigated the cytotoxic effect using DNA fragmentation, cell cycle arrest, flow-cytometric, morphological analysis using U937 cells. Cynaropicrin when treated in combination with L-cysteine, N-acetyl-L-cysteine, reactive oxygen species scavengers or rottlerin, and reports that it inhibits the cynaropicrin-mediated cytotoxicity and morphological change. Therefore, it is proven that cynaropicrin found to be more cytotoxic to leukocyte-derived cancer cells than fibroblasts (Madhuri et al. 2012). According to Chopra and Vishwakarma (2018), the cytostatic effects of *S. costus* extracts were examined using gastric AGS cancer cells, and recorded that treatment resulted in apoptosis and G2-arrest in a dose- and time-dependent manner. The effects were attributed to the regulation of cyclins and pro-apoptotic molecules and suppression of anti-apoptotic molecules. Ko et al. (2005) reports that *S. costus* root can be a candidate to deal with gastric cancers either by the traditional herbal therapy or by the combination therapy with the conventional chemotherapy. Rasul et al. (2012) reports that costunolide and sesquiterpene lactone isolated from *S. costus* possess potent anticancer properties.

Anti-convulsant Activity

Madhuri et al. (2012) reported that the root extracts of *Saussurea costus* prepared from different solvents such as petroleum, ether and water were investigated for the anticonvulsant activity by pentylenetetrazole, picrotoxin-induced convulsions, and the maximal electroshock tests performed in mice by which it is proved that the petroleum extract at a dose of 100 and 300 mg/kg i.p. showed the potent anti-convulsant activity (Shirishkumar et al. 2009).

Anti-inflammatory Activity

According to Chopra and Vishwakarma (2018), *Saussurea costus* is used in Korean traditional prescriptions for inflammatory diseases. The total methanol extract at 0.1 mg/ml as a final concentration exhibited more than 50% of inhibition on cytokine induced neutrophil chemotactic factor (CINC) induction. The ethanolic extract of *S. costus* roots at a dose range of 50–200 mg/kg, p.o. was studied for the acute and chronic inflammation induced in mice and rats (Madhuri et al. 2012). The plant extract showed anti-inflammatory activity through carrageenan-induced paw edema and peritonitis animal models which showed the anti-inflammatory activity in a dose dependent manner (Gokhale et al. 2002). The preventive effects of dehydrocostus lactone from *S. costus* on NF-Kappa B activation in LPS treated RAW 264.7 macrophages and U937 human monocytic cells were examined to explain the molecular mechanism for the suppression of LPS-induced nitric oxide (NO) production (Jin et al. 2000). The results showed that the suppression of NO production is mediated by the inhibitory action on the inducible nitric oxide synthase (iNOS) gene expression through the inactivation of NF-Kappa B and these lactones can act as pharmacological inhibitors of the NF-Kappa B activation (Jin et al. 2000, Chopra and Vishwakarma 2018). As reported in literatures (Kang et al. 2004, Chopra and Vishwakarma 2018), the costunolide inhibited protein and mRNA expression of interleukin-1b (IL-1b) in LPS stimulated RAW 264.7 cells and even suppresses the AP-1 transcription activity which was confirmed by an electrophoretic mobility shift assay (EMSA). It even inhibited the phosphorylation of mitogen activated protein kinases (MAPK) which is specific inhibitors such as SAPK/JNK and p38 MAP kinase. This proves that costunolide can be used as an anti-inflammatory compound for human.

Anti-oxidant Activity

Thara and Zuhra (2012) studied the different bioactive components of *Saussurea costus* possess antioxidant activity against the gram positive bacteria, *Staphylococcus aureus* MTCC 87 and gram negative *Escherichia coli*. Chang et al. (2012) indicate that n-butanol fractionates from the root of this species @ 1000 ppm have the strongest inhibitory potential on 2,2-diphenyl-1-picrylhydrazyl (DPPH) radical and reducing power at 92.98% and 0.38, respectively. Hence, this data shows that *S. costus* is helpful in preventing the anti-oxidative stresses (Chopra and Vishwakarma 2018, Chang et al. 2012).

Hepatoprotective

According to publication by Chopra and Vishwakarma (2018), constunolide and dehydrocostus lactone isolated from the roots of *Saussurea costus* investigated to have a strong suppressive effect on the expression of the Hepatitis B surface antigen (HBsAg) in human hepatoma Hep 3B cells, but it has no effect on the viability of cells. These two compounds usually suppress HBsAg production by Hep3B cells in a dose-dependent manner with IC_{50} of 1.0 M and 2.0 M, respectively (Pandey et al. 2007). Northern blotting analysis showed the suppression of HBsAg gene expression at the mRNA level. The suppressive effect of costunolide on replication of human liver cells also investigated in another human hepatoma cell line Hep A_2 which derived from Hep G2 cells by transfecting a tandemly repeat hepatitis B virus DNA. This investigation recommends the active constituents of this species, costunolide and dehydrocostus lactone, had the potential to be developed as specific anti-HBV drugs in years to come (Chen et al. 1995).

Gastro-protective Activity

The active molecules such as costunolide, saussureamines (A,B,C) and dehydrocostus lactone isolated from the methanolic extract of the roots of *Saussurea costus* proves the gastro-protective effect on acidified ethanol-induced gastric mucosal lesions in rats in a dose dependent manner, 5 and 10 mg/kg, respectively (Madhuri et al. 2012, Chopra and Vishwakarma 2018). According to another study by Matsuda et al. (2000) the inhibitory effect was also shown by saussureamine A on gastric mucosal lesions induced by water immersion stress in mice and this proves the gastroprotective effect of this species in drug development programme. The acetone extract of costunolide from *S. costus* exhibits cholagogic effect and inhibitory effect on the formation of gastric ulcer in mice (Pandey et al. 2007). According to Chopra and Vishwakarma (2018) followed by a study of Yamahara et al. (1985), it is proven that the variation in gastric acidity output, serum gastrin and plasma somatostatin concentration were recorded during *S. costus* decoction perfusion into the stomach of patients with chronic superficial gastritis. This investigation reveals that the decoction of *S. costus* can accelerate the gastric and increase the endogenous motilin release (Chen et al. 1994).

Anti-parasitic Activity

According to Chopra and Vishwakarma (2018), the *Saussurea costus* possess anti-parasitic activity. Rhee et al. (1985) evaluated anti-parasitic activities against *Clonorchis sinensis* and *Trypanosoma cruzi*, and the study proves that the decoction of *S. costus* when administered orally into rabbits infected with *Clonorchis sinensis*, it has been observed that this suppresses the egg laying capacity of rabbits. Lirussi et al. (2004) tested the methanolic extract of *S. costus* in vitro with the epimastigote form of *Trypanosoma cruzi* clone Bra C15 C2 at 27°C in F-29 medium at a concentration of 100 g/ml in axenic cultures; allopurinol was used as reference drug, and observed that this extract exhibited the inhibitory activity upto 100%.

Anti-obesity Activity

It has been investigated that betulinic acid isolated from *Saussurea costus* suppressed the hypothalamic protein tyrosine phosphatase 1B in mice and enhanced the antiobesity effect of leptin in obese rats (Choi et al. 2013). The experiment proves that the combinational treatment of BA and ethanol extract of *Orthosiphon stamineus* would be effective for the treatment of obesity.

Cultivation and Agronomic Practices

Plant genetic resources in agri-horticultural crops and their wild relatives are of immense value (Chopra and Vishwakarma 2018). Wild population of this plant could be in temperate and sub-alpine region of Himalayas. For successful growth, *Saussurea costus* prefers the moist sandy loam soil, rich in organic carbon. Bio-edaphic conditions at an altitude of 1200-1800 m above mean sea level is suitable for the cultivation. According to Kuniyal et al. (2005) propagation through seeds is common and on an average 70% seeds germinate under field conditions and the seeds of *S. costus* can be planted in April or May in nurseries and the seedlings can be transplanted to suitable area when they assume 10-15 cm height. Organic manures such as FYM, and vermivompost can be ideal for better growth. The plants usually require 5-6 irrigations between May-September for better growth and quality establishment. According to Kuniyal et al. (2005), the wild population of *S. costus* was covering approximately 400–600 hactares in early 1960s, and now it has been reduced to only 80 ha in Lahual Valley of Northern Himalayas.

Population Status, Threats and Future Perspectives

Convention on International Trade on Endangered Species put *Saussurea costus* under Appendix 1, where this species has been categorized as critically endangered plant of Himalayas (Anonymous 1973, Arora and Bhojwani 1989, Chopra and Vishwakarma 2018). Based on current data on the wild population, this species has restricted and fragmented range of distribution, and its wild population is declining at a faster rate. Increasing anthropogenic activities, over-exploitation for its wide application in pharmaceutical industries, climate change coupled with man-forced introduction of invasive and exotic species is creating irreversible habitat loss of this species at local, regional and global scales. *S. costus* preservation and conservation can be accomplished through *ex-situ* and *in-situ* conservation. The former includes the protection of wild population of this species outside their natural habitats which involves collection, maintenance and conservation of samples of live plants, seeds, vegetative propagules, tissues or cell cultures. Botanical gardens, seed banks, tissue culture and cryopreservation would be helpful for preservation and conservation of this species. *In-situ* maintenance of *S. costus* through establishment of conservation and multiple-use areas offers several advantages over off-site methods in terms of coverage, viability of the resources, and economic sustainability (Chopra and Vishwakarma 2018). It involves natural habitat management for direct manipulating populations of *S. costus*. Benedetto et al. (2018) and Rao et al. (2013) reported that *Saussurea costus* is spontaneously growing in Jammu and Kashmir State of India between the elevation range of 3200-3800 m above sea level, and therefore, such areas need to be prioritized for the conservation of this and similar species in their natural habitat for conservation of pure gene pool.

LITERATURES CITED

Anonymous. 1973. Convention on International Trade in Endangered Species of Wild Fauna and Flora. Signed at Washington, DC on 3 March, 1973 and amended at Bonn, Germany.

Anonymous. 2001. *The Ayurvedic Pharmacopoeia of India*, ed.1st, Vol-1. GOI, Ministry of Health & Family Welfare, Department of ISM & H, New Delhi, India.

Arora R, Bhojwani SS. 1989. In vitro propagation and low temperature storage of *Saussurea lappa* C.B.Clarke-an endangered medicinal plant. *Plant Cell Reports* 8(1): 44-47.

Bagheri S, Ebadi N, Taghipour Z, Manayi A, Toliyat T, Ardakani M. 2018. Preparation of *Saussurea costus* traditional oil and investigation of different parameters for standardization. *Research Journal of Pharmacognosy* 5(2): 51-56.

Bedi YS, Dutta HC, Kaur H. 2011. Plants of Indian Systems of Medicine, Vol. II: Monographs on twenty commercially important medicinal plants. Lambert Academic Publishing, United State of America.

Benedetto C, Auria MD, Mecca M, Prasad P, Singh P, Singh S, Sinisgalli C, Milella L. 2018. Chemical and biological evaluation of essential oil from *Saussurea costus* (Falc.) Lipsch. from Garhwal Himalaya collected at different harvesting periods. *Natural Product Research* https://doi.org/10.1080/14786419.2018.1440219.

Butola JS, Samant SS. 2010. *Saussurea* species in Indian Himalayan Region: diversity, distribution and indigenous uses. *International Journal of Plant Biology* 1(1): 43-51.

Chang KM, Choi SI, Chung SJ, Kim GH. 2011. Anti-microbial activity of *Saussurea lappa* C.B.Clarke roots. *Preventive Nutrition and Food Science* 16(4): 376-380.

Chang KM, Choi SI, Kim GH. 2012. Anti-oxidant activity of *Saussurea lappa* C.B.Clarke roots-research note. *Preventive Nutrition and Food Science* 17(4): 306-309.

Chen HC, Chou CK, Lee SD, Wang JC, Yeh SF. 1995. Active compounds from *Saussurea lappa* Clarks that suppress hepatitis B virus surface antigen gene expression in human hepatoma cells. *Antiviral Research* 27 (1-2): 99-109.

Chhabra BR, Gupta S, Jain M, Kalsi PS. 1998. Sesquiterpene lactones from *Saussurea lappa*. *Phytochemistry* 49: 3801-3804.

Cho JY, Kim AR, Jung JH, Chun T, Rhee MH, Yoo ES. 2004. Cytotoxic and pro-apoptotic activities of cynaropicrin, a sesquiterpene lactone, on the viability of leukocyte cancer cell lines. *European Journal of Pharmacology* 92: 85-94.

Choi EJ, Gun-Hee K. 2010. Evaluation of anticancer activity of dehydrocostus lactone in vitro. *Molecular Medicine Reports* 3(1): 185-188.

Choi EM, Kim GH, Lee YS. 2009. Protective effects of dehydrocostus lactone against hydrogen peroxide-induced dysfunction and oxidative stress in osteoblastic MC3T3-E1 cells. *Toxicolology In Vitro* 23: 862–867.

Choi JY, Na MK, Hyun Hwang I, Ho Lee S, Young Bae E, Yeon Kim B, Seog Ahn J. 2009. Isolation of betulinic acid, its methyl ester and Guaiane sesquiterpenoids with protein Tyrosine phosphatase 1B inhibitory activity from the roots of *Saussurea lappa* C.B.Clarke. *Molecules* 14: 266–272.

Choi YJ, Park SY, Kim JY, Won KC, Kim BR, Son JK, Lee SH, Kim YW. 2013. Combined treatment of betulinic acid, a PTP1B inhibitor, with orthosiphon stamineus extract decreases body weight in high-fat-fed mice. *Journal of Medicinal Food* 16(1): 2-8.

Chopra VL, Vishwakarma RA. 2018. Plants for Wellness and Vigour. New India Publishing Agency, New Delhi, India.

de Kraker JW, Franssen MC, de Groot A, Shibata T, Bouwmeester HJ. 2001. Germacrenes from fresh costus roots. *Phytochemistry* 58: 481-487

Dhillon RS, Kalsi PS, Singh WP, Gautam VK, Chhabra BR. 1987. Guaianolide from Saussurea lappa. *Phytochemistry* 26: 41209-41210.

FRLHT. 2009. Environmental Information system. www.frlht.org.

Gokhale AB, Damre AS, Kulkarni KR, Saraf MN. 2002. Preliminary evaluation of anti-inflammatory and anti-arthritic activity of *S. lappa, A. speciosa* and *A. aspera*. *Phytomedicine* 9 (5): 433-437.

Govindan SV, Bhattacharaya SC. 1977. Alantolides and cyclocostunolides from Saussurea lappa. *Indian Journal of Chemistry* 15: 956.

Gwari G, Bhandari U, Andola HC, Lohani H, Chauhan N. 2013. Volatile constituents of *Saussurea costus* roots cultivated in Uttarakhand Himalayas, India. *Pharmacognosy Research* 5: 179-182.

Hajra PK, Rao RR, Singh DK, Uniyal BP. 1995. Flora of India, Vols. 12 & 13: Asteraceae. Botanical Survey of India, Calcutta, India.

Jin M, Lee HJ, Ryu JH, Chung KS. 2000. Inhibition of LPS-induced NO Productioon and NF-kappa B activation by a serquiterpene from *Saussurea lappa*. *Archives of Pharmacal Reseasrch* 23: 54-58.

Jung IH, Kim Y, Lee CO, Kang SS, Park JH, Im KS. 1998. Cytotoxic constituents of Saussurea lappa. *Archives of Pharmacal Reseasrch* 21: 153-156.

Kaewsomboon T, Sawangkan K, Satirapipathkul C. 2012. Characterization of konjac glucomannan film containing the extracts of *Atractylodes lancea* and *Saussurea lappa*. *Advanced Material Research* 506: 401-404.

Kalsi P, Gurdeep K, Sunila S, Talwar KK 1984. Dehydrocostuslactone and plant growth activity of derived guaianolides. *Phytochemistry* 23: 2855–2861.

Kalsi PS, Kumar S, Jawanda GS, Chhabra BR. 1995. Guaianolides from *Saussurea lappa*. *Phytochemistry* 40: 1713-1715.

Kim EJ, Hong JE, Lim SS, Kwon GT, Kim J, Kim JS, Lee KW, Park JHY. 2012. The hexane extract of *Saussurea lappa* and its active principle, dehydrocostus lactone, inhibit prostate cancer cell migration. *Journal of Medicinal Food* 15(1): 24-32.

Kim EJ, Lim SS, Park SY, Shin HK, Kim JS, Park JHYP. 2008. Apoptosis of DU145 human prostate cancer cells induced by dehydrocostus lactone isolated from the root of *Saussurea lappa*. *Food and Chemical Toxicology* 46(12): 3651-3658.

Ko SG, Kim HP, Jin DH, Bae HS, Kim SH, Park CH, Lee JW. 2005. *Saussurea lappa* induces G2-growth arrest and apoptosis in AGS gastric cancer cells. *Cancer Letters* 220(1): 11-19.

Kokate CK, Purohit AP, Gohkale SB. 2002. Pharmacognosy. In: Terpenoids, 21st edn. Nirali Prakasan, India, pp 377–378

Kuniyal CP, Rawat YS, Oinam SS, Kuniyal JC, Vishvakarma SCR. 2005. *Saussurea costus (S. lappa)* cultivation in the cold desert environment of the Lahaul valley, northwestern Himalaya, India: arising threats and need to revive socio-economic values. *Biodiversity Conservation* 14 (5): 1035-1045.

Li Y, Xu C, Zhang Q, Liu JY, Tan RX (2005) In vitro anti-helicobacter pylori action of thirty chinese herbal medicines used to treat ulcer diseases. *Journal of Ethnopharmacology* 98: 329–333

Lirussi D, Li J, Prieto JM, Gennari M, Buschiazzo H, R'ýos JL, Zaidenberg A. 2004. Inhibition of Trypanosoma cruzi by plant extracts used in Chinese medicine. *Fitoterapia* 75: 718-723.

Madhuri K, Elango K, Ponnusankar S. 2012. *Saussurea lappa* (Kuth root): review of its traditional uses, phytochemistry and pharmacology. *Oriental Pharmacy and Experimental Medicine* 12(1): 1-9.

Matsuda H, Kageura T, Inoue Y, Morikawa T, Yoshikawa M (2000) Absolute stereo structures and syntheses of Saussureamines (A,B,C,D,E), amino acid-sesquiterpene conjugates with gastroprotective effect from the roots of *Saussurea lappa*. *Tetrahedron* 56: 7763–7777.

Nadkarni AK. 1982. Indian Materia Medica, Vol-1. Bombay popular prakashan, Mumbai, India.

Pandey MM, Govindarajan R, Rawat AKS, Pangtey YPS, Mehrotra S. 2004. High performance liquid chromatographic method for quantitative estimation of an antioxidant principle chlorogenic acid in *Saussurea costus* and *Arctium lappa*. *Natural Product Sciences* 10: 40-42.

Pandey MM, Rastogi SR, Rawat AKS. 2007. *Saussurea costus*: botanical, chemical and pharmacological review of an ayurvedic medicinal plant. *Journal of Ethnopharmacology* 110(3): 379-390.

Pusalkar PK, Singh DK. 2009. *Saussurea forrestii* (Asteraceae)-a new record for Indian Flora. *Rheedea* 19 (1 & 2): 61-63.

Qiu J, Wang J, Luo H, Du X, Li H, Luo M, Dong J, Chen Z, Deng X. 2011. The effets of subinhibitory concentrations of costus oil on virulence factor production in *Staphylococcus aureus*. *Journal of Applied Microbiology* 10(1): 333-340.

Rao AS, Kelkar GR, Bhattacharyya SC. 1960. Terpenoids-XXI, The structure of costunolide, a new sesquiterpene lactone from costus root oil. *Tetrahedron* 9: 275-283. Rao RN, Raju SS, Babu KS, Vadaparthi PRR. 2013. HPLC determination of costunolide as a marker of *Saussurea lappa* and its herbal formulations. *International Journal of Biochemistry* 3: 99-107.

Rasul A, Yu B, Yang L, Arshad M, Khan M, Ma T, Yang H. 2012. Costunolide, a sesquiterpene lactone induces G2/M phase arrest and mitochondria-mediated apoptosis in human gastric adenocarcinoma SGC-7901 cells. *Journal of Medicinal Plants Resesarch* 6(7): 1191-1200.

Rawat GS, Pangtey YPS. 1987. A contribution to the ethnobotany of alpine regions of Kumaon. *Journal of Economic and Taxonomic Botany* 11: 139-148.

Rhee JK, Back BK, Ahn BZ. 1985. Structural investigation on the effects of the herbs on *Clonorchis sinensis* in rabbits. *The American Journal of Chinese Medicine* 13: 119-125.

Robinson A, Kumar TV, Sreedhar E, Naidu VG, Krishna SR, Babu KS, Srinivas PV, Rao JM. 2008. A new sesquiterpene lactone from the roots of *Saussurea lappa*: structure–anticancer activity study. *Bioorganic and Medicinal Chemistry Letters* 18: 4015-4017.

Ruke DD, Traas PC, Heide RT, Boelen H, Takken HJ 1978. Acidic components in essential oils of costus root, patchouli and olibanum. *Phytochemistry* 17:1664–1666.

Salooja KC, Sharma VN, Siddiqi S. 1950. Chemical examination of the roots of *S. lappa* Part I, on the reported isolation of the alkaloid Saussurine. *Journal of Scientific and Industrial Reseasrch* 9: 1.

Shah R. 2006. Nature s Medic inal plants of Uttaranc hal: (Herbs, Grasses & Ferns). Vol. I & II. Gyanodaya Prakashan, Nanital, Uttarakhand, India.

Siddique MAA, Wafai BA. 1993. New bases number for *Saussurea lappa* C.B.Clarke an important endangered medicinal plant of Kashmir Himalayas. *Chromosome Information Series* 54: 5-6.

Singh H, Singh T, Handa KL. 1957. A note on costus oil from Kashmir costus roots. *The Indian Forester* 83: 606.

Singh IP, Talwar KK, Arora JK, Chhabra BR, Kalsi PS. 1992. A biologically active guaianolide from *Saussurea lappa*. *Phytochemistry* 31: 2529–2531.

Sun CM, Syu WJ, Don MJ, Lu JJ, Lee GH. 2003. Cytotoxic sesquiterpene lactones from the root of *Saussurea lappa*. *Journal of Natural Products* 66: 1175-1180.

Talwar KK, Singh IP, Kalsi PS. 1991. A serquiterpenoid with plant growth regulatory activity from *Saussurea lappa*. *Phytochemistry* 31: 1336-1338.

Taniguchi M, Kataoka T, Suzuk H, Uramoto M, Ando M, Arao K, Magae J, Nishimura T, Otake N, Nagai K. 1995. Costunolide and dehydrocostus lactone as inhibitors of killing function of cytotoxic T lymphocytes. *Bioscience, Biotechnology and Biochemistry* 59: 2064–2067.

TPL. 2018. The Plant List 2013, version 1.1. Published on internet: http://www.theplantlist.org/ [Accessed on 15 August 2018].

Tsarong T. 1994. Tibetan Medicinal Plants. Tibetan Medical Publications, India.

VirJee DU, Kachroo P. 1985. Chromosomal conspectus of some alpine-subalpine taxa of Kashmir Himalaya. *Chromosome Information Series* 39: 33-35.

Wang F, Xie ZH, Gao Y, Xu Y, Cheng XL, Liu JK. 2008. Sulfonated guaianolides from *Saussurea lappa*. *Chemical and Pharmaceutical Bulletin* 56: 864–865

Wani PA, Ganaie KA, Nawchoo IA, Wafai BA. 2006. Phenological episodes and reproductive strategies of *Inula racemosa* (Asteraceae)-a critically endangered medicinal herbs of Northwest Himalaya. *International Journal of Botany* 2(4): 388-394.

Yamahara J, Kobayashi M, Miki K, Kozuka M, Sawada T, Fujimura H. 1985. Cholagogic and antiulcer effect of Saussurea radix and its components. *Chemical and Pharmaceutical Bulletin* 33: 1285-1288.

Yang H, Xie J, Sun H. 1997. Study on chemical constituents of *Saussurea lappa* I. *Acta Botanica Yunnanica* 19: 85-91.

Yoshikawa M, Hatakeyama S, Inoue Y, Yamahara J. 1993. Saussureamines A, B, C, D and E, new anti-ulcer principles from Chinese *Saussurea radix*. *Chemical and Pharmaceutical Bulletin* 41: 214-216.

Yu HH, Lee JS, Lee KH, Kim KY, You YO. 2007. *Saussurea lappa* inhibits the growth, acid production, adhesion, and water-insoluble glucan synthesis of *Streptococcus mutans*. *Journal of Ethnopharmacology* 111(2): 413-417.

Zahara K, Tabassum S, Sabir S, Rahmatullah MA. 2014. A review of therapeutic potential of *Saussurea lappa*-an endangered plant from Himalaya. *Asian Pacific Journal of Tropical Medicine* 7: 60-69.

24

Bringing Innovation and Digitization of Web Accessible Herbarium Database to Build a National Resource in India: A Case Study of Plant Inventory Biodiversity Data of Janaki Ammal Herbarium

Bikarma Singh[1,3], Sumit G. Gandhi[2,3], Harish Chander Dutt[4]
Abdul Rahim[5], Rajan Sachdev[6] and Yashbir Singh Bedi[2,3]

[1]*Biodiversity and Applied Botany Division,* [2]*Plant Biotechnology Division*
CSIR-Indian Institute of Integrative Medicine Jammu Tawi-180001
Jammu & Kashmir, INDIA
[3]*Academy of Scientific and Innovative Research, Anusandhan Bhawan*
New Delhi-110001, INDIA
[4]*Department of Botany, University of Jammu, Jammu-180006*
Jammu & Kashmir, INDIA
[5]*PME Division, CSIR-Indian Institute of Integrative Medicine*
Jammu Tawi-180001, Jammu & Kashmir INDIA
[6]*Plus Automation, 328-New Plot, Jammu-180005, Jammu & Kashmir, INDIA*
Corresponding authors: drbikarma@iiim.ac.in, sumit@iiim.ac.in

ABSTRACT

Systematic studies through stored specimens in herbaria and museum of defined geographical areas provide a set of baseline data for understanding the vegetation composition of a defined ecosyetm. A stored plant collection in herbaria offers a rich portrait of biodiversity, making innumerable contribution to science and human society. Digitization coupled with hosting of such information on world wide web (www) system has been accepted as an important tool for the dissemination of the botanical specimens in the form of databasing, imaging and geo-referencing of voucher specimens. In the present communication, we describe the development of the Janaki Ammal Herbarium (JAH) database that contains a comprehensive and searchable data of voucher snaps, historical background and statistical analysis of 23,225 voucher specimens. The information on JAH database is easy to extract data on botanical names, author citation, variety (if any), family, collection number, specific location, collection date, habitat, collector's name and remarks on morphology. The goal of the present investigation is to digitize (database, image, georeference) all specimens in JAH, enabling them to be made available through a single portal. At concluding, all vouchers of JAH made publicly accessible *via* world wide web user interfaces, and also linked to PubMed, NCBI Taxonomy, where the site provide information related to Entrez records, comments and references of the particular species, while Wiki species site is linked to the free species directory. It also provide link of online accessible herbarium-based databases of the world.

Keywords: Digitization, Taxonomy, Biodiversity, Website data basing, Janaki Ammal Herbarium, CSIR, India

INTRODUCTION

Thousands of plant collections catalogued and systematically stored in different herbaria across the world are the major frontier for the species description, discovery and rediscovery (Bebber et al. 2010, Singh and Bedi 2017). These biological specimens constitute an irreplaceable wealth for education and provide baseline information for systematic research (Carranza-Rojas et al. 2017, Tschope et al. 2013). Species reaction to habitat loss and fragmentation, biological invasions, pollution, and the consequences of global climate alteration were the main areas of study related to species decline (Barnosky et al. 2011), and the loss of biodiversity have become catastrophe disciplines and depend heavily on the baseline information that herbarium and museum collections offers (Carvalho et al. 2008). Most of the biological specimens resulting from the biodiversity inventories are stored in the herbaria, museums, or in private collection, and contains the main primary occurrence data that is used as the basis of many biodiversity scientific studies (Suarez and Tsutsui 2004, Baird 2010, Tulig et al. 2012). Studies conducted by Lavoie (2012) reports 264,543,834 herbarium specimens stored in 733 herbaria across the world, and could be used as tools for biogeographical and environmental studies. In the context of global climate change, the importance of herbarium institutions and museum specimens are day-by-day increasing (Johnson et al. 2011).

The storage and maintenance of authentic specimens of plants are quite inexpensive as compared to the benefits which they offer to the end users by serving as the central referral facility for the correct identification of raw material for research and development purposes. To keep pace with times, botanists and curators, who have, historically managed and worked with live or/and preserved collections are anxious to move toward the digitalization of their collections. Hence, there is need for proactive the approach to convert the prized collections into digital form for both preservation and dissemination purposes, make them a prime resource for research in many different disciplines (Pupulin and Gonzalez 2003). Efforts in digitization ranges from the creation of digital images and photographs to historical documents. The organization of such data into digital, easily accessible, searchable form adds a great value to the knowledge of biodiversity. Internet provides an excellent mechanism for the dissemination of specimen data and taxonomic information (Wang et al. 2009, Knapp et al. 2002, Rao et al. 2012). One of the reasons that these collections have moved online is that they are all too often hidden from the public eye (Matthew and Christopher 2002). Not only would digitization of the collection make it easier to access, it would also allow researchers to look at the specimens before the start of a new research area. The digital collection would therefore offers advantages like increased visibility through the WWW, create access to a previously hidden collection abode in museum or in herbarium, reduce the need for the physical handling, support access regardless of user skill and could be used as an inventory of specimens in the collection (Schmidt 2007). Access to public domain information on plant will advance research in bioinformatics, ethnobotany, taxonomy, biogeography and phytochemistry (Gaikwad et al. 2008).

Keeping in view the needs of digital herbarium specimens in accurate species recognition and describing the new taxa, and to make the raw data available online for researchers, CSIR-IIIM Jammu has built a custom-designed internet-accessible database for JAH. This herbarium is recognized internationally under the acronym RRLH and is registered in Index Herbariorum-A Global Directory of Public Herbaria and Associated Staff at the New York Botanical Garden, USA (Holmgren et al. 1990). JAH serve as both the internal laboratory information management system and the public portal for accessing the biodiversity survey data.

STUDY AREA

Background of Janaki Ammal Herbarium

Janaki Ammal Herbarium at the CSIR-Indian Institute of Integrative Medicine (IIIM) is named after Edavaleth Kakkat Janaki Ammal, the first Indian woman to receive a Doctorate Degree in Science and a Founding Fellow of the Indian National Science Academy. She worked at IIIM (formerly, known by the name of Regional Research Laboratory) in Jammu and made the most significant contribution in shaping the field of plant genetics in India. Till to-date, accessions at this herbarium includes a total of 23,225 herbarium specimens of 3,267 species under 1,168 genera and 219 families. The specimens comprised of angiosperm, gymnosperm and pteridophyte groups of plants mainly of medicinal, aromatic and otherwise economic values from Himalayan regions, adjoining parts of India and even some specimens housed are of foreign origin. RRLH herbarium sheets are prepared following standard methods of Jain and Rao technique (Jain and Rao 1977) and specimens pigeoned as per Bentham and Hooker's System (Bentham and Hooker 1862-1983) of plant classification. The authors of scientific plant names and abbreviations used are as per Brummitt and Powell (1992), and follow rules of International Code of Botanical Nomenclature (ICBN) while incorporating the latest changes in naming of plants.

Statistical Analysis of Accession

Data were assembled from accession register maintained at JAH for 23,225 herbarium specimens collected and accessioned between 1958-2017. The data reveals that the maximum number (34.03%) of the total herbarium were filed and incorporated during 1963-1967, followed by 20.82% between 1958-1962 (Table 1).

Table 1: Year-wise herbarium sheets accessioned at Janaki Ammal Herbarium

Sl. No.	Period	RRLH Accession Nos.	Registered Accession Nos.	Total sheets	% of Accession
1.	1958-1962	00001-4836	I	4836	20.82
2.	1963-1967	4837-12740	I, II	7904	34.03
3.	1968-1972	12741-14087	II	1347	5.08
4.	1973-1977	14088-15594	II	1507	6.49
5.	1978-1982	15595-16380	II	786	3.38
6.	1983-1987	16381-16960	II	580	2.50
7	1988-1992	16951-17457	II, III	507	2.18
8.	1993-1997	17458-18004	III	547	2.36
9.	1998-2002	18005-19794	III	1780	7.66
10.	2003-2007	19795-21581	III	1787	7.69
11.	2008-2012	21582-22280	III	699	3.01
12.	2013-2017	22281-23225	IV	945	4.07
	Total			23,225	100.00

Both these intervals include the specimens donated to this herbarium by distinguished and renowned botanist working in the western Himalayan region, and their collection even dates back to 1867. The lowest number of accession recorded was between the year 1988-1992 and the reason being a large number herbarium specimens collected during this session were under pipeline for incorporation. The bar diagram presented in Fig. 1 shows the details of accessions filed in five years of interval period between 1958 to 2017.

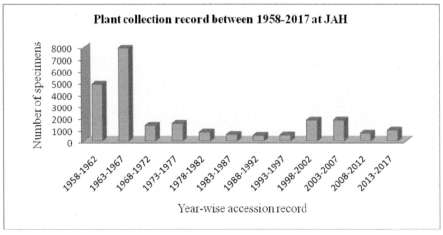

Fig. 1: Dynamics of specimens accessioned in the herbarium collection at Janaki Ammal Herbarium

Many prominent and distinguished scientists in the form of botanist, ecologist and ethnobotanist of India contributed in the enrichment of Janaki Ammal Herbarium. The prominent angiosperm collectors were A.K. Dutta, Baldev Singh, B.K. Arbol, B.K. Kapahi, Des Raj, D.R. Sharma, J.N. Naskar, K.S. Ahluwalia, L.D. Kapoor, N.M. Sood, P.K. Dutta, P.S. Jamwal, Rattan Chand, Santosh Gupta, S.N. Sobti, S.K. Malhotra, T.N. Dhar, Tajub Lapcha and Y.K. Sarin, however, the housed pteridophyte collections belong to E.N. Trotter, H.S. Kirn, R.R. Sterwat and Vir Singh. The year-wise details of collectors who contributed to to collection of herbarium specimens of Janaki Ammal Herbarium is presented in table 2.

Table 2: Botanists who contributed in the enrichment of Janaki Ammal Herbarium

Sl. No.	Accessioned period	Botanist contributed in field survey
1.	1958-1962	B.K. Arbol, K.S. Ahluwalia, L.D. Kapoor, P.S. Jamwal, R.Chand, R.L. Badhwar, S.N. Sobti and T.N. Dhar.
2.	1963-1967	B. Singh, B.K. Arbol, D. Raj, J.N. Naskar, K.S. Ahluwalia, L.D. Kapoor, N.M. Sood, P.K. Dutta, P.S. Jamwal, R. Chand, Santosh Gupta, S.N. Sobti, S.K. Malhotra, T. Lapcha, T.N. Dhar, V. Singh and Y.K. Sarin,
3.	1968-1972	A.K. Dutta, B.K. Arbol, B.K. Kapahi, B.S. Sethi, D.R. Sharma, J.P. Singh, K.S. Ahluwalia, L.D. Kapoor, Shamlal, P.S. Jamwal, P. Singh, R. Chand, R.L. Badhwar, S.N. Sobti, S. Gupta, T. Lapcha, and Y.K. Sarin.
4.	1973-1977	A.K. Dutta, B.K. Arbol, B.K. Kapahi, D.R. Sharma, E.N. Trotter, J.P. Singh, K.L. Pandoo, L.D. Kapoor, P.S. Jamwal, R.L. Badhwar, R.R. Stewart, S.K. Kapur, T.N. Dhar, and Y.K. Sarin.
5.	1978-1982	C.K. Atal, O.P. Suri, P.S. Jamwal, P. Singh, R. Bedi, S.K. Kapur, and S.N. Sobti.
6.	1983-1987	D.R. Sharma, D.S. Singh, J.P. Singh, S.K. Kapur, S.K. Tandon, Y.K. Sarin, and V. Shah.
7	1988-1992	B.K. Kapahi, D.R. Sharma, R.S. Dixit, S.K. Tandon, S.C. Sharma, and T.N. Srivastava
8.	1993-1997	B.K. Kapahi, H.S. Kirn, R.S. Kapil, T.N. Srivastava, and S.K. Tandon.
9.	1998-2002	B.M. Sharma, B.K. Kapahi, Deepak, H.S. Kirn, P. Baleshwar, S. Jamwal and T.N. Srivastava.
10.	2003-2007	B.M. Sharma, B.K. Kapahi, K.L. Gupta, M. Ayoab, N. Ali, P. Baleshwar, S. Jamwal, S.N. Sharma, S. Kitchlu, and S. Paul
11.	2008-2012	Arun Kumar, O. Baleshwar, Kuldeep, P. Sharma, S.N. Sharma, S. Kitchlu, S. Ram, S.G. Gandhi, Y.S. Bedi, and V.K. Gupta
12.	2013-2017	B.Singh, S. Gairola, S. Kitchlu, and Y.S. Bedi.

DIGITIZED JANAKI AMMAL HERBARIUM

Procedure

The protocol and procedures for the digitization workflow for JAH were discussed among the core group members and detailed SOP was prepared. This involved determining how specimens would be transported to and from the IT section where digitization of the specimens takes place. Hebarium sheets were carefully transported from JAH to dedicated scanning set up, and were returned to the original position in the herbarium cupboards after scanning. Scanned images were appropriately named and stored as separate file on hard drive as well as in CD-ROM. For each herbarium scanned sheet matching passport data was prepared and entered in excel spreadsheet. The compelete information was passed on to the IT (Information Technology) expert from Plus Automation, Jammu, for developing digitized sheets for individual specimens.

Data Content

Herbarium database presented on JAH website encompassing 3,267 plant taxa having 23,225 vouchers. All the information presented on the herbarium specimen label, viz. species name, author citation, sub-species (if any), variety (if any), family, subfamily (if any), collection number, location, date of collection, habitat and collector's name were first entered into Microsoft Excel spreadsheets and an updated catalogue prepared. To make the data available through internet, all the specimens were digitalized at a minimum resolution of 300 dpi to construct a web-accessible digital herbarium using Hyper Text Markup Language (HTML).

Families were arranged alphabetically and navigation bar at the footer of each page provide links for each taxonomic family. Within the page for each family, the individual species along with their synonyms list with links to individual pages of each accession. These individual web pages contain a scanned image of the herbarium sheet along with detail date of collections, identification and taxonomy. Moreover, these pages were also provided links to external databases (PubMed, NCBI Taxonomy, Wikispecies). For each plant, clicking on any of these links would open their respective pages on these external databases. Fig. 2 provides snapshots of web-accessible database available for public access at World Wide Web (WWW) (http://www.iiim.res.in/herbarium/index.htm).

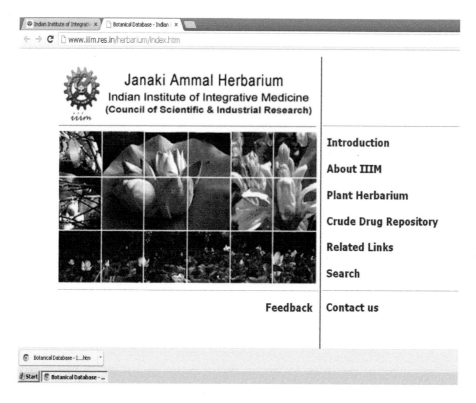

Fig. 2: Home page of the digital Janaki Ammal Herbarium database

Data Construction

The JAH has been digitalized using HTML, the publishing language of the WWW under HTML 4.01, which is a subversion of HTML 4. In addition to the text, multimedia, and hyperlink features of the previous versions of HTML (HTML 3.2 [HTML32] and HTML 2.0 [RFC1866]), HTML 4 supports more options, scripting languages, style sheets, better printing facilities, and documents that are more accessible to users with disabilities. HTML 4 is an SGML application confirming to International Standard ISO 8879 - Standard Generalized Markup Language. As a result, the web page displays correctly in any browser that confirm to standard. Every HTML document has a context-rich TITLE element in the head section to identify the contents of a document, since users often consult documents out of context. For reasons of accessibility, user agents always make the content of the title element available to users (including title elements that occur in frames).

The web interface for the digital herbarium make use of cascading style sheets (CSS) and is designed to be usable in browsers that do not support graphics. Style sheets supplant the limited range of presentation mechanisms in HTML. Style sheets make it easy to specify the way a data will appear with respect to amount of white space between text lines, the amount lines are indented, the colors used for the text and the backgrounds, the font size and style, and a host of other details. Style sheets of the digital herbarium have been placed in separate files so as to make them easy to reuse and manage. The website information architecture allows image data stored within the relevant family of the website structure so as to have an ease of manageability for the administrator/ webmaster for routine reference and upgradation as well as quick loading of web pages for the online visitors. Images are computed in JPEG (jpg) format for compression to photographic images so as to support downloading at a reasonably faster speed. The degree of compression has been adjusted, with a selectable trade-off between the storage size and the image quality achieving 10:1 compression. For future scalability, caching can be implemented to speed up the image loading process. There is currently no noticeable lag due to the degree of compression used in the digitalization.

Web Design

Web template system of JAH consists of three main parts, *viz.* (i) template engine i.e. the primary processing element, (ii) content resource means each individual plant species data stream along with the picture, and (iii) template resource, i.e. the web templates specified according to the template language. This has been deployed in JAH database software for creation of content and to deliver the web pages over the internet. The template and the content resources were processed, and combined by the template engine for the mass production of the web documents. These were also deployed a modular web page structured with components such as header, footer, global navigation bar (GNB), local navigation bar and content. These components were modified independently of each other, and gaves the flexibility to the admin team/ webmaster to focus on technical maintenance. The content of the template system give ease of the design change, ease of interface localization (menus and other presentation standards were easy to make uniform for users browsing on the site) and provision to work separately on design and code by different team members at the same time. Herbarium data were stored in database and allows rapid indexed searches to be carried out and the content to be generated dynamically. Non-text content such as high resolution digital image files and the documents requiring storage were stored on the file system instead of inside the database to provide greater efficiency in search. A schematic design detailed of the digital JAH herbarium followed is presented in Fig. 3 and the detailed architectural flow chart depicted in Fig. 4.

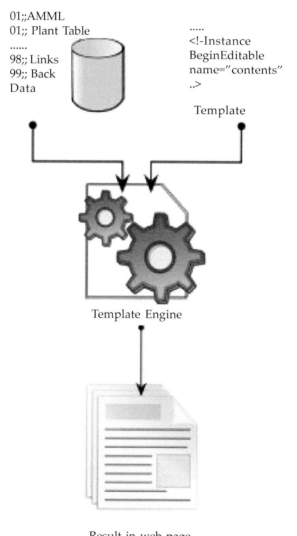

Result in web page

Fig. 3: Design details of the digital Janaki Ammal Herbarium

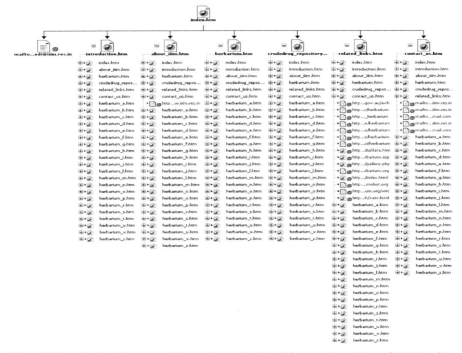

Fig. 4: Detailed architectural flow chart depicted for construction of JAH online

Internet User

Users are presented with a simple search interface. This appears on the right side of JAH web page. For convenience of the visitors for this website, the contents are grouped into six types, viz. introduction, about IIIM, plant herbarium, crude drug repository, related links, and contact us. Detailed instructions of these groups appear upon placing and clicking the mouse cursor over the web interface. Users can search for history of the JAH herbarium, key role of IIIM organization in the field scientific research and development, plant herbarium details, about housed crude drug samples, important links related to different known herbaria of the world, *viz.* Australia's Virtual Herbarium, Australian Natural Herbarium at Canberra, Queensland Herbarium at Mt. Coottha, National Herbarium of Victoria at Melbourne, The Charless Sturt University Virtual Herbarium, Southern Cross University, University of Florida Herbarium, The University of North Carolina Herbarium, The New York Botanical Garden Virtual Herbarium, Louisiana State University Herbarium Image Gallery, University of South Carolina Herbarium, The University of Georgia Herbarium, Missouri Botanical Garden, ENVIS Digital Herbarium,

oriaceae/dioscorea_deltoidea.htm

Fig. 5: Online herbarium sample sheet (example *Dioscorea deltoida*) of JAH

Neotropical Herbarium Specimens, W3 TROPICOS Nomenclatural database and associated authority.

The JAH website provides the basic and the advanced search capabilities. The image of plant herbarium provided allows species to be identified by browsing the digital images. Each image hyperlinked to corresponding family and species webpage. Users can click on the relevant alphabet of the family name. The search can be performed by putting several criteria's including nomenclature fields like flora of, botanical name, synonym(s), family, plant stage, place of collection, altitude (m), date of collection, accession number or locality. However individual species vouchar can be found by searching with the simple alphabetical search interface. The result of the web search provide user with 13 rows of information. These includes flora of the region, botanical name, basionym(s) and synonym(s) (if any), family name, stage at which plant was collected (flowering, fruiting or vegetative condition), accurate place of collection, altitude/height of the place at which the specimen was collected in meter (m), collection date, accession number recorded at RRLH register, total accession sheets of the particular species housed, cupboard number/pigeon hole number in which species were preserved, brief remarks on the specimens recorded by the collector and name of the person responsible for the identification of the particular specimen. The detail of an example herbarium sample sheet (*Dioscorea deltoida* Wall.) provided Fig 5. The herbarium species of JAH is linked to PubMed (http://www.ncbi.nlm.nih.gov), NCBI Taxonomy, where the site provide information related to entrez records, comments and references of the particular species, NCBI LinkOut, and Wiki species, where the site linked to the free species directory.

UTILITY OF JAH DATABASE

1. *Species Information of Himalayan Region*

Indian Himalayas is a part of global biodiversity hotspot region in the world[23] and information related to plant species, family, scientific name, author citation, synonym, common name, habit, habitat, taxonomy and distribution record abode in Himalayan region can be extracted from JAH. The study on ecology and taxonomy can be easily undertaken from the herbarium specimens and associated data. The scientific name and other data of the species can be used for cross-linking to external data portals such as PubMed, NCBI taxonomy and Wiki species.

2. *Medicinal and Aromatic Information*

Botanical collections housed in JAH were mainly of medicinal and aromatic plants collected for research in Natural Product Chemistry and for product development in different system of medicine such as Ayurveda, Unani, Sidda, Allopathy and Amchi. Hence, this herbarium database can be useful for researchers in accessing different products prepared from botanical collections, their accession status and herbarium images for easy authentication.

3. *Biogeography and Distribution Information*

Information related to locality, latitude and longitude in decimal units, district, state and country can be obtained from JAH, which will be helpful in studying plant distribution and biogeography of a particular species. JAH provide the original observational data on the particular species written on the herbarium label.

4. *Information on Invasive Species*

Botanists, ecologists, and conservation biologists widely believe that invasions by non-native species are a leading cause of recent extinctions of species. The loss of biodiversity and extinction are major ongoing crisis in terms of conservation. Invasive exotic species are causing dramatic changes in many ecological systems worldwide (Gurevitch and Padilla 2004). Herbarium and museum collection can be used to measure evolution in invasive species. Therefore, JAH database will be helpful in studying the migration, colonization and evolution of invasive species particularly in hotspot regions like Himalayas and Indo-Myanmar regions. This database will be useful in working out the first record of the invasive species, and their associated impacts in different ecosystem.

5. *Information on Phenology*

Recorded information on flowering and fruiting period help to investigate the taxonomy of plant. JAH provide information on phenological season of the particular species recorded by collectors and botanist from different eco-regions, which can be access online by looking at the specimens as well as the data presented.

6. *Global Climate Change*

There is a widespread agreement that global climate change threatens the survival of biological communities and individual species. By examining the collections housed at herbaria and museum specimens, scientists have documented the effects of climate change on a variety of organisms. In future, JAH database will serve as a baseline resource data for predicting the climate change and would be helpful in providing information on distribution of species through time, and the biology of particular species in response to climate change.

7. *Information on Habitat Loss and Loss of Biodiversity*

Habitat loss in the form of fragmentation and degradation considered as to be the greatest threats to biodiversity. Herbarium collections allow investigators to documents the time scale of these changes and their ecological consequences. JAH database will be useful to show the loss of biological diversity of Indian Himalaya, species decline or local extinction of plants.

8. *Monetary Importance*

Herbarium collections act as centralized storehouse of reference material and accumulated resources that eliminate the need for time-consuming fieldwork. Given the costs of traveling to remote localities to collect specimens, it is very easy to believe that herbarium collections save the research organization a lot of mony. Another importance is that the collected herbarium eliminates the wastefulness of duplication and redundancy. Therefore, online JAH database will provide direct financial benefits to research organization and social community.

ONLINE HERBARIUM DATABASES

Herbarium databases accessible online aimed to convert the morphological characteristic of plant collections into digital information and store the whole content in network environment. Highly developed websites are convenient for browsing and querying the digital images of plant specimens, has become an important tool as users spot their potential for their research. Some well known internet accessible herbarium-based database along country, location and website link is provided in table 3.

Table 3: Web accessible authentic herbarium-based database across the world

Sl.No.	Herbarium name	Country	Website approach
1.	Australia's Virtual Herbarium at Perth	Australia	http://www.Avh.chah.org.av
2.	BGCI- Botanic Gardens Conservation International at London	England	http://www.bgci.org
3.	C.V. Star Virtual Herbarium from New York Botanical Garden's International Plant Science Centre	United States	http://sciweb.nybg.org/science2/Virtual Herbarium.asp
4.	Harvard University Herbarium at Cambridge	United States	http://www. Kiki.huh. harvard.edu
5.	Herbarium Berolinense of Botanical Garden and Botanical Museum (BGBM) at Berlin-Dahlem	Germany	http://ww2.bgbm.org/Herbarium/default.cfm
6.	Herbarium Catalogue of the Royal Botanic Gardens (RGB) at Kew	England	http://apps.kew.org/herbcat/navigator.do
7.	Janaki Ammal Herbarium at Jammu	India	http://www.iiim.res.in/herbarium/herbarium.htm

8.	Kerala Forest Research Institute Herbarium at Dehradun	India	http://www.kfri.res.in
9.	LWG -Virtual Herbarium of the National Botanical Research Institute at Lucknow	India	http://www.nbri. res.in/herbarium
10.	MBLWHOL Library Digital Herbarium at Cape God	United States	http://www. mblwhoil brary.org
11.	National Institute of oceanography (NIO) - Digital Herbarium	India	http://www.nio.org
12.	New Zealand Virtual herbarium	New Zealand	http://www. virtualher barium.org.nz
13.	Oregon State University Herbarium at Corvallis	United States	http://www. oregonstate. edu/dept/botany/ herbarium
14.	Plant Database of Northern Ontario	Canada	http://www. orthernontario flora.ca
15.	Regional Plant Resource Centre (RPRC digital her barium) at Odisha	India	http://www.rprcbbsr.com/ herbarium
16.	Tropicos, the interactive database of Missourie Botanical Garden	United States	http://www.tropicos.org
17.	University and Jepson Herbaria at University of California	United States	http://ucjeps.berkeley.edu
18.	Utah Valley University Herbarium at Orem	United States	http://www. herbarium. ovu.edu
19.	vPlants-AVirtual Herbarium of Chicago region	United	States http://www. vplants. org
20.	WTU Herbarium Image Collection: Plants of Washington	United States	http://www.biology. burkey. washington. edu/ herbarium/ imagecollection.php

CONCLUSION

Voucher specimens are critical for the proper authentication and certification of the identity of research and commercial botanicals. The collection represent the work of botanists who collect specimens from field, press and dry them, and bring them back to the herbarium to mount on papers following specific procedures as a mechanism of recording botanical diversity (Baskauf and Kirchoff 2008, Dutta and Bedi 2013, Singh et al. 2013), but the lack of adequate storage of voucher represents a mortal mistake in a scientific study, and serious legal exposure in a commercial perspective (Gropp 2003). It is anticipated that the ecologist and the taxonomist of India and around the world will find JAH database constructive for collaborative and comparative work on large biogeographic scales. The data will be constantly scrutinized by the experts and will be updated accordingly. On the whole, JAH is developed as a single knowledge-base for the plant derived therapeutic substances and can be used as an integrated resource for scientists, researchers, policy makers, students and tribal communities through a click of mouse. As the database grows, JAH can be used for research in the areas such as Geographical Information System (GIS) studies, chemo-informatics and biodiversity informatics. Furthers, this aims to assist global and national priority of biodiversity conservation, better human and smart use of information using information technology. In coming time, it will serve as a storage of holistic knowledge-base with data on taxonomy, ethnobotany, ecology, anatomy, morphology, biodiversity, conservation, biogeography, phytochemistry, and bioactivity of medicinal and aromatic plants abode in the Himalayan regions. Comprehensive images and habitat data endow with supplementary information to authenticate DNA based identification and make possible the assessment of habitat distinctiveness from different regions of the world.

During the last few centuries, scientific collections were considered as an essential components of research, particularly for taxonomist, ecologist, ethnobotanist and evolutionary biologist. We are still in this stage of species discovery, and a majority of the species that exists on the earth has not yet been described and discovered. Contemporary collections need to be well curated and maintained, and this will require assurance to support and train taxonomists to maintain this modern facilities. The benefit of these collections to society must be maximized by stepping up the rate at which this information will entered in the database and made easy accessible. Optimistically, digitalization and web accessible system will help those fighting to save their collected flora and fauna from dismemberment. Nothing will ever replace the taxonomist knowledge and training that the botanical collection provides, therefore, funding in such areas should focused a national priority.

Availability and Requirements

Janaki Ammal Herbarium of Indian Institute of Integrative Medicine at Jammu is freely available and accessible online at http://www.iiim.res.in /herbarium/ index.htm.

List of Abbreviations Used

BGCI: Botanic Gardens Conservation International

CSIR: Council of Scientific and Industrial Research

CSS: Cascading Style Sheets

ENVIS: Environmental Information System

GIS: Geographical Information System

GNB: Global Navigation Bar

HTML: Hyper Text Markup Language

ICBN: International Code of Botanical Nomenclature

IIIM: Indian Institute of Integrative Medicine

ISO: International Organization for Standardization

JAH: Janaki Ammal Herbarium

JPEG: Joint Photographic Expert Group

MBL: Marine Biological Laboratory

NCBI: National Centre for Biotechnology Information

NIO: National Institute of Oceanography

RPRC: Regional Plant Resource Centre

RRLH: Regional Research Laboratory Herbarium, an Internationally recognized acronym of Janaki Ammal Herbarium, formal known as Regional Research Laboratory

SGML: Standard Generalized Markup Language

USA: United States of America

WWW: World Wide Web.

Competing Interests
All authors declare that they have no competing interests.

ACKNOWLEDGEMENTS

Authors are thankful to Council of Scientific and Industrial Research (CSIR) for funding under CSIR Network Project SMM0009. It represents institutional publication number IIIM/2149/2017.

REFERENCES CITED

Baird R. 2010. Leveraging the fullest potential of scientific collections through digitization. *Biodiversity Informatics* 7: 130-136.

Barkworth ME, Murrell ZE. 2012. The US Virtual Herbarium: working with individual herbaria to build a national resource. *Zookeys* 209: 55-73. doi:10.3897/zookeys.209.3205.

Barnosky AD, Matzke N, Tomiya S, Wogan GOU, Swartz B, Quental T, Marshall C, McGuire JL, Lindsey EL, Maguire KC, Mersey B, Ferrer EA. 2011. Has the Earth's sixth mass extinction already arrived? *Nature* 471: 51-57. doi: 10.1038/nature09678.

Baskauf SJ, Kirchoff BK. 2008. Digital plant images as specimens: toward standards for photographing living plants. *Vulpia* 16-30.

Bebber DP, Carine MA, Wood JRI, Wortley AH, Harris DJ, Prance GT, Davidse G, Paige J, Pennington TD, Robson NKB, Scotland RW. 2010. Herbaria are a major frontier for species discovery. *Proceeding National Academy of Sciences USA*, 107 (51): 22169-22171. doi: 10.1073/pnas.1011841108.

Bentham G, Hooker JD. 1862-1983. Genera Plantarum: ad exemplaria imprimis in herbariis kewensibus servata definite. London, Soho, Lovell Reeve & Co., doi: http://dx.doi.org/10.5962/bhl.title.747.

Brummitt RK, Powell CE. 1992. Authors of Plant Names. Kew, Royal Botanic Gardens 1-732.

Carranza-Rojas J, Goeau H, Bonnet P, Mata-Montero E, Joly A. 2017. Goining deeper in the automated identification of herbarium specimens. *BMC Evolutionary Biology* 17: 181.

Carvalho MR, Bockmann FA, Amorim DA, Branda~O CRF. 2008. Systematics must embrace comparative biology and evolution, need speed and automation. *Evolutionary Biology* 35 (2): 150-157. doi: 10.1007/s11692-2008-9018-7.

Dutta HC, Bedi YS. 2013. Problems associated with studying spatial distribution of plants through herbarium anthology: A case study of family Berberidaceae in North-West Himalaya. *Proceeding National Academy of Sciences India. Sect. B Biological Sciences* 83: 1-7. doi: 10.1007/s40011-013-0227-1.

Gaikwad J, Khanna V, Vemulpad S, Jamie J, Kohen J, Ranganathan S. 2008. CMKb: a web-based prototype for integrating Australian Arboriginal Customary medicinal plant knowledge. *BMC Bioinformatics* 9 (12): S25. doi: 10.1186/1471-2105-9-S12-S25.

Gropp RE. 2003. Are university natural science collections going extinct?. *BioScience* 53: 550.

Gurevitch J, Padilla DK. 2004. Are Invasive species a major cause of extinction? *Trends in Ecology and Evolution* 19(9): 470-474. doi: 10.1016/j.tree.2004.07.005.

Holmgren PK, Holmgren NH, Barnett LC. 1990. Index Herbariorum, Pt 1: The herbaria of the world, 8th ed. New York, Bronax, New York Botanical Garden 10: 1-693.

Jain SK, Rao RR. 1977. *A handbook of field and herbarium methods.* India, New Delhi, Today and Tomorrow's Printers and Publishers 16: 1-157.

Johnson KG, Brooks SJ, Fenberg PB, Glover AG, James KE, Lister AM, Michel E, Spencer M, Todd JA, Valsami-Jones E, Young JR, Stewart JR. 2011. Climate change and biosphere response: unlocking the collection vault. *Bioscience* 61, 147–153.

Knapp S, Bateman RM, Chalmers NR, Humphries CJ, Rainbow PS, Smith AB, Taylor PD, Vane-Wright RI, Wilkinson M. 2002. Taxonomy needs evolution, not revolution. *Nature* 419: 559-560.

Lavoie C. 2013. Biological collection in an area changing world-Herbaria as tools for biogeographical and environmental studies. *Perspectives in Plant Ecology, Evolution and Systematics* 15: 68-76. doi: org/10.1016/J. ppees.2012.10.002.

Matthew SD, Christopher P. 2002. vPlants-A virtual Herbarium of the Chicago region. *First Monday* 5: 3.

Pupulin F, González GAR. 2003. Costa Rican Orchidaceae Types (Crotypes) Digital Imaging Documentation at Ames, Harvard University. *Lankesteriana* 7: 11-16.

Rao KS, Sringeswara AN, Kumar D, Pulla S, Sukumar R. 2012. A digital herbarium for the flora of Karnataka. *Current Science* 102 (9): 1268-1271.

Schmidt L. 2007. Digitization of herbarium Specimens, a collaborative project. *ACRL 13th National Conference* 13: 64-69.

Singh B, Adhikari D, Barik SK, Chettri A. 2013. *Pterocymbium tinctorium* (Merrill, 1901) (Magnoliophyta: Malvales: Sterculiaceae: Sterculioideae): New record from mainland India and extension of geographic distribution. *Check List* 9(3): 622-625.

Singh B, Bedi YS. 2017. Eating from raw wild plants in Himalaya traditional knowledge documentary on Sheena tube in Kashmir. *Indian Journal of Natural Products and Resources* 8(3): 269-275.

Singh B, Borthakur SK, Phukan SJ. 2013. A survey on ethnomedicinal plants utilized by the indigenous people of Garo Hills with special reference to the Nokrek Biosphere Reserve (Meghalaya), India. *Journal of Herbs, Spices and Medicinal Plants* 20(1): 1-30. doi: 10.1080/10496475.2013.819476.

Suarez AV, Tsutsui ND. 2004. The value of museum collections for research and society. *BioScience* 54(1): 66-74.

Tschope O, Macklin JA, Morris RA, Suhrbier L, Berendsohn WG. 2013. Annotating biodiversity data via the internet. *Taxon* 62(6): 1248-1258.

Tulig M, Tarnowsky N, Bevans M, Kirchgessner A, Thiers BM. 2012. Increasing the efficiency of digitization workflows for herbarium specimens. *ZooKeys* 209: 103-113.

Wang N, Sherwood AR, Kurihara A, Conklin KY, Sauvage T, Presting GG. 2009. The Hawaiian Algal Database-A laboratory, Lims and Online resource for biodiversity data. *BMC Plant Biolology* 9: 117. doi: 10.1186/1471-2229-9-117.

25

Distribution, Conservation and Pharmacological Status of Genus *Aconitum* L. with Special Reference to Western Himalaya

Sabeena Ali[1,3], Sumit G. Gandhi[2,3], Yashbir Singh Bedi[2] and Qazi Parvaiz Hassan[1, 3]*

[1]*Microbial Biotechnology Division, CSIR-Indian Institute of Integrative Medicine Srinagar-190005, Jammu & Kashmir, INDIA*
[2]*Plant Biotechnology Division, CSIR-Indian Institute of Integrative Medicine Jammu Tawi-180001, Jammu & Kashmir, INDIA*
[3]*Academy of Scientific and Innovative Research, Anusandhan Bhawan New Delhi-110001, INDIA*
Corresponding author: qphassan@iiim.ac.in

ABSTRACT

The genus *Aconitum* is in the limelight of the current scenario of the pharmacopeia and has became the centre of attention in the field of herbal medicines, as is manifested with various biological and pharmacological cores which make the genus most imperative. For vast decades some species of the genus *Aconitum* and its derivatives were considered as vizulently fatal but were liquidated as intact drug elites through traditional as well as modern system of medicine. The retrospection of the validated information in account of the genus *Aconitum* represent a pioneering approach to corroborate the scientific studies on almost every aspect of the herbal genus *Aconitum*, as a number of species belonging to this genus has been enlisted in the Red Data Book of IUCN which have been renowned for their medicinal benefits in various systems of medicine. The tubers some species of *Aconitum* contain lethal diterpene alkaloids such as aconitine, mesaconitine, and hypaconitine, which easily can be directed into screened harmless alkaloids such as benzoylaconine, aconine, and pyraconine by heating or alkaline treatment. Currently, the processed tubers are widely and safely practiced for the remedy of pain, neuronal disorders, inflammation, and rheumatism and have a huge therapeutic index for curing ailments like hysteria, throat infection, dyspepsia, abdominal pain, diabetes and also as a cardiac depressant in massive arterial tension of cardiac origin. An important component extracted from some species of *Aconitum*, Lycaconitine has been found to be effective against multi-drug resistant cancers. The di-terpenoid alkaloids of aconite source and their increasing demand for the active compounds used in different systems of medicine has led to over-harvesting of the tubers, resulting in rapid depletion of the natural stocks of this valuable plant. As a consequence, the genus is now on the list of rare and threatened species. Since the commercial demand of this important medicinal plant is very high and hence needs to be conserved as the species density of this genus has been threatned due to many factors. As such diverse efficient and standardized protocols for its regeneration like micropropagation, somatic embryogenesis and some other conventional efforts like storage of seeds as a means of *ex-situ* conservation must be attributed on the aspects of conservation of genus *Aconitum*. The exploitation of population genetic diversity will provide useful information for the biological conservation of these esteemed species of *Aconitum* in the regions . Since a complete revision of the genus *Aconitum* in the region has not yet been carried out and the high level of endemism and a lack of relevant information of the genus *Aconitum* have made its various scientific studies a very interesting object. The scientific and systematic information about the genus *Aconitum* generated demands validation as only a little attention has been done on its various aspects from cultivation, conservation, phytochemistry and pharmacological evaluation to the molecular characterization. Consequently species specific measures are needed and a lot of important attributes has to be assigned to the species of this genus which will ultimately help to ensure simultaneously both conservation as well as sustainability in raw material production of the genus.

Keywords: *Aconitum*, endemism, conservation, diversity, bioprospecting

INTRODUCTION

Medicinal plants are an important element of various systems of medicine throughout the world. Recent global drift towards recovery of interest in plant ascertained drugs are the compelling reasons for evolving a mission mode approach towards medicinal plants. The prevailing research in drug discovery from medicinal plants involves a multifaceted approach combining the botanical, phytochemical, biological and molecular techniques. The drug disclosure in the medicinal plant factory projects to procure advanced and paramount leads against various pharmacological targets. Although the drug breakthrough from medicinal plants continues to provide an important source of new drug leads, numerous challenges are encountered including the procurement of plant materials, the selection and the implementation of appropriate high-throughput screening bioassays, and the scale-up of bioactive compounds. In India, Jammu & Kashmir state has a rich diversity of medicinal plants because it provides a diverse type of habitats for their growth. Kashmir valley represents the temperate region, Jammu area represents sub-tropical and tropical and Ladakh as cold desert region. Conjointly 50% of medicines of plant source prescribed in British pharmacopoeia are growing in the J&K state. As many as 570 plant species are of medicinal importance. Owing to the high medicinal value, the genus *Aconitum* is one of the most exploited plants in the indigenous drug industry. *Aconitum* L. is a representative genus of the buttercup family, Ranunculaceae, and is comprised by more than 300 species (Utelli et al. 2000). The plants are usually perennial or biennial herbs growing in moisture imbibing well drained soils and densely shaded areas of alpine meadows. Leaves are mostly cauline, lobed, rarely divided and dentate. Flowers are simple or branched recemes (Sajan et al. 2011). *Aconitum* is widely distributed across North Asia and North America. The Hengduan Mountain eous region of China comprises of about 166 species of *Aconitum* (Jabbour and Renner 2012, Luo et al. 2005, Luo and Yang 2005) and among them 76 species have been medicinally used (Xiao et al. 2006). In Pakistan this genus is represented by seven species (Riedle et al. 1991). In India, the genus comprised of 27 species, mainly distributed in subtropical, alpine and subalpine regions of Himalaya (Sharma et al. 1993), out of which the existence of few species like *Aconitum balfourii* Holmes ex Stapf., *A. bisma* Rap., *A. chasmanthum* Stapf ex Holmes, *A. deinorrhizum* Holmes ex Stapf., *A. elewesii* Stapf., *A. ferox* Wall ex Secinge, *A. falconeri* Stapf., *A. heterophyllum* Wall ex Royle, *A. leave* Royle, *A. lethale* Stapf., *A. laciniatum* Stapf., *A. luriderm* Munz., *A. moschatum* Stapf., *A. palmatum* Rap., *A. spicatum* Stapf. and *A. violaceum* Jack have been reported as critically. The crude extracts of some of these species have been reported to contain extremely lethal substances; however after detoxification these can be utilized therapeutically (Chopra 1984). *Aconitum balfourii* Stapf endemic to alpine and subalpine belts of Indian Himalayan region, an important highly prized herb of this genus *Aconitum* is facing severe threats and has been listed among 37 Himalayan medicinal herbs under priority for in-situ and ex-situ conservation. Its tuberous roots are rich sources of pseudoaconitine (0.4 to 0.5%) and some other alkaloids (Sharma et al. 2012). In Kashmir Himalayas, presently the genus is represented by about 5 to 6 species, among them the most commonly existing species are; *Aconitum heterophyllum* Wall ex Royle *Aconitum heterophyllum* var. *Bracteatum* Stapf, *Aconitum chasmanthum* Stapf ex Holmes, *Aconitum laeve* Royle, *Aconitum violaceum* Jacquem ex Stapf. The genus *Aconitum* finds a key position in the field of research and as many species of this have been enlisted as endangered in Red Data Book (Dar et al. 2001). A number of species of this genus are known for their medicinal values, among these *Aconitum heterophyllum* commonly known as "Ativisha" in Ayurveda is commonly used in Indian System of Medicine and hold prime position in showing vital medicinal assets. A variety now called *A. heterophyllum* var. *bracteatum*, which is sometimes raised as a separate species as *Aconitum kashmiricum* Stapf ex Coventry (Stapf et al. 1905) and locally named as 'Pevak' is considered as critically endangered medicinal herb endemic to Kashmir Himalaya, & reported to be the adulterant of *Aconitum heterophyllum* (Sinha et al. 2008). In recent years, *Aconitum heterophyllum* is described as species aggregate and needs further delimitation. *Aconitum chasmanthum* Stapf ex Holmes grows at high altitudes in Kashmir and its roots resembles to *Aconitum napellus* for which it has been sometimes wrongly identified. Indian *Aconitum chasmanthum* is about seven times more potent in alkalod content than the European variety, *Aconitum napellus* and hence greatly demanded in the world markets (Bhattacharya 1961). Its roots are used to treat rheumatism. Rhizomes are reported to be used in fever, skin diseases, enlargement of spleen and neuralgia (Dubey et al. 2012). *Aconitum laeve* Royle, one of the representative of the genus *Aconitum* is native to North West Himalaya has been found to be a rich source of diterpenoid and non-diterpenoid alkaloids that are generally of aconitine and lycoctonine types and some other type of alkaloids has been isolated (Ulubelen et al. 2002). Its tubers are frequently used for treatment of rheumatism and neuralgia (Wang et al. 1985). *Aconitum violaceum* Jacquem an important medicinal species has been used for various health ailments including renal pain, rheumatism and high fever. A reproducible *in vitro* regeneration system for *Aconitum violaceum* has been developed providing a basis for germplasm conservation and harnessing the medicinally active compounds of *Aconitum violaceum* (Rawat et al. 2013). Globally the molecular front as well as phytochemical analysis of the various species of this genus has been explored in many research institutes (Srivastava et al. 2010). The current archives in account of the genus *Aconitum* represents the progression in studies on scientific aspects its various under different broader research areas are highlighted as follows:

Historical Overview of Aconitum

During the 3rd century, a naturalist once analyzed the plant *Aconitum* from the Greek city Acona. Veritably from the previous literature, the name *Aconitum* had come from a Greek word akonitos, meaning "without struggle" (Yi-Zhun 2007). In ancient Greek mythology, *Aconitum* were said to have routed from the saliva of Cerberus, the three headed dog that guards the entrance to Hades. Saliva from this creature dripped onto the plant, rendering them extremely poisonous (Tripp 1970; Anonymous 1958 and Le Strange 1977). Other historical sources suggest that the name came from the hill Aconitus, a hill in Greek mythology where Hercules fought with Cerberus. In one of the studies it has been stated that the *Aconitum* takes its name from the ancient Greek as a purpose for poisoning the weapons like arrows for killing the animals especially the wolves, hence the English common name as "arrow poison" or "Wolfsbane". The plant have played an important role in Roman history, as it is assumed that Nero ascended to the throne after poisoning Claudius by tickling his throat with a feather dipped in monkshood. The emperor Trajan (98e117 AD) banned widening of this plant in all Roman domestic gardens (Trestrail 2007). One of the most remarkable pieces describing the role played by this plant in ancient Roman society was described by the writer Ovid. He referred to aconite as the "Step-Mother's Poison". In Romeo and Juliet novel, the potency of this herb has been highlighted by Shakespeare where he stated that Romeo committed suicide using this poison (Yi-Zhun 2007).

Trade of Aconites in India

Aconites may be considered as breakdown to society as they are highly poisonous in nature but may be used as medicine under proper guidance of the herbal practitioners after detoxification. Most of the medicinal plant species which possess very effective therapeutic properties have become rare, threatened, vulnerable and endangered in the wild, either due to over-exploitation or due to loss of their natural habitats. The main poisonous species are *Aconitum chasmanthum* and *Aconitum ferox* which are marketed under trade name Vikh, Vish' or

Vatsanabha. The rest of the species are considered as substitutes that are generally used after mitigation. Indian Aconite or *Aconitum ferox*, also called as The King of Poisons' due to the presence of highly poisonous alkaloids in its roots is more poisonous than European *A. napellus* and used in medicine due to its narcotic and sedative properties (Sharma 1993). Some medicinally used non poisonous species of *Aconitum* like *A. heterophyllum* Wall. ex Royle are made into Ayurvedic and Unani formulations and also traded under the trade names Atees, Atish, Atibish. The products obtained from the species of the genus are used worldwide and exported to other countries also.

The Govt. of India Ministry of Commerce through their circular public notice no. 47 (PN)/92-97 dated 30th March 1994 prohibited the export of 56 plant species, where *Aconitum* species and their derivatives and extracts occupied the first items in the list. Such plants have been enlisted under the title 'Negative List of Exports' in 1998 by Government of India, so as to regulate their trade/export on one hand and on the other to check unplanned and illegal exploitation from wild sources. The list was further amended through Notification no. 24 (RE-98)/1997-2002, in which 29 plant species were prohibited for export. *Aconitum* species were included as item no.19 in that list (SHAH 1997). All the 27 species of Indian *Aconitum* have been placed under Negative List of Exports in India and the trade of these species collected from the wild sources has been banned. The ban would aid in conserving these plant species in their natural habitats to certain extent. In short, the legal ban on Aconites by Government of India may be considered as an asset, since it indirectly protects the wild population of Aconites to certain extent.

Polyherbal Formulation of Aconites: Adulteration and Substitution

From the last several years to the present age, endangered medicinal plants have been used in traditional as well as alternative system of medicine and are becoming aware of the side effect and adverse reaction of synthetic drugs and unavailability of such medicinal plants has led to arbitrary substitution and adulteration in raw drug market (Tewari 1991). Adulteration actually is a practice of substituting the original crude drug partially or fully with other substances which is either free from or inferior in therapeutic and chemical properties or addition of low grade or spurious drugs or entirely different drug similar to that of original drug substituted with an intention of enhancement of profits (Sunita 1992). Formulations having *Aconitum* roots as an ingredient are highly effective in various diseases. Practioners while prescribing such medicines should be aware of the quantity of *Aconite* in a formulation and prescribe such drugs only in recommended dose. *Trikarshika* is an Ayurvedic polyherbal formulation of equal quantities of dried rhizome of *Zingiber officinale* Roscoe, root of *Aconitum heterophyllum* wall and rhizome of *Cyperus pangorei* Linn used for enhancing appetite, digestion, febiruge actions (Jayanta et al. 2012). A traditional substitution of roots of *Aconitum heterophyllum* with rhizomes of *Cryptocoryne spiralis* has been found to show promising antidiarrhoeal potential (Prasad et al. 2014). Other formulations used in Ayurveda are *Mahashankha vati, Sanjivani vati, Tamra parpati, Kaphketu ras, Tribhuvankirti ras, Saubhagya vati, Rambana ras, Anand bhairava, Hinguleshvara rase, Panchamrita rasa, Vatavidhvamsani rasa* etc (Tomar et al. 2015). *Kutajghan Vati,* a classical Ayurvedic preparation is a combination of *Holarrhena pubescens* Wall. ex G.Don (Kurchi), *Aconitum napellus* L. and *Aconitum heterophyllum* Wall and helps in reducing inflammation of small and large intestine and hence promotes healing in colonic ulcers and restores proper digestion (Amit et al. 2010).

Distribution of Aconites

The herbaceous genus, *Aconitum* has major centers of diversity in the mountains of East and Southeastern Asia and Central Europe (Kadota 1987) and also comprises ca. 100 species, distributed majorly in North temperate mountainous regions of the world (Mabberley 2008). A small group is also found in Western North America and Eastern United States of America. South-West China and the Eastern Himalaya comprise (Liangqian and Kadota 2001). In Himalayan regions, it is distributed in Pakistan, India, Nepal, Bhutan and South Tibet, where *Aconitum* has been used in local and traditional system of medicine (Shah 2005). In India, the genus is restricted only to the Himalayan region, and interestingly the species found in eastern Himalayas are not known from western Himalayas and vice-versa (Chaudhary and Rao 1998). In India, the genus is represented by about 27 species that is mainly distributed in sub alpine and alpine zones of Himalayas from Kashmir to Uttarakhand and to the hills of Assam (Sharma et al. 1993). Out of 300 species, a total of 33 species are found in great Himalaya. The distribution of *Aconitum* species has been studied by various botanists and taxonomists. Stapf (1905), Rau (1981-82) (medicinal and non-medicinal), and Chaudhari & Rao (1998a,b) described the threatened taxa and their distribution in the Western and Central and

Northeastern Himalayas. Polunin and Stainton (1984) reported the distribution in Nepal and in other parts of the Himalayas, and Stewart (1972) studied their distribution in West Pakistan and Kashmir.

Aconitum heterophyllum is a herbaceous, perennial, rhizomatous plant (Nayar et al. 1990) also known as aconite, monkshood, devil's helmet, blue rocket, is widely distributed over temperate parts of Western Himalayas and its chief habitat is in the alpine and sub-alpine areas at an elevation of 3000-3500 m above sea level such as Gulmarg, Khilanmarg, Sonamarg (Uniyal et al. 2002). *A. heterophyllum* var. *bractateum* (Pevak, Kashmir Monkshood) is distributed over loose soiled, less pebbled, moist and open alpine slopes at an altitude range of 3000-3800m (Dar et al. 2006). *Aconitum chasmanthum* (Indian napellus, Mohri), also known as *Bal-bal-nag* in Kashmiri is widely distributed in alpine & sub-alpine zones of Western Himalaya from Chitral & Hazara to Kashmir and Chamba in Himachal Pradesh at an altitude range of 3500-4000m (Anonymous 2005). *Aconitum laeve* Royle is widely distributed in North West Himalaya from Chitral to Kumaon, mostly in forests at an altitude range of 2500-3500m, Jammu & Kashmir, Himachal Pradesh, Uttar Pradesh, Pakistan, and Nepal. *Aconitum violaceum* Jaquem (Violet Monkshood, Dudhia-bis, Mitha telia, Telia) is distributed over alpine slopes of Himalayan region of India, Pakistan and Nepal, Jammu & Kashmir, Himachal Pradesh and Uttar Pradesh at an altitude range of 3500-4000m (Chaudhary et al. 1998, CAMP 2003). *Aconitum balfourii* Stapf is an endemic to the alpine and subalpine belts of Indian Himalayan region mostly widespread in Kumaon and Garhwal Himalayas on shady slopes from 3000 to 4200 m (Chopra et al. 1984, Samant et al. 1998). It is also found in Nepal. The genus *Aconitum* is usually found in wet alpine zone, however, in Kashmir Himalayas, these species has been reported in temperate zones also.

Table 1: Distribution of genus *Aconitum* growing in Kashmir

Species name	Vernacular name	Life form	Phenology	Habitat	Specimens examined
Aconitum chasmanthum	Bal-bal-nag	Biennial herb	Aug.-Sept.	Sub-alpine & alpine zones of Western Himalaya from Chitral & Hazara to Kashmir and Chamba in Himachal Pradesh in an altitude range of 3500-4000m	Chorwan, A. R. Naqshi, Showkat and Kachroo 19897,19929 (KASH); Between Chorwan & Achoor, A. R. Naqshi, Showkat and Kachroo 21505 (KASH); Gurais, Khupri, A. R. Naqshi 22400 (KASH), Dahinala, Showket Ara 21705 (KASH); Disou meadow, Gurais, Showkat Ara 21703 (KASH); Gurez proper, G. M. Dar 9104 (KASH); Gurez, Razdani Pass, Neelofar 001(KASH); Gulmarg,Rashid 002 (KASH)
Aconitum heterophyllum	Aconite, Monkshood	Biennial herb	July-Aug	W. Himalaya of Jammu & Kashmir among shrubs on grassy slopes and in pasture lands (3000-3500m), Himachal Pradesh, Uttar Pradesh. Pakistan and Nepal (up to 84°E long).	Jawahar Tunnel Hill Top and Upper Munda, 2400m, 27.09.2000, G.H. Khandey; Kargil Ladakh 10.09.05, K.S. Khan, Khilanmarg, 2950m, 09.08.06, A.H. Malik; Aparwhat Gulmarg 3000m, 21.07.07, A.H. Malik, G.H. Dar; Sonamarg Thajwas, 3700m, 16.07.08, Neelofar (KASH)

Contd.

Aconitum heterophyllum var. *bracteatum*	Aconite, Monkshood	Biennial herb	July-Sept	Alpine subalpine habitats, ranging in altitude from 2125 - 3665m.	Zojila, A. R. Naqshi 2153 (KASH); Apharwat, A.R.Naqshi 2268, 2355, 6610 (KASH); Apharwat, Rashid 003 (KASH); Doodhpathri Budgam, Neelofar 005 (KASH).
Aconitum laeve	Grape-leaved monkshood, Jangali Atees	Perennial herb	May-Oct.	W. Himalaya from Chitral to Kumaon, mostly in forests of Jammu & Kashmir (2500-3500m), Himachal Pradesh, Uttar Pradesh, Pakistan, Nepal.	Sarbal, Uppeandra Dhar 7525 (KASH); Sonamarg, Indra Aima 2971 (KASH); Dawar, Naqshi, Showkat & Kachroo 21500 (KASH); Gulmarg, A. R. Naqshi 1036 (KASH); Sonamarg Jyoti 2992 (KASH); Gulmarg, A. R. Naqshi 2455, 4521, 6716 (KASH)
Aconitum violaceum	Violet Monkshood, Dudhia-bis	Biennial herb	July-Sept	Alpine slopes of Himalayan region of India, Pakistan and Nepal, Jammu & Kashmir, Himachal Pradesh and Uttar Pradesh in an altitude range of 3500-4000m.	Pir Panjal, Kashmir, Chadwell & Ramsey 12596-97 (KASH); Yusmarg, A.R. Naqshi 6959 (KASH); Yusmarg, Vinod Gupta 9262 (KASH); Pranshur, G.H. Dar 11749 (KASH)

Endemism in Ranunculaceae: with special reference to genus Aconitum

In Kashmir Himalaya the family Ranunculaceae is represented by 98 taxa in 17 genera ranging from 350-5,500 m above sea level. *Ranunculus* is the largest genus (28 taxa). Of the 98 taxa, 20 are endemic (20.4 %) in which 13.25 % shows specific and 7.14 % intra-specific endemism. In general 40 taxa (40.71 %) of the total Kashmir Himalayan Ranunculaceae remain confined to the Himalaya (including Afghanistan and Tibet). *Aconitum* shows maximum dominance in Central Asia (51 taxa) than Europe (14 taxa) but the reverse is true in so far as the endemic percentage is concerned. However, it is noteworthy that 58.82 % of the endemic Central Asian species of *Aconitum* are either new species or new combinations. In Kashmir Himalaya the endemic percentage is no less significant (50.00) while it reaches maximum in the Eastern Himalaya. The interesting features of affinities revealed by Kashmir Himalayan Ranunculaceae have shown that *Aconitum* shares 80 % wides with Pakistan; 40 % with USSR and 20 % each with Afghanistan and Europe. No doubt 60 % of the wides extend up to China but none beyond this area (Jee et al. 2014).

Table 2: Species appendix along with altitudinal range and geographical distribution of Indian *Aconitum* with threat status (with slight modification after Jee et al. 2014)

Species Name	Altitudinal Range (m)	Endemism					
		SR	BR	WCR	ESR	MDR	SJR
Aconitum chasmanthum	2100-3900	-	-	+	+	+	-
Aconitum deinorrhizum	3400	+	-	-	-	-	-
Aconitum heterophyllum	2280-3600	-	+	-	-	-	-
Aconitum kashmiricum	2400-3300	+	-	-	-	-	-
Aconitum laeve	2400-4000	-	-	-	-	-	+
Aconitum moschatum	3300-3900	+	-	-	-	-	-
Aconitum rotundifolium	3000-4800	-	-	+	-	-	+
Aconitum soongaricum	3000-3300	-	-	+	-	-	+
Aconitum violaceum	3000-4500	+	-	-	-	-	-

SR=Short Range Endemics, BR=Broad Range Endemics WCR=Western and Central Asian Region, ESR=Euro-Siberian Region, MDR=Mediterranean Region, SJR=Sino-Japanese Region

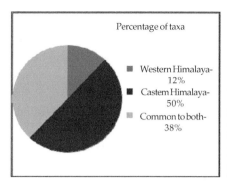
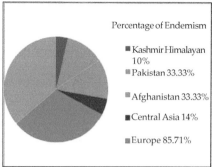

Fig. 1: Extent of Endemism in Western Himalaya and other adjoining areas (I-Total number of taxa; II- total endemism percentage)

Classification of Aconites

The taxonomic classification of aconites is very difficult, since the group exhibits a wide range of geographic and ecological variation wherever it occurs (Kadota et al. 1981, Luo et al. 2005, Sutkowska et al. 2013)). However, based on the life span of the root (annual, biennial and perennial), and by the external/internal structural features of the individual *Aconitum* species, Stapf have classified the Indian *Aconitum*/ Aconites into three main sections which are represented in the table below. Though Stapf's arrangement of Aconites had been criticized, it still serves as good baseline information to taxonomists and anatomists to identify the species to a greater extent (Chakravarty et al. 1954). Aconites have also been classified as *poisonous* and *non-poisonous* based on the content of aconitine present in the species. Shah (2005) reported that the non-poisonous species include *A. heterophyllum, A. laeve* and *A. routndifolium* and the poisonous are *A. chasmanthum, A. ferox, A. deinorrhizum, A falconeri, A. balfourii, A. moschatum, A. violaceum, A. spicatum, A. bisma* and *A. laciniatum.*

Table 3: Categories of *Aconitum* (Stapf et al. 1905)

Section I	Section II	Section III		
Gymaconitum	Lycoctonum	Napellus		
Having annual Roots	Possessing perennial Roots	With biennial roots Sub-category		
		True Napellus with continuous cambium	Anthora six discontinuous concentric rings/ irregular batches of cambium embedded in uniform tissue	Deinorrhizum isolated bands of cambium embedded in secondary phloem

Biological and Pharmacological Aspects

The therapeutic importance of herbal medicine is remarkably increasing since the last few decades, therefore, it is very essential to obtain a proper quality control profile for various medicinal plants used in traditional system of medicine (Nasreen et al. 2010). Since the genus *Aconitum* is reputed for its medicinal and pharmaceutical value and is a rich source of diterpene alkaloids and flavanoids which has been reported to possess significant antipyretic and analgesic properties and a high therapeutic index (Jabeen et al. 2006). Lycaconitine, obtained from several Aconitum species has been found to be effective against multi-drug resistant cancers (Kim et al. 1998). The studies on traditional system of medicine has been showed that the genus *Aconitum* has been used in curing hysteria, throat infection, dyspepsia, abdominal pain, diabetes and is considered as a valuable febrifuge nervine tonic especially combating debility after malaria and in haemoplageia (Dar etal. 2001). Aconitum has been also shown to exhibit antipyretic, analgesic, cytotoxic activities and is used to treat diseases of nervous system, digestive system, rheumatism and fever. The crude methanolic extracts of *Aconitum* species possess pharmacological activities such as antifungal, antibacterial, and insecticidal properties (Anwar et al. 2003). It has also been shown that the plant possesses a good antiviral, antidiarrheal and immunostimulant properties (Venkatasubramaniam et al. 2010). Alkaloids of this genus exhibited anti-inflammatory, antinociceptive, hypotensive, bradycardic, analgetic, and cardiotonic activities (Ameri et al. 1998). It has been used as a cardiac depressant in massive arterial tension of cardiac origin (Chopra et al. 1958). Aconitine containing herbal extract have been found to possess antitumor activity (Solyanik et al. 2004). The genus *Aconitum* is rich in elite compounds having potential biological significance such as benzoylmesaconine, mesaconitine, aconitine, hypaconitine, heteratisine, heterophyllisine, heterophylline, heterophyllidine, atidine, isotisine, hetidine, hetisinone and benzoylheteratisine and other nutrients (Zhaohong et al. 2006). The crude methanolic extracts of Aconitum species possess pharmacological activities such as antifungal, antibacterial, and insecticidal properties (Anwar et al. 2003), antiviral (Zaidi et al. 1988; Pandey et al. 2004) and immune-stimulant properties (Atal et al. 1986). Other compounds isolated from *Aconitum* heterophyllum include *i* flavonoids, tannins, saponins and sugars (Pelliter et al. 1968). The crude extracts of aconites contain extremely lethal substances however after detoxification can be utilized therapeutically (Chopra et al. 1984).

Table 4. Medicinal Aspects of *Aconitum*

Species Name	Vernacular Name	Medicinal Uses
Aconitum heterophyllum Wall ex Royle	Aconite, Monkshood	Anti-inflammatory, anti-pyretic, anti-rheumatic (Uniyal et al. 2002), Anti-viral against spinach mosaic virus (SMV) (Zaidi et al. 1988, Pandey et al. 2004) Antifungal, cytotoxic (Anwar et al. 2003), and immune-stimulant (Atal et al.1986) Diarrheae, vomiting, cough, cold astringent (Shah 2005) Dysentery, dyspepsia, chronic enteritis, and as a febrifuge and combating debility after malaria (Phurpa et al. 2010).
Aconitum heterophyllum var. *Bracteatum* Stapf	Aconite, Monkshood	Anti-inflammatory, anti-pyretic, anti-bacterial and anti-oxidant (Rameashkannan M.V 2014). Diarrheae, vomiting, cough, cold, astringent (Shah 2005)
Aconitum chasmanthum Stapf ex Holmes	Bal-bal-nag	Anti-rheumatic, Anti-pyretic, neurasthenic, diaaphoretic, diuretic, anti-diabetic, diarrhea, dyspepsia, sprue and bronchitis (Chunekar et al. 1988), antifungal (Anwar et al. 2003), skin diseases , enlargement of spleen and neuralgia (Dubey et al. 2012).
Aconitum laeve Royle	Grape-leaved monkshood	Same as *Aconitum heterophyllum*
Aconitum violaceum Jaquem	Violet Monkshood, Dudhia-bis	Anti –inflammatory anti-pyretic, anti-rheumatic, abdominal pain, antidote, febrifuge, renal pain (Janhvi et al. 2013).

Chemical Constituents

The presence of secondary metabolites in plants explains the various uses of plants for various systems of medicine. The most commonly encountered secondary metabolites of plants are saponins, tannins, flavonoids, alkaloids, anthraquinones, cardiac glycosides and cyanogenic glycosides. The genus *Aconitum* has been shown to contain the alkaloids mesaconitine, aconitine, hypoaconitine, heteratisine, heterophyllisine, heterophylline, heterophyllidine, atidine, hetidine, isoatisine, hetsinone and benzoylheteratisine (Wang et al. 2006, Pelltier et al. 1968). *Aconitum heterophyllum* rich in substances having potential biological significance, such as benzoylmesaconine, mesaconitine, aconitine, hypaconitine, heteratisine, heterophyllisine, heterophylline, heterophyllidine, atidine, isotisine, hetidine, hetisinone and benzoylheteratisine and other nutrients (Zhaohong et al. 2006) have been manifested with different biological and pharmacological properties (Anwar et al. 2003; Braca et al. 2003; Carlo et al. 1999, Williams et al. 2004, Pandey et al. 2004, Atal et al. 1986) which make the genus most imperative among the other entire genus. Other compounds isolated from *A. heterophyllum* include flavonoids, tannins, saponins and sugars (Pelliter et al. 1968). In recent years, a number of active constituents are present in this plant, which includes alkaloids, glycosides, flavonoids and sterols (Zhaohong et al. 2006). In *Aconitum chasmanthum*, the main component found is the alkaloid aconitine. The alkaloid content of the roots ranges from 2.98 to 3.11 per cent. The other alkaloids found in *A. chasmanthum* are indaconitine, chasmaconitine, chasmanthinine, chasmanine, and homochasmaconitine (Dubey et al. 2012). From the aerial parts of the *Aconitum laeve* Royle, two new lycoctonine type C19-diterpenoid alkaloids; swatinine-A and swatinine-B along with four known C19-diterpenoid alkaloids foresticine, neoline, delvestine and chasmanine has been isolated so for (Seema et al. 2014). Several diterpenoid alkaloids isolation and detection through HPLC protocols from the genus *Aconitum* have been standardized (Hikino et al. 1983; Jiang et al. 2005, Wang et al. 2006). Other than alkaloids and flavanoids some other chemical constituents like free fatty acids (FFA) in the genus *Aconitum* has been reported (Tornaino et al. 2001; Skonberg and Perkins 2002, Martin et al. 2005, Sajid et al. 2008, Phillip and Matt 2008). Another species *Aconitum balfourii* Stapf dominant to Kumaon and Garhwal Himalayas contains pseudoaconitine, a highly toxic alkaloid as a principal component, and aconitine, benzylaconitine, picroaconitine, and haemonepellene in traces (Anonymous, 1985). Nine other norditerpenoid alkaloids condelphine, bullatine, neoline, isotalatizine, 1-O-methyldelphisine, pseudaconitine, yunaconitine, bikhaconitine, and indaconitine from the aerial parts of *A. balfourii* Stapf has also been isolated (Khetwal 2007). Three compounds 1, 15-dimethoxy-3-hydroxy-14-benzoyl-16 ketoneoline, benzoylaconine and aconitine from Chinese medicinal herb *Aconitum kusnezoffii* Reichb has been isolated (Xu. et al. 2011). The list of chemical compounds isolated from various species of *Aconitum* found from different ecogeographical areas are shown in Table 5

Table 5: Details of chemical constituents isolated from some species of *Aconitum*

A. heterophyllum	Anthorine	Lawson et al. (1937)
	Atidine	Pelletier et al. (1968)
	Atisenol	Pelletier et al. (1982)
	6-Benzoylheteratisine	Aneja et al. (1973)
	11,13:11,16-Diepoxy-16,17-	Gonzlez-Coloma et al. (2004)
	dihydro-11,12-secohetisan-2-ol	Pelletier et al. (1978)
	Dihydroatisine	Pelletier et al. (1968)
	Heterophyllidine, Heterophylline,	Pelletier et al. (1965)
	Hetidine, Hetisinone,	Jacobs et al. (1942)
	O-Methylheterophylline	Ulubelen et al. (2002)
	Isoatisine	Ross et al. (1992)
	Hetisine	Shaheen et al. (2005)
	N-Deethyl-N-formyllyaconitine	Wang et al. (2006)
	O-Methylaconitine	
	Methyl-N-succinoylant	
	hranilatemesaconitine, aconitine,	
	hypoaconitine, heteratisine,	
	heterophyllisine, heterophylline,	
	benzoylheteratisine	
A. chasmanthum	Aconitine, Indaconitine,	Dubey 2012
	Chasmoaconitine,	
	Chasmanthinine, Chasmanine,	
	Homochasmaconitine	
A. laeve	Swatinine, foresticine, neoline,	Seema et al. (2014)
	delvestine chasmanine	
A. violaceum	Indaconitine	Miana et al. (1971)
A. soongaricum	Acetylsongorine	Zhamerashvili et al. (1981)
	Songoramine	Yusunov et al. (1970)
	Songorine	Sultankhozhaev et al. (1982)
	Songorinine	Samatov et al. (1965)
	12-Acetyl-12-epinapelline	Salimov et al. (2004)
	Aconine	Pelletier et al. (1979)

Contd.

A. balfourii	pseudoaconitine, benzylaconitine, picroaconitine, and haemonepellene Balfourine, 8-O-Methylveratroylpseudaconine 9-Hydroxysenbusine A condelphine, bullatine, neoline, isotalatizine, 1-O-methyldelphisine, pseudaconitine, yunaconitine, bikhaconitine, and indaconitine	Anonymous, (1985) Khewal et al. (1992) Khewal et al. (2006) Khetwal et al. (2007)

Assessment of Threat Status of Genus Aconitum

In the past, a large number of studies have been carried out to assess the threat status of various plant species (Soehartono and Newton 2000, Jenkins 1997, Garcia et al. 2002, Peng et al. 2008, Haruntyunuan et al. 2010, Hamayun et al. 2006, Ali and Qaisar 2010, Alam and Aliu 2010; Ali et al. 2012, Kala 2005, Nautiyal et al. 2000, Goraya 2010, Dhar et al. 2002, Kaul 1997, Dar and Naqshi, 2000 and Malik et al. 2011) including some species of genus *Aconitum* also (Baig etal. 2014). The colossal constraints on the genus *Aconitum* due to the annual high demand domestication of raw drugs in India market has made it threatened in different states of Himalayan range (Ved and Goraya 2007) as assessment of some of the species of *Aconitum* like *A. heterophyllum*, *A. falconeri*, *A. ferox*, *A. deinorrhizum*, *A. violaceum*, *A. spicatum* by various researchers have placed into different threat categories of IUCN (Sharma 2012, Chhetri et al. 2005, Srivastava et al. 2010). In the last three decades, the Aconites of Indian Himalaya has been mercilessly exploited and the consequence is that many species of this genus has been enlisted as an endangered taxon (Shah 1983). Since the aconites are rich resource of the important chemical constituents like alkaloids, the extracts have been used in different systems of medicine, so the commercial demand is very high. Also the demand of excessive illegal collection and sale of *Aconitum* by farmers has been continuously carried out. Also the low germination percentage and the cultivation of *Aconitum* species are done in a very small scale because of low availability of land for cultivation of medicinal plants (Srivastava et al. 2010). In addition, under natural conditions, seed germination and seedling establishment in this genus is very rare. Moreover, destruction of natural habitat and the earlier mentioned reasons are collectively responsible for its endangered status. As the species density of this genus has been threatned due to many factors like overexploitation and habitat destructions because of which genus *Aconitum* has found a key position in the Red Data Book of IUCN and degree of consistency (occurrence) has been used to allocate the status of some of the species of *Aconitum*. The non-scientific and ruthless collection of *A. heterophyllum* for extraction purpose in large quantities has led this species towards rarity and is now identified as critically endangered species (IUCN 1993, Nautiyal et al. 2002, CAMP 2003). In some studies the population study of some aconites like *Aconitum heterophyllum*, *Aconitum violaceum* and *Aconitum balfourii* has been found seldom in one or only a few sites and then recommended that it might be due to its specific habitat requirement or due to continuous removal of plants for medicinal uses. On the basis of this study, these three species of this genus has been assigned as endangered species (Nautiyal et al. 2002). The studies have shown that *Aconitum violaceum* in near threatened category and its status is vulnerable, whereas *Aconitum heterophyllum* considered as critically endangered and its status is endangered globally (Rana et al. 2010). As per the data assessed against the IUCN Red List Categories and Criteria (2010), *Aconitum heterophyllum* qualifies for the threat category of Endangered (EN) category. Kaul (1997) assigned the species as Vulnerable in Kashmir Himalaya. IUCN (1998) assigned the plant species as Critically Endangered for North-western Himalaya and also in 2003, IUCN categorized Critically Endangered threat category to the species for Jammu and Kashmir (Kala 2004).The main threats to this genus includes harsh and fragile habitats, squeezed populations, restricted distribution, extensive herbivory of flowering portions, exploitation of tubers for medicinal purposes (Sinha et al. 2008, Dar et al. 2006).

Future Strategies

A number of valuable plant species of this genus attributing the enormous important perspectives are being used in huge amounts by people and the adequate measures has not been taken yet to multiply and domesticate these plant species in order to conserve them. In order to overcome the extinction strategies various field stations has been established in Kashmir Himalaya region. These field stations are concentrating on the maintenance of germplasm and farm cultivation of the elite material. Based on the genetic profiling studies, elites of various species of the genus *Aconitum* are now being taken up for mass multiplication. Besides, the use of conventional propagation methods, application of *in vitro* propagation techniques offers an additional alternative for recovery as well as multiplication of endangered species. Therefore, attempts have to be taken to develop effective *in vitro* propagation protocols for this genus. A gene bank has been established at RRL, Jammu to cover Western Himalayan Region which is fully equipped with state-of- the art facilities for conservation of seeds, live plants and in vitro material of rare, threatened and economically important species. Genetic diversity among various species of the genus *Aconitum* growing in the Himalayan region has to be taken by incorporating various molecular marker analysis techniques that will help in tagging high-yielding genotypes using DNA markers for micropropagation and mass multiplication for conservation of these important elite's in the region.

Diversity Profile of Genus Aconitum

DNA analysis has been proved as an important tool in herbal drug standardization. This is especially useful in case of those that are frequently substituted or adulterated with other species or variants that are morphologically and/or phytochemically indistinguishable. Various types of DNA based molecular techniques are utilized to evaluate DNA polymorphism. Since the phylogenetic relationship of different species of *Aconitum* is ambiguous, and various molecular markers have been used to tackle this issue. The review provides a brief account of a few DNA-based technologies that are useful in genotyping and quick identification of genetically diverse genotypes of some species of genus *Aconitum* of Indian Himalaya.

Molecular Markers as a Quality Control Tool

Molecular markers have been proved to be a valuable tool in the characterization and the evaluation of genetic diversity within the species (Zoghlami et al. 2007). It has been shown that different markers reveal different classes of variation (Powell et al. 1996). Such markers are not typically influenced by environmental conditions and therefore, can be used to describe patterns of genetic variation among plant populations, and to identify duplicate accessions within germplasm collections. Various approaches are available for DNA fingerprinting such as amplified fragment length polymorphism (AFLP) (Zabeau et al. 1993), restriction fragment length polymorphism (RFLP) Botstein et al. (1980) Simple Sequence Repeats (SSRs) Tautz et al. 1989, and random amplified polymorphic DNA (RAPD) Williams et al. 1990. Among these, RAPD is an inexpensive and rapid method not requiring any information regarding the genome of the plant and has been widely used to ascertain the genetic diversity in several plants (Deshwall et al.

2000). There is limited information about the evolution of *Aconitum* in India, because studies are lacking the evolutionary aspects of *Aconitum* in India. Many researchers have studied the evolutionary aspects of *Aconitum* from different biogeographically zones reticulate evolution of high alpine *Aconitum* on the basis of chromosomal and molecular polymerase chain reaction-inter-simple sequence repeat (PCR-ISSR) patternMitka et al. (2007). The study of population genetic diversity is another method that provides useful information for biological conservation. Four populations of *Aconitum* species have been studied through polypeptide pattern and isoenzyme markers to locate the genetic diversity. On the basis of isoenzyme variations and banding pattern, seven polymorphic loci in four populations have been identified (Pathak et al. 2011). In Hengduan Mountains of Southwest China, the phylogenetic studies of *A. delavayi* complex have been investigated by using nuclear ITS (internal transcribed spacer) Zhang et al. 2003. The chloroplast (cp) psbA-trnH has been practiced to study the characterization of *Aconitum* species at its molecular level (He et al. 2010). Different types of molecular markers like ITS (Wang et al. 2009), RAPD (Zhang et al. 2005), and ISSR markers (Luo et al. 2006) or allozyme variation (Zhang et al. 2003) for studying the population variations of *Aconitum* species have shown a high population diversity, and the polymorphism observed in these studies indicates that more investigations are needed at population or species level to ensure a better characterization of *Aconitum* species. Moreover, the phylogenetic relationship of 185 species of the tribe Delphinieae, including 57 *Aconitum* species have been exhibited by utilizing chloroplast trnL-F spacer, trnL intron, and nuclear ITS sequences (Jabbour and Renner 2012). Genetic diversity and variations of *A.* leucostomum have also been examined by ISSR and to supply essential characteristics for identifying *Aconitum* crude drugs (Gao et al. 2014). In ayurvedic raw drug source and prepared herbal products, nrDNA ITS sequence-based SCAR marker have been developed to authenticate *A. heterophyllum* (Seethapathy et al. 2014). In Yunnan, China, two other species of *Aconitum* viz. *A. vilmorinianum* and *A. austroyunnanense*, distributed and used as herbal drug have been identified by their ITS sequences (Zhang et al. 2012).

Some studies have shown that the high polysaccharides and polyphenolics hinder good quality of genomic DNA isolation from *Aconitum*. Therefore, the DNA isolation protocols have been standardized for the genetic diversity analysis and also other molecular biology experiments (Hatwal et al., 2011; Srivastava et al. 2010). Regarding the Kashmir Himalayan Range no such type of studies as genotyping and identification of genetically diverse genotypes of genus *Aconitum* have been carried out.

Conservation Strategies

The goal of conservation is to support sustainable development by protecting and using biological resources in ways that do not diminish the world's variety of genes and species or destroy important habitats and ecosystems. In general, it involves activities such as collection, propagation, characterization, evaluation, disease indexing and elimination, storage and distribution. The conservation of plant genetic resources has long been realized as an integral part of biodiversity conservation. There are two methods for the conservation of plant genetic resources, namely *In-Situ & Ex-Situ* conservation (Kasagana et al. 2011). The best method of conservation for medicinal plants is to grow and evolve in wild and their natural habitats. However, *in situ* conservation is achieved by setting nature reserves and national parks. Plants should also be conserved *Ex-Situ* that is, to grow outside their habitat in controlled environment. In recent years, there has been an increased interest in *in-vitro* culture techniques which offer a viable tool for mass multiplication and germplasm conservation of rare, endangered and threatened medicinal plants (Ajithkumar and Seeni 1998, Prakash et al. 1999). Commercial exploitation and elimination of natural habitat consequent to urbanization has led to gradual extinction of several plant resources. As such efficient aspects for regeneration of these valuable medicinal plants like micropropagation, somatic embryogenesis must be attributed for conservation of the valuable elites like *Aconitum* species and some other conventional efforts like seed storage should also be employed as a means of ex situ conservation. There have been a few reports to date on micropropagation of using shoot tips and nodal explants, as micropropagation proves effective approach to conserve such germplasm. A protocol has been standardized for sterilization of nodal segments and seeds of *Aconitum heterophyllum* for its micropropagation intended for its mass propagation and conservation (Srivastava1 et al. 2010).

Agrotechnology for Rapid Multiplication of Aconites

Seed based multiplication is the most effective, realistic and convenient means for most of the species (Sharma et al. 2006). Some species of

Aconitum are dormant in nature. Cultivation through seed in most of the species of *Aconitum* is difficult due to the poor seed availability and lack of superior germplasm (Nautiyal et al. 2009). Under natural conditions, seed germination and seedling establishment in these species is very difficult and hence cold stratification has been found to be the most effective treatment for breaking seed dormancy in Ranunculaceae growing in temperate and alpine climates (Forbis et al. 2002). Some chemical stimulants such as Giberralic acid, nitrogen containing compounds including nitrate, nitrite and cyanide and treatment of thiourea and ammonium nitrate also helped to improve the germination of seeds of *A. balfourii* and *A. heterophyllum* (Pandey et al. 2000).The effect of pre sowing of seeds in different chemicals and temperature has been studied in *A. heterophyllum* and has been found that low temperature and 0.5 mg/lit IAA concentration was suitable for seed germination (Srivastava et al. 2011). Some physiological factors such as light and temperature have also been shown to affect the germination behavior in *Aconitum deinorrhizum* (Sood et al. 2011). Since there is little information of cultivation of this genus, however, propagation study through tuber segments in some species of *Aconitum* like *Aconitum atrox* has been carried out in alpine region at higher altitude (Rawat et al. 1992) and also at lower altitude by treating tuber segments with a combination of GA3, IBA, and Kinetin (Kuniyal et al. 2006).

Ex-Situ Conservation

Ex-situ conservation involves the process of protecting an endangered species and developing it outside its natural habitat. For medicinal plants, ex-situ Conservation aims at the conservation concern by way of raising of nurseries, seedling supply, and plantations and by establishing medicinal plant gardens. It includes simple seed collection, storage, field plantings or more intensive plant breeding and improvement approaches. It involves three methods namely, field gene banks, seed banks and *in vitro* storage. Of these, seed banks are the most efficient and effective method of conservation for seed. The important aspect of ex-situ conservation is to maintain a wide range of phenotypic and genotypic range of diversity of a species and to propagate the species outside its original natural provenance in a more controlled way. The plants grown under control or protected system (inside greenhouse) instead of natural conditions is like a revelation to get the maximum plants yield in respect of their biomass and concentration of active contents of the *Aconitum* species. So, the cultivation of these valuable plant species with suitable cultivation practices may protect these species in their natural habitat and also fulfill the demand of herbal drug industry (Bahuguna et al. 2013). High polysaccharides and polyphenolics hinder good quality of genomic DNA isolation from *Aconitum*, therefore, DNA isolation protocols have been standardized for the genetic diversity analysis and other molecular biology experiments (Hatwal et al., 2011, Srivastava et al. 2010). All these efforts are included in *ex situ* conservation. Conventional seed storage is an efficient method of *ex situ* conservation but the seed dormancy of *Aconitum* species limits this approach. Therefore, *in vitro* propagation is a more suitable way for conservation of *Aconitum* species (Srivastava et al. 2010).

In-Situ Conservation

In situ conservation aims at either enhancement of existing populations or creation of self-supported new populations via reintroductions and translocations, using sampled or propagated material (Bottin et al. 2007, Menges 2008). In-situ Conservation is usually the preferred conservation strategy for capturing and conserving medicinal plant pockets in their natural habitats. Stress is laid on identification of medicinal plant areas having rich biodiversity of genetic resources that have priority, usually at the species level on the basis of present or potential socio-economic value of the species and their conservation status in the ecosystem. The area-specific action plan and networking of natural sites has to be considered to be the most important aspects of in-situ conservation activities. The ecological requirement of many of the species is complex. Hence moving them out of their own area of comfort to new area may sometime prove counterproductive. Hence by improving the protection, removing all kind of threats is one of the important steps towards *in-situ* conservation.

Plant Tissue Culture

Tissue culture opens up new area for conserving threatned plant species like *Aconitum* as small amount of plant material can generate large number of disease free propagules which can be reintroduced in their native habitat. It can enable the mass propagation of *Aconitum* from a minimum of a plant material so that the large quantities of biomass required for extraction of active constituents can be made available throughout the year, without causing further endangerment of the species (Jabeen et al., 2006). In addition it also overcomes the problems of *ex situ* conservation where the seed availability is mandatory. *In vitro* propagation of plants hold tremendous potential for the production of the high quality plant based medicine (Murch et al., 2000).

Micropropagation as an Important Conservative Tool

Micropropagation is an effective approach to conserve such germplasm. In the area of micropropagation, efficient protocol for regeneration of some species of *Aconitum* has been reported (Bist et al. 2011, Pandey et al. 2004, Jabeen et al. 2006). Successful clonal multiplication through tip tissue culture method has been done in which multiple shoot formation (Murashige and Skoog 1962) and root has been achieved on MS medium (Hatano et al., 1988). The *in vitro* propagation of *Aconitum heterophyllum* Wall by callus induction has been employed (0.5 mg L^{-1}) and BAP (0.25 mg L^{-1}) (Jabeen et al., 2006). Somatic embryogenesis in *Aconitum heterophyllum* has also been attained. A method for the production of hairy roots of *A. heterophyllum* wall has been reported successfully by using *A. rhizogenes* strains such as LBA 9402, LBA 9360, and A4. It has been found that total alkaloid (aconite) content of transformed roots was 2.96%, which was 3.75 times higher compared to 0.79% in the non transformed roots (Giri et al. 1997, 1993). Also the callus induction of small leaf segment of *in vitro* sprouted axillary bud on MS medium fortified with 4.5 mM BAP and 26.9 mM NAA and obtained shoot induction on some concentration of BAP with lowered concentration of NAA (1.1 mM). *In vitro* shoot multiplication and rooting has been observed at 1.1 mM BAP and 12.3mM. IBA, respectively (Pandey et al. 2004). Analysis of the somatic embryos in callus and of sexual embryo *in situ* has been done, (Batygina 2004). These studies has been shown a little effective breeding programmes and enhancing different agro-techniques that could be undertaken for better *in situ and ex situ* conservation and cultivation for commercial purpose for this critically endangered medicinal plant.

Bioprospecting of Novel Bioactive Compounds

Plant species are critically challenged for survival owing to the biotic and abiotic stress that has eventually strengthened their secondary metabolic machinery to synthesize a plethora of small molecules of therapeutic and agricultural importance. Some of the naturally available secondary metabolites possess various pharmacological properties, *viz.*, pesticidal, anti-microbial, anti-viral and anti-cancer properties. Owing to the insect stress and the microbial load, many of the medicinal plants have adapted themselves to combat and survive by producing an array of molecules that are pesticidal and microbicidal. The strategies are focused on purification and characterization of the phytochemicals that could be formulated for mass-utility applications. So the identification of bioactive compounds in plants, their isolation, purification and characterization of active ingredients in crude extracts by various analytical methods is important.

Herbal Drug Identification and Characterization of *Aconitum*

For developing drug standardization, the quality of base material used for formulating the herbal products is a prerequisite. Since the materials used in herbal drugs are traded mostly as roots, bark, twigs, flowers, leaves, fruits and seeds, visible authentication of the material used is difficult and has led to a high level of adulteration. To identify and authenticate the materials, the availability of detailed morphological, histological and pharmacognostic information is essential. Identification of active principles or a biologically active marker compound requires their standardization using appropriate chemical procedures such as TLC, HPLC and LCMS.

A drug sold under the name of *Vatsanabha* in Indian market has been proved to be the mixture of eight different species of *Aconitum* by the help of macroscopic and microscopic studies and by TLC profile, six of which have been identified as *A. spicatum, A. falconeri, A. chasmanthum, A. laciniatum, A. deinorrhizum*, and *A. balfourii* and its authentic species as *A. chasmanthum* Stapf. ex Holmes (Sarkar et al., 2012). From the crude alkaloid extracts of roots and leaves of *A. nagarum*, a high aconitine content has been found by thin layer chromatography (TLC) profiling (Sinam et al., 2011). A High Performance Liquid Chromatography (HPLC) technique has been used for the quantification of the aconitine in the samples of *Aconitum chasmanthum*. Hollow fiber liquid-phase microextraction (HF-LPME) coupled with high-performance liquid chromatography has been also used to simultaneously determine three *Aconitum* alkaloids, including aconitine (AC), hypaconitine (HA) and mesaconitine (MA) in human urine sample (Yang et al., 2010). A high-performance liquid chromatographic (HPLC) method has been developed and validated and found to be accurate and precise, thus, can be successfully applied for the determination of aconitine in marketed ayurvedic oil formulations containing *Aconitum chasmanthum* (Dubey et al. 2012 and for the determination of five principal alkaloids (benzoylmesaconine, mesaconitine, aconitine, hypaconitine and deoxy aconitine) in some species of genus *Aconitum* (Wang et al. 2015). In *A. nagarum* and *A. elwesii* the aconitine alkaloid content has been found to be highest in roots in the month of November and in leaves in the month of August using high performance liquid chromatography (HPLC) (Sinam et al. 2011). LC-MS has become method of choice in many stages of drug development. Hypaconitine (HA), an active and highly toxic constituent derived from *Aconitum* species, has been widely used to treat rheumatism. The metabolism of HA in vitro using male human liver microsomes (MHLMS) Liquid chromatography–high resolution mass spectrometry (LC–MS) has been used to detect with a total of 11 metabolites identified in MHLMS incubations, where mesaconitine (MA), another active and highly toxic constituent of *Aconitum* has been identified (Ling et al. 2011).

CONCLUSION

This investigation remarks the current informative and scientific knowledge on various attributes viz. distribution, cultivation, conservation, phytochemical and pharmacological analysis to the molecular characterization of this medicinally valuable genus *Aconitum*. Although, there are almost more than 300 species in this genus, and the phytochemical and molecular attributes of only a few of the species of this genus have been carried out. The root part of the plants have been used for curing various health ailments as the drug component derivatives of the crude extract exhibit anti-pyretic, anti-inflammatory, antifungal, antiviral, antibacterial, antinociceptive potential. In traditional system of medicine, different species of this genus *Aconitum* is practiced in curing rheumatism, hysteria, throat infection, dyspepsia, abdominal pain, diabetes and renal pain. In present systematics of medicine, *Aconitum* is used as a cardiac depressant in high arterial tension, effective in curing cancerous tumours and also having neuroprotective activity. The preliminary phytochemical investigations on this genus have proved that the genus *Aconitum* is a rich source of diterpenoid alkaloids, flavonoids, saponins, tannins, sugars and can be incorporated into the drug pharmacopeia in order to fulfill the demand of various chemical compounds of aconite source used in various systems of medicine. As such the repository of genus *Aconitum* in the era has been depleted and confronts a severe threat due to over-exploitation. Hence, it is necessary to take a step forward for the conservation of these valuable aconites by incorporating modified approaches to save this genus from extinction which will ultimately prove to satisfy the demands of pharmaceutical industries.

This communication reveals that the genus *Aconitum* has unfolded as a power source and plays an important role in the field of safer herbal medicines. The key barrier for the wide medical utilization of *Aconitum* species is usually their extremely high toxicity, however can be brought into usage by mitigation and other detoxification processes. Only little work has been done for conserving this genus, so it confronts a severe threat due to over-exploitation. So, there is a need to incorporate some modified approaches to save this genus from extinction. Another problem which has to be taken into consideration is that the low level of alkaloid content in some wild species of *Aconitum* is not able to satisfy the demands of pharmaceutical industries, hence the attempts has to be conquered to combat the shortcomings. The studies on its various aspects has provided new acuity for future promising and propitious investigations on the novel components isolated from the different species of the genus, especially alkaloids like aconitine, mesaconitine, hypaconitine, and other nutrients to find novel therapeutics in aiding the drug discovery. In addition for introduction of *Aconitum* species into the era of modern pharmacopeia in order to overcome or treat various diseases, other supplementary biological and pharmacological studies are prerequisite to find the mechanism of actions, safety and efficacy of these valuable medicinal plant species before introducing and implementing into the clinical trials. Summarizing all the scientific knowledge of this valuable medicinal plant, the need of the scenario is to analyze the complete phytochemical version and other pharmacological effects of manifold gamut of compounds of different species of *Aconitum* in order to gauge the taxonomic value of diterpenoid alkaloids and to infer their biosynthetic pathways which will ultimately lead to the conservation and sustainable exploitation of pharmaceutical resources.

REFERENCES CITED

Ameri A. 1998. The effect of *Aconitum* alkaloids on the central nervous system. *Progress in Neurobiology* 56: 211-235.

Anonymous. 1985. Wealth of India-raw material (supplement). Publication and Information Directorate, Council of Scientific and Industrial Research, New Delhi, India.

Anonymous. 1999. Quality evaluation of herbal drugs and validation of ethnobotanical claims. *National Botanical Research Institute Newsletter* 26: 22-26.

Anonymous. 2005. The Wealth of India: Revised edition Vol: I Council of Scientific and Industrial Research, New Delhi 57-60.

Anwar S, Ahmad B, Muhammad SM, Gul W, Nazar-ul-Islam. 2003. Biological and Pharmacological Properties of *Aconitum chasmanthum*. *Journal of Biological Sciences* 3: 989-993.

Atal CK, Sharma ML, Koul A, Khajuria A. 1986). Immunomodulating agents of plant origin. I: Preliminary screening. *Journal of Ethnopharmacology* 18: 133-141.

Bahuguna , PrakashV, Bisht H. 2013. Quantitative enhancement of active content and biomass of two aconitum species through suitable cultivation technology. *International Journal of Conservation Science.* 1: 101-106.

Begum S, Ali M, Latif A, Ahmad W, Alam S, Nisar M, Zeeshan M, Khan MTH, Shaheen F and Ahmad M. 2014. Pharmacologically Active C-19 Diterpenoid Alkaloids from the Aerial parts of *Aconitum laeve* Royle. ACG publications. *Records of Natural Products* 8: (2)83-92

Bhattacharya IC. 1961. A note on Aconitum chasmanthum Stapf. ex Holmes. *Indian Journal of Pharmaceutical Sciences* 23(10): 276-278.

Bist R, Pathak K, Hatwal D, Punetha H, Gaur AK 2011. In vitro propagation of Aconitum balfourii Stapf: An rare medicinal herb of the alpine Himalayas. *Indian Journal of Horticulture* 68(3): 394-398.

CAMP 2003. Threat assessment and management priorities of selected medicinal plants of Western Himalayan states, India. 2003. *Proceedings of the Conservation assessment of medicinal plants workshop.* May 22-26, Shimla, FRLHT, Banglore India.CAMP.

CAMP workshop. 22-25th May, Shimla, Himachal Pradesh.Chakrabarti, L. & Varshney, V. 2001. *Trading in Contraband* 9: 27- 41.

Chakravarty HL, Chakravarti D.1954. Indian Aconites. *Economic Botany* 8: 366-376.

Chaudhary LB, Rao RR. 1998. Notes on the genus *Aconitum* L.(Ranunculaceae) in north west Himalaya (India). *Feddes Repertorium* 109: 527-537.

Chhetri DR Basnet D Chiu PF Kalikotay S Chhetri G and Parajuli S. 2005. Current status of ethnomedicinal plants in the Darjeeling Himalaya. *Current Science.* 89(2): 264 – 268.

Chopra RN. 1984. Poisonous Plants of India. *Academic Publishers,* Jaipur, India, 1: 459-460.

Dar GH, Aman N. 2003. Endemic angiosperms of Kashmir: assessment and conservation, International seminar on recent advances in plant science research,, Department of Botany, University of Kashmir, Srinagar, India, pp:63 (abstract).

Dar GH, Naqshi AR.2001. Threatned plants of the Kashmir Himalaya- a checklist. *Oriental Science* 2: 23-53.

Dar R, Dar GH, Reshi Z. 2006. Recovery and Restoration of some critically endangered endemic angiosperms of the Kashmir Himalaya. *Journal of Biological Sciences* 6 (6): 985-991,.

Dhar U, Kachroo P. 1983. Alpine flora of Kashmir Himalaya. Scientific Publishers. Jodhpur India.

Dhar U, Kachroo P. 1983. Some remarkable features of Endemism in Kashmir Himalaya. In: An assessment of threatned plants of India. 48: 67-71.

Dubey N, Mehta R. 2012. Development and validation of selective High-Performance Liquid Chromatographic method using photodiode array detection for estimation of aconitine in polyherbal ayurvedic taila preparation. *Chromatography Research International.* 2012: 1-5

Forbis T.A., Floyd SK, DeQueiroz A. 2002. The evolution of embryo size in angiosperms and other seed plants: implication for the evolution of seed dormancy. *Evolution.* 56: 2112-2125.

Giri A, Banerjee S, Ahuja PS, Giri C. 1997. Production of hairy roots in Aconitum heterophyllum Wall: using Agrobacterium rhizogenes. In vitro cell and development. *Biologia Plantarum.* 33(4): 280-284.

Giri A, Paramir SA, Kumar APV. 1993. Somatic embryogenesis and plant regeneration from callus cultures of Aconitum heterophyllum Wall. *Plant Cell Tissue and Organ Culture.* 32: 313-218.

Hatano K, Kamur Shayama Y, Nishioka I. 1988. Clonal multiplication of *aconitum* carmichaeli by tip tissue culture and alkoid contents of clonally propagated plants. *Planta Medica* 54: 152-155.

Hatwal D, Bist R, Pathak K, Chaturvedi P, Bhatt JP, Gaur AK.2011. A simple method for genomic DNA isolation for RAPD analysis from dry leaves of *Aconitum balfourii* Stapf. (Ranunculaceae). *Journal of Chemical and Pharmaceutical Research* 3(3): 507-510.

Hikino H, Murakami M, Konno C, Watanabe H. 1983. Determination of Aconitine Alkaloid in Aconitum roots. *Journal of Medicinal Plants Research*

IUCN. 1993. Draft IUCN Red List Categories. IUCN, Gland, Switzerland.

Jabbour F, Renner SS. 2012. A phylogeny of Delphinieae (Ranunculaceae) shows that Aconitum is nested within Delphinium and that Late Miocene transitions to long life cycles in the Himalayas and Southwest China coincide with bursts in diversification. *Molecular Phylogenetics and Evolution* 62 (3): 928–942.

Jabeen N, Shawl AS, Dar GH, Jan A, Sultan P. 2006. Callus induction and Organogenesis from Explants of Aconitum heterophyllum. *Medicinal Plant Biotechnology* 5(3): 287-291.

Jain SK, Rao RR. 1981. Proceedings of the seminar held at Dehradun. BSI, Howrah, pp: 18-22.

Jee V, Dhar U, Kachroo P. 2014. Contribution to the Phytogeographic of Kashmir Himalaya I.Ranunculaceae and Paeoniaceae. *Folia Geobotanica & Phytotaxonomica*. Published by: Springer 4: 387-402.

Kadota Y. 1987. A Revision of Aconitum Subgenus Aconitum (Ranunculaceae) of East Asia. Sanwa Shoyaku Co. Ltd., Utsunomiya, pp. 1-65.

Khetwal S. 2007. Constituents of high altitude Himalayan herbs. A C-19 diterpenoid alkaloid from *Aconitum balfourii*. *Indian Journal of Chemistry* 4: 1364.

Kim DK, Kwon HY,Lee KR, Rhee DK, Zee OP. 1998. Isolation of multidrug resistance inhibitor from Aconitum. *Archives of Pharmacal Research* 21: 344-347.

Kuniyal CP, Bhadula SK, Prasad P. 2006. Flowering, seed characteristics and seed germination behaviour in the populations of a threatened herb Aconitum atrox (Bruhl) Muk. (Ranunculaceae). *Indian Journal of Environmental Sciences* 7(1): 29-36.

Lather A, Gupta K, Bansal P, Singh R, Chaudhary AK. 2010. Pharmacological Potential of Ayurvedic Formulation: Kutajghan Vati-A Review. *Journal of Advanced Scientific Research* 1(2): 41-45.

Luo Y, Yang QE. 2005. Taxonomic revision of Aconitum (Ranunculaceae) from Sichuan, China. *Acta Phytotaxonomica Sinica* 43: 289–386.

Luo Y, Zhang FM, Yang QE.2005. Phylogeny of Aconitum subgenus Aconitum (Ranunculaceae) inferred from ITS sequences. *Plant Systematics and Evolution* 252: 11–25.

Mabberley DJ. 2008 Mabberley's Plant Book. (3rd edn) Cambridge University Press, New York.

Maji JK, Shukla VJ. 2012. Pharmacognosy and phytochemical study of trikarshika churna:a popular polyherbal antioxidant. *International Research Journal of Pharmacy Irjp* 3 (8).

Mitka J, Sutkowska A, Ilnicki T, Joachimiak J. 2007. Reticulate evolution of high alpine aconitum in the Eastern Sudetes and Western Carpathians Central Europe. *Acta Biologica Cracoviensia Botanica* 49/2: 15-26.

Nadeem M, Kumar A, Nandi SK, Palni LMS. 2001. Tissue culture of medicinal plants with particular reference to Kumaun Himalaya. In: Proceedings of the workshop on Himalayan medicnal plants-potential and prospects, Kosi-Katarmal, Almora, 5-7 Nov.

Nautiyal BP, Nautiyal MC, Rawat N, Nautiyal AR. 2009. Reproducrive biology and breeding system of Aconitum balfourii (Benth) Muk: A high altitude endangered medicinal plant of Garhwal Himalaya. *Indian Research Journal of Medicinal Plants* 3(2): 61-68.

Nautiyal BP, Vinay P, Maithani UC, Bisht H, Nautiyal MC. 2002. Population study of three aconites species in Garhwal for the monitoring of species rarity. *Journal of Tropical Ecology* 43(2): 297-303.

Nayar MP, Sastry APK. 1990. Red Data Book of Indian Plants. Published by Botanical Survey of India, Calcutta, India.

Pandey H, Nandi SK, Kumar A, Palni UT, Chandra B, Palni LMS. 2004. In vitro propagation of Aconitum barfourii Stapf; an important aconite of Himalayan alpine. *The Journal of Horticultural Science and Biotechnology* 21: 69-84.

Pandey H, Nandi SK, Nadeem M, Palni LMS. 2000. Chemical stimulation of seed germination in Aconitum heterophyllum wall and A.balfuorii Stapf.: important Himalayan species of medicinal value. *Seed Science and Technology* 28: 39-48.

Pathak K, Hatwal D, Bisht R, Pathak DC, Gaur AK.. 2011. Study of biochemical variability in four populations of Aconitum balfourii by soluble protein and isoenzyme electrophoretic pattern. *Journal of Chemical and Pharmaceutical Research* 3(3): 295-301.

Pelliter SW, Aneja R, Gopinath KW. 1968. The alkaloids of *Aconitum heterophyllum* wall: Isolation and characterization. *Phytochemistry* 7: 625-635.

Polunin O, Stainton A. 1984. Flowers of the Himalaya. pp-580., Oxford University Press, Delhi, India.

Rahman AU, Nuzhat F, Farzana A, Choudhary M, Iqbal M, Khalid A. 2000. New Norditerpenoid Alkaloids from Aconitum falconeri. *Journal of Natural Products* (63(10): 1393-1395.

Rana MS, Samant SS. 2010. Threat categorisation and conservation prioritisation of floristic diversity in the Indian Himalayan region: A state of art approach from Manali Wildlife Sanctuary. *Journal for Nature Conservation* 18: 159-168.

Rau MA. 1975. High Altitude Flowering Plants of West Himalaya.pp-234 Botanical Survey of India, Howrah, India.

Rawat AS, Pharswan AS, Nautiyal MC. 1992. Propagation of Aconitum atrox (Bruhl.) Muk. (Ranunculaceae). A regionally threatened medicinal herb. *Economic Botany*. 46: 337-338.

Rawat JM, Rawat B, Agnihotri RK, Chandra A, Nautiyal S. 2013. In vitro propagation, genetic and secondary metabolite analysis of *Aconitum violaceum* Jacq.: a threatened medicinal herb. *Acta Physiologiae Plantarum* 35(8): 2589-2599

Samant SS, Dhar U, Palni LMS. 1998. Medicinal Plants of Indian Himalaya: Diversity, Distribution and Potential values.HIMAVIKAS. 13, Gyan. Prakash., Nainital.

Sarkar PK, Prajapati PK, Pillai APG, Chauha MGN. 2012. Pharmacognosy of aconite sold under the name Vatsanabha in Indian market. *Indian Journal of Traditional Knowledge* 11(4): 685-696.

Shah NC. 1983. Endangered medicinal and aromatic plants of U.P. Himalaya. In: An assessment of threatened plants of India. Jain, S.K. and R.R. Rao (Eds.). Botanical Survey of India, Howrah, pp. 40-49.

Shah NC. 2005. Conservation aspect of Aconitum species in the Himalayas with special reference to Uttaranchal India. Medicinal Plant Conservation.., 11: 9-15. http://cmsdata.iucn.org/downloads/mpc11.pdf.

Sharma BD, Balakrishnan NP, Rao RR, Hajra PK, Rau MA. 1993 Flora of India. Botanical Survey of India, Calcutta, India.

Sharma E and Gaur AK. 2012. *Aconitum balfourii* Stapf: A rare medicinal herb from Himalayan Alpine. *Journal of Medicinal Plants Research* 6(22): 3810-3817.

Sharma M, Kumar S, Prashar A, Sharma A. 2009. An interesting case of suicidal poisoning. *International Journal of Health & Allied Sciences* 8(4): 14-16.

Sharma RK, Sharma S and. Sharma SS. 2006. Seed germination behaviour of some medicinal plants of Lahaul and Spiti cold desert (Himachal Pradesh): Implication for conservation and cultivation. *Current. Science* 90: 1113-1118.

Sharma SK. 2012. Need for Conservation and Propagation of Medicinal Plants of Himachal Pradesh, India. *Indian Journal of Plant Science* 1(2-3): 217-220.

Shyaula SL. 2011. Phytochemicals, Traditional Uses and Processing of *Aconitum* Species in Nepal. *Nepal Journal of Science and Technology* 12: 171-178.

Sinha PR, Mathur VB, Hussain SA, Chadha. 2008.Special Habitats and Threatened Plants of India. *Envis Bulletin wild life institute of India* 11:1

Sood M, Thakur V. 2011. Effect of light and temperature on germination behavior of Aconitum deinorrhizium Stapf. *International Journal of Farm Sciences*,1(2): 83-87.

Srivastava N, Kamal B , SharmaV, Negi YK , Dobriyal AK , Gupta S, Jadon VS. 2010. Standardization of Sterilization Protocol for Micropropagation of *Aconitum heterophyllum*- An Endangered Medicinal Herb. *Academic Arena* 2(6).

Srivastava N, Sharma V, Dobriyal AK, Kamal B, Gupta S, Jadon VS. 2011. Influence of presowing treatments on *in vitro* seed germination of Ativisha (*Aconitum heterophyllum* Wall) of Uttarakhand. *Biotechnology* 10(2): 215-219.

Srivastava N, Sharma V, Kamal B, Dobriyal AK, Jadon VS. 2010. Advancement in research on *Aconitum* species (Ranunculaceae) under different area: a review. *Biotechnology* Asian network for scientific information.

Srivastava N, Sharma V, Kamal B, Dobariyal AK, Jadon VS. 2010. Polyphenolics free DNA isolation from different types of tissue of *Aconitum* heterophyllum Wall-endangered medicinal species. *Journal of Plant Science* 5(4): 414-419.

Srivastava N, Sharma V, kamal B, Jadon S. 2010. *Aconitum*: Need for sustainable exploitation. *International Journal of Green Pharmacy* 15.249.47.9

Stapf O. 1905. Aconites of India- A monograph. *Annals of the Royal Botanic Garden*10: 115-197.

Stewart RR. 1972. Annotated catalogue of the vascular plants of West Pakistan and Kashmir. Fakhri Printing Press, Karachi, India.

Sunita G. 1992. Substitute and adulterant plants, Periodical Experts Book Agency, New Delhi, India.

Sutkowska A, Boron P , Mitka J. 2013. Natural Hybrid Zone of Aconitum species in the Western Carpathians: Linnaean Taxonomy and ISSR Fingerprinting.*Acta biologica Cracoviensia. Series Botanica* 55(1): 114 – 126.

Tewari NN. 1991. Some crude drugs: source, substitute and adulterant with special reference to KTM crude drug market. *Sachitra Ayurved* 44(4): 284-290.

Tomar BS, Vishen AS. 2015. Vatsnabha-A Wonderful Poisonous Drug. *International Journal of Ayurveda and Pharmaceutical Chemistry* 2350-0204.

Trestrail JH. 2007. Criminal poisoning: investigational guide for law enforcement, toxicologists, forensic scientists, and attorneys. New Jersey: Humana Pressp. 6. ISBN: 158829921X, 9781588299215.

Ulubelen A, Mercli A H, Mericli F, Kolak U, Arfan M, Ahmad M. and Ahmad, H. 2002. Norditerpenoid alkaloids from the roots of *Aconitum* leave Royle. *Die Pharmazie* 57(6): 427-429.

Uniyal BP, Singh PM, Singh DK. 2002. Flora of Jammu and Kashmir. Botanical Survey of India, Kolkata, India.

Utelli A B, Roy A and Baltisberger M. 2000. Molecular and morphological analyses of European Aconitum species (Ranunculaceae). *Plant Systematics and Evolution.* 224: 195- 212.

Venkatasubramaniam P, Subrahmanya KK, Nair VSN. 2010. *Cyperus rotundus* a substitute for *Aconitum heterophyllum*: Studies on the Aurvedic concept of Abhava Pratindh iDravya (drug substitution). *Journal of Ayurveda and Integrative Medicine* 1: 33-39.

Wen ZWJ, Xing J, He Y. 2006. Quantitative determination of diterpenoid alkaloids in four species of *Aconitum* by HPLC. *Journal of Pharmaceutical and Biomedical Analysis* 40(4,3): 1031-1034.

Xiao PG, Wang FP, Gao F, et al. 2006. A pharmacophylogenetic study of *Aconitum* L. (Ranunculaceae) from China. *Acta Phytotaxonomic Sin*ica 44: 1–46.

Xu N, Zhao DF, Liang XM, Zhang H, Xiao YS. 2011. Identification of Diterpenoid Alkaloids from the Roots of *Aconitum* kusnezoffii Reihcb. 16: 3345-3350

Yang Y, Chen J, Shi YP. 2010. Determination of aconitine, hypaconitine and mesaconitine in urine using hollow fiber liquid-phase microextraction combined with high-performance liquid chromatography. *Journal of Chromatography. B, Analytical Technologies in the Biomedical and Life Sciences* 878(28): 2811-2816.

Yoirentomba MS, Devi GAS. 2011. Seasonal variation of bioactive alkaloid content in *Aconitum* spp. From Manipur, India. *International Quarterly Journal of Life Sciences The Biocsan* 6(3): 439-442.

Zhaohong W, Wen J, Xing J, He Y. 2006. Quantitative determination of diterpenoid alkaloids in four species of *Aconitum* by high performance liquid chromatography (HPLC). *Journal of Pharmaceutical and Biomedical Analysis* 40: 1031-1034.

Glossary of Terms

Abscess: Localised collection of pus caused by suppuration in a tissue.

Acne: An inflammatory disease occurring in or around the sebaceous glands.

Actinomorphic: characterized by radial symmetry, such as a starfish or the flower of a daisy.

Acuminate: (of a plant or animal structure, e.g. a leaf) tapering to a point.

Aflatoxin: Poisonous carcinogens that are produced by certain molds which grow in soil, decaying vegetation, hay and grains.

Agronomy: The science of crop production and soil management.

Aliphatic: Relating to organic compounds whose carbon atoms are linked in open chains, either straight or branched, rather than containing a benzene ring. Alkanes, alkenes, and alkynes are aliphatic compounds.

Alkaloid: Class of nitrogenous organic compounds of plant origin which have pronounced physiological actions on humans.

Alpine: Relating to high mountains.

Amino acid: Organic compounds containing amine and carboxyl functional groups, along with a side chain specific to each amino acid.

Anabolism: Synthesis of complex molecules in living organisms from simpler ones together with the storage of energy; constructive metabolism.

Anaemia: Lack of enough blood in the body causing paleness.

Anaesthetic: Inducing loss of feeling or consciousness.

Analgesic: (of a drug) acting to relieve pain.

Analgesic: Relieving pain.

Angiogenesis: Formation of new blood vessels. This process involves the migration, growth, and differentiation of endothelial cells, which line the inside wall of blood vessels.

Annual: A type of flower or plant that lives for only one year.

Antagonist: Drug that attenuates the effect of an agonist. Can be *competitive* or *non-competitive*, each of which can be *reversible* or *irreversible*. A *competitive antagonist* binds to the same site as the agonist but does not activate it, thus blocks the agonist's action. A *non-competitive antagonist* binds to an allosteric (non-agonist) site on the receptor to prevent activation of the receptor. A *reversible antagonist* binds non-covalently to the receptor, therefore can be "washed out". An irreversible antagonist binds covalently to the receptor and cannot be displaced by either competing ligands or washing.

Anther: Part of a stamen that contains the pollen.

Antibody: Also known as an immunoglobulin (Ig), is a large, Y-shaped protein produced mainly by plasma cells that is used by the immune system to neutralize pathogens such as pathogenic bacteria and viruses.

Anticonvulsive: (chiefly of a drug) used to prevent or reduce the severity of epileptic fits or other convulsions.

Antidiabetic: Diabetes treats diabetes mellitus by lowering the glucose level in the blood.

Antidiarrheal: Preventing or controlling diarrhea.

Antidote: An agent which neutralizes or opposes the action of a poison.

Antifungal: Medication also known as an antimycotic medication, is a pharmaceutical fungicide or fungistatic used to treat and prevent mycosis such as athlete's foot, ringworm, candidiasis (thrush), serious systemic infections such as cryptococcal meningitis, and others.

Antihelmintic: Anthelmintics or antihelminthics are a group of antiparasitic drugs that expel parasitic worms (helminths) and other internal parasites from the body by either stunning or killing them and without causing significant damage to the host.

Antioxidant: Substance that inhibits oxidation, especially one used to counteract the deterioration of stored food products.

Antipyretic: (chiefly of a drug) used to prevent or reduce fever.

Antiseptic: Preventing the growth of disease-causing microorganisms.

Aphrodisiac: An aphrodisiac or love drug is a substance that increases libido when consumed. Aphrodisiacs are distinct from substances that address fertility issues or secondary sexual (dys) function such as erectile dysfunction.

Apoptosis: Death of cells which occurs as a normal and controlled part of an organism's growth or development.

Appetizer: Small dish of food or a drink taken before a meal or the main course of a meal to stimulate one's appetite.

Arthritis: Inflammation of the joints.

Aromatic: Having a pleasant and distinctive smell.

Autophagy: Autophagy (or autophagocytosis) is the natural, regulated mechanism of the cell that disassembles unnecessary or dysfunctional components.

Beverages: Liquid intended for human consumption. In addition to their basic function of satisfying thirst.

Bioactivity: Biological activity describes the beneficial or adverse effects of a drug on living matter.

Biodiversity hotspot: Biogeographic region that is both a significant reservoir of biodiversity and is threatened with destruction. The term biodiversity hotspot specifically refers to 25 biologically rich areas around the world that have lost at least 70 percent of their original habitat.

Biodiversity: Reflects the number, variety and variability of living organisms.

Biological resources: Those components of biodiversity of direct, indirect, or potential use to humanity.

Bioprospection: Bioprospecting is the process of discovery and commercialization of new products based on biological resources. Despite indigenous knowledge being intuitively helpful, bioprospecting has only recently begun to incorporate such knowledge in focusing screening efforts for bioactive compounds.

Bovine: Cattle.

Calibration: the action or process of calibrating something.

Calyx: Sepals of a flower, typically forming a whorl that encloses the petals and forms a protective layer around a flower in bud.

Carbohydrate: Biomolecule consisting of carbon, hydrogen and oxygen atoms, usually with hydrogen-oxygen atom ratio of 2:1.

Carminative: Drug that relieves flatulence.

Catalyst: An agent or compound that is added to a process to make a chemical reaction happen more quickly.

Cellulose: An important structural component of the primary cell wall of green plants, many forms of algae and the oomycetes.

Chemotaxonomic: Method of biological classification based on similarities in the structure of certain compounds among the organisms being classified.

Chlorotic: In botany, chlorosis is a condition in which leaves produce insufficient chlorophyll. As chlorophyll is responsible for the green color of leaves, chloroticleaves are pale, yellow, or yellow-white.

Chromatin: Mass of genetic material composed of DNA and proteins that condense to form chromosomes during eukaryotic cell division.

Chromatography: Laboratory technique for the separation of mixture.

Clinical pharmacology: The study of the effects of drugs in humans.

Coniferous: The conifers are a division of vascular land plants containing gymnosperms, cone-bearing seed plants.

Conservation: Prevention of wasteful use of a resource.

Contraceptive: (of a method or device) serving to prevent pregnancy.

Corolla: Petals are modified leaves that surround the reproductive parts of flowers.

Cosmetics: Cosmetics are substances or products used to enhance or alter the appearance of the face or fragrance and texture of the body. Many cosmetics are designed for use of applying to the face, hair, and body.

Crenulate: (especially of a leaf, shell, or shoreline) having a finely scalloped or notched outline or edge.

Cryopreservation: Cryopreservation is the use of very low temperatures to preserve structurally intact living cells and tissues. Unprotected freezing is normally lethal and this chapter seeks to analyze some of the mechanisms involved.

Crystallization: Chemical solid–liquid separation technique, in which mass transfer of a solute from the liquid solution to a pure solid crystalline phase occurs.

Cutaneous: relating to or affecting the skin.

Cytotoxicity: Quality of being toxic to cells. Examples of toxic agents are an immune cell or some types of venom, e.g. from the puff adder (Bitis arietans) or brown recluse spider (Loxosceles reclusa).

Deciduous: Shedding the leaves annually, as certain trees and shrubs.

Decoction: Concentrated liquor resulting from heating or boiling a substance, especially a medicinal preparation made from a plant.

Deforestation: Deforestation, clearance, or clearing is the removal of a forest or stand of trees from land which is then converted to a non-forest use.

Dehydrated: Cause (a person or their body) to lose a large amount of water.

Desiccation: Removal of moisture from something.

Dieresis: Increased or excessive production of urine.

Distillation: A process of evaporation and re-condensation used for purifying liquids.

Diuretic: Diuretics, also called water pills, are medications designed to increase the amount of water and salt expelled from the body as urine.

Drug: An agent that is used therapeutically to treat diseases. It may also be defined as any chemical agent and/or biological product or natural product that affects living processes

Dyes: Natural or synthetic substance used to add a colour to or change the colour of something.

Dysentery: Infection of the intestines resulting in severe diarrhoea with the presence of blood and mucus in the faeces.

Ecologist: Researcher who study the interrelationships between organisms and their environments.

Ecology: The study of the environment and how living things interact with it.

Ecosystem: A community of living and non-living things that interact by exchanging matter and energy.

Ecosystem: A dynamic complex of plant, animal and micro-organism communities and their non-living environment interacting as a functional unit.

Eco-tourism: Tourism directed towards exotic, often threatened, natural environments, intended to support conservation efforts and observe wildlife.

ED$_{50}$: *In vitro* or *in vivo* dose of drug that produces 50% of its maximum response or effect.

Efficacious: (of something inanimate or abstract) successful in producing a desired or intended result; effective.

Efficacy: Describes the way that agonists vary in the response they produce when they occupy the same number of receptors. High efficacy agonists produce their maximal response while occupying a relatively low proportion of the total receptor population. Lower efficacy agonists do not activate receptors to the same degree and may not be able to produce the maximal response.

Elliptic: Relating to or having the form of an ellipse.

Emaciation: The state of being abnormally thin or weak.

Emollient: Emollients are non-cosmetic moisturisers which come in the form of creams, ointments, lotions and gels.

Endangered: (of a species) seriously at risk of extinction.

Endemic: Distribution restricted to a particular area: used to describe a species or organism that is confined to a particular geographical region, for example, an island or river basin.

Environment: Physical surroundings; all that is around you.

Enzymes: Proteins that start a chemical reaction.

Epidermal: The protective outer layer of the skin. In invertebrate animals, the epidermis is made up of a single layer of cells.

Epiphytic: An epiphyte is an organism that grows on the surface of a plant and derives its moisture and nutrients from the air, rain, water (in marine environments) or from debris accumulating around it.

Essential Oil: Volatile perfumery material derived from a single source of vegetable or animal origin by a process, such as hydrodistillation, steam distillation, dry distillation or expression.

Ethanol: Form of natural gas that can be produced from corn.

Ethnobotanical: The study of a region's plants and their practical uses through the traditional knowledge of a local culture and people.

Expectorant: A medicine which promotes the secretion of sputum by the air passages, used to treat coughs.

Extinction: the state or process of being or becoming extinct.

Extract: A concentrate of dried, less volatile aromatic plant part obtained by solvent extraction with a polar solvent.

Extraction: The process of isolating essential oil with the help of a volatile solvent.

Fatty acid: A carboxylic acid consisting of a hydrocarbon chain and a terminal carboxyl group, especially any of those occurring as esters in fats and oils.

Febrifuge: A medicine used to reduce fever.

Filariasis: Parasitic disease caused by an infection with roundworms of the Filarioidea type. These are spread by blood-feeding diptera as black flies and mosquitoes.

Flavonoid: Class of plant and fungus secondary metabolites.

Flavonoids: any of a large class of plant pigments having a structure based on or similar to that of flavone.

Flavour: Refers to that characteristic quality of a material as affects the taste or perception.

Folklore: the traditional beliefs, customs, and stories of a community, passed through the generations by word of mouth.

Fragile: (of an object) easily broken or damaged.

Galactagogue: a food or drug that promotes or increases the flow of a mother's milk.

Gastritis - Inflammation of the stomach.

Genotype: The set of genes in our DNA which is responsible for a particular trait. The phenotype is the physical expression, or characteristics, of that trait.

Germplasm: Living genetic resources such as seeds or tissues that are maintained for the purpose of animal and plant breeding, preservation, and other research uses.

Glabrous: (chiefly of the skin or a leaf) free from hair or down; smooth.

Glandular: relating to or affecting a gland or glands.

Glutamic acid: An á-amino acid that is used by almost all living beings in the biosynthesis of proteins. It is non-essential in humans, meaning the body can synthesize it.

Gonorrhoea: Sexually transmitted infection (STI) caused by bacteria called Neisseria gonorrhoeae or gonococcus.

Gregarious: (of a person) fond of company; sociable.

Habitat: In ecology, a habitat is the type of natural environment in which a particular species of organism lives.

Haemagglutinating: Hemagglutination, or haemagglutination, is a specific form of agglutination that involves red blood cells (RBCs). It has two common uses in the laboratory: blood typing and the quantification of virus dilutions in a haemagglutination assay.

Haematemesis: Hematemesis or haematemesis is the vomiting of blood. The source is generally the upper gastrointestinal tract, typically above the suspensory muscle of duodenum.

Half-life: Half-life (t½) is an important pharmacokinetic measurement. The metabolic half-life of a drug *in vivo* is the time taken for its concentration in plasma to decline to half its original level. Half-life refers to the duration of action of a drug and depends upon how quickly the drug is eliminated from the plasma. The clearance and distribution of a drug from the plasma are therefore important parameters for the determination of its half-life.

Heart palpitations: Abnormally rapid and irregular beating of the heart

Hepatocytes: A hepatocyte is a cell of the main parenchymal tissue of the liver. Hepatocytes make up 70-85% of the liver's mass.

Hermaphrodite: A person or animal having the both male and female sex organs or other sexual characteristics, either abnormally or (in the case of some organisms) as the natural condition.

Homeopathy: A system of complementary medicine in which ailments are treated by minute doses of natural substances that in larger amounts would produce symptoms of the ailment.

Humidity: The amount of water vapour present in air. Water vapour, the gaseous state of water, is generally invisible to the human eye.

Humus: The organic component of soil, formed by the decomposition of leaves and other plant material by soil microorganisms.

Hydro-Distillation: Distillation of a substance carried out in direct contact with boiling water.

Hydrophobic: The hydrophobic effect is the observed tendency of nonpolar substances to aggregate in an aqueous solution and exclude water molecules.

Hyperglycaemic: An excess of glucose in the bloodstream, often associated with diabetes mellitus.

Hypodermis: In arthropods, the hypodermis is an epidermal layer of cells that secretes the chitinous cuticle.

IC$_{50}$: In a functional assay, the molar concentration of an agonist or antagonist which produces 50% of its maximum possible inhibition. In a radioligand binding assay, the molar concentration of competing ligand which reduces the specific binding of a radioligand by 50%.

Immunomodulatory: a chemical agent (as methotrexate or azathioprine) that modifies the immune response or the functioning of the immune system (as by the stimulation of antibody formation or the inhibition of white blood cell activity) Other Words from immunomodulator.

In vitro: Taking place in a test-tube, culture dish or elsewhere outside a living organism.

In vivo: Taking place in a living organism.

Inflorescence: the complete flower head of a plant including stems, stalks, bracts, and flowers.

Infusion: A process of treating a substance with water or organic solvent, with or without heating.

Inhibition: A feeling that makes one self-conscious and unable to act in a relaxed and natural way.

Insecticidal: Substances used to kill insects. They include ovicides and larvicides used against insect eggs and larvae, respectively.

Insecticide: A type of chemical used to kill insects.

Invasive: Tending to spread very quickly and undesirably or harmfully.

Keystone species: A dominant predator whose removal allows a prey population to explode and often decreases overall diversity.

Laxative: Substances that loosen stools and increase bowel movements. They are used to treat and prevent constipation.

Leucoderma: Leucoderma, sometimes known as vitiligo, is a rare skin disease characterized by white spots and patches.

Liana: Any of various long-stemmed, woody vines that are rooted in the soil at ground level and use trees, as well as other means of vertical support, to climb up to the canopy to get access to well-lit areas of the forest.

Lithophytic: Plants that grow in or on rocks. Those that grow on rocks are also known as epipetric or epilithic plants.

Macrophages: Large phagocytic cell found in stationary form in the tissues or as a mobile white blood cell, especially at sites of infection.

Mass Spectro Photometry: An analytical technique that ionizes chemical species and sorts the ions based on their mass-to-charge ratio. In simpler terms, a mass spectrum measures the masses within a sample.

Mastitis: Inflammation of the breast.

Meristem: Meristem is the tissue in most plants containing undifferentiated cells (meristematic cells), found in zones of the plant where growth can take place.

Mesocarp: The middle layer of the pericarp of a fruit, between the endocarp and the exocarp.

Microorganisms: Tiny living things that can only be seen with a microscope.

Midrib: A large strengthened vein along the midline of a leaf.

Migraine: A periodic condition with localised headaches, frequently associated with vomiting and sensory disturbances

Monitoring: The performance and analysis of routine measurements aimed at detecting changes in the environment or health status of populations.

Morbidity: The condition of being diseased.

Mountaineous: (of a region) having many mountains.

Mushroom: A mushroom, or toadstool, is the fleshy, spore-bearing fruiting body of a fungus, typically produced above ground on soil or on its food source.

Nutraceutical: A food stuff that is held to provide health or medicinal benefits in addition to its basic nutritional value also called functional food.

Nutraceutical: A nutraceutical is a food or food component that claims to have health benefits, including treatment and prevention of disease.

Odour: That property of a substance which stimulates and is perceived by the olfactory sense.

Orbicular: Having the shape of a flat ring or disc.

Orchard: An orchard is an intentional planting of trees or shrubs that is maintained for food production.

Organic acid: An organic compound with acidic properties.

Organic matter: Dead plants, animals and manure converted by earthworms and bacteria into humus.

Osteoporotic: A disease where increased bone weakness increases the risk of a broken bone. It is the most common reason for a broken bone among the elderly.

Pedunculate: Having, supported on, or growing from a peduncle.

Perennial: Lasting or existing for a long or apparently infinite time; enduring or continually recurring.

Perfume: A suitably blended composition of various materials of synthetic and/or natural origin to give a desired odour effect. It is carried in a suitable medium to the extent of not more than 20 percent.

Perfumery: Mixture of fragrant essential oils or aroma compounds, fixatives and solvents, used to give the human body, animals, food, objects, and living-spaces an agreeable scent.

Permeability: Permeability is the measure of the ability of a material to support the formation of a magnetic field within itself otherwise known as distributed inductance in Transmission Line Theory.

Phagocytic: Phagocytes are cells that protect the body by ingesting harmful foreign particles, bacteria, and dead or dying cells.

Pharmaceutical: Relating to medicinal drugs, or their preparation, use, or sale.

Pharmacology: The study of the effects of drugs. The branch of biology concerned with the study of drug action, where a drug can be broadly defined as any man-made, natural, or endogenous (from within the body) molecule which exerts a biochemical or physiological effect on the cell, tissue, organ, or organism (sometimes the word pharmacon is used as a term to encompass these endogenous and exogenous bioactive species).

Phenol: An aromatic organic compound with the molecular formula C6H5OH.

Phonological: Phenology is the study of periodic plant and animal life cycle events and how these are influenced by seasonal and inter-annual variations in climate, as well as habitat factors (such as elevation).

Phthisis: Pulmonary tuberculosis or a similar progressive wasting disease.

Phytochemistry: Study of phytochemicals, which are chemicals derived from plants. Those studying phytochemistry strive to describe the structures of the large number of secondary metabolic compounds found in plants.

Phytoconstituent: Synthetic intensifies that happen actually in plants.

Plant: Multicellular predominantly photosynthetic eukaryotes of the kingdom Plantae. Historically, plants were treated as one of two kingdoms including all living things that were not animals, and all algae and fungi were treated as plants.

Plateaus: An area of fairly level high ground.

Pollination: The act of transferring pollen grains from the male anther of a flower to the female stigma. The goal of every living organism, including plants, is to create offspring for the next generation.

Polymorphic: Occurrence of two or more clearly different morphs or forms, also referred to as alternative phenotypes, in the population of a species.

Polypetalous: A flower in which the corolla consists of separate petals.

Polysaccharides: A carbohydrate (e.g. starch, cellulose, or glycogen) whose molecules consist of a number of sugar molecules bonded together.

Population: The number of living things that live together in the same place. In biology, a population is all the organisms of the same group or species, which live in a particular geographical area, and have the capability of interbreeding.

Potency: A measure of the concentrations of a drug at which it is effective.

Prebiotics: A non-digestible food ingredient that promotes the growth of beneficial microorganisms in the intestines.

Precipitation: In meteorology, precipitation is any product of the condensation of atmospheric water vapor that falls under gravity.

Precognitive: Having or giving foreknowledge of an event.

Preservation: The action of preserving something.

Primitive: Relating to, denoting, or preserving the character of an early stage in the evolutionary or historical development of something.

Propagation: The breeding of specimens of a plant or animal by natural processes from the parent stock.

Pubescent: Relating to or denoting a person at or approaching the age of puberty.

Purgative: Strongly laxative in effect.

Pyrexia: Raised body temperature; fever.

Raceme: A flower cluster with the separate flowers attached by short equal stalks at equal distances along a central stem. The flowers at the base of the central stem develop first.

Resin: A sticky flammable organic substance, insoluble in water, exuded by some trees and other plants (notably fir and pine).

Rheumatism: Any disease marked by inflammation and pain in the joints, muscles, or fibrous tissue, especially rheumatoid arthritis.

Rhizome: Modified subterranean plant stem that sends out roots and shoots from its nodes. Rhizomes are also called creeping rootstalks or just rootstalks.

Saponins: A toxic compound which is present in soapwort and makes foam when shaken with water.

Sclerenchymatous: A supportive tissue of vascular plants, consisting of thick-walled, usually lignified cells. Sclerenchyma cells normally die upon reaching maturity but continue to fulfill their structural purpose in the plant.

Screening: The presumptive identification of unrecognized disease or defect by the application of tests, examinations or other procedures which can be applied rapidly. Screening is an initial examination only and positive responders require a second diagnostic examination.

Sessile: (of an organism, e.g. a barnacle) fixed in one place; immobile.

Shrub: Small- to medium-sized woody plant. Unlike herbs, shrubs have persistent woody stems above the ground.

Side effect: Any unintended effect of a pharmaceutical product occurring at doses normally used in humans which is related to the pharmacological properties of the drug.

Signal: Reported information on a possible causal relationship between an adverse event and a drug, the relationship being unknown or incompletely documented previously. Usually more than a single report is required to generate a signal, depending upon the seriousness of the event and the quality of the information.

Solitary: Existing alone.

Somaclonal variation: The variation seen in plants that have been produced by plant tissue culture. Chromosomal rearrangements are an important source of this variation.

Species: A group of living organisms consisting of similar individuals capable of exchanging genes or interbreeding. The species is the principal natural taxonomic unit, ranking below a genus and denoted by a Latin binomial, e.g. *Homo sapiens*.

Specificity: The ability of a method, system or tool to correctly classify the proportion of persons who truly do not have a characteristic, as not having it.

Spectroscopy: The study of the interaction between matter and electromagnetic radiation.

Splenocytes: Any one of the different white blood cell types as long as it is situated in the spleen or purified from splenic tissue.

Stamen: Male reproductive organ of a flower. It produces the pollen. The stamen has two parts: anther and stalk.

Steam Distillation: Distillation of a substance by bubbling steam through it.

Steroid: Synthetic hormones that can boost the body's ability to produce muscle and prevent muscle breakdown.

Stigma: The sticky stem of the pistil of the female reproductive system in a plant. It is the portion of the ovary where pollen germinates and is essential for plant reproduction.

Stimulant: Making a body organ active.

Stomatitis: Inflammation of the mouth and lips. It refers to any inflammatory process affecting the mucous membranes of the mouth and lips, with or without oral ulceration.

Subtropical: A humid subtropical climate is a zone of climate characterized by hot and humid summers, and mild winters.

Succulent: A plant (especially a xerophyte) having thick fleshy leaves or stems adapted to storing water.

Susceptible: Likely or liable to be influenced or harmed by a particular thing.

Synonyms: A word or phrase that means exactly or nearly the same as another word or phrase in the same language.

Tannin: A yellowish or brown bitter tasting organic substance present in some galls , barks and other plant tissues.

Taxon (plural taxa): In biology, a taxon is a group of one or more populations of an organism or organisms seen by taxonomists to form a unit. Although neither is required, a taxon is usually known by a particular name and given a particular ranking, especially if and when it is accepted or becomes established.

Taxonomist: A taxonomist is a biologist that groups organisms into categories.

Technology: Instruments, tools or inventions developed through research to increase efficiency.

Temperate: relating to or denoting a region or climate characterized by mild temperatures.

Thin Layer Chromatography: Thin-layer chromatography (TLC) is a chromatography technique used to separate non-volatile mixtures Thin-layer chromatography is performed on a sheet of glass, plastic, or aluminium foil, which is coated with a thin layer of adsorbent material, usually silica gel, aluminium oxide (alumina), or cellulose.

Thymol: Phenol obtained from thyme oil or other volatile oils used as a stabilizer in pharmaceutical preparations, and as an antiseptic (antibacterial or antifungal) agent.

Topographic: Relating to the arrangement of the physical features of an area.

Traditional Knowledge: The term traditional knowledge generally refers to knowledge systems embedded in the cultural traditions of regional, indigenous, or local community.

Trifoliolate: Having three leaflets a trifoliolate leaf.

Tuber: A much thickened underground part of a stem or rhizome, e.g. in the potato, serving as a food reserve and bearing buds from which new plants arise.

Urbanization: The growth of the city into rural areas.

Validation: Establishing documented evidence which provides a high degree of assurance that a specific process will consistently produce a product meeting its pre-determinant specifications and quality attributes.

Valley: Low area between hills or mountains often with a river running through it. In geology, a valley or dale is a depression that is longer than it is wide.

Vinegar: An aqueous solution of acetic acid and trace chemicals that may include flavorings. Vinegar typically contains 5–20% by volume acetic acid.

Volatile: A material is said to be volatile when it has the property of evaporating at room temperature when exposed to atmosphere.

Vulnerable: Exposed to the possibility of being attacked or harmed, either physically or emotionally.

Xenograft: A tissue graft or organ transplant from a donor of a different species from the recipient.

Yield: The amount of a crop produced in a given time or from a given place.

Index

A

B

C